1983

Differential Equations
An Introduction

David A. Sanchez
Richard C. Allen, Jr.
Walter T. Kyner

The University of New Mexico

Addison-Wesley Publishing Company
Reading, Massachusetts · Menlo Park, California
London · Amsterdam · Don Mills, Ontario · Sydney

Sponsoring Editor: Wayne Yuhasz
Copy Editor: Rima Zolina
Production Editor: Martha K. Morong

Text Designer: Marie E. McAdam
Illustrator: ANCO/Boston
Cover Designer: Ann Scrimgeour Rose
Art Coordinator: Robert Trevor

Production Manager: Sherry Berg

The text of this book was composed in Century Schoolbook by University
Graphics, Inc.

Library of Congress Cataloging in Publication Data

Sanchez, David A.
   Differential equations.

   Includes index.
   1. Differential equations.   I. Allen, Richard C.
II. Kyner, Walter T.   III. Title.
QA372.S17 1983   515.3′5   82-16326
ISBN 0-201-07760-4

ISBN 0-201-07760-4
ABCDEFGHIJ-DO-898765432

Preface

This text is based on a course, successfully given at the University of New Mexico for the last eight years, which is a one-semester introduction to ordinary differential equations for mathematics majors, engineering students, and majors in the physical sciences. It differs from the traditional introductory course in that numerical methods are used throughout as a tool to analyze the qualitative behavior of solutions as well as to approximate them.

The primary emphasis of the book is on ordinary differential equations, as well as on the analysis of a number of the mathematical models described by them. A course taught with this book should be one with mathematical depth, but attractive to an audience with a large proportion of engineering and physical sciences students. To this end, we have provided a thorough discussion of first order linear systems and nonlinear systems using phase plane and potential plane methods. We develop the Laplace transform, using it to analyze linear systems, and introduce series solutions and special functions. The existence and uniqueness theorem for the initial value problem is thoroughly discussed in an appendix. We feel that with an appropriate selection of topics a course as "pure" or as "applied" as one may wish can be taught using this book.

The traditional introductory text in ordinary differential equations generally contains one chapter devoted to numerical techniques. There, three or four algorithms are introduced and possibly a brief analysis of errors is given. But this ends the subject, and nowhere else in the text are numerical methods used to assist in the study of solutions of differential equations, especially nonlinear ones. Quite often the algorithms given and the discussions presented do not reflect the recent advances in numerical analysis and the state of computing, and one gets the impression that the chapter on numerical techniques was inserted as an unexciting and unimportant, yet required, part of any acceptable introductory text. Students must surely wonder what value there is in approximating $y(1)$ for the problem $y' = y$, $y(0) = 1$, using four different algorithms and two different step sizes, neither of which provides much accuracy.

It is our opinion, substantiated in practice here at the University of New Mexico, that this is not the most effective way to teach ordinary differential equations and to introduce numerical methods. Instead various numerical methods, beginning with the most elementary ones, can be introduced throughout the course and used to complement the discussions at hand. The computer can be used to give really significant accuracy, to compare different algorithms and changes of step size, and to give the student a true appreciation for numerical methods and the behavior of solutions of ordinary differential equations.

The prerequisites for a student using this book are knowledge of the calculus as developed in a standard two- or three-semester course. To be able to use effectively the numerical techniques developed, the student should have some knowledge of a programming language (Fortran or Pascal) and some experience with computers. Although the simple numerical algorithms can be implemented with a hand calculator, the more sophisticated ones require access to a computer.

We note here some of the principal features of each of the chapters. Chapter 1 treats first order scalar differential equations with specific applications to circuit theory, population biology, and mixing problems with a feedback control law. The Euler and improved Euler algorithms are introduced and used in some of the applications. Chapter 2 discusses second order linear differential equations. The notion of solution vectors is used to develop the variation of parameters method. Although this is not traditional, the geometrical notions introduced will be advantageous when the student studies two-dimensional nonlinear systems. The Euler and improved Euler methods are extended to second order equations.

Chapter 3 is the first full chapter devoted to numerical methods. In it the various types of errors in numerical solutions are discussed and the Taylor series and classical Runge–Kutta methods are introduced. A program for the Runge–Kutta method is given. Chapter 4 is a fairly standard discussion of Laplace transforms with Heaviside's formulas emphasized as a method for partial fraction decompositions. A discussion of the transfer function in systems analysis is given.

Chapter 5 deals with linear systems, with primary emphasis on two- and three-dimensional systems in which the eigenvalues and eigenvectors can be found easily. The fundamental matrix is computed using both the eigenvector method and the more efficient Laplace transform. The beginning of the chapter includes a brief introduction of those concepts from linear algebra necessary for the study of linear systems of differential equations.

Series solutions of second order linear equations and some of the special functions of mathematical physics are discussed in Chapter 6. The method of reduction of order is used to obtain the second solution in the special cases of solutions near a regular singular point. In Chapter 7 nonlinear systems are discussed and both the phase plane and the potential plane are used to analyze stability and the existence of periodic solutions. A detailed analysis of competitive and predator-prey models of two populations is also provided.

Chapter 8 is a further discussion of numerical methods. The use of two algorithms of different order to analyze and control local error is demonstrated. The program RKF45, which uses simultaneously a fourth- and fifth-order Runge–Kutta method to adjust step size, is introduced and some numerical examples are given. It is not necessary to understand the details of the code to be able to use it to solve realistic problems.

The book contains more material than can be used in a one-semester course. The course given here at the University of New Mexico is primarily for engineers and is based on the following outline:

Chapters 1, 2, and 3 (Runge–Kutta methods only);

Chapters 4, 5, and 7 (selected topics).

The RKF45 program is introduced and several assignments based on the topics from Chapter 7 are given. A special section of this course for mathematics and physical sciences majors uses

Chapters 1, 2, and 3 (Runge–Kutta methods only),

Chapters 5 and 7 (selected topics), and Chapters 6 or 8,

where the choice is dependent on the desired degree of emphasis on numerical methods. The outline

Chapters 1, 2, 5, 6, and 7

is appropriate for a course for pure mathematics students. The sections on numerical methods can be deleted easily without loss of continuity and, if desired, summarized later.

For schools with an academic calendar based on the quarter system most of the book could probably be covered in a two-quarter sequence. For a one-quarter introductory course Chapters 1, 2, and 5 and selected topics from Chapter 7 are suitable.

In conclusion, we wish to emphasize that this book is not intended as an introduction to numerical methods for ordinary differential equations. As the reader can see, our aim is to give the important topics and analytical tools needed to study ordinary differential equations, and to show on occasion how numerical techniques and the high-speed computer can complement them. One merely needs to examine the graph of the numerical solutions given in Chapter 8 for the problem of the precession of the perihelia of the planet Mercury to see the utility (and beauty) of numerical methods.

We wish to thank our many students and colleagues who have helped in the development of the course on which this book is based. Their ideas, comments, and criticisms over the years have helped us in the focus of our task. We wish to thank in particular W. J. Kammerer (Georgia Institute of Technology), F. H. Mathis (Baylor University), R. E. Plant (University of California, Davis), L. F. Shampine (Sandia National Laboratories), R. E. Showalter (The University of Texas at Austin), A. P. Stone (University of New Mexico), D. D. Warner (Clemson University), K. Schmitt (University of Utah), and T. Bowman (University of Florida). Our special thanks go to Elizabeth Poor for her technical typing and preparation of the manuscript and to Margaret Poor for working many of the problems and examples. Finally, we wish to thank the staff of Addison-Wesley for their support and advice during the writing of this text, with special appreciation to Martha Morong for her assistance in the technical preparation of this book.

Albuquerque, New Mexico　　　　　　　　　　　　　　　　　D.A.S.
November 1982　　　　　　　　　　　　　　　　　　　　　　R.C.A.
　　　　　　　　　　　　　　　　　　　　　　　　　　　　　　W.T.K.

Contents

Elementary Numerical Methods 3

The Laplace Transform 4

Linear Systems of Differential Equations 5

Nonconstant Coefficient Second Order Linear Equations and Series Solutions 6

Nonlinear Differential Equations 7

More on Numerical Methods 8

First Order Differential Equations

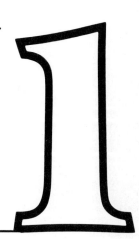

1.1 GENERAL REMARKS

An *ordinary differential equation* is a relationship between an unknown function of a *single variable* and one or more of its derivatives. Such relationships often result when expressing scientific laws connecting physical quantities and their rates of change in mathematical terms. For example, Newton's law stating that the rate of change of the momentum of a particle is equal to the force acting on it or the statement that a population is growing exponentially can both be translated into mathematical language as ordinary differential equations.

Our goal in studying differential equations is to determine both the qualitative and quantitative properties of those functions which satisfy them; such functions are called *solutions*. This can sometimes be done by representing the solutions as elementary functions, e.g., polynomials or trigonometric functions. More often, geometric arguments are needed to determine the qualitative properties, and numerical methods are needed to determine the quantitative properties. These techniques are especially important when an explicit representation of the solution cannot be found.

Three differential equations that arise quite often in applications are:

1. $\dfrac{dy}{dt} = k(t)F(y)$, the growth equation;

2. $L\dfrac{d^2Q}{dt^2} + R\,\dfrac{dQ}{dt} + \dfrac{Q}{C} = E(t)$, the LCR oscillator equation; and

3. $\dfrac{d^2\theta}{dt^2} + \dfrac{g}{l}\sin\theta = 0$, the pendulum equation.

These equations will be examined later in more detail, and we note here only that the standard notation is imprecise, but clear. In all three equations t is called the *independent variable* and the unknown functions y, Q, and θ, whose derivatives explicitly appear, are called the *dependent variables*. It is assumed that each equation is defined in an appropriate domain. A solution is a function $y(t)$, $Q(t)$, or $\theta(t)$ that satisfies the equation in a subset of the domain of definition that includes a nonempty interval of the real numbers. Hence the solutions y, Q, and θ are functions of t but, in contrast with the explicit notation $k(t)$ and $E(t)$, this functional dependence is to be understood from the context.

Our study is made easier by grouping together those differential equations that have a significant common property. The most important property is that of *order*. The *order of a differential equation* is the order of the highest derivative of the dependent variable which appears in the equation. For example, the oscillator and pendulum equations are second order, while the growth equation is first order. In the remainder of this chapter, first order differential equations will be studied and in later chapters higher order equations will be discussed.

A *partial differential equation* is a relationship between an unknown function of *at least two* variables and one or more of its derivatives. For instance, a partial differential equation, which describes the temperature $u(x, t)$ in a thin heated wire as a function of position x and time t, is

$$\frac{\partial u}{\partial t} = k\,\frac{\partial^2 u}{\partial x^2} \qquad \text{or} \qquad u_t = ku_{xx},$$

where k is a constant dependent on the physical properties of the wire. Since the mathematical ideas and methods in partial differential equations are sufficiently different from those of ordinary differential equations, the two subjects are best studied separately. In this book we concern ourselves with ordinary differential equations and usually delete the adjective "ordinary."[1]

---

[1] It is not clear why "ordinary" ever became standard terminology in a subject that motivated the invention of the calculus, that has such a wide applicability to science and technology and that contains so many fascinating ideas and methods.

EXERCISES 1.1

1. Which of the following equations are ordinary differential equations and which are partial differential equations?

a) $y'' + 3y' + 2y = \sin t$

b) $\left(\dfrac{dy}{dt}\right)^2 + 4 = t^2$

c) $u_{xx} - 9u_{tt} = 3 \sin x \cos t$

d) $\dfrac{\partial u}{\partial x} + t^2 \dfrac{\partial u}{\partial t} = 0$

e) $\dfrac{d^3 y}{dt^3} - 4 \dfrac{dy}{dt} + 7y = 4e^{-t}$

f) $u_t = 3t - 5$

2. Determine the order of each of the following differential equations:

a) $y' + 4y = 0$

b) $\dfrac{d^2 y}{dt^2} + 4y^2 = 7$

c) $y'' + \epsilon y'(y^2 - 1) + 4y = 3 \sin t$

d) $\dfrac{d}{dt}(p(t)y') + q(t)y = 0$

e) $y'y'' + 4yy' = 0.$

f) $y(t) = \displaystyle\int_0^t e^s y(s)\,ds.$ This is not a differential equation. Convert it to one!

1.2 FIRST ORDER DIFFERENTIAL EQUATIONS IN GENERAL

Many of the important concepts and techniques of the theory of differential equations can be introduced by studying first order differential equations, i.e., equations of the form

$$\frac{dy}{dt} = f(t, y). \tag{1.2.1}$$

It is assumed that $f(t, y)$ is a real-valued continuous function defined on a domain D in the t, y plane. A *solution* of equation (1.2.1) is a differentiable function $y(t)$ defined on a nonempty t-interval I such that

$$\frac{d}{dt} y(t) = f(t, y(t))$$

for all t in I. The graph of the solution $y(t)$ must also lie in D.

In this book we usually consider open rectangular domains D which are specified by an interval of t values and an interval of y values. For example, D could be the whole t, y plane, or D could be the set

$$\{0 < t < 2, -1 < y < 1\},$$

an open rectangle in the t, y plane, or the set

$$\{-1 < t < 1, -\infty < y < \infty\},$$

an open strip in the t, y plane.

In the following examples some first order differential equations and their solutions are presented. Techniques for obtaining solutions will be given later.

Example 1 Solve $\dfrac{dy}{dt} = -\dfrac{1}{2} ty$.

Here $f(t, y) = -\frac{1}{2}ty$ and its domain of definition is $D = \{-\infty < t < \infty, -\infty < y < \infty\}$. The function $y(t) = 2 \exp(-\frac{1}{4}t^2)$ is a solution on the interval $I = \{-\infty < t < \infty\}$ and the values of $y(t)$ are in the interval $0 < y \le 2$. To verify that it is actually a solution, we substitute the function $y(t)$ into the differential equation to obtain

$$\frac{d}{dt}\left[2 \exp\left(-\frac{1}{4} t^2\right)\right] = -\frac{1}{2} t \left[2 \exp\left(-\frac{1}{4} t^2\right)\right],$$

and so

$$\frac{d}{dt} y(t) = -\frac{1}{2} ty(t)$$

as required. Note that if K is an arbitrary constant, then the function $y(t) = K \exp(-\frac{1}{4}t^2)$ is also a solution. We suggest that the reader verify that any multiple of $\exp(-\frac{1}{4}t^2)$ is a solution and, in particular, $y(t) = 0$ is a solution. □

Example 2 Solve $\dfrac{dy}{dt} = -y^2$.

Here $f(t, y) = -y^2$ is a function which depends only on y with domain of definition $D = \{-\infty < t < \infty, -\infty < y < \infty\}$. If c is any real number, the function $y(t) = 1/(t - c)$ is a solution to the differential equation on any interval I not containing c. To verify this, we calculate

$$\frac{d}{dt}\left(\frac{1}{(t - c)}\right) = \frac{-1}{(t - c)^2}$$

and so

$$\frac{d}{dt} y(t) = -y(t)^2$$

on any interval that does not contain c. For instance, if $c = 2$, then we have that $y(t) = 1/(t - 2)$ is a solution on the interval $I_1 = \{-\infty < t < 2\}$ or $I_2 = \{2 < t < \infty\}$, but not on both. In this example, however, $y(t) = K/(t - c)$ is not a solution unless $K = 0$ or $K = 1$, so any nonzero multiple of $1/(t - c)$ is not a solution, but $y(t) = 0$ is again a solution. □

Example 3   Solve $\dfrac{dy}{dt} = \dfrac{1}{t}$, $t \neq 0$.

Here $f(t, y) = 1/t$, a function that depends only on t and is not defined at $t = 0$. There are two different domains of definition for f:

$$D_1 = \{-\infty < t < 0, -\infty < y < \infty\}$$

and

$$D_2 = \{0 < t < \infty, -\infty < y < \infty\}.$$

Two solutions of the differential equation are, for instance,

$$y_1(t) = \ln(-t) + 340$$

for t in $I = \{-\infty < t < 0\}$, and

$$y_2(t) = \ln t + 60$$

for t in $I = \{0 < t < \infty\}$, since

$$\frac{d}{dt} y_1(t) = \frac{d}{dt} [\ln(-t) + 340] = \frac{1}{-t}(-1) = \frac{1}{t}, \quad -\infty < t < 0$$

and

$$\frac{d}{dt} y_2(t) = \frac{d}{dt} (\ln t + 60) = \frac{1}{t}, \quad 0 < t < \infty.$$

Note that the graph of $y_1(t)$ lies in D_1, that of $y_2(t)$ lies in D_2, and no solution can be defined on any interval containing $t = 0$.   □

We saw in Example 2 that it is not always easy to predict the interval of definition of a solution of a differential equation by examining the differential equation itself. In fact, how can one be sure that a solution even exists? To be more precise, is it always true that the differential equation (1.2.1) has at least one solution whose graph contains (t_0, y_0), an arbitrary point in the domain of definition of $f(t, y)$? The answer to this is given by the following theorem.

The Basic Existence Theorem   Let $f(t, y)$ be a real-valued continuous function defined on a domain D in the ty plane and let (t_0, y_0) be an arbitrary point in D. Then there exists a solution $y(t)$ of the differential equation

$$\frac{dy}{dt} = f(t, y)$$

which is defined on some interval I containing t_0 and which satisfies the condition

$$y(t_0) = y_0.$$

Since the graph of $y(t)$ contains the point (t_0, y_0), we say that the solution *passes through the point* (t_0, y_0). The differential equation together with the *initial condition* $y(t_0) = y_0$ is an *initial value problem,* and the Basic Existence Theorem states that under very reasonable conditions on $f(t, y)$, it will always have a solution. The condition $y(t_0) = y_0$ is frequently a mathematical translation of the statement "at the start of an experiment the time is t_0 and the initial state of the system of interest is y_0." For instance, if the differential equation describes the motion of a particle, then y_0 could represent the initial displacement from a fixed reference point. Or if the differential equation describes the growth of a population, then y_0 could be the size of the population at time t_0.

The proof of the Basic Existence Theorem is difficult and will be discussed in Appendix 1 along with related theorems.

Although the above theorem asserts that under very simple conditions there are solutions to the differential equation (1.2.1), it does not suggest a method for constructing one. Furthermore, it does not address the question of uniqueness, i.e., can there be more than one solution passing through the point (t_0, y_0)? This is an important question. For any physical system modeled by an initial value problem, nonuniqueness would mean lack of predictability, and consequently the model should probably be scrapped.

Continuity of the function f alone, as required in the Basic Existence Theorem, is not sufficient to guarantee that only one solution of a given differential equation passes through a given point (t_0, y_0). One simple condition is that, in addition, $\partial f/\partial y$ be continuous in the domain D. This is summarized in the following theorem which is also proved in Appendix 1.

The Basic Uniqueness Theorem Let $f(t, y)$ and $\partial f(t, y)/\partial y$ be continuous in D and let $y_1(t)$ and $y_2(t)$ be any two solutions of

$$\frac{dy}{dt} = f(t, y)$$

on $a < t < b$. If $y_1(t_0) = y_2(t_0)$ for some t_0, $a < t_0 < b$, then $y_1(t) = y_2(t)$ for all t, $a < t < b$.

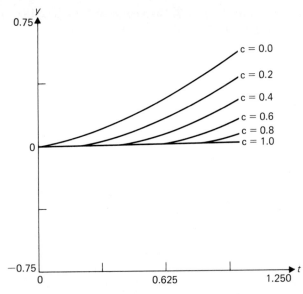

Figure 1.1   Some solutions of $dy/dt = y^{1/3}$, $y(0) = 0$.

A spectacular example of nonuniqueness is the initial value problem

$$\frac{dy}{dt} = y^{1/3}, \qquad y(0) = 0,$$

which has an infinite number of solutions, all defined on the interval $-\infty < t < \infty$. They are for any constant $c \geq 0$,

$$y(t) = \begin{cases} 0 & \text{for } t \leq c, \\ \tfrac{2}{3}(t - c)^{3/2} & \text{for } t > c. \end{cases}$$

A graph of $y(t)$ for some values of c is given in Fig. 1.1. The Basic Uniqueness Theorem does not apply since $\partial f/\partial y = \tfrac{1}{3}y^{-2/3}$ is not continuous in any domain D containing the initial point $(0, 0)$.

In order to examine the geometric meaning of the differential equation (1.2.1), it is helpful to introduce some terminology. An *integral curve* of the differential equation is the graph of a solution. A *tangential line segment* is a line segment through the point (t, y) with slope $m = f(t, y)$. The set of tangential line segments in the domain of $f(t, y)$ is called the *direction field* of the differential equation.

If $y(t)$ is a solution to (1.2.1), defined on an interval I, then its integral curve is simply the set of points $\{(t, y(t)) \mid t \text{ in } I\}$. The identity

$$\frac{dy}{dt}(t) = f(t, y(t)) \quad \text{for all } t \text{ in } I$$

can be interpreted as the geometric condition that at $(t, y(t))$ the slope of the integral curve and the slope of the tangential line segment must be equal. Note that to construct the tangential line segment through a point (t_0, y_0), one need only calculate the value $m = f(t_0, y_0)$ and does not need to know the solution of the differential equation itself.

For instance, in Example 1,

$$\frac{dy}{dt} = -\frac{1}{2} ty$$

and if $(t_0, y_0) = (0, 2)$, then $m = f(0, 2) = -\frac{1}{2}(0)(2) = 0$. Hence the slope of the integral curve at $(0, 2)$ is zero and the tangential line segment at $(0, 2)$ is horizontal. But as we have seen in Example 1, $y(t) = 2 \exp(-\frac{1}{4}t^2)$ is a solution and its graph (an integral curve) contains the point $(0, 2)$. Also dy/dt at $t = 0$ equals zero, as the reader may verify by direct differentiation. A sketch of the direction field of this differential equation can be constructed by drawing short line segments at selected points $\{(t_j, y_j)\}$ with slopes $m_j = -\frac{1}{2}t_j y_j$. Direction fields and several integral curves for Examples 1, 2, and 3 are sketched in Figs. 1.2, 1.3, and 1.4.

The direction field of Fig. 1.2 shows that the integral curves move toward the t-axis as t increases, i.e., toward the integral curve $y = 0$. This integral curve is said to be *stable*, a concept which will be defined and discussed in subsequent chapters. In contrast, the direction field of Fig. 1.3 shows that the integral curves move away from the t-axis as t increases, i.e.,

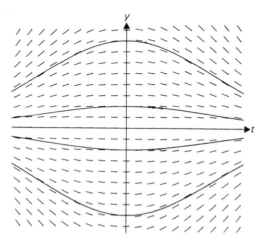

Figure 1.2 Direction field and several integral curves for Example 1: $dy/dt = -\frac{1}{2}ty$.

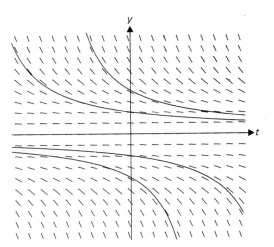

Figure 1.3   Direction field and several integral curves for Example 2: $dy/dt = -y^2$.

away from the integral curve $y = 0$. This integral curve is said to be *unstable*. The direction field of Fig. 1.4 shows that the integral curves move away from the y-axis as t increases, but in contrast to the first two examples, the y-axis is not an integral curve. These examples suggest that the direction field contains more geometric information about the analytic behavior of solutions than might be expected.

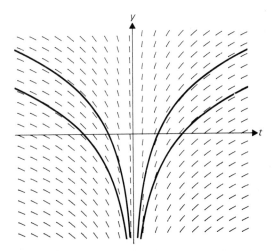

Figure 1.4   Direction field and several integral curves for Example 3: $dy/dt = 1/t, t \neq 0$.

EXERCISES 1.2

For each of the following differential equations, verify by direct substitution that the given function $y(t)$ is a solution. State the interval I on which it is defined.

1. a) $\dfrac{dy}{dt} = 10(y - t), \quad y(t) = t + 0.1$

b) $\dfrac{dy}{dt} = 4ty^2, \quad y(t) = \dfrac{-1}{(1 - 2t^2)}$

c) $\dfrac{dy}{dt} = -2y + e^t, \quad y(t) = \dfrac{1}{3}e^t + \dfrac{2}{3}e^{-2t}$

d) $\dfrac{dy}{dt} = \dfrac{t(3 - 2y)}{(t^2 - 1)}, \quad y(t) = \dfrac{1}{(t^2 - 1)} + \dfrac{3}{2}$

e) $\dfrac{dy}{dt} = \dfrac{3}{4}y^{5/3}, \quad y(t) = \left(1 - \dfrac{t}{2}\right)^{-3/2}$

2. Show that a solution to $dy/dt = e^{t-y}$ is given implicitly by $e^y = e^t + c$, where c is an arbitrary constant.

3. Show by implicit differentiation that a pair of functions $(x(t), y(t))$ satisfying $x^2 + y^2 = c^2$ also satisfy the differential equation $dy/dx = -x/y$.

4. Use the Basic Theorems to deduce that equations of the form $y' = g(t)y + h(t)$, with $g(t)$, $h(t)$ continuous on some interval I, always possess unique solutions on I.

5. Find points (t_0, y_0) where there might not be a unique solution satisfying $y(t_0) = y_0$ for the following differential equations:

a) $\dfrac{dy}{dt} = -\ln|y - 1|$

b) $\dfrac{dy}{dt} = \dfrac{1}{t}\sec y$

c) $\dfrac{dy}{dt} = \sqrt{(y - 1)(y - 2)}$

d) $\dfrac{dy}{dt} = \dfrac{y^2}{y - t}$

6. Show that the functions

$$y(t) = 0, \quad -\infty < t < \infty,$$

and

$$y(t) = (t - t_0)^3, \quad -\infty < t < \infty,$$

are both solutions of the initial value problem

$$\frac{dy}{dt} = 3y^{2/3}, \quad y(t_0) = 0.$$

Graph both solutions for $t_0 = -1, 0, 1$ and explain why the Basic Uniqueness Theorem does not apply.

7. Consider the problem $dy/dt = \sqrt{|1 - y^2|}$, $y(0) = 1$. Verify that

a) $y(t) = 1$ is a solution on any interval containing $t = 0$;

b) $y(t) = \cosh t$ is a solution on any interval $0 \le t \le b$ for any $b > 0$;

c) $y(t) = \cos t$ is a solution on a suitable interval. What is the largest interval containing $t = 0$ on which $\cos t$ is a solution?

8. Sketch the direction field and graph the given integral curves for the differential equation in the specified region D.

a) $\dfrac{dy}{dt} = 2y, D = \{0 < t < 1, -1 < y < 1\}$, $y(t) = Ke^{2t}, K = 0, \pm\frac{1}{4}, \pm\frac{1}{2}$.

b) $\dfrac{dy}{dt} = y - t, D = \{0 < t < 1, -1 < y < 2\}$, $y(t) = Ke^t + t + 1, K = -1, -\frac{1}{2}, -\frac{1}{4}, 0$.

c) $\dfrac{dy}{dt} = \dfrac{1}{y^2}, D = \{0 < t < 2, -2 < y < 2\}$, $y(t) = [3(t - c)]^{1/3}, c = 0, \pm\frac{1}{2}, \pm 1$.

9. A given function can be the solution to more than one differential equation. Show that $y(t) = t^2$ is a solution to both

$$\frac{dy}{dt} = 2t, \quad y(1) = 1,$$

and

$$\frac{dy}{dt} = \frac{2y}{t}, \quad y(1) = 1.$$

10. A weaker condition, which insures uniqueness of solutions of $y' = f(t, y)$ in D, is that $f(t, y)$ be continuous in D and there exist a constant K such that

$$|f(t, y_1) - f(t, y_2)| \leq K|y_1 - y_2|$$

for any points (t, y_1), (t, y_2) in D.

a) For the function $f(t, y) = |y|$ with $D = \{-\infty < t < \infty, -\infty < y < \infty\}$ show that $\partial f/\partial y$ is not continuous at $y = 0$ but f does satisfy the above condition.

b) Show that the functions

$$y(t) \equiv 0, \qquad y(t) = Ce^{-t}, C < 0,$$
$$y(t) = Ce^{t}, C > 0$$

for $-\infty < t < \infty$ are all possible solutions of $y' = |y|$. Show also that only one of these solutions will pass through any point in D of (a).

c) If $D = \{0 < t < 2, -\infty < y < \infty\}$ and if $f(t, y) = \sin|y|$, what is a possible choice of K?

1.3 FIRST ORDER DIFFERENTIAL EQUATIONS
OF THE FORM $dy/dt = g(t)$

In this section we restrict our attention to differential equations of the form

$$\frac{dy}{dt} = g(t), \tag{1.3.1}$$

where $g(t)$ is a continuous function defined on some t-interval. If $\int g(t)dt$ is an antiderivative of $g(t)$ and c is an arbitrary constant, then, by definition,

$$G(t) = \int g(t)dt + c$$

satisfies $G'(t) = g(t)$. Since antiderivatives can only differ by a constant, any solution of (1.3.1) is given by

$$y(t) = \int g(t)dt + c,$$

where c is an arbitrary constant.

Example 1 Solve $\dfrac{dy}{dt} = t^2 + 1$, $-\infty < t < \infty$.

A solution is $y(t) = t^3/3 + t + c$. □

Example 2 Solve $\dfrac{dy}{dt} = 4t \exp 2t$, $-\infty < t < \infty$.

A solution is $y(t) = (2t - 1) \exp 2t + c$. □

Example 3 Solve $\dfrac{dy}{dt} = \dfrac{1}{\sqrt{1-t^2}}$, $-1 < t < 1$.

A solution is $y(t) = \arcsin t + c$. □

It should be noted that in each of the above examples the differential equation has many solutions, in fact, a different solution for each choice of the constant of integration c. To solve an initial value problem, that is, to find the solution passing through a given point (t_0, y_0), an appropriate value of c must be chosen.

Example 4 Solve $\dfrac{dy}{dt} = \sec^2 t$, $y\left(\dfrac{9\pi}{4}\right) = 13$.

To find the particular solution of $dy/dt = \sec^2 t$ that passes through the point $(9\pi/4, 13)$, one first integrates to obtain the general solution $y(t) = \tan t + c$. Setting $t = 9\pi/4$, we have

$$y\left(\frac{9\pi}{4}\right) = \tan\frac{9\pi}{4} + c = 1 + c = 13.$$

Hence $c = 12$ and the solution is

$$y(t) = \tan t + 12, \quad \frac{3\pi}{2} < t < \frac{5\pi}{2}.$$

Note that the domain of definition of $\tan t$ must be chosen so as to contain the initial point $t_0 = 9\pi/4$. □

Let us now consider the initial value problem

$$\frac{dy}{dt} = g(t), \qquad y(t_0) = y_0, \tag{1.3.2}$$

where $g(t)$ is a continuous function defined on an interval I containing t_0 in its interior. The Fundamental Theorem of Calculus states that the function

$$G(t) = y_0 + \int_{t_0}^{t} g(s)\,ds$$

is differentiable on I and satisfies

$$\frac{dG(t)}{dt} = g(t), \qquad G(t_0) = y_0.$$

We conclude that

$$y(t) = y_0 + \int_{t_0}^{t} g(s)\,ds \tag{1.3.3}$$

is the solution of the initial value problem (1.3.2). Since $g(t)$ is continuous on I, it is the unique solution.

Example 5 Solve the initial value problem $\dfrac{dy}{dt} = \dfrac{1}{t(1-t)}$, $y\left(\dfrac{1}{4}\right) = 1, 0 < t < 1.$

Using partial fractions we have that

$$\frac{1}{t(1-t)} = \frac{1}{t} + \frac{1}{1-t}.$$

From (1.3.3), the solution is therefore given by

$$y(t) = 1 + \int_{1/4}^{t} \left(\frac{1}{s} + \frac{1}{1-s}\right) ds = 1 + \left(\ln \frac{s}{1-s}\right)\Big|_{1/4}^{t}$$

$$= 1 + \ln\left(\frac{t}{1-t}\right) - \ln\frac{1}{3} = (1 + \ln 3) + \ln\left(\frac{t}{1-t}\right).$$

Note that the solution is undefined at the endpoints of the interval $0 < t < 1$. □

The advantage of using the formula (1.3.3), which incorporates the initial conditions, is that if an antiderivative of $g(t)$ cannot be found, one can use numerical quadrature (e.g., the trapezoidal rule or Simpson's rule) to approximate values of $y(t)$ for $t \neq t_0$.

EXERCISES 1.3

Solve each of the following differential equations.

1. $\dfrac{dy}{dt} = t^4 + 2$

2. $\dfrac{dy}{dt} = e^{3t} + \cos \pi t$

3. $\dfrac{dy}{dt} = \dfrac{3t - 2}{t^2 - 2t - 8}$

4. $\dfrac{dy}{dt} = e^t \cos 2t$

5. $\dfrac{dy}{dt} = t^{1/3}\sqrt{t^{4/3} - 1}$

6. $\dfrac{dy}{dt} = \dfrac{2t + 1}{t^2 - 2t + 1}$

7. $\dfrac{dy}{dt} = 2t \sec t$

Solve each of the following initial value problems.

8. $\dfrac{dy}{dt} = t^2 - 4t + 1,\ y(1) = 1$

9. $\dfrac{dy}{dt} = \dfrac{1}{2} + \sin t,\ y(0) = 2$

10. $\dfrac{dy}{dt} = \dfrac{4}{t},\ y(e) = 0$

11. $\dfrac{dy}{dt} = \dfrac{t}{t^2 + 4t - 5},\ y(0) = 1$

12. $\dfrac{dy}{dt} = te^{-t^2}$, $y(0) = 1$

13. $\dfrac{dy}{dt} = t \ln 2t$, $y(\frac{1}{2}) = 0$

14. $\dfrac{dy}{dt} = \sqrt{1 + 2t}$, $y(4) = 10$

15. $\dfrac{dy}{dt} = t \cos(2t + 1)$, $y(-\frac{1}{2}) = 1$

16. If a ball is thrown vertically (up or down) from a building s_0 feet high with an initial velocity of v_0 ft/sec, show that its position (measured from the ground) at the end of t seconds is

$$s(t) = s_0 + v_0 t - \tfrac{1}{2}gt^2.$$

Use the fact that

$$\frac{dv}{dt} = -g = -32 \text{ ft/sec}^2 \quad \text{and} \quad \frac{ds}{dt} = v,$$

and assume the upward direction is the positive direction.

17. In Exercise 16, assume that $s_0 = 100$ ft and $v_0 = 64$ ft/sec.

a) How high will the ball go?

b) What is the velocity of the ball as it passes the building on its way down?

c) When does it hit the ground and with what velocity?

18. Let N_0 be the number of atoms of carbon-14 per gram of carbon present when a tree is felled ($t = 0$) and N be the number of atoms per gram of carbon that decays at an approximate rate of

$$\frac{dN}{dt} = -kN_0 e^{-kt} \text{ atoms/g/year,}$$

where $k \simeq 1.24 \cdot 10^{-4}$. A piece of wood found in an Egyptian tomb contained approximately 58% of the amount of carbon-14 present in a live tree. How old is the tomb?

19. A spherical mothball loses mass by evaporation at a rate proportional to its surface area. If it loses half its mass in 75 days, show that it will take approximately one year to disappear completely.

20. Approximate the value of $y(1)$ by using the trapezoidal or Simpson's rule for the following initial value problems:

a) $\dfrac{dy}{dt} = \cos t^2$, $y(0) = 1$

b) $\dfrac{dy}{dt} = e^{t^2}$, $y(0) = 0$

c) $\dfrac{dy}{dt} = \dfrac{1}{\ln(t + 2)}$, $y(0) = -2$

1.4 SEPARABLE EQUATIONS

One of the most important classes of first order differential equations consists of equations which can be written as

$$\frac{dy}{dt} = q(t)f(y), \tag{1.4.1}$$

where q and f are continuous functions. They are called *separable equations* because the key step in their solution is the separation of the independent and dependent variables. Separable equations serve as mathematical models for a variety of interesting problems, e.g., the determination of the velocity of an object falling in a gravitational field, the description of the growth of populations, and the analysis of certain nonlinear electric circuits.

The first step in constructing solutions of the differential equation (1.4.1) is so simple that it is frequently overlooked; namely, one should seek constant solutions. If $y(t) = a$ is a constant solution on an interval, then its derivative is zero. Therefore, by substitution we get

$$0 = q(t)f(a),$$

and, unless $q(t) = 0$ on the interval, which is unlikely in problems of physical interest, we must have $f(a) = 0$. Conversely, if $f(a) = 0$, then $y(t) = a$ is a constant solution. Hence, the zeros of f are constant solutions, and, in general, they are the only constant solutions.

Now assume that $y(t)$ is a nonconstant solution of (1.4.1) and that $f(y(t))$ is nonzero on an interval I. We divide both sides of Eq. (1.4.1) by $f(y(t))$ to obtain the *separated form* of the differential equation:

$$\frac{1}{f(y(t))} \frac{d}{dt} y(t) = q(t). \tag{1.4.2}$$

If $Q(t)$ is an antiderivative of $q(t)$ and $F(y)$ is an antiderivative of $1/f(y)$, then we can integrate both sides of (1.4.2) to obtain

$$F(y(t)) = Q(t) + C. \tag{1.4.3}$$

The constants of integration associated with F and Q have been combined in the single constant C. Note that the construction of the composite function $F(y(t))$ from an antiderivative of $1/f(y)$ is based on the chain rule. To see this, we differentiate $F(y(t))$:

$$\frac{d}{dt} F(y(t)) = \frac{d}{dy} F(y)|_{y=y(t)} \frac{d}{dt} y(t) = \frac{1}{f(y(t))} \frac{d}{dt} y(t),$$

which is the left side of Eq. (1.4.2).

Equation (1.4.3) is not a solution of the differential equation (1.4.1), but is a functional relationship between the nonconstant solution $y(t)$ and the independent variable t. It is called an *implicit solution*. One would like to take an implicit solution and solve for $y(t)$ and hence obtain an *explicit solution*. In theory, this can be done; however, it is not always possible to represent an explicit solution in closed form, i.e., in terms of familiar functions.[2] We shall return to this problem after studying several examples.

Example 1 Solve the initial value problem $\dfrac{dy}{dt} = 3t^2 e^{-y}$, $y(0) = 1$.

---

[2] It is not easy to give a satisfactory definition of a closed form solution. We have waffled by appealing to a reader-dependent concept, *familiar function*.

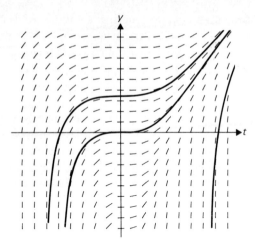

Figure 1.5 Direction field and several integral curves for $dy/dt = 3t^2 e^{-y}$.

Since $f(y) = e^{-y}$, there are no constant solutions and we separate variables to obtain

$$\exp[y(t)] \frac{d}{dt} y(t) = 3t^2.$$

Integrating both sides gives the implicit solution

$$\exp[y(t)] = t^3 + C. \tag{1.4.4}$$

The explicit solution is found by taking the natural logarithm of both sides of the above equation yielding

$$y(t) = \ln(t^3 + C), \tag{1.4.5}$$

which is defined on an interval where $t^3 + C$ is positive. To solve the initial value problem, C must be chosen so that $y(0) = 1$. Substituting $y = 1$ and $t = 0$ into (1.4.4) gives $C = e$. Therefore, the explicit solution to the initial value problem is

$$y(t) = \ln(t^3 + e).$$

The interval of definition of the solution is determined by the initial data. Since $\ln u$ is defined only for $u > 0$, this requires that $t^3 + e > 0$ or, equivalently, $t > -\exp(\frac{1}{3}) \approx -1.396$. The direction field and several integral curves for various initial values for Example 1 are sketched in Fig. 1.5. $\square$

Example 2   Solve $\dfrac{dy}{dt} = -\dfrac{y^2}{t}$, $t \neq 0$.

First we note that since $f(y) = -y^2$, $y(t) = 0$ is a constant solution defined on any interval that does not contain $t = 0$. Also, since this differential equation satisfies the hypotheses of the Basic Uniqueness Theorem of Section 1.1, in any domain that does not include points on the y-axis, we know that any nonconstant solution must be either always positive or always negative. Stated another way, any nonconstant solution cannot intersect the solution $y(t) = 0$ (otherwise uniqueness would be violated), and so it must always be positive or negative. Therefore, if $y(t)$ is a nonconstant solution, we can divide by $-[y(t)]^2$ to obtain the separated form of the differential equation:

$$-\frac{1}{[y(t)]^2} \frac{d}{dt} y(t) = \frac{1}{t}, \quad t \neq 0.$$

From this an implicit solution is obtained by integrating both sides to get

$$\frac{1}{y(t)} = \ln|t| + C,$$

and one can now write the solution as

$$y(t) = \frac{1}{\ln|t| + C},$$

where C is to be determined from given initial conditions. For example, $y(1) = 2$ implies that

$$2 = \frac{1}{\ln 1 + C} = \frac{1}{C};$$

hence $C = \frac{1}{2}$ and the solution is

$$y(t) = \frac{1}{\ln t + \frac{1}{2}}.$$

But $y(t)$ is only defined for an interval I that contains the initial point $t = 1$ and for which $\ln t + \frac{1}{2} \neq 0$. Since $\ln(e^{-1/2}) = -\frac{1}{2}$, the correct expression for the solution is

$$y(t) = \frac{1}{\ln t + \frac{1}{2}}, \quad t > e^{-1/2} \approx 0.6065.$$

Note that the same expression for $y(t)$ would satisfy the initial conditions $y(e^{-5/2}) \approx y(0.0821) = -\frac{1}{2}$, but in this case $y(t)$ would be defined for $0 < t < e^{-1/2}$ instead. The direction field and several integral curves for Example 2 are sketched in Fig. 1.6.   □

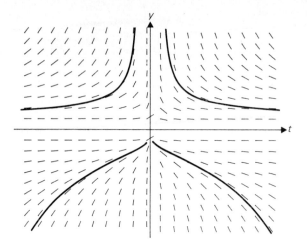

Figure 1.6 Direction field and several integral curves for $dy/dt = -y^2/t,\ t \neq 0$.

Example 3   Solve $\dfrac{dy}{dt} = \dfrac{t^4}{y^4}$.

Since there is no constant solution, we separate variables

$$[y(t)]^4 \frac{d}{dt} y(t) = t^4,$$

and integrate to obtain an implicit solution,

$$[y(t)]^5 = t^5 + C.$$

An explicit solution is

$$y(t) = (t^5 + C)^{1/5}, \tag{1.4.6}$$

where the constant C can be specified if initial conditions are given. Note that $y'(t) = t^4(t^5 + C)^{-4/5}$.

What is the interval of definition of the solution? This question requires a careful answer. The function $y(t)$ given by (1.4.6) is defined and continuous for all t, but it is not differentiable at the point $t = b$, where $b^5 + C = 0$. Therefore, one solution will be defined on the interval $b < t < \infty$ and a second solution on the interval $-\infty < t < b$. This is not just a technical nicety; a solution is a *continuously differentiable* function which satisfies the differential equation on an interval. It is important that $(t, y(t))$, for t in the interval, be in the domain of definition of the differential equation. In this example, $y(b) = 0$ but since $f(t, y) = t^4/y^4$, the differential equation is

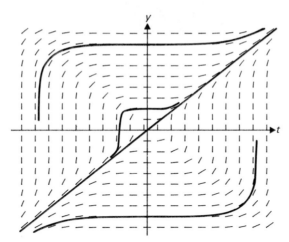

Figure 1.7   Direction field and several integral curves for $dy/dt = t^4/y^4$.

not defined for $y = 0$, and the function (1.4.6) is not a solution on any interval containing $t = b = (-C)^{1/5}$.

The direction field and several integral curves for Example 3 are sketched in Fig. 1.7 for various values of C. For instance, if $C = 0$, we have two distinct solutions: $y(t) = t$, $t > 0$, and $y(t) = t$, $t < 0$.   ☐

Example 4   Solve $\dfrac{dy}{dt} = \dfrac{\cos t}{y}$.

Since $f(y) = 1/y$, there are no constant solutions. We separate variables:

$$y(t) \frac{d}{dt} y(t) = \cos t,$$

and integrate to obtain the implicit solution

$$\tfrac{1}{2}[y(t)]^2 = \sin t + C. \tag{1.4.7}$$

Since the left-hand side of the equation is nonnegative and the differential equation is not defined when $y = 0$, C must satisfy

$$\sin t + C > 0.$$

Since $-1 \le \sin t \le 1$, this implies that $C > -1$. So we have the two explicit solutions:

$$y_1(t) = \sqrt{2(\sin t + C)} \tag{1.4.8}$$

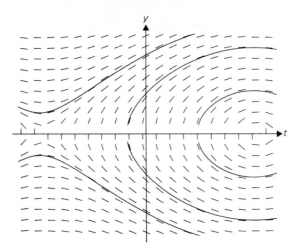

Figure 1.8 Direction field and several integral curves for $dy/dt = (\cos t)/y$.

if the initial conditions require that $y(t)$ be positive, and

$$y_2(t) = -\sqrt{2\,(\sin t + C)},$$

if the initial conditions require that $y(t)$ be negative. Note that $y(t) = 0$ is not admissible since the differential equation is not defined for $y = 0$.

Equation (1.4.7) represents a family of pairs of solutions. If $-1 < C \leq 1$, each pair $y_1(t)$ and $y_2(t)$ is defined on an interval of length π, and $y_1(t)$ will be positive while $y_2(t)$ will be negative. Neither will be defined at the endpoints of the intervals since they vanish there and the differential equation is not defined when $y = 0$. If $C > 1$, then each pair is defined for $-\infty < t < \infty$ since $\sin t + C$ never vanishes. The direction field and several integral curves for Example 4 are sketched in Fig. 1.8. $\quad\square$

Example 5 Solve $\dfrac{dy}{dt} = \dfrac{t^2}{1 + y^5}.$

In the last four examples it has been possible to solve explicitly for the solution $y(t)$. This is not the case here since separating variables gives

$$[1 + y(t)^5]\frac{d}{dt}\,y(t) = t^2,$$

and integrating gives the implicit solution

$$y(t) + \tfrac{1}{6}y(t)^6 = \tfrac{1}{3}t^3 + C.$$

The last expression cannot be solved for $y(t)$, and therefore for each value of C it defines an implicit solution.

However, given initial conditions, the constant C can be specifically determined. For instance, if it is required that $y(-1) = 2$, then letting $t = -1$ and $y = 2$ gives

$$2 + \tfrac{1}{6}2^6 = \tfrac{1}{3}(-1)^3 + C.$$

The solution is $C = 13$ and

$$y(t) + \tfrac{1}{6}y(t)^6 = \tfrac{1}{3}t^3 + 13$$

implicitly defines the solution of the initial value problem. □

We summarize the steps for solving separable equations in the following algorithm.

The separable equation algorithm.

1. Find all constant solutions.

2. Separate variables and integrate to obtain an implicit solution with possible restrictions on the additive constant.

3. If possible, find explicit solutions from the implicit solution. Determine their intervals of definition.

4. If initial data are given, use them in step 2 to determine the additive constant and again in step 3 to insure that you have the desired explicit solution.

Example 6   Solve $\dfrac{dy}{dt} = y(1 - y)$, $y(0) = 2$.

1. Since $f(y) = y(1 - y) = 0$ when $y = 0, 1$, there are two constant solutions, $y(t) = 0$ and $y(t) = 1$.

2. To construct the nonconstant solutions, we separate variables and use partial fractions to obtain

$$\frac{1}{y(t)[1 - y(t)]}\frac{d}{dt}y(t) = \left[\frac{1}{y(t)} + \frac{1}{1 - y(t)}\right]\frac{d}{dt}y(t) = 1.$$

The implicit solution obtained by integration is

$$\ln|y(t)| - \ln|1 - y(t)| = t + C$$

or

$$\ln\left|\frac{y(t)}{1 - y(t)}\right| = t + C.$$

Since $-\infty < \ln|u| < \infty$, this implies that C is unrestricted.

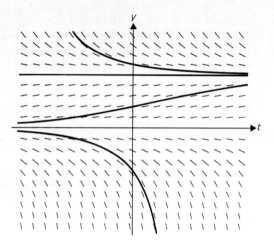

Figure 1.9 Direction field and several integral curves for $dy/dt = y(1 - y)$.

4. To solve the initial value problem, set $t = 0$ and $y = 2$:

$$\ln\left|\frac{2}{1 - 2}\right| = \ln 2 = 0 + C,$$

and so $C = \ln 2$. Therefore, the implicit solution is

$$\ln\left|\frac{y(t)}{1 - y(t)}\right| = t + \ln 2.$$

Since $y(0) = 2$, $1 - y(t) < 0$ (why?). So the implicit solution becomes

$$\ln\left[\frac{y(t)}{y(t) - 1}\right] = t + \ln 2$$

or

$$\frac{y(t)}{y(t) - 1} = 2e^t.$$

3. The above expression can be solved to find the explicit solution $y(t)$:

$$y(t) = \frac{-2e^t}{1 - 2e^t} = \frac{2e^t}{2e^t - 1}.$$

Note that the expression is valid for all $t > \ln \frac{1}{2}$ and $y(t)$ is always larger than 1 for $t \geq 0$. If we multiply the numerator and denominator by $\exp(-t)$, we obtain

$$y(t) = \frac{2}{2 - e^{-t}},$$

and it is seen that $y(t)$ approaches the constant solution $y(t) = 1$ as t approaches infinity. The direction field and several integral curves for Example 6 are sketched in Fig. 1.9.

EXERCISES 1.4

Solve each of the following differential equations:

1. $\dfrac{dy}{dt} = \dfrac{t^2}{y}$

2. $\dfrac{dy}{dt} = \dfrac{t^2\sqrt{y}}{1 + t^3}$

3. $\dfrac{dy}{dt} = e^{t+y}$

4. $\dfrac{dy}{dt} = \dfrac{t - ty^2}{y + t^2 y}$

5. $t\dfrac{dy}{dt} = y^2 - 3y + 2$

6. $\dfrac{dy}{dt} = \dfrac{t + 1}{y^4 + 1}$

7. $\dfrac{dy}{dt} = \dfrac{te^t}{2y}$

8. $\dfrac{dy}{dt} = \dfrac{1}{\tan t \cos^2 y}$

9. $t^2 y \dfrac{dy}{dt} = y - 1$

10. $\dfrac{dy}{dt} = \dfrac{\ln t \cos y}{t \sin 2y}$

Solve each of the following initial value problems:

11. $\dfrac{dy}{dt} = \dfrac{\sin t}{y}$, $y(0) = -1$; for what values of t is $y(t)$ defined?

12. $\dfrac{dy}{dt} = t^2 e^{-y}$, $y(0) = 2$

13. $3\dfrac{dy}{dt} \sin t + 2y \cos t = \cos t$, $y\left(\dfrac{\pi}{2}\right) = \dfrac{1}{2}$

14. $\dfrac{dy}{dt} = \dfrac{y^2 + y}{t}$, $y(1) = 1$

15. $\dfrac{dy}{dt} = \dfrac{3t^2 + 2}{2(y - 1)}$, $y(0) = -1$. What other initial condition does the implicit solution satisfy at $t = 0$?

16. $\dfrac{dy}{dt} = \sqrt{1 - y^2}$, $y\left(\dfrac{\pi}{2}\right) = 0$

17. $t^2 \dfrac{dy}{dt} - \dfrac{1}{2} \cos 2y = \dfrac{1}{2}$, $y(1) = \dfrac{\pi}{4}$. What is $\lim_{t \to \infty} y(t)$?

18. $\dfrac{dy}{dt} = ty^4 \sqrt{1 + 3t^2}$, $y(1) = -1$

19. $\dfrac{dy}{dt} = -3y^{4/3} \sin t$, $y\left(\dfrac{\pi}{2}\right) = \dfrac{1}{8}$. For what values of t is $y(t)$ defined? What if $y(\pi/2) = 8$?

20. $\dfrac{dy}{dt} + \dfrac{4y}{1 - t^2} = 0$, $y(2) = 9$. What is $\lim_{t \to \pm\infty} y(t)$?

21. In Exercise 16 of Section 1.3, the motion of a ball thrown upwards with initial velocity v_0 and subject only to the force of gravity was discussed. If it is also subject to air resistance proportional to its velocity $v(t)$, then its upward motion is described by the initial value problem

$$\frac{dv}{dt} = -g - kv, \qquad v(0) = v_0,$$

where $k > 0$ is the constant of proportionality.

a) Show that the solution of the initial value problem is

$$v(t) = \left(\frac{g}{k} + v_0\right) e^{-kt} - \frac{g}{k}.$$

b) Show that the ball reaches its maximum height when

$$t = t_{\max} = \frac{1}{k} \ln\left(1 + \frac{k}{g} v_0\right).$$

c) If $g = 32$ ft/sec^2, $v_0 = 128$ ft/sec, and $t_{\max} = 3.5$ sec, then estimate the maximum height attained by the ball. (*Hint:* To find k you may need the approximation $\ln(1 + x) \simeq x - \frac{1}{2}x^2 + \frac{1}{3}x^3$ with error less than 0.016 for $|x| < \frac{1}{2}$.

22. If a falling object is subject to gravity and an opposing force $f(v)$ of air resistance, then its velocity satisfies the initial value problem

$$\frac{dv}{dt} = g - f(v), \qquad v(0) = v_0.$$

a) If $f(v) = kv^2$, $k > 0$, that is, if the air resistance is proportional to the square of the velocity, show that $v(t)$ approaches $\sqrt{g/k}$ as t approaches infinity. This limiting or free-fall velocity is independent of v_0 but you may assume $v_0 = 0$.

b) If the limiting velocity of a human in free fall is 50 m/sec, estimate at what time a parachutist should open his parachute if he falls from an altitude of 2000 meters and the parachute must be opened when he reaches an altitude of 500 meters. Assume $v_0 = 0$ and $g = 9.8$ m/sec^2.

23. The law of radioactive decay states that the rate of disintegration of a radioactive substance is proportional at any instant to the amount of the substance present. If we denote the amount of a radioactive substance present after t years by $A(t)$, then the differential equation governing the above process is

$$\frac{dA}{dt} = -kA,$$

where k is a positive constant of proportionality. Suppose we know that 0.5 percent of radium disappears in 12 years (this condition can be used to evaluate k). What percentage will disappear in 100 years? What is the half-life of radium? (The half-life of a radioactive substance is the time it takes for half of the substance to decay.)

24. A family of curves which intersects each member of a given family of curves $F(x, y, C) = 0$, where C is a parameter, at an angle of $\pi/2$ is called the *orthogonal trajectories* of F. One can find them by differentiating $F(x, y, C)$ implicitly with respect to x to obtain

$$\frac{\partial F}{\partial x} + \frac{\partial F}{\partial y}\frac{dy}{dx} = 0$$

and then replacing dy/dx with $-1(dy/dx)$ and solving the resulting differential equation. Find

the orthogonal trajectories of

a) The family of parabolas $x^2 - 4Cy = 0$;

b) The family of circles $x^2 + y^2 - C^2 = 0$;

c) The family of ellipses $x^2 + 4y^2 - C = 0$;

d) The family exponentials $e^{2x}y - C = 0$.

In each case, graph a few members of each family so as to show the orthogonality.

25. Water escapes from a tank through a small hole in the bottom; the rate of escape is given by

$$\frac{dV}{dt} \text{ (ft}^3\text{/sec)} = -4.8 \cdot \text{Area of hole (ft}^2) \cdot \sqrt{h},$$

where $V(t)$ is the volume of water in the tank after t seconds and the depth of the water is h feet. If by reasons of symmetry one can express the volume as $V(h)$, one obtains the differential equation

$$V'(h)\frac{dh}{dt} = -4.8 \cdot \text{Area of hole (ft}^2) \cdot \sqrt{h}.$$

This equation can be solved to determine the time T when the tank will be empty: $h(T) = 0$.

a) Find the time required to empty a cylindrical tank of radius 4 ft and height 12 ft, if water is leaking from the bottom through a square hole with a 1-inch side.

b) Find the time required to empty a conical funnel 12 inches high and with base diameter 3 inches, if the apex hole has a diameter of ½ inch.

c) Using the fact that 1 foot ≃ 0.305 meters, modify the constant -4.8 to find the metric equivalent of the differential equation.

1.5 LINEAR FIRST ORDER DIFFERENTIAL EQUATIONS

A first order differential equation is linear if it can be written in the form

$$\frac{dy}{dt} + p(t)y = q(t). \tag{1.5.1}$$

We shall assume in the discussions to follow that $p(t)$ and $q(t)$ are continuous functions of t. If $q(t) = 0$, the differential equation is said to be *homo-*

geneous; otherwise, the equation is *inhomogeneous* and $q(t)$ is called the *inhomogeneous* or *forcing term.* The differential equations

$$\frac{dy}{dt} + t^2 y = 0 \tag{1.5.2a}$$

and

$$\frac{dy}{dt} + y = \sin t \tag{1.5.2b}$$

are linear, while the equations

$$\frac{dy}{dt} + t^2 y^2 = 0 \quad \text{and} \quad y\frac{dy}{dt} + y = \sin t$$

are not. Equation (1.5.2a) is homogeneous and (1.5.2b) is inhomogeneous.

We begin our analysis of (1.5.1) by considering the associated homogeneous equation obtained by setting $q(t) = 0$:

$$\frac{dy}{dt} + p(t)y = 0. \tag{1.5.3}$$

But by rewriting the equation as

$$\frac{dy}{dt} = -p(t)y,$$

one sees immediately that it is a separable equation. The only constant solution is $y(t) = 0$, and separating the variables gives

$$\frac{1}{y(t)} \frac{d}{dt} y(t) = -p(t).$$

If $P(t) = \int p(t) \, dt$ is an antiderivative of $p(t)$ we can integrate the last relation to get

$$\ln(y(t)) = -P(t) + C.$$

The explicit solution is therefore

$$y(t) = C \exp[-P(t)], \tag{1.5.4}$$

where C is an arbitrary constant. Note that if the constant C is chosen to be zero, one also obtains the solution $y(t) = 0$ from (1.5.4).

The expression (1.5.4) is called the *general solution* of the homogeneous equation (1.5.3). To justify this terminology we first need to observe an important property of linear homogeneous equations. If $y(t)$ is a solution of

$$\frac{dy}{dt} + p(t)y = 0,$$

then $u(t) = Cy(t)$ is also a solution for any constant C. This last statement is verified by noting that

$$\frac{d}{dt}u(t) + p(t) = \frac{d}{dt}Cy(t) + p(t)\,Cy(t)$$

$$= C\left[\frac{d}{dt}y(t) + p(t)\,y(t)\right] = 0,$$

since $y(t)$ is a solution. But this implies that $u(t)$ is also a solution.

To justify the statement that $y(t) = C\exp[-P(t)]$ is a general solution it must be shown that if $u(t)$ is *any* solution of (1.5.3), it can be written as some constant times $\exp[-P(t)]$. It is not obvious how to prove this statement, so we invoke a mathematical adage: "When stuck, reformulate." With this in mind, we pose an equivalent problem: Show that

$$\frac{d}{dt}u(t)\exp[P(t)] = 0.$$

The equivalence of the two statements is clear because the last relation holds if and only if $u(t)\exp[P(t)] = C$, a constant. Since $\exp[P(t)] > 0$, the last equation is equivalent to $u(t) = C\exp[-P(t)]$, the desired statement. But

$$\frac{d}{dt}\{u(t)\exp[P(t)]\} = \exp[P(t)]\frac{d}{dt}u(t) + u(t)\exp[P(t)]\frac{d}{dt}P(t)$$

$$= \exp[P(t)]\left[\frac{d}{dt}u(t) + p(t)\,u(t)\right] = 0$$

since $u(t)$ is a solution to the homogeneous equation (1.5.3). Therefore $y(t) = C\exp[-P(t)]$ is a general solution.

Example 1   Find the general solution to $\dfrac{dy}{dt} + (\cos t)y = 0$.

Here $p(t) = \cos t$ and $P(t) = \sin t$ is an antiderivative. Using (1.5.4), we get the general solution

$$y(t) = C\exp(-\sin t).$$

Another antiderivative of $\cos t$ is $P(t) = \sin t + 10$, and so

$$y(t) = C\exp(-\sin t - 10)$$
$$= C\exp(-10)\exp(-\sin t)$$
$$= C^*\exp(-\sin t),$$

where $C^* = C\exp(-10)$, is also a solution. The two solutions clearly have the same degree of arbitrariness and hence are equivalent. Actually, more can be said. If $P(t)$ is an antiderivative of $p(t)$, so is $P(t) + c$ for any constant c. Therefore, the general solution has the form

$$y(t) = C \exp[-P(t) - c] = C^* \exp[-P(t)]$$

with $C^* = C \exp(-c)$. In other words, the arbitrary additive constant of integration c is absorbed into the arbitrary multiplicative constant of the general solution of the linear homogeneous differential equation (1.5.3). □

Example 2 Solve $\dfrac{dy}{dt} + 2ty = 0$, $y(10) = 3$.

To solve this initial value problem, we first find the general solution of the differential equation. Since $p(t) = 2t$, an antiderivative is $P(t) = t^2$, and so

$$y(t) = C e^{-t^2}$$

is the general solution. To satisfy the initial condition $y(10) = 3$, we let $t = 10$ and $y = 3$ to obtain

$$3 = C e^{-100}.$$

Hence $C = 3 \exp(100)$, and so the solution to the initial value problem is

$$y(t) = 3e^{100} e^{-t^2} = 3e^{(100-t^2)}. □$$

Let us now return to (1.5.1), retaining the forcing function $q(t)$, and solve the resulting inhomogeneous equation

$$\frac{dy}{dt} + p(t)y = q(t). \tag{1.5.5}$$

The basis of our method for solving (1.5.5) is the observation that if $P(t) = \int p(t)\,dt$ is an antiderivative of $p(t)$, then by the product rule and the chain rule of differentiation we get

$$\frac{d}{dt}\{\exp[P(t)]\,y(t)\} = \exp[P(t)]\frac{d}{dt}y(t) + p(t)\,y(t)\exp[P(t)]$$

$$= \exp[P(t)]\left[\frac{d}{dt}y(t) + p(t)\,y(t)\right],$$

since $dP(t)/dt = p(t)$. Now, multiplying both sides of (1.5.5) by $\exp[P(t)]$ and rearranging, we obtain

$$\exp[P(t)]\left[\frac{d}{dt}y(t) + p(t)\,y(t)\right] = \exp[P(t)]\,q(t),$$

or, by our observation,

$$\frac{d}{dt}\{\exp[P(t)]\,y(t)\} = \exp[P(t)]\,q(t). \tag{1.5.6}$$

Equation (1.5.6) is of the form studied in Section 1.2. It can be solved by simple integration, i.e., by finding antiderivatives. Because of this, the function $\exp[P(t)]$ is called an *integrating factor*.

We now integrate both sides of (1.5.6) to get

$$\exp[P(t)]\, y(t) \;=\; \int^t \exp[P(s)]\, q(s)\, ds \;+\; C,$$

where C is an arbitrary constant. Dividing by $\exp[P(t)]$ yields

$$y(t) \;=\; \exp[-P(t)] \int^t \exp[P(s)]\, q(s)\, ds \;+\; C \exp[-P(t)]. \quad (1.5.7)$$

Before continuing the discussion, it would be useful to look at an example with $p(t) = k$, a fixed constant, and use formula (1.5.7).

Example 3 $\quad \dfrac{dy}{dt} + ky = \sin t.$

Since $p(t) = k$, $P(t) = kt$ is an antiderivative and $\exp[P(t)] = e^{kt}$. From integral tables we know that

$$\int^t e^{ks} \sin s\, ds \;=\; e^{kt}\,\frac{k \sin t - \cos t}{1 + k^2} \;+\; H,$$

where H is a constant of integration. After substituting this into (1.5.7), we obtain

$$y(t) \;=\; e^{-kt}\left(e^{kt}\,\frac{k \sin t - \cos t}{1 + k^2} + H \right) + C e^{-kt}$$

$$=\; \frac{k \sin t - \cos t}{1 + k^2} \;+\; C^* e^{-kt},$$

where $C^* = C + H$. Once again we see that there is a one-parameter family of solutions parameterized by the constant C^*. It is called the *general solution* of the differential equation.

Observe that the above solution can be written as the sum of two functions, one of which, $C^* \exp(-kt)$, is the general solution of the homogeneous equation

$$\frac{dy}{dt} + ky = 0,$$

and the other $(k \sin t - \cos t)/(1 + k^2)$, is a *particular solution* to the inhomogeneous equation

$$\frac{dy}{dt} + ky = \sin t.$$

We suggest that the reader verify these statements. □

Let us return to the expression (1.5.7) for the solution of the inhomogeneous differential equation and consider the first term

$$y_p(t) = \exp[-P(t)] \int^t \exp[P(s)] \, q(s) \, ds.$$

It satisfies

$$\frac{d}{dt} y_p(t) = \frac{d}{dt} \exp[-P(t)] \int^t \exp[P(s)] \, q(s) \, ds$$

$$+ \exp[-P(t)] \frac{d}{dt} \int^t \exp[P(s)] \, q(s) \, ds$$

$$= -p(t) \exp[-P(t)] \int^t \exp[P(s)] \, q(s) \, ds$$

$$+ \exp[-P(t)] \exp[P(t)] \, q(t)$$

$$= -p(t) y_p(t) + q(t),$$

and therefore it is a particular solution of the inhomogeneous differential equation. The expression (1.5.7) is a sum of a particular solution and the general solution $C \exp[-P(t)]$ of the homogeneous equation and hence is the general solution of the linear inhomogeneous differential equation (1.5.5). The constant C can be determined if initial conditions are given.

We now summarize all of this in the following algorithm:

The integrating factor algorithm.

1. Write the equation in the form

$$\frac{dy}{dt} + p(t)y = q(t).$$

2. Find $P(t)$, an antiderivative of $p(t)$, and the integrating factor $\exp[P(t)]$.

3. Rewrite the differential equation as

$$\frac{d}{dt} \{\exp[P(t)] \, y(t)\} = \exp[P(t)] \, q(t).$$

4. Integrate to obtain

$$\exp[P(t)] \, y(t) = \int^t \exp[P(s)] \, q(s) \, ds + C.$$

5. Find the general solution $y(t)$:

$$y(t) = \exp[-P(t)] \int^t \exp[P(s)] \, q(s) \, ds + C \exp[-P(t)].$$

6. If initial data are given, substitute into the above equation and solve for C.

Example 4 Solve $\dfrac{1}{t}\dfrac{dy}{dt} + \dfrac{2}{1 + t^2}y = 3$, $y(1) = 4$.

Employing the above algorithm step by step we can write:

1. $\dfrac{dy}{dt} + \dfrac{2t}{1 + t^2}y = 3t$

2. $p(t) = \dfrac{2t}{1 + t^2}$,

so let

$$P(t) = \int \frac{2t}{1 + t^2}\,dt = \ln(1 + t^2)$$

and the integrating factor is

$$\exp[\ln(1 + t^2)] = 1 + t^2.$$

3. Hence,

$$\frac{d}{dt}[(1 + t^2)\,y(t)] = (1 + t^2)\,3t.$$

4. Therefore,

$$(1 + t^2)\,y(t) = \int^t (1 + s^2)\,3s\,ds + C = \tfrac{3}{4}t^4 + \tfrac{3}{2}t^2 + C.$$

5. The general solution is

$$y(t) = \frac{3}{4}\frac{t^4 + 2t^2}{1 + t^2} + \frac{C}{1 + t^2}.$$

6. To determine C set $t = 1$, $y = 4$ to obtain

$$4 = \frac{3}{4}\cdot\frac{3}{2} + \frac{C}{2},$$

and so $C = 23/4$. The solution to the initial value problem is

$$y(t) = \frac{3}{4}\frac{t^4 + 2t^2}{1 + t^2} + \frac{23}{4}\frac{1}{1 + t^2} = \frac{3}{4}(1 + t^2) + \frac{5}{1 + t^2}. \quad \square$$

Example 5 Solve $\dfrac{dy}{dt} + 2ty = \sin t$, $y(0) = 2$.

1. Not needed.
2. $p(t) = 2t$,

so let

$$P(t) = \int 2t \, dt = t^2$$

and the integrating factor is $\exp(t^2) = e^{t^2}$.

3. Hence

$$\frac{d}{dt}[e^{t^2} y(t)] = e^{t^2} \sin t.$$

4. Therefore,

$$e^{t^2}y(t) = \int^t e^{s^2} \sin s \, ds + C.$$

The integral cannot be expressed in terms of elementary functions, but the expression for the solution is still valid. For convenience we write

$$e^{t^2}y(t) = \int_0^t e^{s^2} \sin s \, ds + C,$$

since the initial data are given at $t = 0$.

5. The general solution is

$$y(t) = e^{-t^2} \int_0^t e^{s^2} \sin s \, ds + Ce^{-t^2}.$$

6. To determine C, let $t = 0$ and $y(0) = 2$ to obtain

$$2 = 1 \cdot \int_0^0 e^{s^2} \sin s \, ds + C \cdot 1 = 0 + C$$

or $C = 2$. The solution is

$$y(t) = e^{-t^2} \int_0^t e^{s^2} \sin s \, ds + 2e^{-t^2}.$$

The trick in step 4 was to choose the antiderivative that vanishes at $t = 0$ where the initial data were given. We generalize this trick to obtain the solution of the initial value problem

$$\frac{dy}{dt} + p(t)y = q(t), \qquad y(t_0) = y_0$$

as

$$y(t) = \exp[-P(t)] \int_{t_0}^t \exp[P(s)] \, q(s) \, ds + y_0 \exp[-P(t)],$$

where $P(t)$ is an antiderivative of $p(t)$. As in the previous example, it may not be possible to evaluate the integral in the above expression, but it is still a valid expression for the solution. Values of $y(t)$ can be numerically estimated, for instance. $\square$

Another method for solving the inhomogeneous linear differential equation (1.5.1) is the *variation of parameters method*. It is equivalent to the integrating factor method for first order differential equations, but, unlike the integrating factor method, it can be extended to higher order equations as we shall see in later chapters. The variation of parameters method originated in celestial mechanics which studies the motion of planets, comets, etc.; it will be presented here as the result of a clever observation.

The general solution to the homogeneous differential equation

$$\frac{dy}{dt} + p(t)\, y = 0$$

is $y(t) = C \exp[-P(t)]$, a constant multiple of $y_h(t) = \exp[-P(t)]$. The assumption of the variation of parameters method is that the solution of the inhomogeneous differential equation

$$\frac{dy}{dt} + p(t)\, y = q(t) \tag{1.5.8}$$

can also be expressed as a multiple of $y_h(t)$. However, the coefficient will not be a constant but a function of t which must be determined.

Therefore assume that $y(t) = y_h(t)\, v(t)$ is a solution of (1.5.8) and substitute it into the differential equation to get

$$\frac{d}{dt}\,[y_h(t)\, v(t)] + p(t)\, y_h(t)\, v(t) = q(t)$$

or

$$\left[\frac{d}{dt}\, y_h(t)\right] v(t) + y_h(t) \left[\frac{d}{dt}\, v(t)\right] + p(t)\, y_h(t)\, v(t)$$

$$= \left[\frac{d}{dy}\, y_h(t) + p(t)\, y_h(t)\right] v(t) + y_h(t) \left[\frac{d}{dt}\, v(t)\right] = q(t).$$

But $y_h(t)$ is a solution of the homogeneous equation, hence the bracketed expression is zero. Therefore $v(t)$ must satisfy

$$y_h(t)\,\frac{d}{dt}\, v(t) = q(t)$$

or

$$\frac{d}{dt} v(t) = y_h(t)^{-1} q(t) = \exp[P(t)] q(t).$$

The last equation is of the type studied in Section 1.2 and can be solved directly by integration to obtain

$$v(t) = \int^t \exp[P(s)] q(s) \, ds + C.$$

Finally,

$$y(t) = y_h(t) v(t) = \exp[-P(t)] \left[\int^t \exp[P(s)] q(s) \, ds + C \right],$$

which is the answer previously obtained by using integrating factors.

The following example shows the use of the method of variation of parameters directly.

Example 6 Find a general solution to $\dfrac{dy}{dt} + \dfrac{2}{t^2} y = \dfrac{6}{t^2}$.

A solution $y_h(t)$ of $(dy/dt) + (2y/t^2) = 0$ is $y_h(t) = \exp(2/t)$ since $P(t) = -2/t$ is an antiderivative of $p(t) = 2/t^2$. Let $y(t) = \exp(2/t) v(t)$; substitute it directly into the differential equation:

$$\frac{d}{dt} [\exp(2/t) v(t)] + \frac{2}{t^2} \exp\left(\frac{2}{t}\right) v(t) = \frac{6}{t^2}$$

or

$$-\frac{2}{t^2} \exp\left(\frac{2}{t}\right) v(t) + \exp\left(\frac{2}{t}\right) \left[\frac{d}{dt} v(t) \right] + \frac{2}{t^2} \exp\left(\frac{2}{t}\right) v(t) = \frac{6}{t^2}.$$

Canceling and multiplying by $\exp(-2/t)$ gives

$$\frac{d}{dt} v(t) = \exp\left(-\frac{2}{t}\right) \frac{6}{t^2}$$

or

$$v(t) = \int^t \exp\left(\frac{-2}{s}\right) \frac{6}{s^2} \, ds + C = 3 \exp\left(\frac{-2}{t}\right) + C.$$

A general solution is therefore

$$y(t) = \exp\left(\frac{2}{t}\right) v(t) = 3 + C \exp\left(\frac{2}{t}\right).$$

EXERCISES 1.5

In each of the Exercises 1 through 12, find a general solution of the given differential equation.

1. $\dfrac{dy}{dt} + y = 0$ **2.** $\dfrac{dy}{dt} - 4y = 0$

3. $\dfrac{dy}{dt} + t^2 y = 0$ **4.** $\dfrac{dy}{dt} - \dfrac{1}{t} y = 0$

5. $\dfrac{dy}{dt} - 3y = 0$ **6.** $\dfrac{dy}{dt} + 2y = t$

7. $\dfrac{dy}{dt} - y = \sin 2t$ **8.** $\dfrac{dy}{dt} + 2ty = 4t$

9. $t^2 \dfrac{dy}{dt} + ty + 1 = 0$

10. $t \dfrac{dy}{dt} - 4y = t^3 + 3t^2$

11. $\dfrac{dy}{dt} + (\cot t)y = 3 \sin t \cos t$

12. $t \dfrac{dy}{dt} + \dfrac{1}{\ln(t)} y = 1$

In each of the Exercises 13 through 26, find a general solution and the solution satisfying the given initial conditions.

13. $\dfrac{dy}{dt} + 2y = 0,\ y(0) = 4$

14. $\dfrac{dy}{dt} - \dfrac{1}{2} y = 0,\ y(2) = -1$

15. $\dfrac{dy}{dt} = (\sec t)y,\ y\left(\dfrac{\pi}{4}\right) = \dfrac{\sqrt{2}}{2}$

16. $\dfrac{dy}{dt} + |t|y = 0,\ y(0) = 4$

17. $\dfrac{dy}{dt} + y = 2t + 5,\ y(0) = 4$

18. $\dfrac{dy}{dt} + 4y = 6 \sin 2t,\ y(0) = -\dfrac{3}{5}$

19. $\dfrac{dy}{dt} - 2y + e^{2t} = 0,\ y(0) = 2$

20. $\dfrac{dy}{dt} - 2y = t^2 e^{2t},\ y(0) = 2$

21. $\dfrac{dy}{dt} - 2ty - t = 0,\ y(0) = 1$

22. $e^y \dfrac{dy}{dt} + e^y = 4 \sin t,\ y(0) = 0$

Hint: Let $u = e^y$.

23. $(t^2 + 1) \dfrac{dy}{dt} - 2ty = t^2 + 1,\ y(1) = \pi$

24. $\dfrac{dy}{dt} - (\tan t)y = e^{\sin t},\ 0 < t < \pi/2,\ y(1) = 0$

25. $\dfrac{dy}{dt} + 2ty = e^{-t^2},\ y(0) = 1$

26. $\dfrac{dy}{dt} - 2y = t^2 e^{2t},\ y(0) = 2$

27. $\dfrac{dy}{dt} + \dfrac{2}{t} y = \dfrac{\cos t}{t^2},\ y(\pi) = 0,\ t > 0$

28. Show that the solution of

$$\dfrac{dy}{dt} + 2ty = 1,\ y(0) = 0$$

is

$$y(t) = e^{-t^2} \int_0^t e^{s^2}\, ds.$$

Use Simpson's rule with four and eight subdivisions to find $y(0.5)$.

The equation $dy/dt + p(t)y = q(t)y^n$ is known as Bernoulli's equation. It is linear when $n = 0$ or 1. Show that it can be reduced to a linear equation for any other value of n by change of variables $z = y^{1-n}$. Apply this method to solve the following equations.

29. $\dfrac{dy}{dt} - y = ty^2$ **30.** $\dfrac{dy}{dt} - \dfrac{t}{2} y = ty^5$

31. $\dfrac{dy}{dt} + \dfrac{y}{t} = t^3 y^3$

32. $2 \dfrac{dy}{dt} - \dfrac{1}{t} y + (\cos t)y^3 = 0$

33. Newton's law of cooling states that the rate of change of temperature of a body is proportional to the difference of temperature between a cooling body and its surroundings. If $T(t)$ represents the temperature of the body at time t and S is the constant temperature of the surroundings, this leads to the differential equation

$$\frac{dT}{dt} = -k(T - S),$$

where k is a constant of proportionality. Suppose a body cools in 10 minutes from 100°C to 60°C. If the surroundings are at a temperature of 20°C, when will the body cool to 25°C?

34. A ball is dropped with zero initial velocity from a building and encounters air resistance proportional to its velocity (see Exercise 22 of Section 1.4). Let its steady state $(t \to \infty)$ velocity be 24 ft/sec.

a) Find its velocity at the end of 3 seconds and how far it has dropped.

b) If the building is 222 ft high, estimate at what time the ball hits the ground.

35. A simple model for the growth of a population is exponential growth where, if $P(t)$ is the size of a population (always nonnegative) at time t, then

$$P'(t) = rP(t), \quad r > 0, \quad P(0) = P_0,$$

where P_0 is the initial size. If the population is being harvested, say, by a predator or disease, at a constant rate $H > 0$, then a possible model is

$$P'(t) = rP(t) - H, \quad P(0) = P_0.$$

a) If $r = 0.01$, $P_0 = 4000$, show that the population will expire in finite time, that is, $P(T) = 0$ for some $T > 0$, if $H > 40$, but will increase without bound if $H < 40$.

b) If $H = 50$, find the time T at which the population expires.

c) What is the solution if $H = 40$?

d) For the general case, show that $H = rP_0$ is the critical harvest rate above which the population expires in finite time.

36. Consider the linear equation

$$y' = -ky + \phi(t), \quad k > 0, \quad y(0) = y_0,$$

where $\phi(t)$ is periodic with period T, that is, $\phi(t + T) = \phi(t)$ for all t. The solution can be written in the form

$$y(t) = y_0(t) + y_p(t),$$

where $y_0(t) \to 0$ as $t \to \infty$ and is called a *transient solution* and $y_p(t)$ is periodic with period T and is called the *steady state solution*. Find the steady state solution of the following:

a) $y' = -y + 4 \cos t$

b) $y' = -2y + 1 + \sin 2t$

c) $y' = -y - 2 \cos^2 t$. *Hint:* Express $\cos^2 t$ in terms of $\cos 2t$.

d) $y' = -2y - 5 \sin t$, $y(0) = 3$. Find a time T for which $|y(t) - y_p(t)| < 10^{-5}$ for $t > T$.

1.6 SOME ELEMENTARY NUMERICAL METHODS

In the previous two sections we have concentrated on methods for finding explicit solutions for certain differential equations, that is, solutions expressible in terms of elementary functions. Some differential equations, however, have solutions that cannot be written down in this manner, e.g.,

$$\frac{dy}{dt} = e^{t^2}.$$

In fact, many of the differential equations encountered in mathematical models of physical phenomena cannot be solved explicitly, or data are only known approximately. So we are led to study methods for obtaining approximations to solutions. Such solutions are usually called simply *numerical solutions*.

In this section we will study some fundamental ideas, which serve as the basis for numerical methods, and present two examples of such methods. In Chapters 3 and 8, numerical methods will be discussed in more detail.

We begin our discussion of numerical methods by considering the initial value problem

$$\frac{dy}{dt} = f(t, y), \qquad y(t_0) = y_0, \tag{1.6.1}$$

defined on some interval $t_0 \leq t \leq b$. The numerical method to be studied will generate a table of approximate values for $y(t)$, the solution to (1.6.1). For now, we suppose that the entries in the table are for equally spaced values of the independent variable t. If we choose an integer n and let

$$h = (b - t_0)/n,$$

our task is to find approximations to $y(t)$ at the equally spaced points $t_i = t_0 + ih$, $i = 1, 2, \ldots, n$. In what follows, the expression $y(t_i)$ will always represent the actual solution of (1.6.1) evaluated at $t = t_i$, whereas y_i will represent an approximation to that value. The reader should keep this notational distinction in mind.

The simplest example of a numerical method for obtaining an approximate solution to (1.6.1) is *Euler's method*. In a sense, Euler's method produces a solution which follows the direction field of the differential equation.

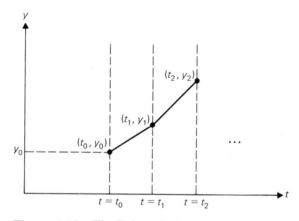

Figure 1.10   The Euler solution.

Consider a ty coordinate system and construct the vertical lines $t = t_0$, $t = t_1, \ldots$, as in Fig. 1.10.

The exact solution to (1.6.1) passes through the point (t_0, y_0) and has the slope $f(t_0, y_0)$ there. Draw a straight line through (t_0, y_0) with slope $f(t_0, y_0)$. This line intersects the vertical line $t = t_1$ at some point with $y = y_1$. Draw a straight line through (t_1, y_1) with slope $f(t_1, y_1)$, and it will intersect the line $t = t_2$ at some point with $y = y_2$. Continuing in this fashion, we shall generate a set of points (t_i, y_i), $i = 1, 2, \ldots, n$, such that the y_i values are approximations to $y(t_i)$.

To derive an analytical expression for the above, proceed as follows. The slope of the direction field at (t_0, y_0) is $f(t_0, y_0)$ and, by using the point-slope form for the equation of a straight line, we get the equation of the first line segment:

$$\frac{y - y_0}{t - t_0} = f(t_0, y_0).$$

Hence, letting $t = t_1$, we obtain the following expression for y_1, the approximation to $y(t_1)$:

$$y_1 = y_0 + (t_1 - t_0) f(t_0, y_0),$$

or, since $h = t_1 - t_0$,

$$y_1 = y_0 + hf(t_0, y_0).$$

The slope at (t_1, y_1) is $f(t_1, y_1)$, and so the equation of the second line segment is

$$\frac{y - y_1}{t - t_1} = f(t_1, y_1).$$

Letting $t = t_2$ and remembering that $t_2 - t_1 = h$, we obtain the expression for y_2, the approximation to $y(t_2)$:

$$y_2 = y_1 = hf(t_1, y_1).$$

In general, if we are at the point (t_i, y_i), we calculate the slope $f(t_i, y_i)$, then use the equation of the ith line segment to obtain the approximation y_{i+1} to the value $y(t_{i+1})$ of the actual solution:

$$y_{i+1} = y_i + hf(t_i, y_i).$$

This is summarized in the following algorithm.

The Euler algorithm. An approximate solution to the initial value problem (1.6.1) at the equally spaced points $t_0, t_1, t_2, \ldots, t_n$ is given by

$$y_{i+1} = y_i + hf(t_i, y_i), \qquad t_{i+1} = t_0 + (i + 1)h, \quad i = 0, \ldots, n - 1, \quad (1.6.2)$$

where h, t_0, and y_0 are given.

Example 1   Find an approximate solution to the initial value problem

$$y' = y, \qquad y(0) = 1$$

at the point $t = 1$ by using the Euler algorithm with step $h = 1/4$.

Here $f(t, y) = y$, so (1.6.2) simplifies to

$$y_{i+1} = y_i + hy_i = (1 + h)y_i,$$

or, since $h = 1/4$,

$$y_{i+1} = \frac{5}{4} y_i.$$

Starting then with $t_0 = 0$, $y_0 = 1$, we compute

$$y_1 = \frac{5}{4} y_0 = \frac{5}{4} \cdot 1 = \frac{5}{4},$$

$$t_1 = t_0 + 1 \cdot \frac{1}{4} = 0 + \frac{1}{4} = \frac{1}{4},$$

$$y_2 = \frac{5}{4} y_1 = \frac{5}{4} \cdot \frac{5}{4} = \frac{25}{16},$$

$$t_2 = t_0 + 2 \cdot \frac{1}{4} = 0 + \frac{1}{2} = \frac{1}{2},$$

$$y_3 = \frac{5}{4} y_2 = \frac{5}{4} \cdot \frac{25}{16} = \frac{125}{64},$$

$$t_3 = t_0 + 3 \cdot \frac{1}{4} = 0 + \frac{3}{4} = \frac{3}{4},$$

$$y_4 = \frac{5}{4} y_3 = \frac{5}{4} \cdot \frac{125}{64} = \frac{625}{256},$$

$$t_4 = t_0 + 4 \cdot \frac{1}{4} = 0 + 1.$$

Therefore the approximation is $y_4 = 625/256 = 2.441$ (correct to three decimal places). The actual solution is $y(t) = e^t$ and $y(1) = 2.718$ (again correct to three decimal places). The absolute error in the numerical solution is $|y_4 - y(1)| = 0.277$. The two solutions are represented graphically in Fig. 1.11, where the approximating points y_0, y_1, y_2, y_3, and y_4 are joined by straight line segments.

   A point worth noting here is that the first segment of the Euler approximation is part of the tangent line to the exact solution. This is not true in general for subsequent segments. The ith segment is tangent to a solution of the differential equation going through the point (t_i, y_i), but not necessarily tangent to the solution of interest, namely the one passing through the initial

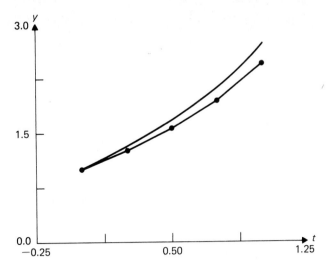

Figure 1.11 Plot of the Euler solution and exact solution to $y' = y$, $y(0) = 1$.

point (t_0, y_0). Consequently, it is possible and usual for the "Euler solution" to drift away from the true solution after many steps of the algorithm. ☐

With $h = \frac{1}{4}$ in Example 1, Euler's method gave an approximation that had an absolute error of 0.277. Our intuition should tell us that we would probably make a smaller error by using a smaller value of h. Computing with $h = \frac{1}{8}$ and $t_0 = 0$ and $y_0 = 1$ as before, we get

$$y_1 = (1 + h)y_0 = \frac{9}{8} \cdot 1 = \frac{9}{8},$$

$$t_1 = t_0 + 1 \cdot \frac{1}{8} = \frac{1}{8},$$

$$y_2 = \frac{9}{8}y_1 = \frac{9}{8} \cdot \frac{9}{8} = \left(\frac{9}{8}\right)^2,$$

$$t_2 = t_0 + 2 \cdot \frac{1}{8} = \frac{1}{4},$$

$$\vdots$$

$$y_8 = \frac{9}{8}y_7 = \frac{9}{8}\left(\frac{9}{8}\right)^7 = \left(\frac{9}{8}\right)^8,$$

$$t_8 = t_0 + 8 \cdot \frac{1}{8} = 1.$$

The approximation is $y_8 = 2.566$, which has an absolute error

$$|y_8 - y(1)| = 0.152$$

(correct to three decimal places). This result is slightly better. If we are willing to do more work, the result can be improved even more as shown in Table 1.1.

TABLE 1.1

n	$h = 1/n$	y_n	$\|y_n - y(1)\|$
16	0.063	2.638	0.080
32	0.031	2.677	0.041
64	0.016	2.697	0.021
128	0.008	2.708	0.010
256	0.004	2.713	0.005
512	0.002	2.716	0.002
1024	0.001	2.717	0.001

Note that if we make the step size half as big, the absolute error will also be approximately half as big. This suggests that the error in the numerical approximation produced by Euler's method is roughly proportional to the step size h. We shall see later that this is true for most problems.

A Fortran program that implements Euler's method is given in Fig. 1.12. It is set up to solve the initial value problem

$$\frac{dy}{dt} = ty^{1/3}, \qquad y(1) = 1. \tag{1.6.3}$$

The solution is computed and printed at 10 equally spaced points in the interval $1 \le t \le 2$, so $N = 10$, and the final time is TF $= 2.0$. This results in a step size $H = (2.0 - 1.0)/10 = 0.1$. The output of the program is reproduced in Table 1.2.

Equation (1.6.3) is separable and the solution is

$$y(t) = \left(\frac{t^2 + 2}{3}\right)^{3/2},$$

so $y(2) \simeq 2.828$. With 10 steps, the above program produced 2.724, a value in error by about 0.105. As in Example 1, the step size h could be decreased to obtain a better result.

The above discussions clearly indicate that Euler's method is not very practical for realistic computation. One must take a large number of steps to

```
C  PROGRAM USES THE EULER METHOD TO SOLVE THE
C  INITIAL VALUE PROBLEM
C
C      DY/DT = F(T,Y), Y(T0) = Y0
C
C  DEFINE FUNCTION F(T,Y) AND INITIAL CONDITIONS.
C
       F(T,Y) = T*Y**(1.0/3.0)
       T0 = 1.0
       Y0 = 1.0
C
C  SET FINAL T AND NUMBER OF STEPS N TO BE TAKEN.
C
       TF = 2.0
       N = 10
C
C  EULER LOOP
C
       T = T0
       Y = Y0
       H = (TF - T0)/FLOAT(N)
       WRITE(6,2)T,Y
     2 FORMAT(1X,F6.2,F10.5)
       DO 1 I = 1,N
          Y = Y + H*F(T,Y)
          T = T + H
          WRITE(6,2)T,Y
     1 CONTINUE
       STOP
       END
```

Figure 1.12 Euler program to solve $y' = ty^{1/3}$, $y(1) = 1$.

TABLE 1.2	t	y	t	y
	1.00	1.000000	1.60	1.821700
	1.10	1.099999	1.70	2.017109
	1.20	1.213550	1.80	2.231903
	1.30	1.341546	1.90	2.467134
	1.40	1.484922	2.00	2.723868
	1.50	1.644643		

achieve even modest accuracy for most problems. How then can we generate a better approximation? One approach is the following. First note that in going from t_0 to t_1, Euler's method uses only information about the slope of the solution at t_0, namely $f(t_0, y_0)$. The fact that $f(t, y)$ may change in the interval is ignored. A more accurate slope value to use might be the average of the slopes at (t_0, y_0) and (t_1, y_1). However, $f(t_1, y_1)$, the slope at (t_1, y_1), cannot be calculated until we know y_1. If this value is approximated by Euler's method, we obtain the following approximation to $y(t_1)$:

$$y_1 = y_0 + \frac{h(s_1 + s_2)}{2},$$

where

$$s_1 = f(t_0, y_0),$$

$$s_2 = f(t_0 + h, y_0 + hs_1).$$

Note that s_2 is the value of f at t_1 and a value of y given by Euler's method. This idea easily extends to the other line segments and leads to

The improved Euler algorithm. An approximate solution to the initial value problem (1.6.1) at the equally spaced points $t_0, t_i, \ldots, t_n$ is given by

$$s_1 = f(t_i, y_i),$$

$$s_2 = f(t_i + h, y_i + hs_1),$$

$$y_{i+1} = y_i + h(s_1 + s_2)/2,$$

$$t_{i+1} = t_0 + (i + 1) h, \quad i = 0, \ldots, n - 1,$$

where h, t_0, and y_0 are given.

Example 2 Find an approximate solution to the initial value problem in Example 1 at the point $t = 1$ by using the improved Euler algorithm with $h = \frac{1}{4}$. Starting with $t_0 = 0$ and $y_0 = 1$, we compute

$$s_1 = y_0 \qquad\qquad\qquad = 1.000$$

$$s_2 = y_0 + \frac{1}{4} s_1 \qquad\quad = 1.250$$

$$y_1 = y_0 + \frac{1}{8} (s_1 + s_2) = 1.281$$

$$t_1 = t_0 + \frac{1}{4} \qquad = 0.250$$

$$s_1 = y_1 \qquad = 1.281$$

$$s_2 = y_1 + \frac{1}{4} s_1 \qquad = 1.601$$

$$y_2 = y_1 + \frac{1}{8} (s_1 + s_2) = 1.641$$

$$t_2 = t_0 + 2 \cdot \frac{1}{4} \qquad = 0.500$$

$$s_1 = y_2 \qquad = 1.641$$

$$s_1 = y_2 + \frac{1}{4} s_1 \qquad = 2.052$$

$$y_3 = y_2 + \frac{1}{8} (s_1 + s_2) = 2.103$$

$$t_3 = t_0 + 3 \cdot \frac{1}{4} \qquad = 0.750$$

$$s_1 = y_3 \qquad = 2.103$$

$$s_2 = y_3 + \frac{1}{4} s_1 \qquad = 2.629$$

$$y_4 = y_3 + \frac{1}{8} (s_1 + s_2) = 2.695$$

$$t_4 = t_0 + 4 \cdot \frac{1}{4} \qquad = 1.000$$

In this case the absolute error to three decimal places is $|y_4 - y(1)| = 0.023$, which is approximately 10% of the error produced by using the Euler algorithm. $\square$

    A Fortran program implementing the improved Euler algorithm is given in Fig. 1.13. Again, it is set up to solve the Eq. (1.6.3). Its output is presented in Table 1.3.

    For the improved Euler algorithm, $y_{10} \approx 2.828$, which differs from the true solution by 0.0008. This is a considerable improvement over the Euler prediction, $y_{10} \approx 2.724$.

```
C   PROGRAM USES THE IMPROVED EULER METHOD TO SOLVE
C   THE INITIAL VALUE PROBLEM
C
C       DY/DT = F(T,Y), Y(T0) = Y0
C
C   DEFINE FUNCTION F(T,Y) AND INITIAL CONDITIONS.
C
        F(T,Y)=T*Y**(1.0/3.0)
        T0=1.0
        Y0=1.0
C
C   SET FINAL T AND NUMBER OF STEPS N TO BE TAKEN.
C
        TF=2.0
        N=10
C
C   IMPROVED EULER LOOP
C
        T=T0
        Y=Y0
        H=(TF-T0)/FLOAT(N)
        WRITE(6,2)T,Y
      2 FORMAT(1X,F6.2,F10.5)
        DO 1 I=1,N
           S1=F(T,Y)
           S2=F(T+H,Y+H*S1)
           Y=Y+H*(S1+S2)/2.0
           T=T+H
           WRITE(6,2)T,Y
      1 CONTINUE
        STOP
        END
```

Figure 1.13   Improved Euler program to solve $y' = ty^{1/3}$, $y(1) = 1$.

	t	y	t	y
TABLE 1.3	1.00	1.000000	1.60	1.873597
	1.10	1.106775	1.70	2.080564
	1.20	1.227788	1.80	2.307834
	1.30	1.363985	1.90	2.556483
	1.40	1.516345	2.00	2.827600
	1.50	1.685873		

TABLE 1.4	n	h	y_n	$|y_n - \tan(1)|$
	10	10^{-1}	1.553790	$0.4 \cdot 10^{-2}$
	100	10^{-2}	1.557369	$0.4 \cdot 10^{-4}$
	1000	10^{-3}	1.557407	$0.4 \cdot 10^{-6}$

Example 3 Approximate the solution to the initial value problem

$$\frac{dy}{dt} = 1 + y^2, \qquad y(0) = 0,$$

at the point $t = 1$ by using the improved Euler method with $h = 0.1, 0.01$, and 0.001. The actual solution to this problem is $y(t) = \tan t$. Using the improved Euler program with

$$F(T,Y) = 1.0 + Y**2$$

$$T0 = 0.0$$

$$Y0 = 0.0$$

$$TF = 1.0$$

$$N = 10, N = 100, \text{ and } N = 1000$$

we obtain Table 1.4. These data suggest that the error in the improved Euler algorithm might be proportional to h^2, rather than to h, as in the Euler algorithm. Again for most problems, this is the case. $\square$

EXERCISES 1.6

In Exercises 1, 2, and 3, perform the indicated computations by hand, retaining only four significant digits at each step of the calculations.

1. Consider the initial value problem

$$\frac{dy}{dt} = -ty^2, \quad y(0) = 2.$$

a) Use the Euler algorithm to compute an approximate solution at $t = 1.0$. Use the values $n = 2, 4, 8$ which correspond to step sizes of $h = \frac{1}{2}, \frac{1}{4}, \frac{1}{8}$.

b) Repeat part (a) using the improved Euler algorithm.

c) Determine the actual solution and compare

the value of $y(1)$ with your results in (a) and (b).

2. Repeat Exercise 1 for the initial value problems

a) $\dfrac{dy}{dt} + 2y = t, y(0) = 1$

b) $\dfrac{dy}{dt} = t + 1, y(0) = 1$

c) $\dfrac{dy}{dt} = t^3 e^{-2y}, y(0) = 0$

d) $\dfrac{dy}{dt} = \dfrac{e^t}{y}, y(0) = 1$

3. Use the improved Euler algorithm with $h = 1$ to obtain an approximate solution at $t = 1$ for

Exercise 2(b) above with the initial condition $y(0) = -\frac{1}{4}$. Why is the answer so accurate?

4. Use the Euler and improved Euler programs to obtain an approximate solution at $t = 0.5$ in Exercises 1 and 2. What is the smallest value of n (or largest value of h) that will produce three significant digits of accuracy in each case?

5. Consider the initial value problem

$$\frac{dy}{dt} = y + t^2, \qquad y(0) = 1.$$

a) Find the solution $y(t)$ and evaluate it for $t = 0.2, 0.4, \ldots, 1.0$.

b) Using the improved Euler program with a step size of $h = 0.2$, find approximate values for the solution at the t values in part (a).

c) Repeat part (b) by using $h = 0.1$.

d) Compare the results of part (c) with those of part (b) and the exact values. The differences in the results for $h = 0.2$ and $h = 0.1$ tend to indicate whether a smaller step size must be used for this range of t values. (A good rule of thumb is to use the solution corresponding to the smaller step size if the two solutions agree to the required accuracy for all t values of interest. If they do not agree, reduce h and repeat. This procedure gives an *indication* but not a proof of the accuracy of the result.)

6. Modify the programs in the text so that the values of t and y are printed out at every pth step, where p is to be specified. (If many integration steps are used, one normally does not want to print out t and y at every step.)

7. Consider the problem

$$\frac{dy}{dt} = y^2, \quad y(0) = 1.$$

a) Find the solution $y(t)$ by using the methods in Section 1.4.

b) Use the Euler algorithm to approximate the solution at $t = 0.5$ by using h values of 0.05 and 0.025. Compare with the actual solution.

c) Reduce h until you have achieved three significant figures of accuracy.

8. Use the improved Euler algorithm to approximate the solution at $t = 0.5$ in Exercise 7. What is the largest value of h that will produce comparable results?

9. The initial value problem

$$\frac{dy}{dt} = y^2 - t^2, \quad y(0) = \frac{1}{2},$$

cannot be solved by analytical methods. Use the improved Euler program to approximate the solution at $t = 1$ to three significant figures.

10. Dawson's integral

$$y(t) = e^{-t^2} \int_0^t e^{x^2}\, dx$$

is the solution to the initial value problem

$$\frac{dy}{dt} = -2ty + 1, \quad y(0) = 0.$$

Approximate $y(0.5)$ to four significant figures by using the improved Euler program. The actual result (correct to five decimal places) is $y(0.5) = 0.42444$. Compare your results with those obtained in Problem 28 of Section 1.5, if you worked it.

11. Apply the improved Euler program to each of the following initial value problems to approximate the solution at the indicated value of t correct to three significant figures. Solve the differential equation analytically and compare with your numerical solution.

a) $\dfrac{dy}{dt} = \dfrac{1}{1 + t^2}$, $y(0) = 0$; $t = 1$

b) $\dfrac{dy}{dt} = 2y$, $y(0) = 1$; $t = 0.5$

c) $\dfrac{dy}{dt} = \dfrac{2y}{t}$, $y(1) = 1$; $t = 2$

d) $\dfrac{dy}{dt} = y - t$, $y(0) = 2$; $t = 1$

e) $\dfrac{dy}{dt} = t^3 e^{-y}$, $y(1) = 0$; $t = 2$

f) $\dfrac{dy}{dt} = y - e^{-t}$, $y(0) = 1$; $t = 1$

g) $\dfrac{dy}{dt} = e^{y+t}$, $y(0) = 1$; $t = 0.25$

h) $\dfrac{dy}{dt} = \dfrac{2t}{y + t^2 y}$, $y(0) = -2$; $t = 1$

i) $\dfrac{dy}{dt} = 4y + 1 - t$, $y(0) = 1$; $t = 0.5$

12. Approximate the solution to the initial value problem

$$\frac{dy}{dt} = 1 + y^2, \quad y(0) = 0$$

at $t = 0.1, 0.2, \ldots, 1.0$ by using the improved Euler program with $h = 0.1$. On one sheet of graph paper plot your numerical solution and the actual solution $y(t) = \tan t$. What do you think would happen if you tried to approximate the solution near $t = \pi/2$?

13. Verify that the error in the Euler algorithm appears to be proportional to h and the error in the improved Euler algorithm proportional to h^2 by using the initial value problem

$$\frac{dy}{dt} = -y, \quad y(0) = 1,$$

and examining the numerical solution at $t = 1$. Use the h values: ½, ¼, ⅛, ⅟₁₆, ⅟₃₂, and ⅟₆₄.

14. Estimate the time required to empty a conical funnel 12 inches high with a 3-inch base diameter if the apex hole has a diameter of ½ inch. Use the improved Euler program with TF = 5.0, modified with a command to STOP if the height is not positive. (See Exercise 25 of Section 1.4.)

1.7 IMPLICIT SOLUTIONS OF FIRST ORDER EQUATIONS

A separable first order differential equation

$$\frac{dy}{dt} = -\frac{g(t)}{h(y)}$$

can be written as

$$g(t)\, dt + h(y)\, dy = 0,$$

and an implicit solution

$$G(t) + H(y) = c$$

can be constructed by finding antiderivatives $G(t)$ and $H(y)$ of $g(t)$ and $h(y)$. If we set

$$F(t,y) = G(t) + H(y),$$

then

$$dF(t, y) = \frac{\partial F}{\partial t}(t, y)\, dt + \frac{\partial F}{\partial y}(t, y)\, dy$$

$$= G'(t)\, dt + H'(y)\, dy$$

$$= g(t)\, dt + h(y)\, dy = 0.$$

This says that the separable equation has an implicit solution $F(t, y) = c$.

But must a differential equation be separable to have an implicit solution? Clearly not. We can take any continuously differentiable function

$F(t, y)$ and define a differential equation

$$dF = \frac{\partial F}{\partial t}(t, y)\, dt + \frac{\partial F}{\partial y}(t, y)\, dy = 0$$

that has an implicit solution $F(t, y) = c$. But it does not follow that the differential equation

$$\frac{dy}{dt} = -\frac{\partial F}{\partial t}(t, y) \bigg/ \frac{\partial F}{\partial y}(t, y)$$

is separable. For instance, let $F(t,y) = ye^{ty} = c$ implicitly define $y(t)$. Then $y(t)$ will be a solution of

$$\frac{dy}{dt} = -\frac{y^2 e^{ty}}{e^{ty} + tye^{ty}} = -\frac{y^2}{1 + ty},$$

which is clearly not separable.

In this section we shall consider the problem of constructing implicit solutions to an important class of equations, *exact differential equations*. We begin with the differential equation

$$M(t, y)\, dt + N(t, y)\, dy = 0. \tag{1.7.1}$$

Suppose it has an implicit solution, $F(t, y) = c$, defined in a region R of the ty plane. Suppose furthermore that

$$\frac{\partial F}{\partial t}(t, y) = M(t, y), \qquad \frac{\partial F}{\partial y}(t, y) = N(t, y), \tag{1.7.2}$$

and that the second partial derivatives of $F(t, y)$ are continuous. If so, $M dt + N dy$ is an *exact differential* and (1.7.1) is an *exact differential equation*.

Some differential equations are so simple that exactness can be determined by inspection, e.g., the left-hand side of

$$\cos y\, dt - t \sin y\, dy = 0$$

is the differential of $F(t, y) = t \cos y$ and the task of finding a solution is done. For more complicated equations we need a test for exactness and a method for constructing an implicit solution. If the equation (1.7.1) is exact and if (1.7.2) is satisfied and all of the second derivatives of $F(t, y)$ are continuous, then it is known from calculus that the mixed second partial derivatives of $F(t, y)$ are equal:

$$\frac{\partial^2 F}{\partial t\, \partial y}(t, y) = \frac{\partial^2 F}{\partial y\, \partial t}(t, y)$$

or, from (1.7.2),

$$\frac{\partial M}{\partial y}(t, y) = \frac{\partial N}{\partial t}(t, y).$$

Clearly, the converse should hold, and it does, if the region R is nice, say, a rectangle,[3] and the coefficient functions are continuously differentiable in R. This is the content of the following theorem.

Theorem   Let $M(t, y)$ and $N(t, y)$ be continuously differentiable in a rectangle R. The differential equation

$$M(t, y)\, dt + N(t, y)\, dy = 0$$

is exact if and only if

$$\frac{\partial M}{\partial y}(t, y) = \frac{\partial N}{\partial t}(t, y) \text{ in } R. \qquad (1.7.3)$$

Proof   If the equation is exact, then $\partial M/\partial y = \partial N/\partial t$ follows from the equality of the mixed partial derivatives of smooth functions.

If $\partial M/\partial y = \partial N/\partial t$, we shall show that the differential equation is exact by constructing a suitable function $F(t, y)$. To do this, integrate $M(t, y)$ with respect to t while holding y fixed:

$$F(t, y) = \int_a^t M(s, y)\, ds + W(y), \qquad (1.7.4)$$

where $W(y)$ is an arbitrary differentiable function of y and (s, y) is in R for $a \le s \le t$. By construction,

$$\frac{\partial}{\partial y} \int_a^t M(s, y)\, dy + W'(y) = N(t, y)$$

or

$$W'(y) = N(t, y) - \frac{\partial}{\partial y} \int_a^t M(s, y)\, ds. \qquad (1.7.5)$$

---

[3] A precise statement would be that R is a *simply connected region,* of which rectangles or discs are examples. For the definition of this concept the interested reader should consult an advanced calculus text.

But is the right side of (1.7.5) a function of y alone? It is if its derivative with respect to t is zero. Let us check:

$$\frac{\partial}{\partial t}\left[N(t, y) - \frac{\partial}{\partial y}\int_a^t M(s, y)\, ds\right] = \frac{\partial N}{\partial t}(t, y) - \frac{\partial^2}{\partial t\, \partial y}\int_a^t M(s, y)\, ds$$

$$= \frac{\partial N}{\partial t}(t, y) - \frac{\partial}{\partial y}\left[\frac{\partial}{\partial t}\int_a^t M(s, y)\, ds\right]$$

$$= \frac{\partial N}{\partial t}(t, y) - \frac{\partial}{\partial y}M(t, y) = 0,$$

since we have assumed that $\partial M/\partial y = \partial N/\partial t$ in R. The interchanging of derivatives is permitted because of the assumed smoothness of $M(t, y)$.

Now set

$$W(y) = \int_b^y N(t, z)\, dz - \int_b^y \frac{\partial}{\partial z}\int_a^t M(s, z)\, ds\, dz, \qquad (1.7.6)$$

where (s, z) is in R for $b \le z \le y$, $a \le s \le t$. By virtue of the relation $\partial M/\partial y = \partial N/\partial t$, the differential of the function $F(t, y)$ defined by (1.7.4), where $W(y)$ is given by (1.7.6), equals $M(t, y)\, dt + N(t, y)\, dy$. Therefore $F(t, y) = c$ is an implicit solution of the differential equation, which completes the proof of the theorem. ∎

An advantage of the proof just given is that it is constructive inasmuch as it gives us a procedure for finding a solution. This is illustrated in the following example.

Example 1   Construct an implicit solution to

$$(6ty + 4)\, dt + (3t^2 + 10y)\, dy = 0.$$

Let $M(t, y) = 6ty + 4$, $N(t, y) = 3t^2 + 10y$, and since

$$\frac{\partial M}{\partial y}(t, y) = 6t = \frac{\partial N}{\partial t}(t, y),$$

the condition (1.7.3) for exactness is satisfied. This result, together with the smoothness of the coefficient functions $M(t, y)$ and $N(t, y)$ insures that there exists an implicit solution. To construct it, set

$$F(t, y) = \int M(t, y)\, dt + W(y) = 3t^2 y + 4t + W(y).$$

From the fact that

$$N(t, y) = 3t^2 + 10y = \frac{\partial F}{\partial y}(t, y) = 3t^2 + W'(y),$$

we have

$$W'(y) = 10y \quad \text{or} \quad W(y) = 5y^2.$$

Hence,

$$F(t, y) = 3t^2y + 4t + 5y^2,$$

and

$$3t^2y + 4t + 5y^2 = c$$

is an implicit solution to the differential equation.

We could have started instead with $N(t, y)$ and constructed an anti-derivative with an additive function $R(t)$. To do this, set

$$F(t, y) = \int N(t, y) \, dy + R(t) = 3t^2y + 5y^2 + R(t).$$

From

$$M(t, y) = 6ty + 4 = \frac{\partial F}{\partial t}(t, y) = 6ty + R'(t),$$

we have

$$R'(t) = 4 \quad \text{and} \quad R(t) = 4t.$$

Hence,

$$F(t, y) = 3t^2y + 5y^2 + 4t,$$

as before. Note that $W(y)$ and $R(t)$ are determined up to an additive constant that is immaterial in the construction of an implicit solution. □

It is sometimes possible to construct an implicit solution to a nonexact differential equation by multiplying it by an *integrating factor* to convert it into an exact differential equation. The finding of an integrating factor usually rests on a clever observation that an expression, such as $t \, dy - y \, dt$, reminds one of the differential

$$d\left(\frac{y}{t}\right) = \frac{t \, dy - y \, dt}{t^2}.$$

On the other hand, it is also true that

$$d\left(-\frac{t}{y}\right) = \frac{t \, dy - y \, dt}{y^2}$$

and

$$d\left(\ln\frac{y}{t}\right) = \frac{t \, dy - y \, dt}{ty}.$$

This suggests that $1/t^2$, $1/y^2$, or $1/ty$ are all possible integrating factors. Which, if any, will lead to an exact differential equation must be determined by further analysis.

This is illustrated in the following example.

Example 2 Construct an implicit solution to

$$(ty - 1)y \, dt + (ty + 1)t \, dy = 0.$$

Here, $M(t, y) = ty^2 - y$, $N(t, y) = t^2y + t$ and

$$\frac{\partial M}{\partial y}(t, y) = 2ty - 1 \neq 2ty + 1 = \frac{\partial N}{\partial t}(t, y),$$

so the equation is not exact. Hoping for inspiration, we rearrange the terms. The arrangement

$$t \, dy - y \, dt + ty^2 \, dt + t^2y \, dy = 0$$

suggests that $d(y/t)$ might be lurking about. Dividing by t^2 gives

$$d\left(\frac{y}{t}\right) + \frac{y^2}{t} \, dt + y \, dy = 0,$$

which unfortunately is not exact since

$$\frac{\partial}{\partial y}\left(\frac{y^2}{t}\right) = \frac{2y}{t} \neq \frac{\partial}{\partial t}(y) = 0.$$

Dividing by y^2 gives the same result, but after dividing the equation by ty, we obtain

$$\frac{t \, dy - y \, dt}{ty} + (y \, dt + t \, dy) = 0.$$

Since the second term is the differential of ty, therefore

$$\ln \frac{y}{t} + ty = c$$

is an implicit solution in any region where $0 < y/t < \infty$.

This example supports the widely held opinion that the finding of integrating factors is an arcane art that may lead to madness. □

EXERCISES 1.7

Show that the following differential equations are exact and find an implicit solution to each of them.

1. $(2ty + y^2) \, dt + (t^2 + 2ty) \, dy = 0$

2. $\left(\frac{1}{y} - \frac{y}{t^2}\right) dt + \left(\frac{1}{t} - \frac{t}{y^2}\right) dy = 0$

3. $(2ty - y \sin t) \, dt + (t^2 + \cos t) \, dy = 0$

4. $\cosh t \cosh y \, dt + \sinh t \sinh y \, dy = 0$

5. $(y + \sin y \sec^2 t) \, dt + (t + \cos y \tan t) \, dy = 0$

6. $[y \cos(t + y) - ty \sin(t + y)] \, dt +$
$[t \cos(t + y) - ty \sin(t + y)] \, dy = 0$

7. $\left(\dfrac{y}{t} - 2t \ln y \right) dt + \left(\ln t - \dfrac{t^2}{y} \right) dy = 0$

8. $(ye^{ty} + 2ty) \, dt + (te^{ty} + t^2) \, dy = 0$

Implicit solutions to the pair of differential equations

$$M(x, y)dx + N(x, y)dy = 0,$$
$$N(x, y)dx - M(x, y)dy = 0$$

form an orthogonal family of curves since the product of their slopes is -1. For both equations to be exact,

$$\frac{\partial M}{\partial y}(x, y) = \frac{\partial N}{\partial x}(x, y)$$

and

$$\frac{\partial N}{\partial y}(x, y) = -\frac{\partial M}{\partial y}(x, y).$$

These equations are called *Cauchy–Riemann equations;* they play a fundamental role in the theory of two-dimensional flow of incompressible irrotational fluids. Verify that the following pairs of equations are exact and find implicit solutions.

9. $e^x \cos y \, dx - e^x \sin y \, dy = 0,$
$e^x \sin y \, dx + e^x \cos y \, dy = 0$

10. $(2x + 2)dx - 2y \, dy = 0,$
$2y \, dx + (2x + 2)dy = 0$

Multiply the following differential equations by the given integrating factor and show that you obtain an exact equation. Find an implicit solution for each of them.

11. $(2t^3y - y^3) \, dt - 2t^4 \, dy = 0, \quad Z(t, y) = t^{-2}y^{-3}$

12. $(2t^3y - y^3) \, dt + (ty^2 - t^4) \, dy = 0, \quad Z(t, y) = t^{-2}y^{-2}$

13. $(2y^2 + ty) \, dt - (t^2 - 2ty) \, dy = 0, \quad Z(t, y) = -t^{-1}y^{-2}$

14. $(3yt^2 - 2y^3) \, dt - (3t^3 - 2ty^2) \, dy = 0, \quad Z(t, y) = y^{-2}(t^2 + y^2)^{-1}$

Another class of differential equations, for which implicit solutions are the rule rather than the excep-

tion, are called *homogeneous* equations. The nomenclature, which is confusing, refers to the special nature of $f(t, y)$. Linear equations with no forcing term are also called homogeneous, but this is a completely different usage. A function $f(t, y)$ is said to be homogeneous if it has the property that $f(t, y) = f(ct, cy)$ for all t, y, and c for which it is defined.

15. Which of the following functions are homogeneous?

a) $\dfrac{3t + 2y}{t - y}$

b) $\dfrac{y + t}{4t}$

c) $\dfrac{2t + y^2}{ty}$

d) $\dfrac{ty^2 + t^3}{t^3 + t^2y + y^3}$

e) $\dfrac{6t^{1/3}y^{2/3} + ty}{4t - y}$

f) $\dfrac{3 \ln 2^y}{t + \sqrt{ty}}$

16. A differential equation $dy/dt = f(t, y)$, where $f(t, y)$ is a homogeneous function, is called a *homogeneous differential equation.* It can be solved by the substitution

$$y = tu, \qquad \frac{dy}{dt} = u + t\frac{du}{dt},$$

which transforms it to a separable equation. Use this technique to solve the following equations.

a) $\dfrac{dy}{dt} = \dfrac{y^2 + ty + t^2}{t^2}, \quad y(1) = 0$

b) $\dfrac{dy}{dt} = \dfrac{2t + y}{y}$

c) $\dfrac{dy}{dt} = \dfrac{t^2 + y^2}{ty}, \quad y(1) = 4$

d) $\dfrac{dy}{dt} = \dfrac{2t - y}{t + 4y}$

e) $\dfrac{dy}{dt} = \dfrac{2t \sin(y/t) + 3y \cos(y/t)}{3t \cos(y/t)}, \quad y(8) = 4\pi$

f) $\dfrac{dy}{dt} = \dfrac{t + y}{t - y}, \quad y(1) = 0$

17. Equations of the form

$$\frac{dy}{dt} = \frac{At + By + C}{Dt + Ey + F}$$

can sometimes be reduced to homogeneous equations by translation of the ty coordinates.

Show that if the pair of simultaneous equations

$$At + By + C = 0, \qquad Dt + Ey + F = 0$$

has a unique solution $t = h$, $y = k$, then the change of variables $t = x + h$, $y = w + k$ gives the transformed differential equation

$$\frac{dw}{dx} = \frac{Ax + Bw}{Dx + Ew},$$

which is homogeneous. Use this idea to solve the following:

a) $\dfrac{dy}{dt} = \dfrac{3t + 4y - 5}{2t + 2}$

b) $\dfrac{dy}{dt} = \dfrac{t + 2y}{2t + 3y - 1}$

c) $\dfrac{dy}{dt} = \dfrac{5y - 2t - 8}{4y - t - 7}$

1.8 A MATHEMATICAL MODEL OF AN ELECTRIC-CIRCUIT PROBLEM

In the following sections we shall study some aspects of the process of describing a physical problem in mathematical terms. This process is called *mathematical modeling*. It is not a logical exercise, like that of solving a linear differential equation, nor is it an action requiring divine guidance. The process is a combination of physical insight, mathematical reasoning, and common sense. We shall first illustrate it with the problem of finding the flux variation in an iron-core induction coil.

The experimental equipment consists of the coil, a multitester (voltmeter and ohmmeter), a variable resistor, a battery, and a switch. The main properties of the iron-core induction coil used here are that the flux is a function of the current and that the voltage drop across the coil due to inductance is the time rate of change of the flux. Since the flux cannot be measured directly, we shall have to deduce it from measurements of the voltage drop across the coil. If the coil, variable resistor, and battery are connected in a loop, i.e., in a series circuit, then the current in each circuit element is the same and the total resistance of the circuit is the sum of the individual resistances of the coil and the resistor. By Ohm's law, the voltage drop across each circuit element is equal to its resistance times the current flow through it. Therefore, the total voltage drop due to resistance is

$$V_{\text{res}} = R \cdot I,$$

where I denotes the current and R the total resistance of the circuit. The induction coil opposes any change in the current with the result that there is a voltage drop across it

$$V_{\text{ind}} = \frac{d\phi}{dt},$$

where ϕ denotes the magnetic flux of the current. In accordance with Kirchhoff's law, the sum of the voltage drops across the coil and the resistor must equal the voltage of the battery:

$$V_{\text{ind}} + V_{\text{res}} = E$$

or

$$\frac{d\phi}{dt} + RI = E. \tag{1.8.1}$$

We now have part of our mathematical model, an ordinary differential equation. The model is not complete since a relation between the magnetic flux ϕ and the current I is not yet formulated. With such a relation, the differential equation (1.8.1) would become a first order equation in one of the unknowns ϕ or I. An acceptable flux–current relation must be simple and stable. Without launching into an extensive study of the theory of electricity and magnetism, let us try to formulate an acceptable relation. It is known that if the current is zero, there is no flux, and if the current is small, the flux is small. A simple relation consistent with these observations is that of proportionality. Hence, suppose that $\phi = LI$, where L, the proportionality constant, is called the *inductance of the coil*. The variable resistor must be adjusted so that the current is small enough for this linear flux–current relation to be valid. The differential equation for the series circuit is then

$$L\frac{dI}{dt} + RI = E. \tag{1.8.2}$$

The constants R and L are not known and must be estimated from experimental data. Figure 1.14 is a diagram of the simple RL circuit just described.

The estimation of constants (also called *parameters*) is an essential aspect of mathematical modeling. If the parameters of a model cannot be estimated, the model is of little value. In the physical sciences, this restriction is so obvious, it requires little discussion. In the biological and social sciences, where there seems to be a great temptation to construct elaborate and untestable mathematical models, this restriction deserves strong emphasis.

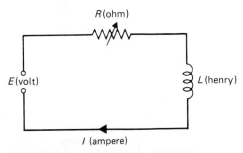

Figure 1.14 A simple RL circuit.

Let us return to the study of the flux variation of the coil. The circuit elements are connected in series with the switch open, and the ohmmeter measures R, the resistance of the coil and variable resistor. Once this is done, the switch is closed ($t = 0$). Since $I(0) = 0$, the solution to the differential equation (1.8.2) is

$$I(t) = \frac{E}{R}(1 - e^{-Rt/L}) \quad \text{and} \quad \phi(t) = LI(t) = \frac{EL}{R}(1 - e^{-Rt/L}). \quad (1.8.3)$$

It follows from (1.8.1) and the last relation that

$$\frac{d\phi}{dt}(t) = E - RI(t) = Ee^{-Rt/L}; \quad (1.8.4)$$

taking logarithms of both sides gives

$$\ln \frac{d\phi}{dt}(t) = \ln E - \frac{Rt}{L}. \quad (1.8.5)$$

This last equation shows that a plot of $\ln(d\phi/dt)$ against t gives a straight line with slope $-R/L$. Values of $(t, d\phi/dt)$ can be found by measuring the voltage drop across the coil with a voltmeter. Since R has been measured earlier, we can estimate the remaining circuit parameter L, the inductance, by estimating the slope of the line (1.8.5). Actual measurements will not lie exactly on the line because of experimental errors or slight deviations from our mathematical model. By plotting some values of $\ln(d\phi/dt)$ against t, the slope R/L can be obtained graphically, or its value can be estimated by linear regression. This last procedure consists of performing a least-squares fit of a straight line to the data and is explained below.

The (simulated) measurements of $d\phi/dt$ in Table 1.5 will be used to illustrate the method of estimating R/L. The data was generated by adding random numbers to values of $d\phi/dt$ computed from (1.8.4) with $R = 10$ and $L = 30$.

TABLE 1.5	t	$d\phi(t)/dt$	$\ln d\phi(t)/dt$	t	$d\phi(t)/dt$	$\ln d\phi(t)/dt$
	1.2	1.467	0.383	7.2	0.197	-1.625
	2.4	0.661	0.414	8.4	0.0855	-2.459
	3.6	0.416	-0.877	9.6	0.0744	-2.598
	4.8	0.361	-1.019	10.8	0.0607	-2.802
	6.0	0.152	-1.884	12.0	0.0329	-3.414

Linear regression is a computational procedure for finding the equation of a line that best describes a given set of data by using the *principle of least squares*. To apply the idea here, let $y = \ln(d\phi/dt)$ and let the line be represented by the equation

$$y = \alpha + \beta t.$$

The data $\{(t_i, y_i),\ i = 1, 2, \ldots, 10\}$ from Table 1.5 do not lie on any line, so we try to choose a line in such a way that the sum of the squares of the deviations $[y_j - (\alpha + \beta t_j)]$, $j = 1, \ldots, 10$, of the data from the line is as small as possible. This leads us to the problem of choosing values of the parameters α and β so that

$$F(\alpha, \beta) = \sum_{j=1}^{10} [y_j - (\alpha + \beta t_j)]^2 \tag{1.8.6}$$

is a minimum; hence the term, principle of least squares. If $F(\tilde{\alpha}, \tilde{\beta})$ is the minimum of $F(\alpha, \beta)$, then

$$y = \tilde{\alpha} + \tilde{\beta} t$$

is the equation of the best-fitting line. From calculus, we know that $\tilde{\alpha}$ and $\tilde{\beta}$ must satisfy

$$0 = \frac{\partial F}{\partial \alpha}(\tilde{\alpha}, \tilde{\beta}) = -2 \sum_{j=1}^{10} [y_j - (\tilde{\alpha} + \tilde{\beta} t_j)],$$

$$0 = \frac{\partial F}{\partial \beta}(\tilde{\alpha}, \tilde{\beta}) = -2 \sum_{j=1}^{10} [y_j - (\tilde{\alpha} + \tilde{\beta} t_j)] t_j.$$

Expanding the two expressions above leads to the following pair of linear equations for $\tilde{\alpha}$ and $\tilde{\beta}$:

$$10\tilde{\alpha} + \left(\sum_{j=1}^{10} t_j \right) \tilde{\beta} = \sum_{j=1}^{10} y_j,$$

$$\left(\sum_{j=1}^{10} t_j \right) \tilde{\alpha} + \left(\sum_{j=1}^{10} t_j^2 \right) \tilde{\beta} = \sum_{j=1}^{10} y_j t_j.$$

Solving for $\tilde{\alpha}$ and $\tilde{\beta}$ gives

$$\tilde{\beta} = \left[\sum_{j=1}^{10} y_j t_j - \frac{1}{10} \left(\sum_{j=1}^{10} t_j \right) \left(\sum_{j=1}^{10} y_j \right) \right] \div \left[\sum_{j=1}^{10} t_j^2 - \frac{1}{10} \left(\sum_{j=1}^{10} t_j \right)^2 \right],$$
$$\tilde{\alpha} = \frac{1}{10} \left(\sum_{j=1}^{10} y_j - \tilde{\beta} \sum_{j=1}^{10} t_j \right). \tag{1.8.7}$$

From the data in Table 1.5,

$$\sum_{j=1}^{10} t_j = 66.00, \qquad \sum_{j=1}^{10} y_j = -16.709,$$

$$\sum_{j=1}^{10} t_j^2 = 554.4, \qquad \sum_{j=1}^{10} y_j t_j = -148.41,$$

and so $\tilde{\beta} = -0.321$ and $\tilde{\alpha} = 0.448$. Since we know that $y = \tilde{\alpha} + \tilde{\beta} t$ approximates $\ln(d\phi/dt) = \ln E - (R/L)t$, we set

$$\ln E \simeq \tilde{\alpha} = 0.448 \qquad \text{or} \qquad E \simeq 1.565$$

and

$$-\frac{R}{L} \simeq \tilde{\beta} = -0.321 \qquad \text{or} \qquad L \simeq \frac{R}{0.321}.$$

If $R \simeq 9.963$ is the measured value of the total resistance, then $L \simeq 31.04$.
We have constructed a mathematical model for the flux variation in an iron-core induction coil and have shown that the parameters of the model can be estimated from measured data. In the next section, a model for population growth is described.

EXERCISES 1.8

1. Solve (1.8.2) when $L = 5$ henries, $R = 15$ ohms and $E = 60$ volts if $I(0) = 0$. At what time will $|I(t) - 4| < 10^{-4}$ amperes?

2. Solve (1.8.2) when $L = 5$ henries, $R = 15$ ohm, and $E(t) = 110 \sin(2\pi \cdot 60t)$ volts, a 60-cycle sine wave of amplitude 110 volts. Assume that $I(0) = 0$.

3. Solve (1.8.2) when $L = 4$ henries, $R = 8$ ohms, $I(0) = 4$ amperes, and $E(t) = 0$ for $0 \leq t < 1$, $E(t) = 6$ volts for $t > 1$. Do this by solving the initial value problem at $t = 0$, then find the general solution for $t > 1$, and finally match the two solutions at $t = 1$, so that $I(t)$ is continuous there. Is $I'(t)$ continuous at $t = 1$?

4. The nationwide average Scholastic Aptitude Test scores in mathematics for selected years from 1963 are given below (range of possible scores is from 200 to 800):

Year	Score
1963	502
1967	492
1970	488
1974	480
1977	470

a) Letting $t = 0$ correspond to 1963, $t = 4$ to 1967, etc., construct a least-squares fit of a straight line to the data. Plot the data and the line.

b) Use the straight line to predict the average SAT mathematics scores for 1979 and 1980 (the actual scores were 467 and 466, respectively).

5. For the data of Table 1.5 plot $(t_j, \ln(d\phi/dt)_j)$ and then with a straightedge draw a best-fitting line. From it estimate R and L. This is called the *eyeballing* method of linear regression.

6. Compute the *residuals*,

$$r_j = (d\phi/dt)_j - E \exp(-Rt_j/L),$$

first with the least-square values of R and L, then with the values you obtained in the preceding problem. Compare the corresponding values of F, $\partial F/\partial\alpha$, and $\partial F/\partial\beta$, recalling that F should be a minimum and $\partial F/\partial\alpha$, $\partial F/\partial\beta$ are equal to zero.

1.9 A MODEL OF POPULATION GROWTH

In the construction of a mathematical model for population growth, the basic difficulty is that there are no physical principles (like Ohm's law or Newton's laws of motion) on which to anchor the construction. Nevertheless, the model should reflect biological reality and this can be verified by an analysis of the qualitative behavior of solutions and, hopefully, by comparison with biological data. Finally, the model should be simple enough that the above analysis be possible—a model with 50 parameters is of little value either mathematically or biologically.

In this section, we will create a model for the growth of a single population. Let $x(t)$ be the size of the total population at time t and let $\dot{x}(t) = dx(t)/dt$ be its rate of growth. Then the per capita rate of growth of the population at time t is represented by

$$\frac{\dot{x}(t)}{x(t)},$$

since it is the rate of growth divided by the total population at time t. To obtain a differential equation, there must be some assumption made about the dependency of $\dot{x}/x$ on the time t, the total population $x(t)$, and possibly on some biological parameters. The simplest assumption would be that the per capita rate of growth is a positive constant, which leads to the equation

$$\frac{\dot{x}(t)}{x(t)} = r \quad \text{or} \quad \dot{x} = rx, \quad r > 0.$$

This is the equation for exponential growth, and if x_0 is the initial size of the population at time $t = 0$, then its solution is $x(t) = x_0 e^{rt}$. In some sense, this model is too simple, since one would expect that, as the population size increases, the per capita rate of growth will decrease (due to overcrowding, competition for food, resources, etc.) At best, one would expect pure exponential growth when there is unlimited space and resources. Therefore, the per capita rate of growth should depend on the population size, and so the model should be of the form

$$\frac{\dot{x}(t)}{x(t)} = f(x(t)) \qquad \text{or} \qquad \dot{x} = xf(x).$$

Now the task is to select a plausible function $f(x)$. A possible set of principles which might guide the selection of $f(x)$ are:

a) For small values of $x = x(t)$, the function $f(x)$ should be positive, reflecting the assumption that when the population is small, its growth is *locally* exponential.

b) For large values of $x = x(t)$, the function $f(x)$ should be negative, meaning that too large a population inhibits the rate of growth.

c) For all values of $x = x(t)$, the function $f(x)$ should be decreasing since the per capita growth rate should decrease as the population increases.

These principles are surely not the only ones that could be devised, but they will suffice for the model we wish to describe and they are biologically plausible.

The next problem is that of selecting a candidate for $f(x)$. Certainly, there is no unique function satisfying all the conditions (a), (b), and (c); however, we must adhere to our principle of simplicity. Too complicated a function $f(x)$ will make the analysis of the ensuing differential equation virtually impossible. The simplest function satisfying conditions (a), (b), and (c) is a linear one:

$$f(x) = r - mx, \quad r, m > 0.$$

We see that

a) $f(x) > 0$ for $x < r/m$,

b) $f(x) < 0$ for $x > r/m$,

c) $f'(x) = -m < 0$.

Hence, the biological guidelines are satisfied. The mathematical model for the population growth is given by the differential equation

$$\dot{x} = x(r - mx), \qquad x(0) = x_0, \quad r, m, x_0 > 0,$$

where $x(t)$ is the total population at time t and x_0 is the initial population. Letting $K = r/m$, we can write it as

$$\dot{x} = rx\left(1 - \frac{x}{K}\right), \qquad x(0) = x_0, \quad r, K, x_0 > 0. \tag{1.9.1}$$

This differential equation is called the *logistic equation* of population growth and has been used successfully to model the growth of yeast cells, fruit flies, the population of Sweden, the height of sunflower plants, and the Pacific Halibut fishery, to name a few examples.

An examination of (1.9.1) shows immediately that $x(t) = K$ is a constant solution (so is $x(t) = 0$ but it has no biological interest) and that

$$\dot{x} = rx\left(1 - \frac{x}{K}\right) \begin{cases} >0 & \text{for} \quad x < K, \\ <0 & \text{for} \quad x > K. \end{cases}$$

This means that if the initial value x_0 satisfies $0 < x_0 < K$, the solution $x(t)$ satisfying $x(0) = x_0$ is a strictly increasing function of t. As long as $0 < x(t) < K$ the population will increase, but since $x(t)$ can never cross the line $x = K$ (why?), and the direction field in $0 \leq t < \infty$, $0 < x < K$ always points upward, $x(t)$ must become asymptotic to $x = K$ as $t \to \infty$. By a similar line of reasoning one can conclude that any solution with initial value $x(0) = x_0 > K$ is a strictly decreasing function of t, which becomes asymptotic to $x = K$ as $t \to \infty$. Without solving the equation (which can be done—see Exercise 1a), it can be concluded that every solution $x(t)$, for which $x(0) = x_0 \neq K$ and $x_0 > 0$, approaches the value K as $t \to \infty$. For this reason the constant K is called the *carrying capacity* of the population and represents the natural limit of its size (Fig. 1.15).

If $x(t)$ is small and positive, then for a short period of time the population grows like the solution of $\dot{x} = rx$, since the term $-rx^2/K$ will be negligible. Hence for small values of x the growth looks exponential, and for that reason the parameter r is called the *intrinsic growth rate* of the population.

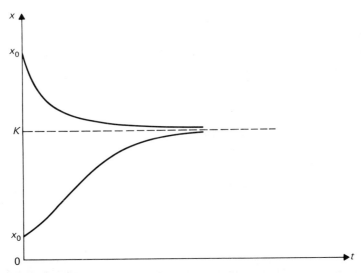

Figure 1.15 Graph of solutions of the logistic equation.

Our mathematical results are biologically plausible since one would expect that a small population will initially grow exponentially. Then, as the competition for food and the effects of overcrowding begin to be felt, its growth rate will diminish. One also would expect that there is a natural limit, K, to the size of the population, which is, in part, determined by the resources available. Of course, if it is believed or observed that the growth of a population under study behaves differently, then this would call for a different set of assumptions and, consequently, a different $f(x)$.

EXERCISES 1.9

1. a) Using the method of separation of variables, show that the solution of the logistic equation with initial data $x(0) = x_0$ is given by

$$x(t) = \frac{Kx_0}{x_0 + (K - x_0)e^{-rt}}.$$

b) Show that $\lim_{t \to \infty} x(t) = K$ and that $x(t)$ is increasing if $0 < x_0 < K$ and decreasing if $x_0 > K$.

2. The function

$$x(t) = \frac{665}{1 + \exp[4.1896 - 0.5355t]}$$

was used by R. Pearl in 1925 to model the growth of a yeast *(Saccharomyces)* population. Here t is measured in hours and $x(t)$ is a number proportional to the quantity of yeast cells.

a) Find the logistic differential equation and the initial conditions that $x(t)$ satisfies.

b) Find a value T for which $K - x(t) < 1$ whenever $t > T$.

c) Graph $x(t)$, $0 \le t \le 16$.

3. The function

$$P(t) = 1.535 + x(t),$$

where $x(t)$ is the solution of the logistic equation

$$\dot{x} = 0.023x \left(1 - \frac{x}{6.336}\right), \qquad x(0) = 0.767,$$

is an excellent fit to the population of Sweden from 1800 to 1920. Here t is measured in years, $x(t)$ is population size in millions, and $t = 0$ corresponds to the year 1800.

Solve the logistic equation and use it to predict the population of Sweden in 1930 (6.142), 1940 (6.372), 1950 (7.041), 1960 (7.495), and 1970 (8.040). The figures in parentheses are the actual populations.

4. A model for the harvesting of a population with logistic growth is

$$\dot{x} = rx \left(1 - \frac{x}{K}\right) - H,$$

where H can be regarded as a constant harvesting rate. Such harvesting could occur as the result of hunting, fishing, or a disease.

a) Show that if $H = 0$, the maximum rate of growth of $x(t)$ is $H_c = rK/4$ and that if $H > H_c$, the population expires in finite time (meaning $x(T) = 0$ for some $T > 0$).

b) Show that if $0 < H < H_c$, the equation has two constant solutions $x(t) = K_1$ and $x(t) = K_2$ with $0 < K_1 < K_2 < K$. Furthermore, all solutions $x(t)$ with $x(0) = x_0$ and $x_0 > K_1$ satisfy $\lim_{t \to \infty} x(t) = K_2$. The effect of harvesting is to reduce the carrying capacity of the population.

5. The sandhill crane *(Grus canadensis)* has a suggested carrying capacity $K = 194.6$ in thousands of birds and an intrinsic growth rate $r = 0.0987$. The sandhill crane is hunted because of its destructive foraging in grain fields.

a) Using the results of Exercise 4, find the critical harvest rate (birds per year) H_c.

b) What is the effect on the carrying capacity if 3000 birds per year ($H = 3.0$) are harvested?

c) Assuming the population is near equilibrium, that is, $x(0) = K = 194.6$, and the harvesting rate is 13,000 birds per year ($H = 13.0$), use the improved Euler method with step size $h = 0.5$ to estimate when the population expires. (This will occur prior to $T = 25$ years.)

d) The same as (c), but harvesting 20,000 birds per year.

6. By writing the expression for the solution of the harvested logistic equation in the form

$$\int_K^0 \frac{1}{rx(1 - x/K) - H} \, dx = \int_0^T dt, \quad H > H_c,$$

one can solve for the time T necessary to reach extinction. Do this and, using the data from Exercise 5, compare your computed results in 5(c) and 5(d) with the actual answer.

7. Below are some other models of population growth that have been suggested by various researchers. Discuss their underlying biological assumptions and qualitative features and compare them to those of the logistic model.

a) $\dot{x} = rx \ln \dfrac{K}{x}$

b) $\dot{x} = rx \left(\dfrac{K}{x} - 1 \right)$

c) $\dot{x} = \dfrac{rx(K - x)}{K + \epsilon x}$, where $\epsilon > 0$ is small

d) $\dot{x} = rx[1 - (x/K)^\alpha]$; consider two cases: $0 < \alpha < 1$ and $\alpha > 1$.

1.10 A SIMPLE MIXING PROBLEM AND FEEDBACK CONTROL

In this section we shall study the problem of adjusting the concentration of salt to a desired level in a brine mixing tank. Simple extensions of this model will apply to a number of problems in dilution or contamination. For instance, a dam or reservoir into which pollutants are flowing via feeder streams can be modeled as a mixing tank.

Let us suppose that at $t = 0$ the tank contains V liters of brine with an initial salt concentration of c_0 grams per liter. Brine is pumped into the top of the tank at a rate of r liters per second and its salt concentration is s grams per liter. The tank contains a mixing device which insures that the concentration of salt c in the tank is uniform throughout the tank. Finally, the mixture in the tank is pumped out through the bottom at the same rate r as the incoming solution.

Our task is to adjust s, the incoming concentration of salt, so that the concentration of salt in the tank attains (and remains at) a predetermined concentration $\bar{c}$. Our mathematical model will show that unless the initial concentration c_0 and the incoming concentration s both equal $\bar{c}$, our task cannot be accomplished in finite time. To overcome this difficulty, we must introduce the notion of a feedback control law which governs s.

The fundamental principle in this and other mixing problems is that the rate of change of the amount of salt in the tank is equal to the rate of change of the incoming salt minus the rate of change of the outgoing salt. Since the amount of salt in the tank is c (grams/liter) $\times$ V (liters), therefore

$$\frac{d}{dt} (cV) = (\text{Rate in}) - (\text{Rate out}).$$

The data given implies that

$$\text{Rate in} = s \text{ (grams/liter)} \times r \text{ (liters/second)},$$

$$\text{Rate out} = c \text{ (grams/liter)} \times r \text{ (liters/second)},$$

and so the governing equation is

$$\frac{d}{dt}(cV) = sr - cr. \qquad (1.10.1)$$

Since the volume V of the tank is constant, the equation (1.10.1) becomes

$$\frac{dc}{dt} = ps - pc, \qquad p = \frac{r}{V}, \qquad (1.10.2)$$

with initial condition $c(0) = c_0$.

Equation (1.10.2) is linear with constant coefficients and it is easily solved to obtain

$$c(t) = s + (c_0 - s)e^{-pt}. \qquad (1.10.3)$$

As t increases, $c(t)$ approaches s in the limit, and therefore we see that we must take $s = \bar{c}$ to solve our problem of attaining a concentration $\bar{c}$. Hence the inflow concentration must equal the desired concentration.

From equation (1.10.3) it is seen that if $s = \bar{c}$ and the initial concentration $c_0 \neq \bar{c}$, an infinite amount of time is required before $c(t) = \bar{c}$. If this is not desirable, the design of our system must be improved, and to do this we will employ a fundamental concept in modern engineering, *feedback control,* the control of a process through monitoring. In this problem, this will be accomplished by changing the incoming concentration s, where the change is governed by the existing concentration in the tank.

The concept of feedback control is neither esoteric nor new. The household thermostat is a feedback control device that turns the furnace on or off depending on the measured temperature. An autopilot mechanism, which measures the actual heading of an airplane and adjusts the control surfaces so as to maintain a fixed heading, is another example. A much older feedback device, not understood until recently, is the control of respiration by measuring the hydrogen-ion concentration in the brain.

Returning to our problem, suppose that the concentration c of salt in the tank can be measured at any instant. This could be accomplished, for example, by a pressure-sensitive device on the bottom of the tank since the mass of fluid is equal to the mass of dissolved salt plus the mass of the water. Suppose furthermore that instrumentation can be devised so that the measurements obtained can be used to control the incoming concentration s.

Then the simplest example of a *feedback control law* is the following one:

$$s = s(c) = \begin{cases} 0 & \text{if} \quad c > \bar{c}, \\ \bar{c} & \text{if} \quad c = \bar{c}, \\ \bar{s} & \text{if} \quad c < \bar{c}, \end{cases}$$

where $\bar{s}$ is some convenient value greater than $\bar{c}$. For instance, the first equation says to pump in fresh water if the concentration is greater than $\bar{c}$. The last equation says to pump in a heavier concentration if $c(t)$ is less than $\bar{c}$.

With this control law, the differential equation (1.10.2) now becomes

$$\frac{dc}{dt} = ps(c) - pc, \tag{1.10.4}$$

and the form of the solution depends on the initial concentration $c(0) = c_0$.

Case 1. If $c_0 > \bar{c}$, then $s(c) = 0$ and the solution of (1.10.4) is

$$c(t) = c_0 e^{-pt}.$$

The desired concentration will be attained at $t = t^*$, where $c(t^*) = \bar{c}$. Hence $t^* = (1/p) \ln(c_0/\bar{c})$.

Case 2. If $c_0 = \bar{c}$, then $s(c) = \bar{c}$, $c(t) = \bar{c}$, and $t^* = 0$.

Case 3. If $c_0 < \bar{c}$, then $s(c) = \bar{s}$ and from (1.10.3) the solution is

$$c(t) = \bar{s} + (c_0 - \bar{s})e^{-pt},$$

where $\bar{s} > \bar{c} > c_0$. The desired concentration will be obtained when $c(t^*) = \bar{c}$, and this gives

$$t^* = \frac{1}{p} \ln \left(\frac{\bar{s} - c_0}{\bar{s} - \bar{c}} \right).$$

On the face of it, our simple control law has solved the problem of attaining $\bar{c}$ in finite time.

We continue our discussion of the mixing problem and begin with the following observation. The simple feedback model has a serious drawback—we have assumed instantaneous mixing of the solution in the tank and the incoming brine. In reality, once equilibrium $\bar{c}$ is attained, the input mechanism will oscillate (or *chatter*, in the control terminology) as it attempts to follow the fluctuating feedback directions.

Therefore, the model should be designed to allow for errors in the adjustments of the input concentration, as well as errors in the measurements of the brine concentration in the tank. One way to do this is to weaken the requirement that the concentration c attain a specific value $\bar{c}$ and instead require that it be close to c. Hence we require that c be in the interval

$$I_\epsilon = \{c \mid |c - \bar{c}| \leq \epsilon\},$$

where $\epsilon > 0$ is some acceptable measure of error.

This last requirement will not eliminate chatter since it is still possible for the concentration c to drift back and forth across the boundary of I_ϵ. A simple way to eliminate chatter is to design a feedback law for the input concentration s with a time delay $\tau > 0$ as follows:

1.
$$s = s(c) = \begin{cases} 0 & \text{if } c > \bar{c} + \epsilon, \\ \bar{c} & \text{if } \bar{c} - \epsilon \leq c \leq \bar{c} + \epsilon, \\ \bar{s} & \text{if } c < \bar{c} - \epsilon, \text{ where } \bar{s} > \bar{c}. \end{cases}$$

2. Change values of $s(c)$ exactly τ seconds after the measurement of the tank concentration c indicates that I_ϵ has been reached.

To guard against chatter, no new measurements (and hence no new instructions) are accepted during the delay period.

As before, the form of the solution will depend on the value c_0 of the initial concentration in the tank.

Case 1. $c_0 > \bar{c} + \epsilon$. In this case $s(c) = 0$ and $c(t) = c_0 e^{-pt}$, and this will remain so until $c(t)$ reaches $\bar{c} + \epsilon$. This will occur at the time t^* which satisfies the relation

$$c_0 e^{-pt^*} = \bar{c} + \epsilon.$$

However, because of the delay, $s(c)$ will not switch from 0 to $\bar{c}$ until $t = t^* + \tau$. At the switching time,

$$c(t^* + \tau) = c_0 e^{-p(t^* + \tau)} = c_0 e^{-pt^*} e^{-p\tau},$$

which implies (from the previous equation) that

$$c(t^* + \tau) = (\bar{c} + \epsilon)e^{-p\tau}.$$

In the last relation it is seen that if the delay time τ is too large, the right-hand side could be very small. It could then occur that $c(t^* + \tau)$ will be less than $\bar{c} - \epsilon$, the lower boundary of I_ϵ. Therefore, in the design of our system τ must be restricted, so that

$$\bar{c} - \epsilon < (\bar{c} + \epsilon)e^{-p\tau}.$$

When $s(c)$ switches to $\bar{c}$, the differential equation and initial conditions are

$$\frac{dc}{dt} = p\bar{c} - pc, \qquad c(t^* + \tau) = (\bar{c} + \epsilon)e^{-p\tau}.$$

The solution is

$$c(t) = \bar{c} + [(\bar{c} + \epsilon)e^{-p\tau} - \bar{c}]e^{-p(t - t^* - \tau)}, \quad t \geq t^* + \tau,$$

and the restriction on τ and a simple estimate give

$$c(t) > \bar{c} + [(\bar{c} - \epsilon) - \bar{c}]e^{-p(t-t^*-\tau)} \geq \bar{c} - \epsilon$$

for $t \geq t^* + \tau$. Furthermore,

$$c(t) - \bar{c} = (\bar{c} + \epsilon)e^{-p(t-t^*)} - \bar{c}e^{-p(t-t^*-\tau)} \leq (\bar{c} + \epsilon)e^{-p(t-t^*)} - \bar{c}e^{-p(t-t^*)} \leq \epsilon;$$

hence $|c(t - \bar{c}| \leq \epsilon$ for $t \geq t^* + \tau$ as desired.

Case 2. $\bar{c} - \epsilon \leq c_0 \leq \bar{c} + \epsilon$. Now $s(c) = \bar{c}$, and for $t > 0$ the solution of (1.10.2) is

$$c(t) = \bar{c} + [c_0 - \bar{c}]e^{-pt}.$$

We then have the estimates

$$c(t) = \bar{c}(1 - e^{-pt}) + c_0 e^{-pt} \leq \bar{c}(1 - e^{-pt}) + (\bar{c} + \epsilon)e^{-pt}$$
$$= \bar{c} + \epsilon e^{-pt} \leq \bar{c} + \epsilon$$

and

$$c(t) \geq \bar{c}(1 - e^{-pt}) + (\bar{c} - \epsilon)e^{-pt} = \bar{c} - \epsilon e^{-pt} \geq \bar{c} - \epsilon.$$

Therefore $|c(t) - \bar{c}| \leq \epsilon$, as required.

Case 3. $c_0 < \bar{c} - \epsilon$. In this case, $s(c) = \bar{s} > \bar{c} > c_0$, and the concentration will increase and reach $\bar{c} - \epsilon$, the lower boundary of I_ϵ at a time $t_* > 0$. At time $t_* + \tau$, $s(c)$ will switch to $\bar{c}$, and to insure that $x(t_* + \tau)$ is less than $\bar{c} + \epsilon$ (the upper boundary of I_ϵ), a further restriction on the delay time τ is needed. We ask the reader (see Exercise 14) to show that this restriction is

$$\bar{s} + (\bar{c} - \epsilon - \bar{s})e^{-p\tau} < \bar{c} + \epsilon,$$

and that with this requirement $|c(t) - \bar{c}| \leq \epsilon$ for $t \geq t_* + \tau$. Note that the last restriction can be written as

$$(\bar{s} - \bar{c}) - \epsilon < [(\bar{s} - \bar{c}) + \epsilon]e^{-p\tau},$$

which is analogous to the requirement on τ in Case 1.

EXERCISES 1.10

1. A tank contains a homogeneous solution of 5 kilograms of salt and 500 liters of water. Starting at $t = 0$, fresh water is poured into the tank at a rate of 4 liters/min. A mixing device maintains homogeneity. The uniform solution leaves the tank at 4 liters/min.

a) What is the differential equation governing the amount of salt in the tank at any time?

b) In how many minutes will the concentration of salt reach a 0.1% level?

2. A tank contains a homogeneous solution of 2 kilograms of salt and 50 liters of water. Pure salt is fed into the tank at the rate of 1 kilogram per minute. The uniform solution leaves the tank at the rate of 2 liters per minute. Find the amount and concentration of the salt after 25 minutes.

3. A tank contains a homogeneous solution of 5 kilograms of salt and 1000 liters of water. A brine solution containing 0.5 kilograms of salt per liter enters at 5 liters per minute. The uniform solution leaves the tank at the same rate. When will the tank contain 100 kilograms of salt?

4. In Exercise 1, change the exit rate to 2 liters/min and assume the tank has a 1000-liter capacity. What is the amount and concentration of salt in the tank at the instant of overflow? *Hint:* Find V as a function of time and use it in the differential equation for c.

5. In Exercise 3 suppose the exit rate is changed to 3 liters/min. and the tank has a 2000-liter capacity. What is the amount and concentration of salt after 250 minutes? At overflow?

6. For the simple feedback control strategy given, suppose that the tank contains 100 liters of brine and the inflow and exit rates are 4 liters/min. If the desired concentration is $\bar{c} = 12$ grams/liter, find the time t^* at which it is achieved if:

   a) The initial concentration in the tank is $c_0 = 20$ grams/liter,

   b) The initial concentration in the tank is $c_0 = 4$ grams/liter and we use an incoming concentration of 16, 13, and 12.1 grams/liter.

7. Two tanks initially contain 200 liters of fresh water each. Starting at $t = 0$, brine containing 5 kg/liter of salt is added to the first tank at a rate of 2 liters/min. The uniform solution from the first tank is transferred to the second tank at a rate of 2 liters/min. The uniform solution in the second tank is removed at the same time. Find the concentration in each tank at any time.

8. A 1000-liter tank contains a homogeneous solution of 5 kilograms of salt and 500 liters of water. In order to attain a concentration of 0.02 ± 0.006 kg/liter, brine is added at a rate of 4 liters/min. The uniform solution leaves at the same rate. The concentration of the incoming brine is governed by the following feedback law:

$$s = s(c) = \begin{cases} 0 & \text{if } c > 0.026, \\ 0.02 & \text{if } 0.0194 \le c \le 0.026, \\ 0.04 & \text{if } c < 0.0194. \end{cases}$$

Let t_* be the time at which the concentration is equal to 0.0194. Find t_* and $c(t_*)$. Find the formula for $c(t)$ with $0 \le t \le t_*$ and with $t_* < t$.

9. Assume that the tank of Exercise 8 initially contains 15 kilograms of salt and 500 liters of water. Let t^* be the time at which the concentration is equal to 0.026. Find the formula for $c(t)$ with $0 \le t \le t^*$ and with $t^* < t$.

10. Suppose a delay of $\tau = 2$ min is introduced in the control and the initial conditions are 5 kilograms of salt dissolved in 1000 liters of water. Find $c(t)$ on the intervals $0 \le t \le t_* + \tau$ and $t_* + \tau < t$, where t_* is the time at which the concentration is equal to 0.0194.

11. With a delay of $\tau = 2$ min in the control and the initial conditions of 30 kilograms of salt and 1000 liters of water, find $c(t)$ on the intervals $0 \le t \le t^* + \tau$ and $t^* + \tau < t$, where t^* is the time at which the concentration is equal to 0.026.

12. With no delay and with the initial concentration in the 1000-liter tank denoted by c_0, find the formulas for $c(t)$ if the control law is

$$s(c) = \begin{cases} 0 & \text{if } c > \bar{c}, \\ \bar{s} & \text{if } c \le \bar{c}. \end{cases}$$

Show that t_* is a decreasing function of $\bar{s}$. Therefore to minimize t_*, take $\bar{s}$ as large as possible. Assume V equals 500 liters.

13. Solve Exercise 12 with a delay τ.

14. Derive the restriction on the delay time τ given in the text for case 3 and show that

$$|c(t) - \bar{c}| \le \epsilon \qquad \text{for } t \ge t_* + \tau.$$

MISCELLANEOUS EXERCISES

1.1 The Riccati differential equation (Count J. Riccati, 1676–1754) is a first order nonlinear differential equation of the form

$$\frac{dy}{dt} = p(t)y + q(t)y^2 + r(t).$$

(If $r(t) = 0$, it is a special case of Bernoulli's equation and can be solved exactly.) If $y_1(t)$ is a solution let

$$y(t) = y_1(t) + \frac{1}{x(t)},$$

and by substitution into the differential equation, show that $x(t)$ satisfies a linear differential equation. Hence, knowing a particular solution, we can find the general solution.

1.2 Use the result of the previous exercise to find the solution of the following initial value problems, given a particular solution $y_1(t)$.

a) $\dfrac{dy}{dt} = y^2 - 1$, $y(0) = 3$; $y_1(t) = 1$

b) $\dfrac{dy}{dt} = y + \dfrac{2}{t^3} y^2 - t^2$, $y(1) = 4$; $y_1(t) = t^2$

c) $\dfrac{dy}{dt} = -y^2 + \dfrac{15}{4t^2}$, $y(1) = 4$; $y_1(t) = \dfrac{a}{t}$

1.3 The Riccati differential equation occurs in the design of a feedback-control law for the linear-regulator problem with quadratic cost. A one-dimensional version of that problem is the following: a system with output $x(t)$ and input or control $u(t)$ is governed by the linear differential equation

$$\frac{dx}{dt} = a(t)x + b(t)u, \qquad x(t)_0 = x_0, \quad t_0 \le t \le t_1.$$

Choose a control $u = u(t)$ so as to minimize the cost or performance measure

$$C[u] = \tfrac{1}{2}kx(t_1)^2 + \tfrac{1}{2} \int_{t_0}^{t_1} [w_1(t)x(t)^2 + w_2(t)u(t)^2]dt,$$

where $w_1(t)$ and $w_2(t)$ are given weighting functions and k is a nonnegative constant.

The theory of this problem shows that the optimal control $u(t)$, which minimizes $C[u]$, is realized through the feedback control law:

$$u(t) = p^*(t)x(t), \qquad p^*(t) = -\frac{b(t)}{w_2(t)} p(t),$$

where $p(t)$ satisfies the Riccati differential equation

$$\frac{dp}{dt} = -2a(t)p + \frac{b(t)^2}{w_2(t)} p^2 - w_1(t), \qquad p(t_1) = k.$$

If one can find $p(t)$, the problem is completely solved. Hence, we find the solution of

$$\frac{dx}{dt} = [a(t) + b(t)p^*(t)]x, \qquad x(t_0) = x_0,$$

and then evaluate

$$C[u] = \tfrac{1}{2}kx(t_1)^2 + \tfrac{1}{2} \int_{t_0}^{t_1} [w_1(t) + w_2(t)p^*(t)^2]x(t)^2 \, dt,$$

which will be a minimum.

For the linear regulator problem

$$\frac{dx}{dt} = x + u, \qquad x(0) = 1,$$

$$C[u] = \tfrac{1}{2}kx(1)^2 + \tfrac{1}{2} \int_0^1 [3x(t)^2 + u(t)^2] \, dt.$$

a) Solve the problem completely for the case $k = 3$, i.e., find the feedback control law, the optimal output $x(t)$ and the value of $C[u]$.

b) Find the feedback control law for the case $k = 0$.

c) For the case $k = 2.75$, find the feedback control law, then use the improved Euler method to approximate $x(t)$ on $0 \le t \le 1$. With the numerical values you can use a numerical quadrature formula (e.g., Simpson's rule) to estimate the minimum cost $C[u]$.

1.4 Given the first order linear equation

$$\frac{dx}{dt} + ax = \phi(t),$$

where $a \ne 0$ and $\phi(t)$ is a continuous periodic function of period T, show that it can have at most one solution of period T. (*Hint:* You could assume it has two solutions and find the differential equation satisfied by their difference.) Show by example that if $a = 0$, there could exist infinitely many periodic solutions or none.

1.5 It began to snow sometime in the evening. At midnight the snowplow started out and by 1:00 A.M. it had plowed two miles down the main highway. By 2:00 A.M. it had plowed only one more mile. Assuming it snows at a constant rate and the plow removes snow at a constant rate, when did it start to snow? (An almost classic problem devised by Prof. R. P. Agnew).

1.6 The fact that $y^2 < y^2 + t^2$ for $t > 0$ implies that the solution of

$$\frac{dy}{dt} = y^2 + t^2, \qquad y(0) = y_0 > 0$$

grows faster than the solution of

$$\frac{dy}{dt} = y^2, \qquad y(0) = y_0.$$

Use this fact to show that the solution of the first differential equation becomes infinite for a finite value of $t > 0$. (A little analysis of this kind prior to doing any numerical work pays off!)

1.7 Following the idea developed in the previous problem, show that the solution of

$$\frac{dy}{dt} = y^{1/2} + t^2, \qquad y(0) = 1,$$

is bounded for $t > 0$ and satisfies

$$1 + t + \frac{t^3}{3} < y(t) < 3e^t - t^2 - 2t - 2.$$

1.8 Show that the solution of

$$\frac{dy}{dt} = y^2 \cos t, \qquad y(0) = y_0 \neq 0,$$

is defined for $-\infty < t < \infty$ if $|y_0|^{-1} > 1$, whereas it is only defined on a finite interval if $|y_0|^{-1} < 1$.

1.9 Find the exact solution for $\epsilon > 0$ of the equation

$$\epsilon \frac{dy}{dt} + y = 1 + t, \qquad y(0) = 0,$$

and sketch its graph for $t \geq 0$ and $\epsilon = 10^{-1}, 10^{-2}$. For each value of ϵ, determine the interval $0 \leq t \leq t_\epsilon$ for which the solution $y(t)$ satisfies

$$(1 + t) - y(t) \geq 2\epsilon.$$

This narrow region is an example of a *boundary layer*—a region of sudden transition. The boundary layer phenomenon occurs in many problems of fluid dynamics.

1.10 Given the initial value problem

$$\frac{dy}{dt} = \frac{-\sqrt{1 - y^2}}{t^2}, \qquad y(a) = 0, \quad a > 0,$$

and the infinite strip

$$S = \{(t, y) \,|\, |y| < 1, \quad t > 0\},$$

show that, as t approaches zero, the solution stays within the strip and hits the lateral boundary an infinite number of times, but never reaches the left-hand boundary.

REFERENCES

There are countless introductory texts on ordinary differential equations. A few studies the reader will find worth examining are:

1. W. E. Boyce and R. C. DiPrima, *Elementary Differential Equations*, 3rd ed., Wiley, New York, 1977.

2. F. Brauer and J. A. Nohel, *Ordinary Differential Equations*, 2nd ed., Benjamin, Menlo Park, 1973.

3. M. Braun, *Differential Equations and Their Applications*, Springer–Verlag, New York, 1975.

4. W. R. Derrick and S. I. Grossman, *Elementary Differential Equations with Applications*, 2nd ed., Addison-Wesley, Reading, 1981.

5. S. L. Ross, *Introduction to Ordinary Differential Equations*, 3rd ed., Wiley, New York, 1974.

For further study of the modeling of populations and ecological systems a classic study is

6. E. C. Pielou, *An Introduction to Mathematical Ecology*, Wiley–Interscience, New York, 1969.

Other interesting discussions of population models as well as the modeling of other biological systems are

7. J. M. Smith, *Mathematical Ideas in Biology,* Cambridge Press, Cambridge, 1968.

8. J. M. Smith, *Models in Ecology,* Cambridge Press, Cambridge, 1974.

9. S. Levin (Ed.), *Studies in Mathematical Biology,* Parts I and II, Mathematical Association of America Studies in Mathematics, Volumes 15 and 16, 1978. This is a collection of articles, some of which are quite advanced in level; it gives an idea of the scope of problems in biology being modeled mathematically.

Linear Second Order Differential Equations

2.1 INTRODUCTION

Second order ordinary differential equations play a central role in the theory of differential equations as well as in their applications to engineering, physics, and other sciences. Their role in the theory is due, in part, to their ability to model concepts, such as linear independence (which occurs in the study of equations of order two or higher, but not in equations of order one) and to their usefulness in studying such topics as the phase plane analysis of nonlinear oscillators (in which the topology of the two-dimensional plane is essential). Their role in applied mathematics is due to the great variety and importance of phenomena which can be described by second order differential equations, such as LCR circuits and mechanical systems with one degree of freedom, and because they are an essential tool in solving many of the equations of mathematical physics.

In this chapter we shall restrict our study to linear second order equations. An introduction to nonlinear second order equations is given in Chapter 7.

2.2 EXAMPLES FROM MECHANICS AND CIRCUIT THEORY

The analysis of simple mechanical systems, which give rise to oscillatory or damped oscillatory motion, occurs in almost every introduction to differential equations because it is based on easily understood concepts of mechanics, namely Newton's and Hooke's laws. Furthermore, it illustrates the effects of damping and the phenomenon of resonance, which are of fundamental importance in the design of vibration absorbing devices. More important, the physical description, and its consequent mathematical translation, goes hand in hand with physical reality, as evidenced by the widespread use of the model and variations of it in engineering, structural analysis, and applied mechanics.

The result of translating the physical description of a simple oscillatory mechanical system into a mathematical model is a linear second order differential equation with constant coefficients. In the opinion of the authors, a thorough understanding of the qualitative behavior of solutions of this equation as parameters vary, or external driving forces are added, is *the most important topic* to be learned from an introductory course in differential equations. We now proceed to construct the mathematical model that gives us the differential equation.

Consider a mass m suspended by means of a light vertical spring from a fixed vertical support (see Fig. 2.1a). If the natural length of the spring is l_0 and it is in equilibrium, then, when the mass is attached, it will be extended by an additional amount l. If we assume Hooke's law, then the tension in the spring is directly proportional to its extension, and this gives the relation

$$kl = mg,$$

where g is the gravitational constant. This means that in an equilibrium state the upward tension in the spring must balance the downward force of gravity. The constant of proportionality k is called the *spring constant*.

Now suppose that at an initial time t_0 the mass is vertically displaced y_0 units from equilibrium and also has a velocity v_0 upon release. If $y = y(t)$ is the distance of the mass (measured downwards) at time t from equilibrium, then the spring extension at time t is $l + y(t)$. Newton's law states that the

Rate of change in downward momentum = Net downward force

or

$$\frac{d}{dt}\left[m \frac{dy}{dt}(t) \right] = mg - k[l + y(t)].$$

Since $kl = mg$, this becomes (letting $y = y(t)$) the second order differential equation

$$m \frac{d^2y}{dt^2} = -ky \qquad \text{or} \qquad m \frac{d^2y}{dt^2} + ky = 0. \tag{2.2.1}$$

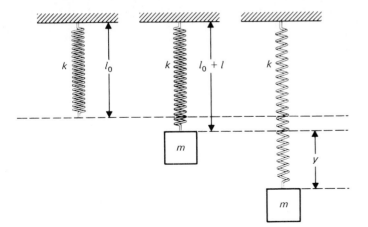

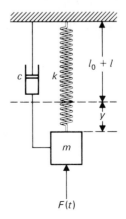

Figure 2.1(a) A spring–mass system with no damping or forcing.

Figure 2.1(b) A spring–mass system with damping and forcing.

In addition we are given the initial conditions of displacement and velocity

$$y(t_0) = y_0, \qquad v(t_0) = \frac{dy}{dt}(t_0) = v_0. \tag{2.2.2}$$

The problem is to solve the constant coefficient second order linear differential equation (2.2.1) subject to the initial conditions (2.2.2).

However, the description of the physical model is quite simple, and we might wish to take into account the effect of damping due, for instance, to air resistance on the mass. A plausible model for damping is to assume that it resists the motion with a force proportional to the velocity. Therefore, the damping force F_α is

$$F_\alpha = -c\,\frac{dy}{dt}(t),$$

where the positive constant c is called the *damping coefficient*. Such damping is called *viscous damping* and is often modeled by the addition of a piston-like device called a *dashpot* (see Fig. 2.1b).

Finally, there could also be a time dependent external force $F(t)$ acting on the mass. This could be the result of a movable, rather than fixed, support, or the mass could be attached to another mass–spring system, as in a shock or vibration absorbing mechanism. Then Newton's law gives

$$\frac{d}{dt}\left(m\,\frac{dy}{dt}\right) = \text{Net downward force} = -ky - c\,\frac{dy}{dt} + F(t)$$

or

$$m \frac{d^2y}{dt^2} + c \frac{dy}{dt} + ky = F(t). \tag{2.2.3}$$

The solution $y = y(t)$ and its derivative $v = y'(t)$ will give the displacement from equilibrium and velocity of the mass at time t subject to the initial conditions

$$y(t_0) = y_0, \qquad y'(t_0) = v_0. \tag{2.2.4}$$

The object in this chapter is to analyze and solve, whenever possible, the initial value problem (2.2.3) and (2.2.4).

The next example from circuit theory shows that the initial value problem (2.2.3) and (2.2.4) is not solely a mathematical model for a mass–spring system. A useful mathematical model described by, say, a differential equation and/or initial value problem will often represent a number of distinct physical situations. One can then study the mathematical model in the abstract and, with an appropriate choice of the parameters, interpret the quantitative and qualitative results obtained in terms of the physical reality under study.

For the circuit theory example, we connect three circuit elements in series, as shown in Fig. 2.2; the symbols for the resistor, capacitor, and inductor are identified by the letters R, C, and L, respectively. The imposed electromotive force, which could be a battery or a signal generator, is represented by a circular symbol and denoted by $E(t)$. One can assume the presence of voltmeters or oscilloscopes that would monitor the process without affecting it.

If the current I is flowing through a resistor, the voltage drop across the resistor is proportional to the current:

$$V_R = RI,$$

where R denotes the magnitude of the resistor. On the other hand, if current is flowing through an inductor, the voltage drop across the inductor is pro-

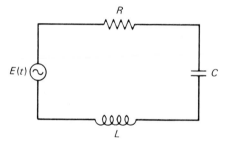

Figure 2.2 An LCR circuit.

portional to the rate of change of the current with time:

$$V_{\mathrm{L}} = L\, dI/dt,$$

where L denotes the magnitude of the inductor. Note that there is no voltage drop due to inductance if the current is constant. Since the inductor consists of a coil of wire, there is always resistance present but, for convenience, it is lumped together with the resistance R of the resistor.

A perfect capacitor does not allow current to flow through it but, instead, it stores electric charges. As the amount of charge Q changes, current flows into and out of the capacitor. The voltage drop across the capacitor is proportional to its charge:

$$V_{\mathrm{C}} = Q/C,$$

where C denotes the magnitude of the capacitor.

The three equations above are related through a fundamental principle of circuit theory, namely, that the sum of the voltage drops in a closed loop is equal to the imposed electromotive force, hence,

$$V_{\mathrm{L}} + V_{\mathrm{R}} + V_{\mathrm{C}} = E(t).$$

Substituting the above equations for V_{L}, V_{R}, and V_{C}, we have

$$L\frac{dI}{dt} + RI + \frac{Q}{C} = E(t). \tag{2.2.5}$$

Since the current is the time rate of change of the charge, this gives the second equation,

$$\frac{dQ}{dt} = I. \tag{2.2.6}$$

These two first order equations are coupled and are equivalent to the single second order equation obtained by setting

$$\frac{dI}{dt} = \frac{d^2Q}{dt^2}, \qquad I = \frac{dQ}{dt}$$

in (2.2.5) to obtain

$$L\frac{d^2Q}{dt^2} + R\frac{dQ}{dt} + \frac{Q}{C} = E(t), \tag{2.2.7}$$

a linear second order ordinary differential equation with constant coefficients whose solution is $Q(t)$, the charge.

Sometimes it is desirable to model the circuit by a differential equation with I as the dependent variable. This is accomplished by differentiating (2.2.5) to obtain

$$L\frac{d^2I}{dt^2} + R\frac{dI}{dt} + \frac{I}{C} = \frac{dE(t)}{dt}. \tag{2.2.8}$$

However, if the impressed electromotive force $E(t)$ is rapidly changing, e.g., as in a square wave or in a saw-toothed wave, equation (2.2.7) is preferable to (2.2.8) since the right-hand side function, *the forcing function,* is smoother.

The pair of differential equations (2.2.5) and (2.2.6) or the equivalent second order differential equation (2.2.7) can be solved for $Q(t)$ and $I(t)$. However, if we are to compare these functions with measured values of the charge and current, more information is needed to specify the solution functions completely. It is sufficient, and for this problem natural, to specify the charge and current at a convenient time, say, $Q = Q_0$ and $I = I_0$ at $t = t_0$ to give the initial conditions

$$Q(t_0) = Q_0, \qquad I(t_0) = Q'(t_0) = I_0. \tag{2.2.9}$$

The reader can easily see that there is no essential *mathematical* difference between the initial value problem (2.2.3), (2.2.4) for the mass–spring system and the initial value problem (2.2.7), (2.2.9) for an LCR circuit.

In the next six sections we shall mathematically analyze the second order linear differential equation and the corresponding initial value problem. The developed solution techniques apply mainly to the case of constant coefficients, but much of the theory applies to the case of time-dependent coefficients as well. In the latter sections we shall return to the two examples above and interpret the mathematical results in terms of the physical reality.

2.3 GENERAL THEORY—INTRODUCTION

We shall study the linear second order differential equations

$$a(t) \frac{d^2y}{dt^2} + b(t) \frac{dy}{dt} + c(t)y = f(t), \tag{2.3.1}$$

where $a(t)$, $b(t)$, $c(t)$, and $f(t)$ are continuous functions on a nonempty interval. In order to insure that the equation is always of second order, the coefficient $a(t)$ of d^2y/dt^2 must never be zero in the interval. If the forcing function $f(t)$ is zero, the differential equation

$$a(t) \frac{d^2y}{dt^2} + b(t) \frac{dy}{dt} + c(t)y = 0 \tag{2.3.2}$$

is called *homogeneous.* (That unfortunate usage again! This should not be confused with homogeneous first order equations discussed in Exercise 16 of Section 1.7.)

The essential property of linear equations is, of course, *linearity.* Namely, if $y_1(t)$ and $y_2(t)$ are solutions of the homogeneous equation (2.3.2), then

$$y(t) = Ay_1(t) + By_2(t)$$

is a solution for any choice of the constants A and B. This follows easily from the linearity of differentiation and the property above:

$$a(t) \frac{d^2}{dt^2} [Ay_1(t) + By_2(t)] + b(t) \frac{d}{dt} [Ay_1(t) + By_2(t)] + c(t)[Ay_1(t) + By_2(t)]$$

$$= A \left[a(t) \frac{d^2}{dt^2} y_1(t) + b(t) \frac{d}{dt} y_1(t) + c(t)y_1(t) \right]$$

$$+ B \left[a(t) \frac{d^2}{dt^2} y_2(t) + b(t) \frac{d}{dt} y_2(t) + c(t)y_2(t) \right].$$

But each term in brackets is zero because both $y_1(t)$ and $y_2(t)$ are solutions. Therefore the expression above equals zero, hence $y(t)$ is a solution.

By analogy with linear first order equations, we find that if $v(t)$ is a solution to the inhomogeneous equation (2.3.1) and $w(t)$ is a solution to the homogeneous equation (2.3.2), then

$$y(t) = v(t) + Kw(t)$$

is a one-parameter family of solutions to the inhomogeneous equation (2.3.1). As usual, the constant (or parameter) K is arbitrary. Substitute the assumed solution into the differential equation (2.3.1) to obtain

$$a(t) \frac{d^2}{dt^2} [v(t) + Kw(t)] + b(t) \frac{d}{dt} [v(t) + Kw(t)] + c(t)[v(t) + Kw(t)]$$

$$= \left[a(t) \frac{d^2}{dt^2} v(t) + b(t) \frac{d}{dt} v(t) + c(t)v(t) \right]$$

$$+ K \left[a(t) \frac{d^2}{dt^2} w(t) + b(t) \frac{d}{dt} w(t) + c(t)w(t) \right]$$

$$= f(t) + 0.$$

The first bracketed term in the preceding equation is equal to $f(t)$ since $v(t)$ is a solution to the inhomogeneous equation (2.3.1), and the second bracketed term is equal to zero since $w(t)$ is a solution to the homogeneous equation (2.3.2). Thus $y(t)$ is a solution to the inhomogeneous equation (2.3.1).

Example 1   Show that $y(t) = \frac{1}{2}t^4 + Kt^{-2}$ is a solution to

$$t^2 \frac{d^2y}{dt^2} + t \frac{dy}{dt} - 4y = 6t^4. \tag{2.3.3}$$

First compute the derivatives

$$\frac{dy}{dt} (t) = 2t^3 - 2Kt^{-3}, \qquad \frac{d^2y}{dt^2} (t) = 6t^2 + 6Kt^{-4}.$$

Then substituting into (2.3.3) gives

$$t^2(6t^2 + 6Kt^{-4}) + t(2t^3 - 2Kt^{-3}) - 4(\tfrac{1}{2}t^4 + Kt^{-2})$$
$$= t^4(6 + 2 - 2) + Kt^{-2}(6 - 2 - 4) = 6t^4,$$

and so $y(t)$ is a solution. $\square$

Given a one-parameter family of solutions $y(t) = v(t) + Kw(t)$ to equation (2.3.1), do we have all possible solutions? In other words, if $z(t)$ is a solution, can K be selected so that $z(t) = v(t) + Kw(t)$? The answer to this question is no. It is easy to check that $z(t) = \frac{1}{2}t^4 + 3t^2$ is also a solution to equation (2.3.3) and we have shown that $y(t) = \frac{1}{2}t^4 + Kt^{-2}$ is a one-parameter family of solutions to the same equation. But for any choice of K, we have

$$\frac{1}{2}t^4 + 3t^2 \neq \frac{1}{2}t^4 + Kt^{-2}$$

on any nonempty interval.

Upon reflection, the attempt to represent all solutions to a second order differential equation by a one-parameter family is unreasonable. As was pointed out in the preceding section, linear second order differential equations can be used to model mechanical or electrical problems whose solution is determined by specifying two quantities, e.g., the displacement and velocity or the charge and current at a given time. One should therefore expect that linear second order differential equations would have a two-parameter family of solutions. For example,

$$y(t) = \frac{1}{2}t^4 + At^2 + Bt^{-2}$$

is a two-parameter family of solutions to the differential equation of Example 1. The following theorem is a mathematical justification for two-parameter families of solutions.

Theorem   Let $a(t)$, $b(t)$, $c(t)$, and $f(t)$ be continuous functions on a nonempty interval I. Then there exists a pair of solutions, $y_1(t)$ and $y_2(t)$, of the homogeneous differential equation (2.3.2) and a solution $z(t)$ of the nonhomogeneous differential equation (2.3.1), defined on the interval I, such that the two-parameter family

$$y(t) = z(t) + K_1 y_2(t) + K_2 y_2(t) \qquad (2.3.4)$$

contains all the solutions of the inhomogeneous differential equation (2.3.1). Furthermore, for any t_0 in I and any numbers r and s, there is a unique choice of K_1 and K_2 such that the solution (2.3.4) satisfies the *initial conditions*

$$y(t_0) = r, \qquad \frac{dy}{dt}(t_0) = s. \qquad (2.3.5)$$

The solution (2.3.4) is called the *general solution*.

A proof of this theorem will be found in Appendix 1. In the case of a homogeneous equation, where $f(t) = 0$, the term $z(t)$ will not appear in the expression (2.3.4).

The task of solving a differential equation subject to constraints of the type (2.3.5) is called an *initial value problem*.

Example 2 Find the solution to the initial value problem

$$t^2 \frac{d^2y}{dt^2} + t \frac{dy}{dt} - 4y = 6t^4,$$

$$y(2) = 3, \qquad \frac{dy}{dt}(2) = 13,$$

given that a two-parameter family of solutions to the differential equation is

$$y(t) = \tfrac{1}{2}t^4 + At^2 + Bt^{-2}.$$

First compute the derivative of $y(t)$ to get

$$\frac{dy}{dt}(t) = 2t^3 + 2At - 2Bt^{-3}.$$

It is required at $t = 2$ that $y = 3$ and $dy/dt = 13$, therefore parameters A and B must satisfy the algebraic equations obtained by substituting these data into the expressions for $y(t)$ and $dy(t)/dt$. Since

$$3 = 8 + 4A + \tfrac{1}{4}B, \qquad 13 = 16 + 4A - \tfrac{1}{4}B,$$

we get $A = -1$, $B = -4$, and the unique solution to this initial value problem is

$$y(t) = \tfrac{1}{2}t^4 - t^2 - 4t^{-2}. \ \square$$

EXERCISES 2.3

1. For the following linear second order differential equations, show that each of the given functions is a solution. What initial conditions does each satisfy at $t = t_0$?

a) $y'' + 3y' + 2y = 0$; $y_1(t) = 4e^{-t}$, $y_2(t) = -3e^{-2t}$; $t_0 = 0$

b) $y'' + 9y = 0$; $y(t) = 5\cos 3t - 2\sin 3t$; $t_0 = \pi/6$

c) $y'' - 2y' + y = t$; $y(t) = t + 2$, $y(t) = t + 2 + 3e^t - te^t$; $t_0 = 0$

d) $y'' + 4y = 3e^{2t}$; $y_1(t) = \tfrac{3}{8}e^{2t}$, $y_2(t) = \tfrac{3}{8}e^{2t} + 4\cos(2t - \pi/4)$; $t_0 = \pi/2$

e) $t^2y'' - 3ty' + 4y = 0$; $y_1(t) = t^2$, $y_2(t) = t^2 \ln t$; $t_0 = 1$

f) $y'' - 2y' + 2y = 3\sin t - \cos t$;
$y_1(t) = \sin t + \cos t$,
$y_2(t) = \sin t + \cos t - 6e^t \sin t$; $t_0 = 0$

g) $t^2y'' + ty' + (t^2 - \tfrac{1}{4})y = 0$;
$y_1(t) = \cos t/\sqrt{t}$,
$y_2(t) = \sin t/\sqrt{t}$; $t_0 = \pi$

2. For the following differential equations from Exercise 1, the given function represents a two-parameter family of solutions. Use it to solve the given initial value problem.

a) From Exercise 1(a): $y(t) = Ae^{-t} + Be^{-2t}$; $y(0) = 1$, $y'(0) = 2$

b) From Exercise 1(b): $y(t) = A \cos 3t + B \sin 3t$; $y(0) = 0$, $y'(0) = 1$

c) From Exercise 1(b): $y(t) = K \cos(3t - \phi)$, where K and ϕ are constants, $K > 0$, $-\pi/2 \le \phi \le \pi/2$; $y(0) = 0$, $y'(0) = 1$

d) From Exercise 1(c): $y(t) = t + 2 + Ae^t + Bte^t$; $y(1) = 3$, $y'(1) = 2$

e) From Exercise 1(e): $y(t) = \frac{5}{8}e^{2t} + A \cos 2t + B \sin 2t$; $y(\pi/4) = 0$, $y'(\pi/4) = 1$

f) From Exercise 1(f): $y(t) = At^2 + Bt^2 \ln t$; $y(2) = 4$, $y'(2) = 8$

g) From Exercise 1(g): $y(t) = \sin t + \cos t + Ke^t \cos(t - \phi)$, where K and ϕ are constants, $K > 0$, $-\pi/2 \le \phi \le \pi/2$; $y(\pi/2) = 1$, $y'(\pi/2) = 3$

2.4 CONSTANT COEFFICIENT HOMOGENEOUS EQUATIONS— THE CASE OF REAL ROOTS

Linear differential equations with constant coefficients are easy to solve and provide a rich class of examples to illustrate the general theory of linear differential equations. They can model a surprisingly large number of interesting physical processes. In this section, we shall study the homogeneous second order linear differential equation

$$a \frac{d^2y}{dt^2} + b \frac{dy}{dt} + cy = 0, \qquad (2.4.1)$$

where a, b, and c are constants with $a \ne 0$. The inhomogeneous equation with constant coefficients will be studied in Sections 2.7 and 2.8.

Recalling that exponential functions are solutions to first order linear equations with constant coefficients, it is natural to ask if they also satisfy second order equations. Let $y(t) = \exp(\lambda t)$; substitute it into (2.4.1) to obtain

$$a\lambda^2 e^{\lambda t} + b\lambda e^{\lambda t} + ce^{\lambda t} = e^{\lambda t}(a\lambda^2 + b\lambda + c) = 0.$$

Since $\exp(\lambda t)$ can never be zero, this equality can only be satisfied if λ is a root of the *characteristic equation*

$$a\lambda^2 + b\lambda + c = 0. \qquad (2.4.2)$$

It is no accident that the characteristic equation is of second degree in λ; each time $\exp(\lambda t)$ is differentiated, another factor of λ appears.

The quadratic formula

$$\lambda = \frac{-b \pm (b^2 - 4ac)^{1/2}}{2a}$$

implies that the roots of the characteristic equation fall into the following three categories:

    I. Real and distinct if $b^2 - 4ac > 0$;

    II. Real and equal if $b^2 - 4ac = 0$;

    III. Complex and distinct if $b^2 - 4ac < 0$.

This classification of the roots will give a classification of all the solutions to the differential equation (2.4.1). In this section, the case of real roots (types I and II) will be discussed.

Type I. $b^2 - 4ac > 0$.

    If the roots of the quadratic equation (2.4.2) are real and distinct, there are two solutions

$$y_1(t) = e^{\lambda_1 t} \quad \text{and} \quad y_2(t) = e^{\lambda_2 t}, \tag{2.4.3}$$

where

$$\lambda_1 = \frac{-b + (b^2 - 4ac)^{1/2}}{2a},$$

$$\lambda_2 = \frac{-b - (b^2 - 4ac)^{1/2}}{2a}.$$

A *general linear combination* of the two solutions

$$y(t) = c_1 e^{\lambda_1 t} + c_2 e^{\lambda_2 t}, \tag{2.4.4}$$

where c_1 and c_2 are arbitrary constants, is also a solution. This follows from the general property of linearity discussed in the previous section.

Example 1 Solve the initial value problem

$$y'' + y' - 6y = 0, \quad y(0) = 5, \quad y'(0) = 0.$$

Set $y(t) = e^{\lambda t}$, substitute, and obtain the characteristic equation

$$\lambda^2 + \lambda - 6 = 0.$$

This is easily factored:

$$(\lambda - 2)(\lambda + 3) = 0,$$

and so the roots are $\lambda_1 = 2$ and $\lambda_2 = -3$. The corresponding solutions are

$$y_1(t) = e^{2t} \quad \text{and} \quad y_2(t) = e^{-3t}.$$

Let c_1 and c_2 be arbitrary constants and let

$$y(t) = c_1 e^{2t} + c_2 e^{-3t}, \quad y'(t) = 2c_1 e^{2t} - 3c_2 e^{-3t}.$$

Since $y(t)$ is a general linear combination of the two solutions e^{2t} and e^{-3t}, it is a solution to the differential equation. The constants c_1 and c_2 can be chosen so that the initial conditions are satisfied. We substitute $t = 0$, $y = 5$, $y' = 0$ in the expressions for $y(t)$ and $y'(t)$ and obtain two linear algebraic equations for c_1 and c_2:

$$5 = c_1 + c_2, \qquad 0 = 2c_1 - 3c_2.$$

Using Cramer's rule, we get

$$c_1 = \frac{\begin{vmatrix} 5 & 1 \\ 0 & -3 \end{vmatrix}}{\begin{vmatrix} 1 & 1 \\ 2 & -3 \end{vmatrix}} = \frac{-15}{-5} = 3,$$

$$c_2 = \frac{\begin{vmatrix} 1 & 5 \\ 2 & 0 \end{vmatrix}}{\begin{vmatrix} 1 & 1 \\ 2 & -3 \end{vmatrix}} = \frac{-10}{-5} = 2.$$

Therefore, the solution to the initial value problem is

$$y(t) = 3e^{2t} + 2e^{-3t}. \quad \square$$

Example 2 Solve the initial value problem

$$y'' + 6y' + 4y = 0, \qquad y(0) = 1, \qquad y'(0) = -3.$$

The characteristic equation

$$\lambda^2 + 6\lambda + 4 = 0$$

has solutions $\lambda_1 = -3 + \sqrt{5}$, $\lambda_2 = -3 - \sqrt{5}$. Therefore,

$$y(t) = c_1 \exp[(-3 + \sqrt{5})t] + c_2 \exp[(-3 - \sqrt{5})t],$$

and so

$$\begin{aligned} y'(t) = (-3 + \sqrt{5})c_1 \exp[(-3 + \sqrt{5})t] \\ + (-3 - \sqrt{5})c_2 \exp[(-3 - \sqrt{5})t]. \end{aligned}$$

The equations for c_1 and c_2 are obtained by setting $t = 0$, $y(0) = 1$, and $y'(0) = -3$ to get

$$1 = c_1 + c_2, \qquad -3 = (-3 + \sqrt{5})c_1 + (-3 - \sqrt{5})c_2.$$

By Cramer's rule,

$$c_1 = \frac{\begin{vmatrix} 1 & 1 \\ -3 & -3 - \sqrt{5} \end{vmatrix}}{\begin{vmatrix} 1 & 1 \\ -3 + \sqrt{5} & -3 - \sqrt{5} \end{vmatrix}} = \frac{-\sqrt{5}}{-2\sqrt{5}} = \frac{1}{2},$$

$$c_2 = \frac{\begin{vmatrix} 1 & 1 \\ -3 + \sqrt{5} & -3 \end{vmatrix}}{\begin{vmatrix} 1 & 1 \\ -3 + \sqrt{5} & -3 - \sqrt{5} \end{vmatrix}} = \frac{-\sqrt{5}}{-2\sqrt{5}} = \frac{1}{2},$$

and so the solution to the initial value problem is

$$y(t) = \frac{1}{2} e^{(-3+\sqrt{5})t} + \frac{1}{2} e^{(-3-\sqrt{5})t} = e^{-3t} \left[\frac{e^{\sqrt{5}t} + e^{-\sqrt{5}t}}{2} \right],$$

or

$$y(t) = e^{-3t} \cosh(\sqrt{5}t). \quad \square$$

Type II. $b^2 - 4ac = 0$.

If the roots of the quadratic equation (2.4.2) are equal, only one exponential solution is obtained: $x_1(t) = e^{\lambda t}$, where $\lambda = -b/2a$. A second solution is needed if we are to solve initial value problems. This second solution can be constructed by one of several mathematical tricks. (A famous applied mathematician once remarked that there is a hierarchy of tricks based on frequency of use: a *cheap trick* is used for one or two problems, a *device* is used for several problems, a *method* is used for many!) We shall use a mathematical device borrowed from perturbation theory, an important idea in applied mathematics.

The device consists of changing the insoluble problem by a small amount, a perturbation, into one that can be solved. We then carefully select a solution to the perturbed problem that can be changed by a limiting process into a previously inaccessible solution to the original problem.[1] This is demonstrated with an example before stating the general result for Type II.

Consider the differential equation

$$y'' + 6y' + 9y = 0, \tag{2.4.5}$$

with characteristic equation

$$\lambda^2 + 6\lambda + 9 = 0 \quad \text{or} \quad (\lambda + 3)^2 = 0.$$

---

[1] The authors' approach to solving the case $b^2 - 4ac = 0$ is not a conventional one, but is intended to introduce the reader to the perturbation approach. For the more common approach, see Exercise 5 of this section.

Clearly, $\lambda = -3$ is the only root, and the corresponding solution is $y_1(t) = e^{-3t}$. Now change the differential equation (2.4.5) so that the resulting characteristic equation has two real distinct roots that are almost equal to -3. Take these roots to be

$$\lambda_1 = -3 + \epsilon \quad \text{and} \quad \lambda_2 = -3 - \epsilon,$$

where ϵ, the perturbation, is a small real number. Note that we are working backwards—from roots of the characteristic equation to the differential equation itself. The new characteristic equation is constructed by multiplying its two factors

$$[\lambda - (-3 + \epsilon)][\lambda - (-3 - \epsilon)] = \lambda^2 + 6\lambda + 9 - \epsilon^2 = 0,$$

and so the corresponding differential equation is

$$y'' + 6y' + (9 - \epsilon^2)y = 0. \tag{2.4.6}$$

We see that as ϵ approaches zero, the differential equation (2.4.6) becomes the original (2.4.5). A two-parameter family of solutions to (2.4.6) is

$$y(t, \epsilon) = c_1 \exp[(-3 + \epsilon)t] + c_2 \exp[(-3 - \epsilon)t]. \tag{2.4.7}$$

If ϵ is set equal to zero,

$$y(t, 0) = (c_1 + c_2)e^{-3t},$$

which is a solution to (2.4.5), but not a new one.

To get a new solution, we employ a cheap trick. Rewrite $y(t, \epsilon)$ as

$$y(t, \epsilon) = e^{-3t}[c_1 e^{\epsilon t} + c_2 e^{-\epsilon t}],$$

and using the power series expansion for $e^{\pm z}$,

$$e^{\pm z} = 1 \pm z + \frac{z^2}{2!} \pm \frac{z^3}{3!} + \cdots,$$

we get the expression for $y(t, \epsilon)$:

$$y(t, \epsilon) = e^{-3t}\left[(c_1 + c_2) + (c_1 - c_2)\epsilon t + (c_1 + c_2) \frac{\epsilon^2 t^2}{2!} + (c_1 - c_2) \frac{\epsilon^3 t^3}{3!} + \cdots \right].$$

$$\tag{2.4.8}$$

The trick is to let

$$c_1 + c_2 = k_1 \quad \text{and} \quad c_1 - c_2 = k_2/\epsilon$$

and substitute

$$y(t, \epsilon) = e^{-3t}\left[k_1 + k_2 t + k_1 \frac{\epsilon t^2}{2!} + k_2 \frac{\epsilon^2 t^3}{3!} + \cdots \right].$$

Therefore, if $t \neq 0$,

$$\lim_{\epsilon \to 0} y(t, \epsilon) = y(t, 0) = k_1 e^{-3t} + k_2 t e^{-3t}, \qquad (2.4.9)$$

and the previously inaccessible solution $y_2(t) = te^{-3t}$ appears.

It is properly labeled inaccessible for there seems to be no way to predict it from knowledge of solutions to differential equations whose characteristic equations have real distinct roots! We now verify that te^{-3t} is a solution by substituting it into the differential equation (2.4.5):

$$\frac{d^2}{dt^2}(te^{-3t}) + 6\frac{d}{dt}(te^{-3t}) + 9(te^{-3t})$$

$$= (-6e^{-3t}+9te^{-3t}) + 6(e^{-3t}-3te^{-3t}) + 9te^{-3t} = 0.$$

As the next example shows, the two-parameter family of solutions (2.4.9) can be used to solve any initial value problem for the differential equation (2.4.5).

Example 3 Solve $y'' + 6y' + 9y = 0, \quad y(t_0) = r, \quad y'(t_0) = s$.

We have

$$y(t) = k_1 e^{-3t} + k_2 t e^{-3t},$$
$$y'(t) = -3k_1 e^{-3t} + k_2(e^{-3t}-3te^{-3t}).$$

To determine k_1 and k_2 we substitute the initial data:

$$r = k_1 e^{-3t_0} + k_2 t_0 e^{-3t_0},$$
$$s = -3k_1 e^{-3t_0} + k_2(e^{-3t_0} - 3t_0 e^{-3t_0}),$$

and solve the resulting equations. Using Cramer's rule, we obtain

$$k_1 = \frac{\begin{vmatrix} r & t_0 e^{-3t_0} \\ s & e^{-3t_0} - 3t_0 e^{-3t_0} \end{vmatrix}}{\begin{vmatrix} e^{-3t_0} & t_0 e^{-3t_0} \\ -3e^{-3t_0} & e^{-3t_0} - 3t_0 e^{-3t_0} \end{vmatrix}} = \frac{(r - 3rt_0 - st_0)e^{-3t_0}}{(e^{-3t_0})^2}$$

$$= (r - 3rt_0 - st_0)e^{3t_0},$$

$$k_2 = \frac{\begin{vmatrix} e^{-3t_0} & r \\ -3e^{-3t_0} & s \end{vmatrix}}{\begin{vmatrix} e^{-3t_0} & t_0 e^{-3t_0} \\ -3e^{-3t_0} & e^{-3t_0} - 3t_0 e^{-3t_0} \end{vmatrix}} = \frac{(3r + s)e^{-3t_0}}{(e^{-3t_0})^2}$$

$$= (3r + s)e^{3t_0}.$$

Hence the solution is

$$y(t) = (r - 3rt_0 - st_0)e^{-3(t-t_0)} + (3r + s)te^{-3(t-t_0)},$$

and consequently

$$y'(t) = -3(r - 3rt_0 - st_0)e^{-3(t-t_0)} + (3r + s)(1 - 3t)e^{-3(t-t_0)}.$$

To check our computations, we substitute $t = t_0$, then

$$y(t_0) = r - 3rt_0 - st_0 + 3rt_0 + st_0 = r,$$

$$y'(t_0) = -3r + 9rt_0 + 3st_0 + 3r + s - 9rt_0 - 3st_0 = s,$$

as required. $\square$

The equation (2.4.5) was discussed in detail since any differential equation of Type II can be solved by the same method. However, once it is known that $e^{\lambda t}$ and $te^{\lambda t}$ are solutions, it is not necessary or even interesting to derive them by a perturbation device again and again. Instead observe that if

$$a\lambda^2 + b\lambda + c = 0, \qquad b^2 - ac = 0,$$

then $\lambda = -b/2a$ and

$$y(t) = k_1 e^{\lambda t} + k_2 te^{\lambda t}$$

is a two-parameter family of solutions to

$$ay'' + by' + cy = 0.$$

The verification of this is straightforward:

$$a\frac{d^2}{dt^2}(k_1 e^{\lambda t} + k_2 te^{\lambda t}) + b\frac{d}{dt}(k_1 e^{\lambda t} + k_2 te^{\lambda t}) + c(k_1 e^{\lambda t} + k_2 te^{\lambda t})$$

$$= k_1(a\lambda^2 e^{\lambda t} + b\lambda e^{\lambda t} + ce^{\lambda t}) + k_2[a(2\lambda e^{\lambda t} + \lambda^2 te^{\lambda t}) + b(e^{\lambda t} + \lambda te^{\lambda t}) + cte^{\lambda t}]$$

$$= k_1 e^{\lambda t}(a\lambda^2 + b\lambda + c) + k_2 e^{\lambda t}[t(a\lambda^2 + b\lambda + c) + (2a\lambda + b)] = 0,$$

since $a\lambda^2 + b\lambda + c = 0$ and $2a\lambda + b = 0$.

Example 4 Solve $y'' - 4y' + 4y = 0$, $y(0) = 3$, $y'(0) = -1$.

The characteristic equation is

$$\lambda^2 - 4\lambda + 4 = (\lambda - 2)^2 = 0$$

with the double root $\lambda = 2$. Hence a two-parameter family of solutions is

$$y(t) = k_1 e^{2t} + k_2 te^{2t}.$$

Therefore,

$$y'(t) = 2k_1 e^{2t} + k_2(e^{2t} + 2te^{2t}),$$

and letting $t = 0$, $y(0) = 3$, $y'(0) = -1$, we obtain

$$3 = k_1, \qquad -1 = 2k_1 + k_2.$$

This can be solved directly to get $k_1 = 3$, $k_2 = -7$, and hence the solution is

$$y(t) = 3e^{2t} - 7te^{2t}.$$

EXERCISES 2.4

1. For each of the following constant coefficient linear differential equations, the characteristic equation has real, distinct roots. Write down the general solution in each case and, if initial data are given, the solution satisfying that data.

a) $y'' - 9y' + 20y = 0$

b) $y'' - 4y = 0$

c) $y'' - y' - 30y = 0$

d) $y'' - 6y' = 0$

e) $y'' - 5y = 0$; $y(0) = 0$, $y'(0) = 1$

f) $y'' - 7y' = 0$; $y(0) = 1$, $y'(0) = 1$

g) $y'' - 4y' - 12y = 0$; $y(0) = 4$, $y'(0) = -4$

h) $y'' - 2y' - 3y = 0$; $y(0) = 0$, $y'(0) = 0$

i) $y'' - \frac{1}{16}y = 0$; $y(0) = 2$, $y'(0) = -1$

j) $y'' + 3y' + 2y = 0$; $y(1) = 1$, $y'(1) = 3$

2. Repeat Exercise 1 for each of the following differential equations whose characteristic equation has real, equal roots.

a) $y'' - 4y' + 4y = 0$

b) $y'' + 2\sqrt{5}y' + 5y = 0$

c) $y'' = 0$

d) $y'' + 2y' + y = 0$

e) $y'' + y' + \frac{1}{4}y = 0$; $y(0) = 1$, $y'(0) = 1$

f) $y'' + 32y' + 256y = 0$; $y(0) = 2$, $y'(0) = 1$

g) $y'' - 6y' + 9y = 0$; $y(0) = 0$, $y'(0) = 3$

h) $y'' = 0$, $y(1) = 4$, $y'(1) = -3$

i) $y'' + \frac{1}{2}y' + \frac{1}{16}y = 0$; $y(1) = 4$, $y'(1) = 0$

j) $y'' + \sqrt{10}\,y' + \frac{5}{2}y = 0$; $y(0) = 10$, $y'(0) = 0$

3. Show that the relation $y(t) = c_1 e^{kt} + c_2 e^{-kt}$ can be written in the form

$$y(t) = k_1 \cosh kt + k_2 \sinh kt,$$

where $k_1 = c_1 + c_2$, $k_2 = c_1 - c_2$. Recall that

$$\sinh x = \frac{e^x - e^{-x}}{2}, \qquad \cosh x = \frac{e^x + e^{-x}}{2}.$$

4. The previous problem shows that if the roots of the characteristic equation are $\lambda_1 = -\lambda_2 = k$, then the general solution can always be written in the form

$$y(t) = c_1 \cosh kt + c_2 \sinh kt.$$

Express the solution of the following initial value problems in this form:

a) $y'' - 16y = 0$; $y(0) = 3$, $y'(0) = 2$

b) $y'' - \frac{1}{9}y = 0$; $y(0) = 1$, $y'(0) = -1$

c) $y'' - 2y = 0$; $y(0) = 0$, $y'(0) = 1$

5. Another way of finding this second solution in the case of equal roots $\lambda_1 = \lambda_2 = k$ follows the line of reasoning given in the discussion of the method of variation of parameters in Chapter 1. Knowing $y(t) = ce^{kt}$ is a solution for any constant c, assume a solution is of the form $y(t) =$

$c(t)e^{kt}$. Substitute it into the differential equation and solve the ensuing differential equation for $c(t)$. Use this device to find the general solution of each of the following.

a) $y'' + 4y' + 4y = 0$

b) $y'' - 10y' + 25y = 0$

c) $y'' + 2ky' + k^2y = 0$

2.5 CONSTANT COEFFICIENT HOMOGENEOUS EQUATIONS— THE CASE OF COMPLEX ROOTS

In this section we will consider the case where the roots of the characteristic equation are complex numbers. If the characteristic equation is

$$a\lambda^2 + b\lambda + c = 0,$$

then its roots will be complex when $b^2 - 4ac < 0$, and since a, b, and c are real, the roots will be of the form

$$\lambda_1 = p + iq \quad \text{and} \quad \lambda_2 = p - iq,$$

where p and q are real numbers. These roots will lead to complex-valued solutions $e^{\lambda_1 t}$ and $e^{\lambda_2 t}$, but we will be able to obtain a real-valued general solution from them.

Let

$$ay'' + by' + cy = 0 \tag{2.5.1}$$

be a differential equation whose characteristic equation, $a\lambda^2 + b\lambda + c = 0$, has the complex distinct roots,

$$\lambda = p + iq = -\frac{b}{2a} + i\frac{(4ac - b^2)^{1/2}}{2a},$$

$$\lambda^* = p - iq = -\frac{b}{2a} - i\frac{(4ac - b^2)^{1/2}}{2a}. \tag{2.5.2}$$

As usual, (*) denotes a complex conjugate, that is, i is replaced by $-i$ (in some texts, the notation $\bar{\lambda}$ is used for the complex conjugate of λ).

To show that the complex-valued exponential

$$y(t) = \exp(\lambda t) = \exp[(p + iq)t] \tag{2.5.3}$$

is a solution, it must be verified that it has the analogous property of real exponentials, namely that

$$\frac{dy}{dt}(t) = \lambda y(t). \tag{2.5.4}$$

But we can show that equation (2.5.4) is valid by using Euler's formula

$$\exp(iqt) = \cos qt + i \sin qt,$$

which the reader may wish to verify (see Appendix 4 and Exercise 4, p. 96). This formula and the law of exponents imply that $y(t) = \exp(\lambda t)$ can be written in the form

$$y(t) = \exp[(p + iq)t] = \exp(pt)\exp(iqt)$$
$$= e^{pt}[\cos qt + i \sin qt].$$

Therefore,

$$\frac{dy}{dt}(t) = \frac{d}{dt}\{e^{pt}[\cos qt + i \sin qt]\}$$
$$= pe^{pt}[\cos qt + i \sin qt] + qe^{pt}[-\sin qt + i \cos qt]$$
$$\doteq pe^{pt}[\cos qt + i \sin qt] + iqe^{pt}[\cos qt + i \sin qt]$$
$$= (p + iq)e^{pt}[\cos qt + i \sin qt] = \lambda y(t).$$

and (2.5.4) is verified. The verification of (2.5.4) with λ replaced by λ^* is similar. These differentiation formulas show that the complex exponential behaves exactly like the real one with respect to differentiation. Therefore,

$$y(t) = c_1 e^{\lambda t} + c_2 e^{\lambda^* t}$$

is a two-parameter family of solutions to the differential equation (2.5.1) since

$$a\frac{d^2}{dt^2}(c_1 e^{\lambda t} + c_2 e^{\lambda^* t}) + b\frac{d}{dt}(c_1 e^{\lambda t} + c_2 e^{\lambda^* t}) + c(c_1 e^{\lambda t} + c_2 e^{\lambda^* t})$$
$$= c_1 e^{\lambda t}(a\lambda^2 + b\lambda + c) + c_2 e^{\lambda^* t}(a\lambda^{*2} + b\lambda^* + c) = 0.$$

The last expression is zero because both λ and λ^* are roots of the characteristic equation.

For many applications of differential equations, particularly in electrical engineering, complex-valued solutions are employed. However, real-valued solutions are also needed. We shall obtain these real solutions from the real and imaginary parts of $y(t)$ (or $y^*(t)$) and then show that a general linear combination of these two real solutions can be used to solve initial value problems with arbitrary data.

If $\lambda = p + iq$, then the solution $y(t)$ can be written as

$$y(t) = c_1 \exp[(p + iq)t] + c_2 \exp[(p - iq)t]$$
$$= e^{pt}[c_1 \cos qt + c_1 i \sin qt + c_2 \cos qt - c_2 i \sin qt].$$

Since $y(t)$ is a solution for *any* choice (real or complex) of the constants c_1 and c_2, let (a) $c_1 = c_2 = \frac{1}{2}$, which gives the real solution

$$y_1(t) = e^{pt} \cos qt,$$

and (b) $c_1 = -c_2 = 1/(2i)$, which gives another real solution

$$y_2(t) = e^{pt} \sin qt.$$

Now observe that since

$$\exp[(p + iq)t] = e^{pt} \cos qt + ie^{pt} \sin qt,$$

then

$$y_1(t) = e^{pt} \cos qt = \text{Re}\{\exp[(p + iq)t]\}$$

and

$$y_2(t) = e^{pt} \sin qt = \text{Im}\{\exp[(p + iq)t]\}.$$

(Recall that if $z = a + ib$ is a complex number, then its real part is $\text{Re}[z] = a$ and its imaginary part is $\text{Im}[z] = b$.) Therefore we have shown that the real and imaginary parts of $e^{\lambda t}$, where $\lambda = p + iq$, are real solutions of (2.5.1). Furthermore,

$$y(t) = k_1 e^{pt} \cos qt + k_2 e^{pt} \sin qt$$

is, for real k_1 and k_2, a real-valued two-parameter family of solutions.

Example 1   Find two-parameter families of real- and complex-valued solutions of $y'' + 4y' + 5y = 0$.

The characteristic equation, $\lambda^2 + 4\lambda + 5 = 0$, has the distinct complex roots $\lambda = -2 + i$ and $\lambda^* = -2 - i$. One two-parameter family of complex-valued solutions is

$$y(t) = c_1 \exp[(-2 + i)t] + c_2 \exp[(-2 - i)t].$$

Since $\lambda = p + iq = -2 + i$, $p = -2$, $q = 1$, and a two-parameter family of real-valued solutions is

$$y(t) = k_1 e^{-2t} \cos t + k_2 e^{-2t} \sin t. \quad \square$$

It remains to be shown that the parameters k_1 and k_2 in the solution to the differential equation (2.5.1),

$$y(t) = k_1 e^{pt} \cos qt + k_2 e^{pt} \sin qt, \tag{2.5.5}$$

can be chosen to satisfy arbitrary initial data $y(t_0) = r$, $y'(t_0) = s$. We first compute

$$y'(t) = k_1[pe^{pt} \cos qt - qe^{pt} \sin qt] + k_2[pe^{pt} \sin qt + qe^{pt} \cos qt];$$

substituting the data gives the pair of equations:

$$r = k_1 e^{pt_0} \cos qt_0 + k_2 e^{pt_0} \sin qt_0$$

$$s = k_1 e^{pt_0}(p \cos qt_0 - q \sin qt_0) + k_2 e^{pt_0}(p \sin qt_0 + q \cos qt_0).$$

Using Cramer's rule to solve for k_1 and k_2 gives

$$k_1 = \frac{\begin{vmatrix} r & e^{pt_0} \sin qt_0 \\ s & e^{pt_0}(p \sin qt_0 + q \cos qt_0) \end{vmatrix}}{W}$$

$$= \frac{e^{pt_0}}{W} [r(p \sin qt_0 + q \cos qt_0) - s \sin qt_0],$$

$$k_2 = \frac{\begin{vmatrix} e^{pt_0} \cos qt_0 & r \\ e^{pt_0}(p \cos qt_0 - q \sin qt_0) & s \end{vmatrix}}{W}$$

$$= \frac{e^{pt_0}}{W} [s \cos qt_0 - r(p \cos qt_0 - q \sin qt_0)],$$

where the denominator is

$$W = \begin{vmatrix} e^{pt_0} \cos qt_0 & e^{pt_0} \sin qt_0 \\ e^{pt_0}[p \cos qt_0 - q \sin qt_0] & e^{pt_0}[p \sin qt_0 + q \cos qt_0] \end{vmatrix}$$

$$= e^{2pt_0} [\cos qt_0(p \sin qt_0 + q \cos qt_0)]$$

$$- e^{2pt_0} [\sin qt_0(p \cos qt_0 - q \sin qt_0)]$$

$$= e^{2pt_0}(q \cos^2 qt_0 + q \sin^2 qt_0) = qe^{2pt_0} \neq 0,$$

because $q = (4ac - b^2)^{1/2}/2a \neq 0$. Since $W \neq 0$, the above expressions for k_1 and k_2 uniquely determine them as functions of t_0, r, and s; hence $y(t)$ and $y'(t)$ will satisfy any given initial conditions.

Example 2 Solve $y'' + 6y' + 13y = 0$, $y(\pi/2) = -2$, $y'(\pi/2) = 8$.

The characteristic equation $\lambda^2 + 6\lambda + 13 = 0$ has complex distinct roots $\lambda = -3 + 2i$ and $\lambda^* = -3 - 2i$, so (2.5.5) becomes

$$y(t) = k_1 e^{-3t} \cos 2t + k_2 e^{-3t} \sin 2t.$$

If we differentiate first the exponential and then the trigonometric terms, we obtain

$$y'(t) = -3y(t) + 2e^{-3t}(-k_1 \sin 2t + k_2 \cos 2t).$$

(The reader will find the trick of writing $y'(t)$ in the form above very useful in the computing of the values of k_1 and k_2.) Substituting the initial data gives

$$-2 = -k_1 e^{-3\pi/2}, \qquad 8 = 6 - 2k_2 e^{-3\pi/2},$$

and hence $k_1 = 2e^{3\pi/2}$, $k_2 = -e^{3\pi/2}$. The solution to the initial value problem is

$$y(t) = 2e^{3\pi/2}e^{-3t} \cos 2t - e^{3\pi/2}e^{-3t} \sin 2t$$
$$= 2e^{-3(t-\pi/2)} \cos 2t - e^{-3(t-\pi/2)} \sin 2t. \quad \square$$

In many applications it is sometimes useful to replace the real two-parameter family of solutions

$$y(t) = k_1 e^{pt} \cos qt + k_2 e^{pt} \sin qt \qquad (2.5.6)$$

by

$$y(t) = A e^{pt} \cos(qt - \phi), \qquad (2.5.7)$$

where the two parameters are A, the amplitude of the solution, and ϕ, the phase angle. For instance, if $p = 0$, the value A immediately tells us the minimum and maximum values of the solution, namely $\pm A$. A relationship between the parameters (k_1, k_2) and (A, ϕ) is easy to derive. Since it is assumed that not both k_1 and k_2 are zero, rewrite (2.5.6) as

$$y(t) = \sqrt{k_1^2 + k_2^2}\, e^{pt} \left[\frac{k_1}{\sqrt{k_1^2 + k_2^2}} \cos qt + \frac{k_2}{\sqrt{k_1^2 + k_2^2}} \sin qt \right].$$

The numbers $\alpha = k_1/\sqrt{k_1^2 + k_2^2}$ and $\beta = k_2/\sqrt{k_1^2 + k_2^2}$ satisfy conditions

$$|\alpha| \leq 1, \qquad |\beta| \leq 1, \qquad \alpha^2 + \beta^2 = 1.$$

Therefore, letting

$$A = \sqrt{k_1^2 + k_2^2}, \qquad \cos \phi = \alpha = \frac{k_1}{A}, \qquad \sin \phi = \beta = \frac{k_2}{A}, \qquad (2.5.8)$$

we can write the solution $y(t)$

$$y(t) = A e^{pt} (\cos \phi \cos qt + \sin \phi \sin qt)$$
$$= A e^{pt} \cos(qt - \phi).$$

The trigonometric equations determining ϕ can also be written as

$$\phi = \arctan\left(\frac{k_2}{k_1}\right), \qquad -\frac{\pi}{2} \leq \phi \leq \frac{\pi}{2},$$

where $\phi = \pi/2$ if $k_1 = 0$, $k_2 > 0$, and $\phi = -\pi/2$ if $k_1 = 0$, $k_2 < 0$.

Example 3   Solve $y'' - 2y' + 26y = 0$, $y(0) = 3/\sqrt{2}$, $y'(0) = -12/\sqrt{2}$.

The characteristic equation is $\lambda^2 - 2\lambda + 26 = 0$, with roots $\lambda = 1 + 5i$,

$\lambda^* = 1 - 5i$, and therefore

$$y(t) = k_1 e^t \cos 5t + k_2 e^t \sin 5t,$$

and

$$y'(t) = y(t) + 5e^t(- k_1 \sin 5t + k_2 \cos 5t).$$

Substituting the initial data gives

$$\frac{3}{\sqrt{2}} = k_1, \qquad \frac{-12}{\sqrt{2}} = \frac{3}{\sqrt{2}} + 5k_2.$$

So $k_1 = 3/\sqrt{2}$, $k_2 = -3/\sqrt{2}$, and the solution is

$$y(t) = \frac{3}{\sqrt{2}} e^t (\cos 5t - \sin 5t).$$

To express the solutions in the amplitude–phase angle form, use equations (2.5.8) to obtain

$$k_1 = \frac{3}{\sqrt{2}} = A \cos \phi, \qquad k_2 = \frac{-3}{\sqrt{2}} = A \sin \phi;$$

hence,

$$A = \sqrt{k_1^2 + k_2^2} = \sqrt{\tfrac{9}{2} + \tfrac{9}{2}} = 3$$

and

$$\phi = \arctan\left(\frac{k_2}{k_1}\right) = \arctan(-1) = -\frac{\pi}{4}.$$

The amplitude–phase angle form of the solution is

$$y(t) = 3e^t \cos\left(5t + \frac{\pi}{4}\right).$$

Note that the phase angle is measured in radians. □

EXERCISES 2.5

1. Write down the general solution of the following linear differential equations and, if initial data are given, the solution satisfying that data. Give only real solutions.

a) $y'' + 3y' + \tfrac{13}{4}y = 0$
b) $y'' + 9y = 0$
c) $y'' - 2y' + 2y = 0$
d) $y'' - y' + 3y = 0$
e) $y'' + \tfrac{49}{4}y = 0$
f) $y'' + 4y' + 5y = 0$; $y(0) = 0$, $y'(0) = 1$
g) $y'' + y = 0$; $y(0) = 2$, $y'(0) = -2$
h) $y'' + 3y' + \tfrac{5}{2}y = 0$; $y(0) = 2$, $y'(0) = -6$
i) $y'' + 4y = 0$; $y(\pi/8) = 1$, $y'(\pi/8) = 2$
j) $y'' - 4y' + 7y = 0$; $y(1) = 0$, $y'(1) = \sqrt{3}$

2. Express the solution to each of the following initial value problems in the form

$$y(t) = Ae^{pt} \cos(qt - \phi),$$

where A is the amplitude of the solution and ϕ is the phase angle.

a) $y'' + y = 0$; $y(0) = 1$, $y'(0) = 1$

b) $y'' + 2y' + 5y = 0$; $y(0) = 0$, $y'(0) = 4$

c) $y'' + 4y = 0$; $y(\pi/4) = 2$, $y'(\pi/4) = -2$

d) $y'' + 6y' + 13y = 0$; $y(\pi/2) = 4$, $y'(\pi/2) = 0$

3. If the solution of an initial value problem is given in the amplitude–phase angle form

$$y(t) = Ae^{pt} \cos(qt - \phi),$$

and the initial data $y(0)$ and $y'(0)$ are given, show that

$$A^2 = y(0)^2 + \left[\frac{y'(0) - py(0)}{q} \right]^2,$$

$$\cos \phi = \frac{y(0)}{A},$$

$$\sin \phi = \frac{y'(0) - py(0)}{qA}.$$

Use this result to find the solutions of the following initial value problems:

a) $y'' + 25y = 0$; $y(0) = 1$, $y'(0) = 5$

b) $y'' + 2y' + 10y = 0$; $y(0) = 4$, $y'(0) = 5$

c) $y'' - y' + \frac{13}{4}y = 0$; $y(0) = -2$, $y'(0) = 2$

4. The series for e^z,

$$e^z = 1 + z + \frac{z^2}{2!} + \cdots + \frac{z^n}{n!} + \cdots = \sum_{n=0}^{\infty} \frac{z^n}{n!},$$

converges for all real or complex z. By letting $z = i\theta$ and rearranging the series into its real and imaginary parts, prove Euler's formula

$$e^{i\theta} = \cos \theta + i \sin \theta.$$

Hint: See Appendix 4 for a sketch of a proof.

5. If initial data for a differential equation are given at more than one point, the differential equation together with the data is called a *boundary value problem*. For example, the differential equation

$$y'' - y' - 2y = 0$$

together with the conditions

$$y(0) = 0, \qquad y(1) = 1$$

is called a *two-point boundary value problem* because a solution $y(t)$ of the differential equation is sought whose graph passes through the two points $(0, 0)$ and $(1, 1)$. Find $y(t)$ in the example.

6. Solve the following two-point boundary value problems:

a) $y'' + y = 0$; $y(0) = 1$, $y(\pi/2) = -1$

b) $y'' + 2y' - 3y = 0$; $y(0) = 0$, $y(1) = 1$

c) $y'' + 2y' + y = 0$; $y(0) = 2$, $y'(2) = -2$

d) $y'' + y = 0$; $y(0) = 1$, $y(\pi) = \alpha$. For what values of α will there be one solution, many solutions, or no solution?

e) $y'' + \omega^2 y = 0$; $y(0) = 0$, $y(\pi) = 0$. For what values of ω will there be one solution, many solutions, or no solution?

2.6 THEORY OF HOMOGENEOUS SECOND ORDER LINEAR DIFFERENTIAL EQUATIONS

In this section we shall study some theoretical results about solutions to linear homogeneous second order differential equations

$$a(t) \frac{d^2y}{dt^2} + b(t) \frac{dy}{dt} + c(t)y = 0, \qquad (2.6.1)$$

where $a(t)$, $b(t)$, and $c(t)$ are continuous functions on a nonempty interval. We have seen that if the coefficients of the equation are constant, it is not difficult to construct a two-parameter family of solutions and to determine the parameters, so that the solution takes on arbitrary initial data. If the coefficients are not constant, there is no general method of constructing solutions. It is therefore important to have criteria for those solutions that can be constructed to insure that initial value problems can be solved. As we shall see, this leads to interesting geometric results about the constant-coefficient solutions.

Let

$$y(t) = c_1 y_1(t) + c_2 y_2(t) \qquad (2.6.2)$$

be a two-parameter family of solutions to (2.6.1). It is called a *general solution* if the coefficients c_1 and c_2 can be chosen to solve an arbitrary initial value problem, i.e., satisfy initial data of the form

$$y(t_0) = r, \qquad y'(t_0) = s. \qquad (2.6.3)$$

To see what conditions this imposes on the solutions $y_1(t)$ and $y_2(t)$, substitute the initial data to obtain the two algebraic equations for c_1 and c_2:

$$r = c_1 y_1(t_0) + c_2 y_2(t_0), \qquad s = c_1 y_1'(t_0) + c_2 y_2'(t_0).$$

Using Cramer's rule, we have

$$c_1 = \frac{\begin{vmatrix} r & y_2(t_0) \\ s & y_2'(t_0) \end{vmatrix}}{\begin{vmatrix} y_1(t_0) & y_2(t_0) \\ y_1'(t_0) & y_2'(t_0) \end{vmatrix}} \qquad c_2 = \frac{\begin{vmatrix} y_1(t_0) & r \\ y_1'(t_0) & s \end{vmatrix}}{\begin{vmatrix} y_1(t_0) & y_2(t_0) \\ y_1'(t_0) & y_2'(t_0) \end{vmatrix}}$$

Clearly, the denominator must be nonzero to be able to solve for c_1 and c_2 if r and s are arbitrary numbers. This is a restriction on the solutions $y_1(t)$ and $y_2(t)$ that can occur in a general solution, namely that

$$W(y_1, y_2)(t) = \begin{vmatrix} y_1(t) & y_2(t) \\ y_1'(t) & y_2'(t) \end{vmatrix}$$

must be nonzero at $t = t_0$. The function W is called the *Wronskian* of the two functions $y_1(t)$ and $y_2(t)$.

The Wronskian has an interesting geometric meaning in terms of the vector product (cross product) of *solution vectors* of second order linear homogeneous differential equations.

Definition   A solution vector is defined as

$$\mathbf{Y}(t) = y(t)\mathbf{e}_1 + \frac{dy}{dt}(t)\mathbf{e}_2,$$

where $y(t)$ is a solution of a second order differential equation, and $\mathbf{e}_1$ and $\mathbf{e}_2$ are orthogonal unit vectors.

Vectors formed from solutions to differential equations are essential in the geometric theory of differential equations and will occur often in this book.

A nonvanishing Wronskian means that the solution vectors

$$\mathbf{Y}_1(t) = y_1(t)\mathbf{e}_1 + y_1'(t)\mathbf{e}_2,$$

$$\mathbf{Y}_2(t) = y_2(t)\mathbf{e}_1 + y_2'(t)\mathbf{e}_2$$

are not parallel. To show this, compute the cross product (vector product)

$$\mathbf{Y}_1(t) \times \mathbf{Y}_2(t) = [y_1(t)y_2'(t) - y_2(t)y_1'(t)]\mathbf{e}_3 = W(y_1, y_2)(t)\mathbf{e}_3,$$

where the vector $\mathbf{e}_3 = \mathbf{e}_1 \times \mathbf{e}_2$. By definition of the cross product, $|W(y_1, y_2)(t)|$ is the area of the parallelogram spanned by $\mathbf{Y}_1(t)$ and $\mathbf{Y}_2(t)$, so if the Wronskian is nonzero, the vectors cannot be parallel.

If two vectors $\mathbf{Z}_1$ and $\mathbf{Z}_2$ are parallel, then scalars c_1 and c_2, not both zero, can be selected so that

$$c_1\mathbf{Z}_1 + c_2\mathbf{Z}_2 = \mathbf{0},$$

where $\mathbf{0}$ is the zero vector with all zero components. The vectors are then said to be *linearly dependent*. This concept is now extended to vector-valued functions.

Definition   Two vector-valued functions $\mathbf{Z}_1(t)$ and $\mathbf{Z}_2(t)$ are *linearly dependent* on an interval $t_1 < t < t_2$ if there exist scalar constants c_1 and c_2, not both zero, such that the linear combination

$$c_1\mathbf{Z}_1(t) + c_2\mathbf{Z}_2(t) = \mathbf{0}, \quad t_1 < t < t_2.$$

The vectors $\mathbf{Z}_1(t)$ and $\mathbf{Z}_2(t)$ are *linearly independent* if they are not linearly dependent.

A solution vector $\mathbf{Y}(t) = y(t)\mathbf{e}_1 + y'(t)\mathbf{e}_2$ has a special structure, since its components are formed from a solution to a second order linear differential equation. Therefore, any linear combination of solution vectors is also a solution vector. For, if $\mathbf{Y}_1(t)$ and $\mathbf{Y}_2(t)$ are as given above, then

$$c_1\mathbf{Y}_1(t) + c_2\mathbf{Y}_2(t) = c_1[y_1(t)\mathbf{e}_1 + y_1'(t)\mathbf{e}_2] + c_2[y_2(t)\mathbf{e}_1 + y_2'(t)\mathbf{e}_2]$$
$$= [c_1y_1(t) + c_2y_2(t)]\mathbf{e}_1 + [c_1y_1'(t) + c_2y_2'(t)]\mathbf{e}_2.$$

But the last expression is a solution vector since $c_1y_1(t) + c_2y_2(t)$ is a solution.

It was shown above that a nonvanishing Wronskian means that the solution vectors are not parallel and consequently linearly independent. But the converse is also true, as the following lemma shows.

Lemma If $\mathbf{Y}_1(t)$ and $\mathbf{Y}_2(t)$ are two linearly independent solution vectors of a homogeneous linear second order differential equation, the Wronskian is nonzero on their interval of definition.

Proof We shall show that if the Wronskian is zero at some point t_0, then the solution vectors are linearly dependent. Suppose

$$W(y_1, y_2)(t_0) = \begin{vmatrix} y_1(t_0) & y_2(t_0) \\ y_1'(t_0) & y_2'(t_0) \end{vmatrix} = 0.$$

Then the linear system of equations

$$c_1y_1(t_0) + c_2y_2(t_0) = 0,$$
$$c_1y_1'(t_0) + c_2y_2'(t_0) = 0,$$

with c_1 and c_2 as unknowns, has a nontrivial solution $c_1 = a$, $c_2 = b$. But this means that the solution vector

$$\mathbf{Y}(t) = a\mathbf{Y}_1(t) + b\mathbf{Y}_2(t)$$
$$= [ay_1(t) + by_2(t)]\mathbf{e}_1 + [ay_1'(t) + by_2'(t)]\mathbf{e}_2,$$
$$t_1 < t < t_2,$$

satisfies

$$\mathbf{Y}(t_0) = \mathbf{0} = 0\mathbf{e}_1 + 0\mathbf{e}_2.$$

Therefore $y(t) = ay_1(t) + by_2(t)$, which is a solution of the differential equation, satisfies $y(t_0) = 0$, $y'(t_0) = 0$. The zero solution satisfies these initial conditions, so by the uniqueness theorem of Section 2.2 we must have that

$y(t) \equiv 0$, $t_1 < t < t_2$, which in turn implies that $\mathbf{Y}(t) \equiv 0$, $t_1 < t < t_2$. Since not both a and b are zero, this implies that $\mathbf{Y}_1(t)$ and $\mathbf{Y}_2(t)$ are linearly dependent, which proves the lemma. ∎

The reader is warned that the cross product of two vector-valued functions that are *not* solution vectors of a linear homogeneous second order differential equation can vanish at a point even though the two vector-valued functions are linearly independent on an interval containing that point. Indeed,

$$\mathbf{Z}_1(t) = t^2\mathbf{e}_1 + 2t\mathbf{e}_2 \qquad \text{and} \qquad \mathbf{Z}_2(t) = t^3\mathbf{e}_1 + 3t^2\mathbf{e}_2$$

are linearly independent on any interval, yet the cross product is zero at $t = 0$. It is easy to show (Exercise 6, p. 104) that the functions $y_1(t) = t^2$ and $y_2(t) = t^3$ are not solutions of any linear homogeneous second order differential equation with continuous coefficients on an interval containing $t = 0$ (see also Exercise 1(d), p. 103).

Theorem A linear combination of two solutions $y_1(t)$ and $y_2(t)$ to a linear homogeneous second order differential equation is a general solution to that differential equation if and only if the two solution vectors are linearly independent.

Proof Let us show that if a linear combination of the two solutions is a general solution, then the two solution vectors $\mathbf{Y}_1(t)$ and $\mathbf{Y}_2(t)$ are never parallel and hence are linearly independent. The proof is by contradiction.

If the two vectors are parallel at $t = t_0$, then we can pick a *nonzero* vector

$$\mathbf{z} = r\mathbf{e}_1 + s\mathbf{e}_2$$

that is perpendicular to both $\mathbf{Y}_1(t_0)$ and $\mathbf{Y}_2(t_0)$, i.e., the scaler (dot) products $\mathbf{Y}_1(t_0) \cdot \mathbf{z}$ and $\mathbf{Y}_2(t_0) \cdot \mathbf{z}$ are zero. The existence and uniqueness theorem of Section 2.3 insures the existence of a solution $y(t)$ satisfying $y(t_0) = r$, $y'(t_0) = s$, and the solution vector corresponding to this solution is

$$\mathbf{Y}(t) = y(t)\mathbf{e}_1 + y'(t)\mathbf{e}_2,$$

which satisfies

$$\mathbf{Y}(t_0) = r\mathbf{e}_1 + s\mathbf{e}_2 = \mathbf{z}.$$

Furthermore,

$$\mathbf{Y}(t_0) \cdot \mathbf{Y}(t_0) = r^2 + s^2 \neq 0,$$

since $\mathbf{z}$ is a nonzero vector.

Now if a linear combination of $y_1(t)$ and $y_2(t)$ is a general solution, it must be possible to pick parameters c_1 and c_2 so that $y(t)$ above can be expressed as

$$y(t) = c_1 y_1(t) + c_2 y_2(t)$$

and

$$y'(t) = c_1 y_1'(t) + c_2 y_2'(t).$$

Then

$$\begin{aligned} \mathbf{Y}(t) &= y(t)\mathbf{e}_1 + y'(t)\mathbf{e}_2 \\ &= c_1[y_1(t)\mathbf{e}_1 + y_1'(t)\mathbf{e}_2] + c_2[y_2(t)\mathbf{e}_1 + y_2'(t)\mathbf{e}_2] \\ &= c_1\mathbf{Y}_1(t) + c_2\mathbf{Y}_2(t). \end{aligned}$$

However, at $t = t_0$, $\mathbf{Y}(t_0) = \mathbf{z}$ and

$$\begin{aligned} r^2 + s^2 &= \mathbf{Y}(t_0) \cdot \mathbf{Y}(t_0) \\ &= [c_1\mathbf{Y}_1(t_0) + c_2\mathbf{Y}_2(t_0)] \cdot \mathbf{z} \\ &= c_1\mathbf{Y}_1(t_0) \cdot \mathbf{z} + c_2\mathbf{Y}_2(t_0) \cdot \mathbf{z} = 0 + 0 = 0, \end{aligned}$$

since $\mathbf{z}$ is perpendicular to both $\mathbf{Y}_1(t_0)$ and $\mathbf{Y}_2(t_0)$. Thus we have a contradiction, since $r^2 + s^2 \neq 0$.

Now let us show that if the two solution vectors $\mathbf{Y}_1(t)$ and $\mathbf{Y}_2(t)$ are linearly independent, then $y(t) = c_1 y_1(t) + c_2 y_2(t)$ is a general solution. Linear independence on an interval implies that the Wronskian $W(y_1, y_2)(t)$ is nonzero on the interval. Therefore c_1 and c_2 can be selected to match arbitrary initial data since the denominator $W(y_1, y_2)(t_0)$ in the Cramer's rule formulas for c_1 and c_2 is nonzero. Hence, a linear combination of $y_1(t)$ and $y_2(t)$ is a general solution. This concludes the proof of the theorem. ∎

Example 1 Find a general solution to

$$t^2 \frac{d^2 y}{dt^2} + t \frac{dy}{dt} - 4y = 0, \quad t > 0.$$

This is an example of a special type of differential equation called *Euler's equation*. It can be solved by setting $y(t) = t^r$ and then deriving a quadratic equation for the undetermined exponent. Substitution gives

$$t^2 \frac{d^2}{dt^2}(t^r) + t \frac{d}{dt}(t^r) - 4t^r = t^2[r(r-1)t^{r-2}] + t(rt^{r-1}) - 4t^r$$

$$= t^r[r(r-1) + r - 4] = t^r(r^2 - 4) = 0$$

on an interval if and only if $r^2 - 4 = 0$, and therefore $r = 2$ or $r = -2$. This implies that $y_1(t) = t^2$ and $y_2(t) = t^{-2}$ are both solutions and a two-

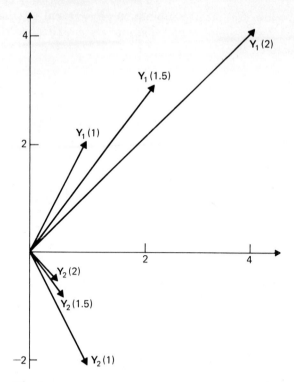

Figure 2.3 Plots of solution vectors for $t = 1, 1.5, 2.$

parameter family of solutions is

$$y(t) = c_1 t^2 + c_2 t^{-2}.$$

This is a general solution since the Wronskian

$$W(y_1, y_2)(t) = \begin{vmatrix} t^2 & t^{-2} \\ 2t & -2t^{-3} \end{vmatrix} = -4t^{-1} \neq 0.$$

The solution vectors $Y_i(t) = y_i(t)\mathbf{e}_1 + y_i'(t)\mathbf{e}_2$, $i = 1, 2$, are

$$\mathbf{Y}_1(t) = t^2 \mathbf{e}_1 + 2t\mathbf{e}_2,$$

$$\mathbf{Y}_2(t) = t^{-2}\mathbf{e}_1 - 2t^{-3}\mathbf{e}_2,$$

For instance, if

$$\mathbf{e}_1 = \begin{bmatrix} 1 \\ 0 \end{bmatrix} \quad \text{and} \quad \mathbf{e}_2 = \begin{bmatrix} 0 \\ 1 \end{bmatrix}$$

are the standard unit vectors parallel to the x and y axes, then

$$\mathbf{Y}_1(t) = t^2 \begin{bmatrix} 1 \\ 0 \end{bmatrix} + 2t \begin{bmatrix} 0 \\ 1 \end{bmatrix} = \begin{bmatrix} t^2 \\ 2t \end{bmatrix}$$

and

$$\mathbf{Y}_2(t) = t^{-2} \begin{bmatrix} 1 \\ 0 \end{bmatrix} - 2t^{-3} \begin{bmatrix} 0 \\ 1 \end{bmatrix} = \begin{bmatrix} t^{-2} \\ -2t^{-3} \end{bmatrix}.$$

They are plotted in Fig. 2.3 for several values of t. $\square$

For the reader's benefit, the results proved in this section can be summarized as follows: Given the homogeneous second order linear equation

$$a(t)y'' + b(t)y' + c(t)y = 0,$$

where $a(t)$, $b(t)$, and $c(t)$ are continuous on the interval $a < t < b$, and two solutions $y_1(t)$ and $y_2(t)$ are defined on $a < t < b$, then

a) The solutions are linearly independent if and only if the Wronskian

$$W(y_1, y_2)(t) = \begin{vmatrix} y_1(t) & y_2(t) \\ y_1'(t) & y_2'(t) \end{vmatrix} \neq 0, \quad a < t < b;$$

b) The two-parameter family of solutions

$$y(t) = c_1 y_1(t) + c_2 y_2(t)$$

is a general solution if and only if $y_1(t)$ and $y_2(t)$ are linearly independent.

EXERCISES 2.6

1. Compute the Wronskian $W(y_1, y_2)(t)$ of the following pairs of functions and determine for what values of t it is nonzero:

a) $y_1(t) = e^{2t}$, $y_2(t) = e^{-2t}$

b) $y_1(t) = e^t$, $y_2(t) = te^t$

c) $y_1(t) = \sin \omega t$, $y_2(t) = \cos \omega t$

d) $y_1(t) = t^3$, $y_2(t) = |t|^3$, $t \neq 0$

e) $y_1(t) = t^2$, $y_2(t) = t^2 \ln t$, $t \neq 0$

f) $y_1(t) = \cos t$, $y_2(t) = \cos(t - \pi)$

g) $y_1(t) = \dfrac{\sin t}{\sqrt{t}}$, $y_2(t) = \dfrac{\cos t}{\sqrt{t}}$, $t \neq 0$

2. Using the methods developed in Sections 2.4 and 2.5 for finding general solutions, find two linearly independent solution vectors of the following differential equations. Use the orthogonal unit vectors

$$\mathbf{e}_1 = \begin{bmatrix} 1 \\ 0 \end{bmatrix}, \quad \mathbf{e}_2 = \begin{bmatrix} 0 \\ 1 \end{bmatrix}.$$

a) $y'' + 5y' + 6y = 0$

b) $y'' - 10y' + 25y = 0$

c) $y'' + 2y' + 5y = 0$. Plot one of the vectors for $t = n\pi/4$, $n = 0, 1, \ldots, 8$.

d) $y'' - 2y' + 5y = 0$. Plot one of the vectors for $t = n\pi/4$, $n = 0, 1, \ldots, 8$.

e) $y'' + 3y' - 10y = 0$

f) $y'' + 9y = 0$. Plot one of the vectors for $t = n\pi/6$, $n = 0, 1, \ldots, 6$.

3. The phase–amplitude form of the general solution of $y'' + w^2y = 0$ is $y(t) = K\cos(wt - \phi)$. Use e_1 and e_2 as in Exercise 2 and show that the corresponding solution vector $Y(t)$ always lies on the ellipse

$$y^2 + \frac{(y')^2}{w^2} = K^2.$$

Find the ellipse for the following initial value problems:

a) $y'' + y = 0$; $y(0) = 4$, $y'(0) = 3$

b) $y'' + 4y = 0$; $y(0) = 1$, $y'(0) = -1$

c) $y'' + \frac{25}{16}y = 0$; $y(0) = 0$, $y'(0) = 7$

4. The *Euler differential equation* of second order is of the special form

$$t^2y'' + bty' + cy = 0, \quad t > 0,$$

where b and c are constants.

a) Assume a solution of the form $y(t) = t^r$ and show that $y(t) = t^\alpha$ is a solution if and only if $r = \alpha$ is a root of the equation

$$r^2 + (b - 1)r + c = 0.$$

b) There are three cases. In each case compute the Wronskian $W(y_1, y_2)(t)$ to show that the given function $y(t) = Ay_1(t) + By_2(t)$ is a general solution.

Case 1: distinct roots $r = \alpha_1$, $r = \alpha_2$; then

$$y(t) = At^{\alpha_1} + Bt^{\alpha_2}.$$

Case 2: a double root $r = \alpha$; then

$$y(t) = At^\alpha + Bt^\alpha \ln t.$$

Case 3: complex roots $r = \alpha + i\beta$, $r^* = \alpha - i\beta$; then

$$y(t) = At^\alpha \cos(\beta \ln t) + Bt^\alpha \sin(\beta \ln t).$$

Note that the identity $t = \exp(\ln t)$ implies $t^{i\beta} = \exp(i\beta \ln t)$.

5. Use the results of Exercise 4 to find the general solution of the following Euler equations. Does the solution approach zero as $t \to \infty$?

a) $t^2y'' + 6ty' + 6y = 0$

b) $t^2y'' + 6ty' - 6y = 0$

c) $t^2y'' + 7ty' + 9y = 0$

d) $t^2y'' - 7ty' + 16y = 0$

e) $t^2y'' + 3ty' + 5y = 0$

f) $t^2y'' - 2ty' + \frac{25}{4}y = 0$

6. Show that

a) $y_1(t) = t^2$ and $y_2(t) = t^3$ cannot both be solutions to the differential equation

$$y'' + b(t)y' + c(t)y = 0,$$

where $b(t)$ and $c(t)$ are continuous at $t = 0$;

b) $y(t) = c_1 + c_2 \cos t$ cannot be the general solution of the differential equation

$$y'' + by' + cy = 0,$$

where b and c are constants.

2.7 THE VARIATION OF PARAMETERS METHOD

In this section we shall study second order inhomogeneous differential equations

$$a(t)y'' + b(t)y' + c(t)y = g(t), \tag{2.7.1}$$

where $a(t)$, $b(t)$, $c(t)$, and $g(t)$ are continuous on some interval and $a(t)$ does not vanish. One method of constructing solutions, the *variation of param-*

eters method, is based on the geometric fact that any vector in a plane can be written as a linear combination of two nonparallel (linearly independent) vectors in the plane. Therefore, for fixed t, any solution vector $\mathbf{Y}(t)$ of the inhomogeneous equation (2.7.1) can be written as a linear combination of any two linearly independent solution vectors $\mathbf{Y}_1(t)$, $\mathbf{Y}_2(t)$ of the corresponding homogeneous differential equation

$$a(t)y'' + b(t)y' + c(t)y = 0. \tag{2.7.2}$$

Hence, we can write

$$\mathbf{Y}(t) = u_1(t)\,\mathbf{Y}_1(t) + u_2(t)\,\mathbf{Y}_2(t), \tag{2.7.3}$$

where the coefficient functions $u_1(t)$ and $u_2(t)$ have yet to be found.

Before we can even start the search for $u_1(t)$ and $u_2(t)$, Eq. (2.7.2) must first be solved for a linearly independent pair of solutions $y_1(t)$ and $y_2(t)$. This is easy to do if the coefficients in the differential equation are constants; otherwise, it is difficult, if not impossible. In any case, suppose that this has been done, so $y_1(t)$ and $y_2(t)$ are known and the two linearly independent solution vectors are

$$\mathbf{Y}_1(t) = y_1(t)\mathbf{e}_1 + y_1'(t)\mathbf{e}_2,$$

$$\mathbf{Y}_2(t) = y_2(t)\mathbf{e}_1 + y_2'(t)\mathbf{e}_2.$$

The solution vector corresponding to the solution $y(t)$ of the inhomogeneous equation (2.7.1) is

$$\mathbf{Y}(t) = y(t)\mathbf{e}_1 + y'(t)\mathbf{e}_2.$$

But the vector equation (2.7.3) also implies that

$$\mathbf{Y}(t) = u_1(t)[y_1(t)\mathbf{e}_1 + y_1'(t)\mathbf{e}_2] + u_2(t)[y_2(t)\mathbf{e}_1 + y_2'(t)\mathbf{e}_2].$$

Hence the vector equation (2.7.3) is equivalent to two scalar equations obtained by equating components in the last two equations:

$$y(t) = u_1(t)\,y_1(t) + u_2(t)\,y_2(t),$$
$$y'(t) = u_1(t)\,y_1'(t) + u_2(t)\,y_2'(t). \tag{2.7.4}$$

The vector equation (2.7.3) is correct since any vector in the plane of two nonparallel vectors can be written as a linear combination of the two vectors; therefore, the scalar equations (2.7.4) must also be correct. But they seem to violate the product rule of calculus. How can the second represent the derivative of the first? This is a valid objection. The answer to it is that $u_1(t)$ and $u_2(t)$ are not arbitrary functions, but must satisfy the constraint

$$u_1'(t)\,y_1(t) + u_2'(t)\,y_2(t) = 0. \tag{2.7.5}$$

on any interval on which the vector equation (2.7.3) is valid. This is because

$$y'(t) = [u_1(t)\ y_1(t) + u_2(t)\ y_2(t)]'$$
$$= [u_1(t)\ y_1'(t) + u_2(t)\ y_2'(t)] + [u_1'(t)\ y_1(t) + u_2'(t)\ y_2(t)]$$
$$= u_1(t)\ y_1'(t) + u_2(t)\ y_2'(t)$$

only if $u_1'(t)$ and $u_2'(t)$ are subject to the constraint (2.7.5). Hence, the constraint (2.7.5) is a consequence of the second component of a solution vector being the derivative of the first component. We will use one of two equations to determine the unknown functions $u_1(t)$ and $u_2(t)$.

A second linear algebraic equation for the derivative of $u_1(t)$ and $u_2(t)$ is found by substituting the expression (2.7.4) into the inhomogeneous differential equation (2.7.1). First compute

$$y''(t) = [u_1(t)\ y_1'(t) + u_2(t)\ y_2'(t)]'$$
$$= u_1(t)\ y_1''(t) + u_2(t)\ y_2''(t) + u_1'(t)\ y_1'(t) + u_2'(t)\ y_2'(t)$$

and then substitute

$$a(t)\ y''(t) + b(t)\ y'(t) + c(t)\ y(t)$$
$$= a(t)[u_1(t)\ y_1''(t) + u_2(t)\ y_2''(t) + u_1'(t)\ y_1'(t) + u_2'(t)\ y_2'(t)]$$
$$\quad + b(t)[u_1(t)\ y_1'(t) + u_2(t)\ y_2'(t)] + c(t)[u_1(t)\ y_1(t) + u_2(t)\ y_2(t)]$$
$$= u_1(t)[a(t)\ y_1''(t) + b(t)\ y_1'(t) + c(t)\ y_1(t)]$$
$$\quad + u_2(t)[a(t)\ y_2''(t) + b(t)\ y_2'(t) + c(t)\ y_2(t)]$$
$$\quad + a(t)[u_1'(t)\ y_1'(t) + u_2'(t)\ y_2'(t)] \overset{?}{=} g(t).$$

The coefficients in the brackets following $u_1(t)$ and $u_2(t)$ are both zero since $y_1(t)$ and $y_2(t)$ are known solutions to the homogeneous equation (2.7.2). Hence, if $y(t)$ is a solution and this procedure is to succeed, another condition must be imposed on $u_1'(t)$ and $u_2'(t)$, namely,

$$a(t)[u_1'(t)\ y_1'(t) + u_2'(t)\ y_2'(t)] = g(t).$$

The representation of a solution vector to the *inhomogeneous* differential equation (2.7.1) as a linear combination of solution vectors to the *homogeneous* differential equation (2.7.2) has led to two linear algebraic equations which must be satisfied by the derivatives of the coefficient functions $u_1(t)$ and $u_2(t)$, namely

$$u_1'(t)\ y_1(t) + u_2'(t)\ y_2(t) = 0, \tag{2.7.6}$$

$$a(t)[u_1'(t)\ y_1'(t) + u_2'(t)\ y_2'(t)] = g(t).$$

Once these equations are solved for $u_1'(t)$ and $u_2'(t)$, one can, in theory, find their antiderivatives $u_1(t)$ and $u_2(t)$ and construct a solution

$$y(t) = u_1(t)\, y_1(t) + u_2(t)\, y_2(t) \qquad (2.7.7)$$

to the inhomogeneous differential equation (2.7.1).

Example 1   Solve the initial value problem $y'' + y = 4 \cos t, \quad y(\pi/2) = 2\pi, \ y'(\pi/2) = -3$.

First solve the homogeneous equation for two linearly independent solutions, $y_1(t) = \cos t$ and $y_2(t) = \sin t$. Then set

$$y(t) = u_1(t) \cos t + u_2(t) \sin t$$

and

$$y'(t) = -u_1(t) \sin t + u_2(t) \cos t.$$

Since $a(t) = 1$ and $g(t) = 4 \cos t$, the equations (2.7.6) imply that $u_1(t)$ and $u_2(t)$ must satisfy

$$u_1'(t) \cos t + u_2'(t) \sin t = 0,$$

$$- u_1'(t) \sin t + u_2'(t) \cos t = 4 \cos t.$$

Since the Wronskian $W(y_1, y_2)(t) = \cos^2 t + \sin^2 t = 1$, these can always be solved for any t to get

$$u_1'(t) = \begin{vmatrix} 0 & \sin t \\ 4 \cos t & \cos t \end{vmatrix} = -4 \cos t \sin t = -2 \sin 2t,$$

$$u_2'(t) = \begin{vmatrix} \cos t & 0 \\ - \sin t & 4 \cos t \end{vmatrix} = 4 \cos^2 t = 2[1 + \cos 2t],$$

where we have used the trigonometric identities

$$2 \cos t \sin t = \sin 2t \qquad \text{and} \qquad 2 \cos^2 t = 1 + \cos 2t.$$

This will simplify the next step which is to find $u_1(t)$ and $u_2(t)$.
  To find $u_1(t)$ and $u_2(t)$, antidifferentiate to get

$$u_1(t) = \int -2 \sin 2t \, dt = \cos 2t$$

$$u_2(t) = \int 2(1 + \cos 2t) \, dt = 2t + \sin 2t,$$

where the constants of integration have been omitted since we are looking for a particular solution of the inhomogeneous problem. From (2.7.4), the solution is

$$y(t) = \cos 2t \cos t + [2t + \sin 2t] \sin t,$$

which can be simplified to obtain

$$y(t) = [\cos 2t \cos t + \sin 2t \sin t] + 2t \sin t$$
$$= \cos (2t - t) + 2t \sin t = \cos t + 2t \sin t.$$

Furthermore,

$$y'(t) = -\cos 2t \sin t + (2t + \sin 2t) \cos t,$$

and a similar simplification gives

$$y'(t) = \sin t + 2t \cos t.$$

This could also be obtained directly by differentiating $y(t)$.

The solution obtained above is a *particular* solution of the inhomogeneous problem since it does not depend on arbitrary parameters. The *general* solution is obtained by adding to the particular solution the general solution of the homogeneous equation (see Exercise 4, p. 113), sometimes called the *complementary* solution. Therefore

$$y(t) = \cos t + 2t \sin t + c_1 \cos t + c_2 \sin t$$

is a two-parameter family of solutions to the inhomogeneous equation. It can be written as

$$y(t) = \underbrace{[\cos t + 2t \sin t]}_{\text{particular solution}} + \underbrace{[c_1 \cos t + c_2 \sin t]}_{\text{complementary solution}}.$$

To find c_1 and c_2 which satisfy the initial conditions, substitute the data

$$2\pi = 0 + \pi + 0 + c_2 \quad \text{and} \quad -3 = 1 + 0 - c_1 + 0,$$

and so $c_1 = 4$ and $c_2 = \pi$. The solution to the initial value problem is therefore

$$y(t) = \cos t + 2t \sin t + 4 \cos t + \pi \sin t$$
$$= 2t \sin t + 5 \cos t + \pi \sin t. \quad \square$$

We see that in the expression for $y(t)$, the first term is the particular solution, and the remainder is a solution of the homogeneous equation. A direct substitution shows that if $y_p(t) = 2t \sin t$, then

$$y_p'' + y_p = (4 \cos t - 2t \sin t) + 2t \sin t = 4 \cos t,$$

so $y_p(t)$ is a particular solution.

As the example shows, the variation of parameters algorithm may produce a term in the particular solution that is a solution to the homogeneous equation. Such a term can always be absorbed in the complementary solution, e.g.,

$$y(t) = \cos t + 2t \sin t + c_1 \cos t + c_2 \sin t$$

can be written as

$$y(t) = 2t \sin t + k_1 \cos t + k_2 \sin t,$$

where $k_1 = c_1 + 1$, $k_2 = c_2$.

Example 2 Find the general solution to

$$t^2 y'' + t y' - 4y = 4t^6, \qquad t > 0.$$

Earlier it was shown that $y_1(t) = t^2$ and $y_2(t) = t^{-2}$ are linearly independent solutions to the corresponding homogeneous differential equation. Let

$$y(t) = u_1(t)t^2 + u_2(t)t^{-2},$$

$$y'(t) = 2u_1(t)t - 2u_2(t)t^{-3};$$

then, since $a(t) = t^2$ and $g(t) = 4t^6$, the constraint equations (2.7.6) are

$$u_1'(t)t^2 + u_2'(t)t^{-2} = 0,$$

$$t^2[2u_1'(t)t - 2u_2'(t)t^{-3}] = 4t^6.$$

The Wronskian is

$$W(y_1, y_2)(t) = \begin{vmatrix} t^2 & t^{-2} \\ 2t & -2t^{-3} \end{vmatrix} = -4t^{-1},$$

and after dividing the second equation by t^2 we have

$$u_1'(t) = \frac{\begin{vmatrix} 0 & t^{-2} \\ 4t^4 & -2t^{-3} \end{vmatrix}}{-4t^{-1}} = t^3,$$

and

$$u_2'(t) = \frac{\begin{vmatrix} t^2 & 0 \\ 2t & 4t^4 \end{vmatrix}}{-4t^{-1}} = -t^7.$$

A simple antidifferentiation gives

$$u_1(t) = \frac{t^4}{4}, \qquad u_2(t) = -\frac{t^8}{8},$$

and a particular solution to the inhomogeneous differential equation is then

$$y_p(t) = \frac{t^4}{4} t^2 + \left(-\frac{t^8}{8}\right) t^{-2} = \frac{t^6}{8}.$$

The general solution

$$y(t) = \frac{t^6}{8} + c_1 t^2 + c_2 t^{-2}$$

is obtained by adding the general solution of the homogeneous equation to $y_p(t)$. ☐

Example 3 Find the solution of

$$y'' + y' - 6y = g(t), \qquad y(t_0) = A, \qquad y'(t_0) = B.$$

Two linearly independent solutions of the corresponding homogeneous equation are $y_1(t) = e^{2t}$, $y_2(t) = e^{-3t}$, and their Wronskian is

$$W(y_1, y_2)(t) = \begin{vmatrix} e^{2t} & e^{-3t} \\ 2e^{2t} & -3e^{-3t} \end{vmatrix} = -5e^{-t}.$$

Therefore,

$$y(t) = u_1(t)e^{2t} + u_2(t)e^{-3t}$$

$$y'(t) = 2u_1(t)e^{2t} - 3u_2(t)e^{-3t},$$

where $u_1(t)$ and $u_2(t)$ satisfy the constraint equations (2.7.6):

$$u_1'(t)e^{2t} + u_2'(t)e^{-3t} = 0,$$
$$2u_1'(t)e^{2t} - 3u_2'(t)e^{-3t} = g(t).$$

Solving this system for $u_1(t)$ and $u_2(t)$ gives

$$u_1'(t) = \frac{1}{5}g(t)e^{-2t}, \qquad u_2'(t) = -\frac{1}{5}g(t)e^{3t}.$$

Since $g(t)$ is not specified and the initial data are given at $t = t_0$, it is convenient (but not necessary) to use the antiderivatives

$$u_1(t) = \frac{1}{5}\int_{t_0}^{t} g(s)e^{-2s} \, ds,$$

$$u_2(t) = -\frac{1}{5}\int_{t_0}^{t} g(s)e^{3s} \, ds,$$

and a particular solution is

$$y_p(t) = \left(\frac{1}{5}\int_{t_0}^{t} g(s)e^{-2s} \, ds\right) e^{2t} + \left(-\frac{1}{5}\int_{t_0}^{t} g(s)e^{3s} \, ds\right) e^{-3t}.$$

The general solution is the particular solution plus the complementary solution:

$$y(t) = y_p(t) + c_1 e^{2t} + c_2 e^{-3t},$$

where the constants c_1 and c_2 are chosen so as to satisfy the initial conditions.

For instance, if $g(t) = 5t$ and $t_0 = 0$, then

$$u_1(t) = \frac{1}{5} \int_0^t 5se^{-2s} \, ds = -\frac{1}{4}e^{-2t}(2t + 1) + \frac{1}{4},$$

$$u_2(t) = -\frac{1}{5} \int_0^t 5se^{3s} \, ds = -\frac{1}{9}e^{3t}(3t - 1) - \frac{1}{9},$$

and

$$y_p(t) = u_1(t)e^{2t} + u_2(t)e^{-3t} = -\frac{5}{6}t - \frac{5}{36} + \frac{1}{4}e^{2t} - \frac{1}{9}e^{-3t}.$$

Note that the choice of antiderivative insures that $y_p(0) = y_p'(0) = 0$. (Why?) On the other hand, if $g(t) = \sec t$ and $t_0 = 1$, then

$$u_1(t) = \frac{1}{5} \int_1^t \sec s \; e^{-2s} \, ds$$

$$u_1(t) = -\frac{1}{5} \int_1^t \sec s \; e^{3s} \, ds,$$

and neither integral can be evaluated by elementary means. Nevertheless, we can write

$$y_p(t) = \frac{1}{5}e^{2t} \int_1^t \sec s \; e^{-2s} \, ds - \frac{1}{5}e^{-3t} \int_1^t \sec s \; e^{3s} \, ds,$$

and the general solution is

$$y(t) = y_p(t) + c_1 e^{2t} + c_2 e^{-3t}.$$

If the initial conditions are, for instance,

$$y(1) = 0 \quad \text{and} \quad y'(1) = 2,$$

then

$$y(1) = y_p(1) + c_1 e^2 + c_2 e^{-3} = 0 + c_1 e^2 + c_2 e^{-3} = 0,$$

$$y'(1) = y_p'(1) + 2c_1 e^2 - 3c_2 e^{-3} = 0 + 2c_1 e^2 - 3c_2 e^{-3} = 2,$$

which can be solved to get $c_1 = \frac{2}{5}e^{-2}$ and $c_2 = -\frac{2}{5}e^3$. The advantage of the choice of antiderivatives is clear in this example. If a different antiderivative (i.e., with a different lower limit than $t_0 = 1$) had been used, then $y_p(1)$ and $y_p'(1)$ would have to be calculated by some numerical technique in order to find c_1 and c_2. $\square$

The method is summarized in the following algorithm.

The variation of parameters algorithm.

1. Solve the homogeneous equation for a linearly independent pair of solutions $y_1(t)$, $y_2(t)$.

2. Solve the pair of linear equations

$$u_1'(t)y_1(t) + u_2'(t)y_2(t) = 0,$$

$$a(t)[u_1'(t)y_1'(t) + u_2'(t)y_2'(t)] = g(t)$$

for $u_1'(t)$ and $u_2'(t)$, then find antiderivatives $u_1(t)$ and $u_2(t)$.

3. A particular solution of the inhomogeneous equation is

$$y_p(t) = u_1(t)y_1(t) + u_2(t)y_2(t)$$

and

$$y_p'(t) = u_1(t)y_1'(t) + u_2(t)y_2'(t).$$

The general solution is

$$y(t) = y_p(t) + c_1y_1(t) + c_2y_2(t),$$

where c_1 and c_2 are arbitrary constants.

4. If initial data $y(t_0) = r$ and $y'(t_0) = s$ are given, select c_1 and c_2 so that

$$r = y_p(t_0) + c_1y_1(t_0) + c_2y_2(t_0),$$

$$s = y_p'(t_0) + c_1y_1'(t_0) + c_2y_2'(t_0).$$

In this case it is convenient to use the antiderivatives

$$u_1(t) = \int_{t_0}^t u_1'(s)\, ds \quad \text{and} \quad u_2(t) = \int_{t_0}^t u_2'(s)\, ds$$

in step 2, so that $y_p(t_0) = y_p'(t_0) = 0$.

EXERCISES 2.7

1. Using the variation of parameters method, find the general solution of the following inhomogeneous differential equations:

a) $y'' - y = 2e^{3t}$

b) $y'' + 9y = 3t + 2$

c) $y'' - 7y' + 12y = 5e^{4t}$

d) $y'' - 4y = 3e^{-2t} - t^2$

e) $y'' + 4y = 4\cos^3 2t$. *Hint:* A useful identity is $\cos^3 \theta = \frac{3}{4} \cos \theta + \frac{1}{4} \cos 3\theta$.

f) $y'' + y = \sec t$

g) $y'' + 6y' + 9y = \dfrac{e^{-3t}}{t^2}$

h) $y'' - 4y = \dfrac{1}{\cosh 2t}$

i) $y'' - 4y' + 3y = \dfrac{1}{1 + e^{-t}}$

2. Using the method of variation of parameters, find the general solution of the following inhomogeneous Euler differential equations (see Exercise 4 of Section 2.6):

a) $t^2y'' - 3ty' + 4y = t + 2$

b) $t^2y'' - 2ty' + 2y = te^t$

c) $t^2y'' + 7ty' + 5y = 10 - 4/t$

d) $t^2y'' - 4ty' + 6y = t^2e^t + \ln t$

3. Solve the following initial value problems.

   a) Exercise 1(a): $y(0) = 1$, $y'(0) = 2$

   b) Exercise 1(b): $y(0) = 0$, $y'(0) = 1$

   c) Exercise 1(f): $y(0) = 2$, $y'(0) = -1$

   d) Exercise 1(g): $y(1) = 1$, $y'(1) = 0$

   e) Exercise 2(b): $y(1) = 0$, $y'(1) = 1$

   f) Exercise 2(c): $y(1) = 1$, $y'(1) = 0$

4. Show that the general solution $y(t)$ of the linear inhomogeneous differential equation

$$a(t)y'' + b(t)y' + c(t)y = g(t)$$

   can be written as a sum $y(t) = y_p(t) + y_h(t)$, where $y_p(t)$ is a particular solution of the inhomogeneous equation and $y_h(t)$ is the general solution of the homogeneous equation. *Hint:* What differential equation is satisfied by $y(t) - y_p(t)$?

5. The solution of

$$y'' + w^2y = f(t); \qquad y(0) = y'(0) = 0$$

can be expressed in the form

$$y(t) = u_1(t) \sin wt + u_2(t) \cos wt,$$

where $u_1(t)$ and $u_2(t)$ are found by the variation of parameters algorithm. By combining $\sin wt$ and $\cos wt$ with the integral expressions for $u_1(t)$ and $u_2(t)$ show that

$$y(t) = \frac{1}{w} \int_0^t \sin[w(t - s)]f(s) \, ds.$$

6. For the given inhomogeneous differential equation, the given functions $y_1(t)$ and $y_2(t)$ are linearly independent solutions of the corresponding homogeneous equation. Use them and the variation of parameters method to find the general solution of the following differential equations:

   a) $ty'' - (2t + 1)y' + (t + 1)y = 4t^2e^t$;
      $y_1(t) = e^t$, $y_2(t) = t^2e^t$

   b) $y'' - \dfrac{2}{t}y' + \left(1 + \dfrac{2}{t^2}\right)y = te^t$;
      $y_1(t) = t \cos t$, $y_2(t) = t \sin t$

   c) $ty'' - y' + 4t^3y = 8t^4$;
      $y_1(t) = \sin t^2$, $y_2(t) = \cos t^2$

2.8   THE METHOD OF UNDETERMINED COEFFICIENTS

An inspection of problems solved by the variation of parameters method reveals that constant coefficient differential equations with exponential, polynomial, or sinusoidal forcing functions usually have exponential, polynomial, or sinusoidal particular solutions. It seems reasonable that particular solutions can be constructed with less effort by deducing the type of the solution from the forcing function and then determining its exact structure by substituting a trial solution into the differential equation. This is called the *method of undetermined coefficients.* It is not as general as the variation of parameters method but for many problems it is much easier. As the reader will see, the method could be nicknamed the "method of judicious guessing." Let us use the method to find the solutions to

$$y'' + y' - 6y = g(t)$$

with four different choices of $g(t)$.

Example 1   a) $y'' + y' - 6y = 5e^{4t}$

Since the derivative of an exponential is again an exponential, and the forcing function is an exponential, let $y(t) = Ae^{4t}$, an exponential of the same type. The undetermined coefficient A will be found by substituting the trial solution into the differential equation

$$(Ae^{4t})'' + (Ae^{4t})' - 6Ae^{4t} = Ae^{4t}(16 + 4 - 6) = 14Ae^{4t} \stackrel{?}{=} 5e^{4t}.$$

Clearly, $A = \frac{5}{14}$ and the particular solution is $5e^{4t}/14$. A general solution needs two linearly independent solutions to the corresponding homogeneous equation. Here $y_1(t) = e^{2t}$ and $y_2(t) = e^{-3t}$, so a general solution is

$$y(t) = \frac{5}{14}e^{4t} + c_1 e^{2t} + c_2 e^{-3t}.$$

b) $y'' + y' - 6y = 5e^{2t}$

With confidence based on the previous example we try $y(t) = Ae^{2t}$. But

$$(Ae^{2t})'' + (Ae^{2t})' - 6Ae^{2t} = Ae^{2t}(4 + 2 - 6) = 0 \neq 5e^{2t},$$

so our guess fails. Fortunately there is an obvious explanation for the failure and an easy modification to prevent such failures. It failed because the trial solution, $y(t) = Ae^{2t}$, was a solution to the corresponding homogeneous equation and its substitution gave us zero. To guard against this difficulty, one should always solve the homogeneous equation first. Its solutions will be needed to construct a general solution or to satisfy initial data, so this is not wasteful. Then check the trial solution to see if it is also a solution to the homogeneous equation. If it is not, proceed; otherwise, the method must be modified. The modification is suggested by the form of the solution constructed by the variation of parameters method:

$$y(t) = te^{2t}.$$

The factor t appears because e^{2t} is a solution to the homogeneous equation, and $u_1'(t) = e^{-2t}g(t)/5$, and $g(t) = 5e^{2t}$, so $u_1'(t) = 1$. Knowing this, we take as our trial solution $y(t) = Ate^{2t}$. Substituting gives us

$$(Ate^{2t})'' + (Ate^{2t})' - 6Ate^{2t} = A(4te^{2t} + 4e^{2t}) + A(2te^{2t} + e^{2t}) - 6Ate^{2t}$$
$$= Ate^{2t}(4 + 2 - 6) + Ae^{2t}(4 + 1) \stackrel{?}{=} 5e^{2t}.$$

Clearly, $A = 1$, and

$$y(t) = te^{2t} + c_1 e^{2t} + c_2 e^{-3t}$$

is a general solution.

c) $y'' + y' - 6y = 5t$

Since the forcing term is a polynomial, this suggests a trial solution $y(t) = At + B$. Substitute to get

$$(At + B)'' + (At + B)' - 6(At + B) = (A - 6B) - 6At \stackrel{?}{=} 5t.$$

Since two polynomials are equal if and only if the coefficients of like powers of t are equal, A and B must satisfy

$$A - 6B = 0, \quad -6A = 5.$$

Hence $A = -\frac{5}{6}$, $B = -\frac{5}{36}$, and the general solution is

$$y(t) = -\frac{5}{6}t - \frac{5}{36} + c_1 e^{2t} + c_2 e^{-3t}.$$

The reader should note that had the guess been simply $y(t) = At$, its substitution would have yielded

$$(At)'' + (At)' - 6At = A - 6At \stackrel{?}{=} 5t,$$

an equation with no solution. Since $g(t) = 5t$ is a polynomial of degree one, the trial solution must be $y(t) = At + B$, the most general polynomial of degree one.

d) $y'' + y' - 6y = 5 \cos t$

The derivative of $\cos t$ is $-\sin t$ and therefore we would expect that both $\cos t$ and $\sin t$ are needed in the trial solution. Choose $y(t) = A \cos t + B \sin t$ and substitute to obtain

$$(A \cos t + B \sin t)'' + (A \cos t + B \sin t)' - 6(A \cos t + B \sin t)$$
$$= (-7A + B) \cos t + (-A - 7B) \sin t$$
$$\stackrel{?}{=} 5 \cos t = 5 \cos t + 0 \sin t.$$

Equating coefficients of $\cos t$ and $\sin t$ we get

$$-7A + B = 5, \quad -A - 7B = 0,$$

which gives $A = -\frac{7}{10}$, $B = \frac{1}{10}$, and the general solution is

$$y(t) = -\frac{7}{10} \cos t + \frac{1}{10} \sin t + c_1 e^{2t} + c_2 e^{-3t}. \quad \square$$

The next problem shows that occasionally a trial solution requires a factor t^2 or even t^3.

Example 2 Find a general solution to

$$y'' + 6y' + 9y = 4e^{-3t}.$$

First solve the homogeneous equation

$$y'' + 6y' + 9y = 0$$

to obtain $y_1(t) = e^{-3t}$, $y_2(t) = te^{-3t}$. Examining the right-hand side would normally suggest a trial solution of the form Ae^{-3t}. But since both Ae^{-3t} and Ate^{-3t} are solutions of the homogeneous equation, we multiply the trial solution by the smallest integer power of t, so that this is no longer the case. Clearly, t^2 will do, and so we try $y(t) = At^2e^{-3t}$. Substitution, some careful differentiation, and collection of terms gives

$$(At^2e^{-3t})'' + 6(At^2e^{-3t})' + 9At^2e^{-3t} = 2Ae^{-3t} \overset{?}{=} 4e^{-3t}.$$

Hence, $A = 2$ and the general solution is

$$y(t) = 2t^2e^{-3t} + c_1e^{-3t} + c_2te^{-3t}. \quad \square$$

The reader should notice that if the forcing function were $4te^{-3t}$, one would normally try a solution of the form $y(t) = (At + B)e^{-3t}$. However, for the above example both Ate^{-3t} and Be^{-3t} are solutions of the homogeneous equation, so the trial solution would be instead

$$y(t) = t^2(At + B)e^{-3t}$$

$$= At^3e^{-3t} + Bt^2e^{-3t}.$$

If the forcing function $g(t)$ is a sum of exponential, polynomial, and sinusoidal functions, then the problem can be split into simpler problems whose solutions can be added to give a solution to the original problem. This procedure is illustrated in the next example.

Example 3   Find a general solution to

$$y'' + y = 6e^t + 6 \sin t.$$

We begin by noticing that $y_1(t) = \cos t$ and $y_2(t) = \sin t$ are two linearly independent solutions of the homogeneous equation. The first equation to be solved is $y'' + y = 6e^t$. A trial solution is $y_p(t) = Ae^t$ and

$$(Ae^t)'' + Ae^t = 2Ae^t \overset{?}{=} 6e^t.$$

Hence $A = 3$ and $y_p(t) = 3e^t$. The second equation to be solved is $y'' + y = 6 \sin t$. Since $6 \sin t$ is a solution to the homogeneous equation, let

$$y_p(t) = t(B \cos t + C \sin t).$$

Then

$$(Bt \cos t + Ct \sin t)'' + (Bt \cos t + Ct \sin t) = -2B \sin t + 2C \cos t$$
$$\overset{?}{=} 6 \sin t + 0 \cos t;$$

hence $B = -3$, $C = 0$, and $y_p(t) = -3t \cos t$. Now add the two particular solutions and a complementary solution to get the general solution

$$y(t) = 3e^t - 3t \cos t + c_1 \cos t + c_2 \sin t. \quad \square$$

The following are guidelines for the method of undetermined coefficients. Unlike the method of variation of parameters, it is difficult to write a specific "algorithm" for the method. Of course, the reader should have carefully gone over the preceding examples before attempting any problems.

Guidelines for the method of undetermined coefficients

Given the *constant coefficient* linear differential equation

$$ay'' + by' + cy = g(t),$$

where $g(t)$ is an exponential, a simple sinusoidal function, a polynomial, or a product of these functions:

1. Solve the homogeneous equation for a pair of linearly independent solutions $y_1(t)$ and $y_2(t)$.

2. If $g(t)$ is *not* a solution of the homogeneous equation, take a trial solution of the same type as $g(t)$ according to the suggestions given in Table 2.1.

3. If $g(t)$ is a solution of the homogeneous equation, take a trial solution of the same type as $g(t)$ multiplied by the lowest power of t for which no term of the trial solution is a solution of the homogeneous equation.

4. Substitute the trial solution into the differential equation and solve for the undetermined coefficients so that it is a particular solution $y_p(t)$.

5. Set $y(t) = y_p(t) + c_1 y_1(t) + c_2 y_2(t)$, where the constants c_1 and c_2 can be determined if initial conditions are given.

6. If $g(t)$ is a sum of the type of forcing functions described above, split the problem into simpler parts. Find a particular solution for each of these, then add the particular solutions to obtain $y_p(t)$.

TABLE 2.1	$g(t)$	**Trial solution**
	ae^{rt}	Ae^{rt}
	$a \sin \omega t$ or $a \cos \omega t$	$A \sin \omega t + B \cos \omega t$
	at^n, where n is a positive integer	$P(t)$, a general polynomial of degree n
	$at^n e^{rt}$, where n is a positive integer	$P(t)e^{rt}$, with $P(t)$ a general polynomial of degree n
	$t^n(a \sin \omega t + b \cos \omega t)$, where n is a positive integer	$P(t)(A \sin \omega t + B \cos \omega t)$, with $P(t)$ a general polynomial of degree n
	$e^{rt}(a \sin \omega t + b \cos \omega t)$	$e^{rt}(A \sin \omega t + B \cos \omega t)$

EXERCISES 2.8

1. Using the method of undetermined coefficients, find the general solution of each of the following inhomogeneous differential equations:

   a) $y'' - y' - 2y = t$

   b) $y'' - 4y = 6e^t$

   c) $y'' - 2y' + y = 4 \sin t$

   d) $y'' + 5y' + 4y = 2e^{-t}$

   e) $y'' = t^2 + t + 1$

   f) $y'' + 2y' + 5y = 10t^2 - e^{-2t}$

   g) $y'' + 4y = t - 2 \sin 2t$

   h) $y'' + y = t \cos 2t$

   i) $y'' + y' - 4y = 2 \sinh t$

   j) $y'' + y = 8 \cos^3 t$

2. Solve the following initial value problems.

   a) Exercise 1(a); $y(0) = 0, \quad y'(0) = 1$

   b) Exercise 1(b); $y(0) = 1, \quad y'(0) = 2$

   c) Exercise 1(c); $y(0) = 1, \quad y'(0) = 0$

   d) Exercise 1(d); $y(0) = 1, \quad y'(0) = 1$

   e) Exercise 1(g); $y(\pi) = 0, \quad y'(\pi) = 1$

   f) Exercise 1(j); $y(\pi/6) = 0; \quad y'(\pi/6) = 0$

3. An alternative method of solving the Euler differential equation

   $$t^2y'' + bty' + cy = 0, \quad t > 0,$$

   is to make the change of independent variable $t = e^z$.

   Show that

   $$\frac{dy}{dt} = \frac{1}{t}\frac{dy}{dz}, \qquad \frac{d^2y}{dt^2} = \frac{1}{t^2}\frac{d^2y}{dz^2} - \frac{1}{t^2}\frac{dy}{dz},$$

   and that the substitution gives the constant coefficient differential equation

   $$\frac{d^2y}{dz^2} + (b - 1)\frac{dy}{dz} + cy = 0$$

   with solutions $y(z) = y(\ln t)$.

4. The change of variables in Exercise 3 transforms an inhomogeneous Euler equation

   $$t^2y'' + bty' + cy = P(t),$$

   where $P(t)$ is a linear combination of powers of t, to the equation

   $$\frac{d^2y}{dz^2} + (b - 1)\frac{dy}{dz} + cy = P(e^z).$$

   The last equation can be solved by the method of undetermined coefficients. Use this trick to find the general solution of

   a) $t^2y'' - 2y = 4t^3$

   b) $t^2y'' + 5ty' - 6y = 4/t^2 - 12$

   c) $t^2y'' + 5ty' + y = 3/t + 5t^2$

   d) $t^2y'' + 5ty' + 4y = 6/t^2$

   e) $t^2y'' + ty' + 9y = At^k, \quad k$ any real number.

2.9 THE PHASE PLANE AND THE OSCILLATOR EQUATION

The *oscillator equation* is the differential equation of an LCR circuit,

$$L\frac{d^2Q}{dt^2} + R\frac{dQ}{dt} + \frac{Q}{C} = E(t), \qquad (2.9.1)$$

and it was derived in Section 2.2 from the two first order equations,

$$\frac{dQ}{dt} = I, \qquad L\frac{dI}{dt} + RI + \frac{Q}{C} = E(t). \qquad (2.9.2)$$

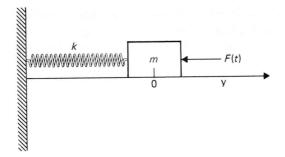

Figure 2.4 A mass–spring system.

Clearly, Eqs. (2.9.1) and (2.9.2) are equivalent: a solution to the pair of first order equations is a solution to the second-order system, and conversely. We found it convenient to use second order equations in the discussion of solution techniques. We shall now find it convenient to use first order systems in the discussion of the geometry of solution curves. By first order systems we mean a set of first order differential equations.

Before starting this discussion, let us consider an important and closely related physical example which is illustrated in Fig. 2.4. The figure depicts a mass supported on a level surface; the mass is attached to a rigid wall by a light spring. If the mass is displaced from its equilibrium position ($y = 0$), the spring either stretches or contracts. The spring then attempts to return to its original length by applying a restoring force to the mass that is proportional to the displacement (Hooke's law). Other forces acting on the mass are a frictional force and, perhaps, an external force. The sum of these forces by Newton's second law equals the time rate of change of the momentum of the system. Hence,

$$m\frac{dy}{dt} = p, \qquad \frac{dp}{dt} = -ky + f\left(y, \frac{dy}{dt}\right) + F(t), \qquad (2.9.3)$$

where y denotes the displacement from equilibrium, p is the momentum, m is the mass, k is the Hooke's constant, $f(y, dy/dt)$ is the friction force, and $F(t)$ is the external force. The first equation defines the momentum, while the second models the forces acting to change the momentum.

The frictional force $f(y, dy/dt)$ is difficult to model. The simplest useful model assumes that the dissipation of energy due to friction is proportional to the velocity and is independent of position, and therefore

$$f\left(y, \frac{dy}{dt}\right) = -c\frac{dy}{dt},$$

where c is a positive constant. The minus sign in this formula is due to the resistance of the surface to the motion of the mass. With this model of the frictional force, the differential equations describing the mass–spring system are

$$m \frac{dy}{dt} = p, \qquad \frac{dp}{dt} = -ky - c \frac{dy}{dt} + F(t), \qquad (2.9.4)$$

and this implies

$$m \frac{d^2y}{dt^2} = \frac{dp}{dt} = -ky - c \frac{dy}{dt} + F(t)$$

or

$$m \frac{d^2y}{dt^2} + c \frac{dy}{dt} + ky = F(t). \qquad (2.9.5)$$

As the reader can see, the differential equation describing the mass–spring system when the mass is sliding on a level surface is the same as when the mass is suspended from a vertical support. In the first case the term $c\, dy/dt$ describes the effect of friction, while in the second case it describes the effect of air resistance. Along with the LCR circuit, all these systems are described by the oscillator equation, and this results in an economy of analysis but also in a transfer of intuition from one discipline to another. For example, electrical-network analogies are used in the analysis of stresses in structures, in underwater acoustics, and in thermodynamics.

With the LCR circuit and the mass–spring system as motivating examples, let us now study the geometry of solution curves in the unforced case when $F(t) \equiv 0$. To do this, consider the example

$$\frac{d^2y}{dt^2} + 2 \frac{dy}{dt} + 17y = 0.$$

An equivalent first order system is obtained by letting $dy/dt = p$; then $d^2y/dt^2 = dp/dt$, which implies

$$\frac{dy}{dt} = p, \qquad \frac{dp}{dt} = -17y - 2p. \qquad (2.9.6)$$

For instance, the solution that satisfies conditions $y(0) = 4$, $p(0) = y'(0) = 0$ is

$$y(t) = e^{-t}(4 \cos 4t + \sin 4t), \qquad p(t) = -17e^{-t} \sin 4t. \qquad (2.9.7)$$

The graphs of the functions $y(t)$ and $p(t)$ are sketched in Figs. 2.5 and 2.6.

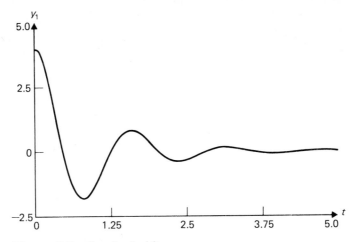

Figure 2.5   Graph of $y(t)$.

One can also regard equations (2.9.7) as the parametric representation of an *integral curve* in the *phase plane,* the yp plane. This integral curve passes through the point $(4, 0)$ at $t = 0$ and spirals into the origin as t increases. In the sketch of the curve shown in Fig. 2.7, the arrow on the curve denotes the direction of increasing time. If the unit vectors $\mathbf{e}_1$ and $\mathbf{e}_2$ are chosen as coordinate vectors, the solution vector

$$\mathbf{Y}(t) = y(t)\mathbf{e}_1 + p(t)\mathbf{e}_2 \qquad (2.9.8)$$

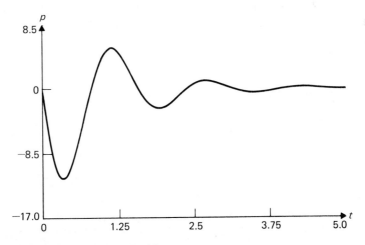

Figure 2.6   Graph of $p(t)$.

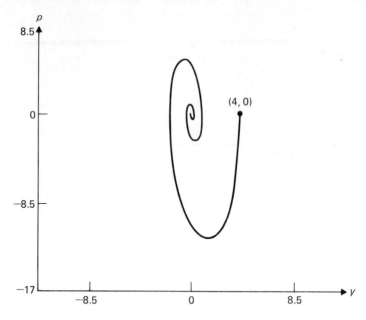

Figure 2.7 Phase plane plot of $p(t)$ versus $y(t)$.

extends from the origin to the integral curve. The derivative of the solution vector,

$$\frac{d\mathbf{Y}}{dt}(t) = \frac{dy}{dt}(t)\mathbf{e}_1 + \frac{dp}{dt}(t)\mathbf{e}_2,$$

is tangent to the integral curve at each point. This is a basic result of vector calculus that can be established by studying the vector

$$\frac{\Delta\mathbf{Y}}{\Delta t} = \frac{\mathbf{Y}(t + \Delta t) - \mathbf{Y}(t)}{\Delta t}$$

as Δt approaches zero.

The points (y, p) in the phase plane can be thought of as moving particles whose velocities are determined by the governing system of first order differential equations. For an integral curve passing through any point (y, p), the velocity vector is given by

$$\frac{d\mathbf{Y}}{dt} = \frac{dy}{dt}\mathbf{e}_1 + \frac{dp}{dt}\mathbf{e}_2,$$

where $\mathbf{e}_1$ and $\mathbf{e}_2$ are unit coordinate vectors. But the components of the velocity vector *at any point* in the phase plane of the system (2.9.6) are

$$\frac{dy}{dt} = p, \qquad \frac{dp}{dt} = -17y - 2p;$$

hence

$$\frac{d\mathbf{Y}}{dt} = p\mathbf{e}_1 + (-17y - 2p)\mathbf{e}_2.$$

For instance, at the point $(y, p) = (2, 1)$ the velocity vector is

$$(1)\mathbf{e}_1 + [-17 \cdot 2 - 2 \cdot 1]\mathbf{e}_2 = \mathbf{e}_1 - 36\mathbf{e}_2.$$

It is the tangent vector to the integral curve passing through $(2, 1)$ at that point.

We see that it is not necessary to solve the differential equation in order to know the direction of the flow of the points in the phase plane. Given a system of differential equations, one can graph the flow by selecting various points (y, p) and then computing the corresponding velocity vector. This is done by substituting the values of y and p into the right-hand sides of the differential equations to obtain the components of the vector. Thus the concept of a *direction field* of a single first order differential equation is extended to a *vector field* of a system of differential equations. To sketch the vector field for the above example, consider the vector

$$p\mathbf{e}_1 + (-17y - 2p)\mathbf{e}_2$$

and compute it for some points (y_0, p_0). Then at each point (y_0, p_0) place a small arrow in the direction of the vector (compare this procedure with the one used to construct direction fields in Chapter 1). The vector field of equation (2.9.6) is sketched in Fig. 2.8.

From a vector field much can be learned about the qualitative behavior of all the solutions of a system of differential equations in a region of the

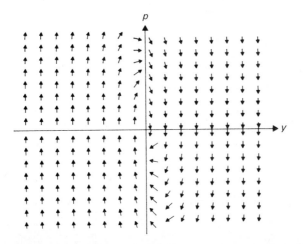

Figure 2.8 The vector field of $\dot{y} = p, \dot{p} = -17y - 2p$.

phase plane. An inspection of the vector field of the system (2.9.6) shows that, with the exception of the constant solution $y(t) = 0$ and $p(t) = 0$, all the solutions spiral into the origin as time increases and the flow is clockwise. This geometric information is of course contained, but hidden, in the formulas for a general solution to the system:

$$y(t) = e^{-t}(c_1 \cos 4t + c_2 \sin 4t);$$

$$p(t) = e^{-t}[-c_1(\cos 4t + 4 \sin 4t) + c_2(4 \cos 4t - \sin 4t)].$$

The displacement y and momentum p are the phase plane variables for the mass–spring system. Other phase plane variables are used with other systems, e.g., in the LCR circuit equation (2.9.2) the charge Q and the current I are the appropriate phase plane variables. We shall use them in a phase plane analysis of the effect of changing the resistance R in an unforced LCR circuit. The governing differential equations,

$$\frac{dQ}{dt} = I, \qquad \frac{dI}{dt} = -\frac{Q}{LC} - \frac{R}{L}I, \qquad (2.9.9)$$

are obtained from equations (2.9.2) with $E(t) \equiv 0$. The terms *flow, position vector,* and *velocity vector* are used in the discussion of the charge–current phase plane since the geometry of the integral curves is the same as that of a corresponding unforced mass–spring system.

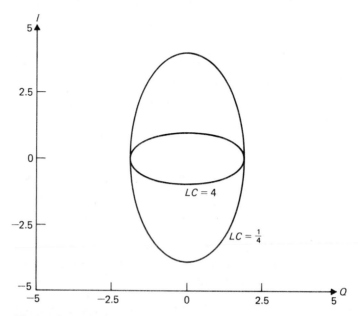

Figure 2.9 Phase plane plot of $I(t)$ versus $Q(t)$.

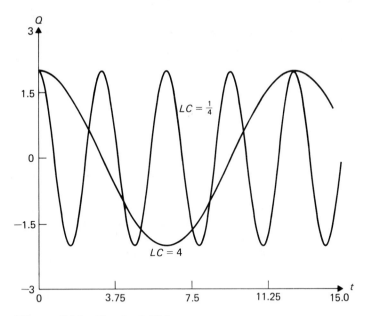

Figure 2.10 Graph of $Q(t)$.

First consider an *undamped* circuit for which the resistance R is zero. Selected integral curves of the system of differential equations

$$\frac{dQ}{dt} = I, \qquad \frac{dI}{dt} = -\frac{Q}{LC}$$

are sketched in Fig. 2.9. The corresponding graphs of $Q(t)$ and $I(t)$ are in Figs. 2.10 and 2.11, respectively. These solutions can be obtained by solving the equivalent second order equation (2.9.1) (with $R = 0$ and $E(t) = 0$) for Q. The solution can be expressed in the phase–amplitude form, and since $I = dQ/dt$, then

$$Q(t) = A \cos (qt - \phi), \qquad I(t) = -Aq \sin (qt - \phi), \qquad q^2 = \frac{1}{LC},$$

are parametric equations of the integral curves. As in Section 2.5, A denotes the amplitude, ϕ the phase, and q the circular, or radial, frequency. The solutions are periodic with period $T = 2\pi/q = 2\pi \sqrt{LC}$.

In this problem, the shape of the integral curves is easily deduced from the general solution since by squaring $Q(t)$ and $I(t)$ one sees that

$$\left[\frac{Q(t)}{A} \right]^2 + \left[\frac{I(t)}{Aq} \right]^2 = \cos^2 (qt - \phi) + \sin^2 (qt - \phi) = 1$$

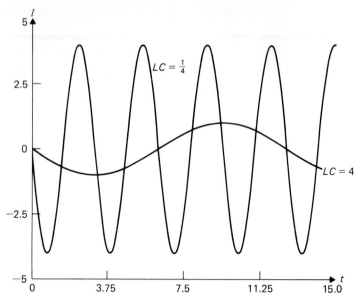

Figure 2.11 Graph of $I(t)$.

or

$$\frac{Q^2}{A^2} + \frac{I^2}{A^2 q^2} = 1,$$

the familiar equation of an ellipse. A moment's reflection will convince the reader that closed curves in the phase plane of a system represent periodic solutions and vice versa. If the period is T and at time t_0 the integral curve passes through a point (Q_0, I_0), then at time $t_0 + T$ it will pass through it again, which means the integral curve is a closed curve.

If the resistance is small but nonzero, that is, $R^2 < 4L/C$, the circuit is said to be *underdamped*. In this case, the integral curves spiral clockwise into the origin; this is illustrated in Figs. 2.12, 2.13, and 2.14. Again, the corresponding solutions can be obtained from (2.9.1); this is similar to the case where the characteristic polynomial has complex roots with nonzero real parts. We have

$$Q(t) = Ae^{pt} \cos(qt - \phi), \qquad (2.9.10)$$

$$I(t) = Ae^{pt}[p \cos(qt - \phi) - q \sin(qt - \phi)],$$

with

$$p = -\frac{R}{2L}, \qquad q = \frac{(4L/C - R^2)^{1/2}}{2L} = \left(\frac{1}{LC} - \frac{R^2}{4L^2}\right)^{1/2}.$$

Note that $p < 0$, so that both $Q(t)$ and $I(t)$ approach zero as t becomes large. Unlike the undamped circuit problem, there is no simple equation satisfied

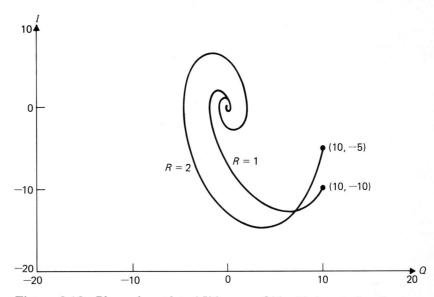

Figure 2.12 Phase plane plot of $I(t)$ versus $Q(t)$ with $L = 1$, $C = \frac{1}{4}$.

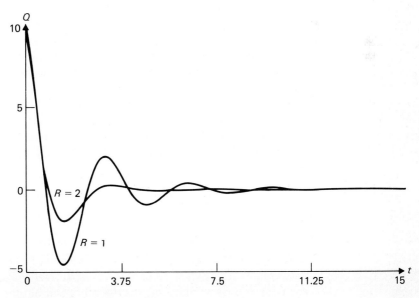

Figure 2.13 Graph of $Q(t)$ with $L = 1$, $C = \frac{1}{4}$.

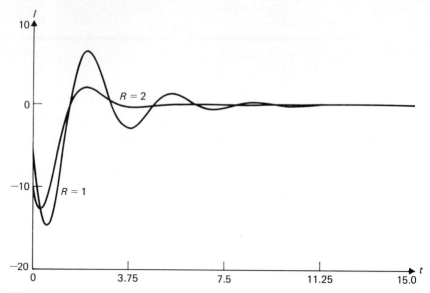

Figure 2.14 Graph of $I(t)$ with $L = 1$, $C = \frac{1}{4}$.

by the solutions to the underdamped circuit problem that demonstrates the spiral nature of the integral curves. The solutions are damped oscillations with *damping factor* Ae^{pt}. The trigonometric factor, for example, $\cos (qt - \phi)$, has period

$$T = 2\pi \left(\frac{1}{LC} - \frac{R^2}{4L^2} \right)^{-1/2}.$$

Because of the damping, the solution is not periodic. However, the time T between successive maximum values is still referred to as the *period of the solution*. In the phase plane it is the time between successive crossings of the positive Q or I axes.

If the resistance is increased so that $R^2 = 4L/C$, we have *critical damping*. A general solution can be obtained by using Eq. (2.9.1); it corresponds to the case where the characteristic polynomial has a double root p. The solution is

$$Q(t) = c_1 e^{pt} + c_2 t e^{pt},$$

$$I(t) = c_1 p e^{pt} + c_2 (pt + 1) e^{pt},$$

with $p = -R/2L$. Two typical integral curves of the system of differential equations are sketched in Fig. 2.15. The corresponding graphs of $Q(t)$ and $I(t)$ are given in Figs. 2.16 and 2.17, respectively. Observe that it is possible for $Q(t)$ to have at most one maximum or minimum value where $dQ/dt = I$

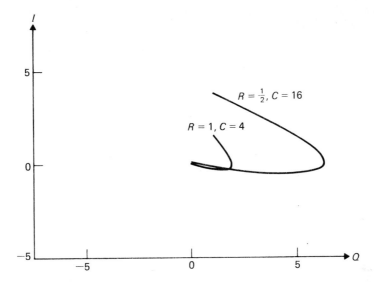

Figure 2.15 Phase plane plot of $I(t)$ versus $Q(t)$ with $L = 1$.

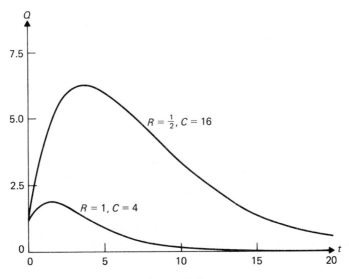

Figure 2.16 Graph of $Q(t)$ with $L = 1$.

$= 0$ before approaching zero monotonically. This can be determined directly from the expressions for the solutions.

   If the resistance is further increased, so that $R^2 > 4L/C$, then $Q(t)$ and $I(t)$ both approach zero monotonically as t increases. The circuit is said to be *overdamped.* For equation (2.9.1) this would be the case where the char-

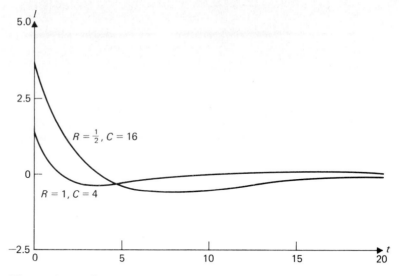

Figure 2.17 Graph of $I(t)$ with $L = 1$.

acteristic polynomial has distinct real negative roots and the solutions are

$$Q(t) = c_1 e^{p_1 t} + c_2 e^{p_2 t},$$

$$I(t) = c_1 p_1 e^{p_1 t} + c_2 p_2 e^{p_2 t}$$

with

$$p_1 = -\frac{R}{2L} - \left(\frac{R^2}{4L^2} - \frac{1}{LC}\right)^{1/2} < p_2 = -\frac{R}{2L} + \left(\frac{R^2}{4L^2} - \frac{1}{LC}\right)^{1/2} < 0.$$

Two typical integral curves are sketched in Fig. 2.18. The corresponding graphs of $Q(t)$ and $I(t)$ are given in Figs. 2.19 and 2.20, respectively.

A visual interpretation of the various types of damping is easily given if the mass–spring–dashpot system of Section 2.2 of this chapter is substituted for the LRC circuit. Therefore the displacement $y(t)$ of the mass replaces the charge $Q(t)$ and the coefficient of damping c replaces the resistance R. The motion of the mass after it is displaced and released can be described in each of the cases as follows:

No damping (c = 0): there is no dashpot present and the motion is a pure oscillation.

Underdamping (c is small): The dashpot is filled with a liquid of very small viscosity and barely resists the motion. The motion is oscillatory but damps out eventually.

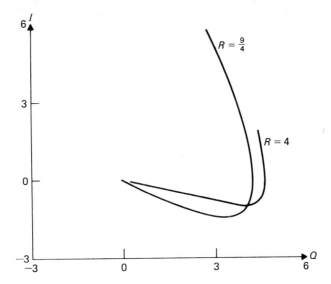

Figure 2.18   Phase plane plot of $I(t)$ versus $Q(t)$ with $L = 1$, $C = 1$.

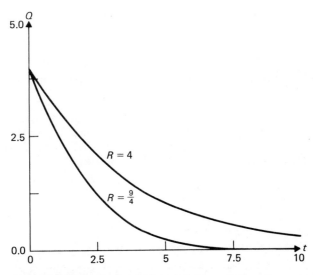

Figure 2.19   Graph of $Q(t)$ with $L = 1$, $C = 1$.

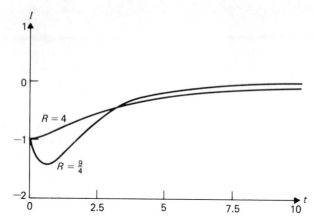

Figure 2.20 Graph of $I(t)$ with $L = 1$, $C = 1$.

Critically damping (c is bigger): the viscosity of the liquid is greater and allows the mass an initial surge but then it slowly comes to rest.

Overdamping (c is very big): the viscosity of the liquid is so great that the mass goes directly to rest with no oscillations or surges.

Any person who has traveled over bumps in the road in automobiles with shock absorbers at various states of efficiency will have direct knowledge of at least one of the above cases.

EXERCISES 2.9

Letting $I = dQ/dt$, write equivalent first order systems for the following second order differential equations and sketch their vector fields in the (Q, I) plane. Find a real general solution of the system by solving the corresponding second order equation. Find a solution such that $Q(0) = 1$, $I(0) = Q'(0) = 0$ and plot it on the same paper used for the vector field of the equation. Classify the equations as not damped, underdamped, critically damped, or overdamped.

1. $Q'' + 4Q = 0$

2. $Q'' + 9Q = 0$

3. $Q'' + Q = 0$

4. $Q'' + 4Q' + 5Q = 0$

5. $Q'' + 4Q' + 13Q = 0$

6. $Q'' + 6Q' + 13Q = 0$

7. $Q'' + 4Q' + 4Q = 0$

8. $Q'' + 6Q' + 9Q = 0$

9. $Q'' + 2Q' + Q = 0$

10. $Q'' + 5Q' + 6Q = 0$

11. $Q'' + 3Q' + 2Q = 0$

12. $Q'' + 5Q' + 4Q = 0$

13. For each of the following undamped oscillator equations find the amplitude and period of the motion and find the equation of the ellipse in the phase plane which the integral curve describes.

a) $Q'' + Q = 0$, $Q(0) = 2$, $Q'(0) = 1$

b) $9y'' + 4y = 0$, $y(0) = 4$, $y'(0) = 2$

c) $y'' + 4y = 0$, $y(0) = 0$, $y'(0) = 8$

d) $Q'' + 10^{-4}Q = 0$, $Q(0) = 1$, $Q'(0) = 1$

14. For each of the following underdamped oscillator equations find the damping factor and period of the motion. Find a time T for which

the motion has magnitude $|y(t)| < 10^{-4}$ for all $t > T$.

a) $y'' + 2y' + 2y = 0$, $y(0) = 2$, $y'(0) = 0$

b) $y'' + \frac{1}{10}y' + 2y = 0$, $y(0) = 2$, $y'(0) = -0.1$

c) $y'' + \frac{19}{5}y' + 2y = 0$, $y(0) = 2$, $y'(0) = 0$

15. For each of the following critically damped oscillator equations find a value $T \geq 0$ for which $|y(T)|$ is a maximum and find that maximum.

a) $y'' + 2y' + y = 0$, $y(0) = 1$, $y'(0) = 1$

b) $y'' + 2y' + y = 0$, $y(0) = 1$, $y'(0) = -3$

c) $y'' + \frac{1}{5}y' + \frac{1}{100}y = 0$, $y(0) = 0$, $y'(0) = 8$

d) $y'' + 200y' + 10^{-2}y = 0$, $y(0) = 2$, $y'(0) = 1$

16. For overdamped oscillator equations, finding a time $T \geq 0$ for which the solution $y(t)$ satisfies $|y(t)| < \epsilon$ when $t \geq T$ involves solving a transcendental equation.

a) Show that for oscillator equations of the form

$$y'' + (n + 1)ay' + na^2y = 0, \quad a > 0,$$

where n is a positive integer, the transcendental equation can be transformed into an algebraic one.

b) Let $\epsilon = 10^{-2}$ and find a time T for which $|y(t)| < \epsilon$ for each of the following:

i) $y'' + 3y' + 2y = 0$, $y(0) = 1$, $y'(0) = 1$

ii) $y'' + 6y' + 8y = 0$, $y(0) = 4$, $y'(0) = 0$

iii) $y'' + 4y' + 3y = 0$, $y(0) = 3$, $y'(0) = -5$

2.10 FORCED OSCILLATIONS

In this section we examine the *forced-oscillator equation* written in the form

$$m \frac{d^2y}{dt^2} + c \frac{dy}{dt} + ky = F(t) \tag{2.10.1}$$

or as an equivalent system

$$m \frac{dy}{dt} = p, \quad \frac{dp}{dt} = -ky - cp + F(t). \tag{2.10.2}$$

It is assumed that the quantities m, c, and k are positive constants. The equation is difficult to analyze because different types of forcing functions produce different types of solutions. We shall restrict our attention here to equations with periodic forcing functions, i.e., functions such that

$$F(t + T) = F(t)$$

for all t, where T, the period, is a positive number.

It can be proved that the damped-oscillator equation ($c > 0$) with a periodic forcing function has a unique periodic solution and that this solution has the same period as the period of the forcing function (3, pp. 100–101). Therefore, a general solution can be written as

$$y(t) = c_1y_1(t) + c_2y_2(t) + w(t), \tag{2.10.3}$$

where $y_1(t)$ and $y_2(t)$ are linearly independent solutions of the corresponding homogeneous differential equation and $w(t)$ is the unique periodic solution of the inhomogeneous differential equation. Since c is positive, then both $y_1(t)$ and $y_2(t)$ tend to zero as t increases and we have $y(t) \simeq w(t)$ for large t. For this reason, $c_1 y_1(t) + c_2 y_2(t)$ is called the *transient solution* and $w(t)$ the *steady state solution*. Clearly, the initial conditions, which are incorporated into the constants c_1 and c_2, will have little effect on the solution when t is large.

In order to exhibit the dependence of the solution on the period of the forcing function, we make the further restriction that $F(t)$ be a finite sum of sinusoidal functions:

$$F(t) = P_0 + \sum_{n=1}^{N} [P_n \cos(n\omega t) + R_n \sin(n\omega t)]. \qquad (2.10.4)$$

The period of $F(t)$ is $T = 2\pi/\omega$. To verify this we notice that

$$\cos\left[n\omega\left(t + \frac{2\pi}{\omega}\right)\right] = \cos(n\omega t + n2\pi) = \cos(n\omega t),$$

$$\sin\left[n\omega\left(t + \frac{2\pi}{\omega}\right)\right] = \sin(n\omega t + n2\pi) = \sin(n\omega t),$$

which implies that $F(t + 2\pi/\omega) = F(t)$ for all t. The unique periodic solution $w(t)$ can now be found by the method of undetermined coefficients.

Let us first solve the case where $P_0 = 0$ and $N = 1$:

$$m\frac{d^2 y}{dt^2} + c\frac{dy}{dt} + ky = P_1 \cos \omega t + R_1 \sin \omega t. \qquad (2.10.5)$$

Set $w_1(t) = A_1 \cos \omega t + B_1 \sin \omega t$ as the trial solution, and after substituting this for y in (2.10.5) one obtains the following equations for A_1 and B_1:

$$(k - m\omega^2)A_1 + c\omega B_1 = P_1, \qquad -c\omega A_1 + (k - m\omega^2)B_1 = R_1.$$

Hence,

$$A_1 = \frac{P_1(k - m\omega^2) - R_1 c\omega}{\Delta}, \qquad B_1 = \frac{R_1(k - m\omega^2) + P_1 c\omega}{\Delta},$$

where $\Delta = (k - m\omega^2)^2 + c^2\omega^2$, and the solution is

$$w_1(t) = \frac{P_1}{\Delta}[(k - m\omega^2)\cos \omega t + c\omega \sin \omega t]$$

$$+ \frac{R_1}{\Delta}[-c\omega \cos \omega t + (k - m\omega^2)\sin \omega t].$$

Using the trigonometric identities

$$\cos Q \cos \omega t + \sin Q \sin \omega t = \cos(\omega t - Q),$$

$$-\sin Q \cos \omega t + \cos Q \sin \omega t = \sin(\omega t + Q),$$

and the fact that $c\omega$, $|k - m\omega^2|$, and $\sqrt{\Delta}$ form the two sides and hypotenuse, respectively, of a right triangle, we can write the solution in the more convenient amplitude–phase angle form:

$$w_1(t) = K(\omega)\{P_1 \cos[\omega t - Q(\omega)] + R_1 \sin[\omega t - Q(\omega)]\}, \quad (2.10.6)$$

where

$$Q(\omega) = \arctan\left(\frac{c\omega}{k - m\omega^2}\right) \qquad (2.10.7)$$

is the *phase displacement* and

$$K(\omega) = [(k - m\omega^2)^2 + c^2\omega^2]^{-1/2} \qquad (2.10.8)$$

is the *amplification factor*. We see that the solution is the original forcing term shifted in time by $Q(\omega)$ and amplified by $K(\omega)$.

The solutions corresponding to the *higher harmonics* of the forcing function, $P_n \cos(n\omega t) + R_n \sin(n\omega t)$, are found by replacing ω by $n\omega$ and P_1, R_1 by P_n, R_n in (2.10.5) and (2.10.6):

$$w_n(t) = K(n\omega)\{P_n \cos[n\omega t - Q(n\omega)] + R_n \sin[n\omega t - Q(n\omega)]\}.$$

The constant term in the forcing function produces a constant solution, $w_0(t) = P_0/k = w_0$. Because of the linearity of the differential equation, the periodic solution corresponding to the periodic forcing solution $F(t)$ is the sum of the $w_n(t)$:

$$w(t) = w_0 + \sum_{n=1}^{N} w_n(t)$$

or

$$w(t) = \frac{P_0}{k} + \sum_{n=1}^{N} K(n\omega)\{P_n \cos[n\omega t - Q(n\omega)]$$

$$+ R_n \sin[n\omega t - Q(n\omega)]\}. \qquad (2.10.9)$$

The amplification factor $K(\omega)$ plays a key role in the analysis of the input–output relationship of oscillator equations. The phase displacement $Q(\omega)$ is of secondary importance in applications. As is seen from Eq. (2.10.9), the amplitude of the nth harmonic of the input function (2.10.4) is multiplied by $K(n\omega)$. The function $K(\omega)$ depends on m, c, and k, the parameters of the oscillator equation, and its graph gives important information on how

$w(t)$, the periodic response of the system, depends on the forcing function $F(t)$. However, a collection of graphs of $K(\omega)$, also called *resonance curves*, would be overwhelming for many choices of the three parameters. This can be avoided by a preliminary study of the amplification factor as a function of the four variables ω, m, c, and k.

First, let us determine $\overline{\omega}$, the *maximum response frequency*, or *resonance frequency*. It is the value of ω at which the amplification factor $K(\omega)$ takes on its maximum. Knowing it would enable us to solve the problem of choosing the input frequency so as to maximize the amplitude of the output. From the expression

$$K(\omega) = [(k - m\omega^2)^2 + c^2\omega^2]^{-1/2} = [m^2\omega^4 - (2km - c^2)\omega^2 + k^2]^{-1/2}$$

it follows that

$$K'(\omega) = -K^3(\omega)[2m^2\omega^3 - (2km - c^2)\omega].$$

Then $K'(\omega) = 0$ implies that

$$\overline{\omega}^2 = \frac{2km - c^2}{2m^2} > 0$$

is the condition for $K(\omega)$ to have a maximum for positive ω. If $2km - c^2 \leq 0$ or equivalently, $c \geq \sqrt{2km}$, the expression for $K'(\omega)$ is negative, hence $K(\omega)$ is a strictly decreasing function of ω. Sketches of several resonance curves are shown in Fig. 2.21.

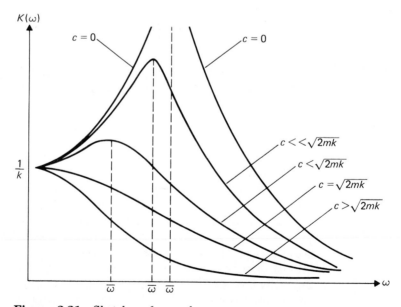

Figure 2.21 Sketches of several resonance curves.

Next set $\omega_0 = (k/m)^{1/2}$, the frequency of the undamped oscillator, and write

$$K(\omega) = \left[m^2 \left(\omega^4 - 2\frac{2km - c^2}{2m^2}\omega^2 + \frac{k^2}{m^2} \right) \right]^{-1/2}$$
$$= m^{-1}[\omega^4 - 2\bar{\omega}^2\omega^2 + \omega_0^4]^{-1/2}.$$

We started with three parameters m, c, and k and still have three parameters m, $\bar{\omega}$, and ω_0. Has anything been gained? The answer depends on the use of the physical system, e.g., the oscillator can serve as a simplified model for microphone diaphragms, seismographs, and oscillator circuits in audio amplifiers. If the maximum response frequency $\bar{\omega}$ is fixed, then ω_0 and m can be selected to reduce distortion for a range of frequencies centered at $\bar{\omega}$ and simultaneously to maximize the sensitivity of the device. It may also be advantageous to fix ω_0 and study $K(\omega)$ for different values of ω, to insure, for instance, that the system will not be damaged at or near the maximum response frequency.

Finally, we wish to discuss briefly the undamped case when $c = 0$. Again consider the situation where $P_0 = 0$ and $N = 1$:

$$m\frac{d^2y}{dt^2} + ky = P_1 \cos \omega t + R_1 \sin \omega t.$$

The analysis is the same as before and (2.10.8) implies that the amplification factor is

$$K(\omega) = [(k - m\omega^2)^2]^{-1/2} = |k - m\omega^2|^{-1}.$$

But $K(\omega)$ is not defined for $\omega = (k/m)^{-1/2} = \omega_0$, the frequency of the undamped oscillator, which is in keeping with the fact that when $\omega = \omega_0$, the forcing function is a solution of the homogeneous equation. In the case where $\omega = \omega_0$, the trial solution must be of the form

$$\omega_1(t) = A_1 t \cos \omega_0 t + B_1 t \sin \omega_0 t, \qquad \omega_0 = \sqrt{k/m},$$

and this is the case of *pure resonance*, where solutions will be oscillatory but their amplitudes become unbounded as t increases.

In a realistic physical situation, the probability that the input frequency ω will be exactly ω_0 is zero, but the problem is that for ω in a neighborhood of ω_0 the amplification factor $K(\omega)$ will be very large, resulting in oscillations of huge amplitude. This may or may not explain the apocryphal tales of certain famous singers shattering wine glasses with high C notes, but it is a serious factor in the design of towers or bridges which are subject to sustained strong winds. Movies taken of the collapse of the Tacoma Narrows Bridge in 1940 due to gale force winds are a vivid example of the destructive force of near-resonance excitations.

EXERCISES 2.10

Find the steady state solution, the transient solution, the phase displacement, and the amplitude factor for the following differential equations:

1. $y'' + 4y' + 13y = 6 \cos 2t - 5 \sin 2t$

2. $y'' + 3y' + 2y = \cos 3t + 2 \sin 3t$

3. $y'' + 4y' + 5y = 7 \cos t + 8 \sin t$

4. $4y'' + 12y' + 9y = 3 \cos t$

5. $3y'' + 13y' + 4y = 4 \cos 2t - \sin 2t$

6. $y'' + cy' + y = A \cos t$, $c = 2, 0.2, 0.002$

7. For the following undamped forced oscillators find the amplitude factor and the phase displacement.

   a) $3y'' + 12y = 5 \cos 3t$

   b) $4y'' + 36y = 3 \sin 2t$

   c) $y'' + \pi^2 y = 2 \sin \frac{22}{7}t$

   d) $y'' + y = A \cos(1 + \epsilon)t$, $\epsilon = 1, 0.1, 0.001$

8. Graph the resonance curves for the forced-oscillator equation and in the first two cases find the maximum response frequency:

$$2y'' + cy' + 9y = A \cos \omega t + B \sin \omega t$$

with $c = 1, 5, 8$.

9. Find an *approximate* solution of the form

$$y(t) = A \sin \epsilon t \sin 2t$$

to the initial value problem

$$y'' + (2 + \epsilon)^2 y = \epsilon F \cos 2t, \quad y(0) = 0, y'(0) = 0.$$

Hint: Find the exact solution and use the trigonometric identity

$$\cos(2 + \epsilon)t = \cos 2t \cos \epsilon t - \sin 2t \sin \epsilon t$$

and the approximation $\cos \epsilon t \simeq 1$. The term $A \sin \epsilon t$ can be thought of as a slowly varying amplitude for the vibration $\sin 2t$. This illustrates the *phenomenon of beats*. Take $A = 2$, $\epsilon = \frac{1}{16}$ and sketch the approximate solution on the interval $0 \le t \le 32\pi$.

10. For the undamped forced oscillator in the case of pure resonance

$$y'' + \omega^2 y = P_1 \cos \omega t + R_1 \sin \omega t$$

show that the particular solution is

$$\omega(t) = \frac{t}{2\omega}\left[P_1 \cos\left(\omega t - \frac{\pi}{2}\right) \right.$$
$$\left. + R_1 \sin\left(\omega t - \frac{\pi}{2}\right) \right].$$

11. A system is described by the forced-oscillator equation with initial data as

$$y'' + 4y = \frac{8}{\pi} \cos 2t,$$

$$y(0) = 0, \quad y'(0) = 0.$$

It becomes overloaded when $|y(t)| \ge 12$. Show that this will first occur somewhere in the interval $6\pi < t < 25\pi/4$.

2.11 ELEMENTARY NUMERICAL METHODS FOR SECOND ORDER EQUATIONS

The writing of a second order differential equation as a pair of coupled first order differential equations permits the application of numerical methods designed for single first order equations to second order equations. This is done by extending these methods to general first order systems and then specializing them to first order systems that are equivalent to second-order equations.

The differential equation

$$a(t) \frac{d^2y}{dt^2} + b(t) \frac{dy}{dt} + c(t)y = q(t) \qquad (2.11.1)$$

with initial data $y(t_0) = r$, $dy(t_0)/dt = s$ can be written as a first order system by introducing a new variable, $z = dy/dt$. Then, since $dz/dt = d^2y/dt^2$, the system

$$\frac{dy}{dt} = z, \qquad \frac{dz}{dt} = \frac{-c(t)y - b(t)z + q(t)}{a(t)} \qquad (2.11.2)$$

with initial data $y(t_0) = r$, $z(t_0) = s$ is equivalent to (2.11.1). For instance,

$$y'' + 4y' + 3y = 4 \sin 2t, \qquad y(0) = 1, \qquad y'(0) = 1$$

is equivalent to

$$y' = z, \qquad y(0) = 1,$$
$$z' = -3y - 4z + 4 \sin 2t, \qquad z(0) = -1,$$

and

$$t^2 y'' - 2y = e^{-t}, \qquad y(1) = 0, \qquad y'(1) = 4$$

is equivalent to

$$y' = z, \qquad y(1) = 0,$$

$$z' = \frac{2y + e^{-t}}{t^2} = \frac{2}{t^2} y + \frac{e^{-t}}{t^2}, \qquad z(1) = 4,$$

as the reader can easily verify.

The system (2.11.2) is a special case of the general pair of coupled first order equations

$$\frac{dy}{dt} = F(y, z, t), \qquad \frac{dz}{dt} = G(y, z, t) \qquad (2.11.3)$$

with initial data $y(t_0) = r$, $z(t_0) = s$. Assuming that $F(y, z, t)$, $G(y, z, t)$ are continuous functions defined on a domain in yzt space, we now extend the previously developed numerical algorithms to this system. These will then be applied to the system of linear differential equations (2.11.2).

We can associate with each equation in (2.11.3) a direction field, one in the ty plane and another in the tz plane. These direction fields are not independent and are coupled in general through the two differential equations. The simplest numerical method for obtaining an approximate solution to the system (2.11.3) is Euler's method which produces a solution by following, in discrete steps, the direction fields in the two planes. Since $F(y, z, t)$ and $G(y, z, t)$ are the slopes of a solution pair $y(t)$ and $z(t)$, Euler's method for a single equation extends to the algorithm

$$y_{i+1} = y_i + hF(y_i, z_i, t_i),$$

$$z_{i+1} = z_i + hG(y_i, z_i, t_i),$$

$$t_{i+1} = t_i + h, \quad i = 0, 1, 2, \ldots, n-1,$$

with the initial data $y_0 = y(t_0)$, $z_0 = z(t_0)$. For the linear system (2.11.2),

$$F(y, z, t) = z,$$

$$G(y, z, t) = \frac{-c(t)y - b(t)z + q(t)}{a(t)}. \tag{2.11.4}$$

This is summarized in the following algorithms.

The Euler algorithm for linear second order differential equations. Given the initial value problem

$$a(t)y'' + b(t)y' + c(t)y = q(t), \quad y(t_0) = y_0, \quad y'(t_0) = z_0,$$

write it as a system

$$y' = z, \quad z' = G(y, z, t), \quad y(t_0) = y_0, \quad z(t_0) = z_0, \tag{2.11.5}$$

with

$$G(y, z, t) = \frac{-c(t)y - b(t)z + q(t)}{a(t)}.$$

An approximate solution to the initial value problem (2.11.5) at the equally spaced points $t_0, t_1, \ldots, t_n$ is given by

$$y_{i+1} = y_i + hz_i,$$

$$z_{i+1} = z_i + hG(y_i, z_i, t_i), \tag{2.11.6}$$

$$t_{i+1} = t_i + h, \quad i = 0, 1, 2, \ldots, n-1,$$

where h, t_0, y_0, and z_0 are given.

Example 1   Use the Euler algorithm with $h = 0.1$ to find an approximate solution to the initial value problem

$$y'' + 2y' + y = -2t, \quad y(0) = 4, \quad y'(0) = 0,$$

at the points $t_i = ih$, where $i = 0, 1, \ldots, 10$.

TABLE 2.2.

t_i	y_i	z_i	t_i	y_i	z_i
0.0	4.0000	0.0000	0.6	3.5086	-1.6457
0.1	4.0000	-0.4000	0.7	3.3440	-1.7874
0.2	3.9600	-0.7400	0.8	3.1653	-1.9044
0.3	3.8860	-1.0280	0.9	2.9748	-2.0000
0.4	3.7832	-1.2710	1.0	2.7748	-2.0775
0.5	3.6561	-1.4751			

Set $y' = z$, $z' = -y - 2z - 2t$, $y(0) = 4$, $z(0) = 0$. Then

$$G(y, z, t) = -(y + 2z + 2t),$$

and (2.11.6) becomes

$$y_{i+1} = y_i + hz_i,$$

$$z_{i+1} = z_i - h(y_i + 2z_i + 2t_i),$$

$$t_{i+1} = t_i + h.$$

Starting with $t_0 = 0$, $y_0 = 4$, $z_0 = 0$, and using $h = 0.1$, we obtained Table 2.2. The solution of the initial value problem is easily found to be

$$y(t) = 2te^{-t} + 4 - 2t, \qquad y'(t) = z(t) = 2(1 - t)e^{-t} - 2,$$

so $y(1) = 2.73576$, $z(1) = -2$. $\square$

It is also easy to extend the improved Euler method. For the general system (2.11.3), set the Euler approximation

$$\tilde{y}_i = y_i + hF(y_i, z_i, t_i),$$

$$\tilde{z}_i = z_i + hG(y_i, z_i, t_i),$$

and define

$$y_{i+1} = y_i + \frac{h}{2} [F(y_i, z_i, t_i) + F(\tilde{y}_i, \tilde{z}_i, t_i + h)],$$

$$z_{i+1} = z_i + \frac{h}{2} [G(y_i, z_i, t_i) + G(\tilde{y}_i, \tilde{z}_i, t_i + h)],$$

$$t_{i+1} = t_i + h, \quad i = 0, 1, 2, \ldots, n - 1,$$

with the initial data $y_0 = y(t_0)$, $z_0 = z(t_0)$. As previously, $F(y, z, t)$ and $G(y, z, t)$ are given by (2.11.4) for the linear system (2.11.2). This is summarized in the following algorithm.

The improved Euler algorithm for linear second order differential equations. Given the initial value problem

$$a(t)y'' + b(t)y' + c(t)y = q(t),$$
$$y(t_0) = y_0, \quad y'(t_0) = z_0,$$

write it as a system

$$y' = z, \quad z' = G(y, z, t), \quad y(t_0) = y_0, \quad z(t_0) = z_0, \quad (2.11.7)$$

with

$$G(y, z, t) = \frac{-c(t)y - b(t)z + q(t)}{a(t)}.$$

An approximate solution to the initial-value problem (2.11.7) at the equally spaced points $t_0, t_1, \ldots, t_n$ is given by

$$\hat{y}_i = y_i + hz_i, \quad \hat{z}_i = z_i + hG(y_i, z_i, t_i),$$

$$y_{i+1} = y_i + \frac{h}{2}(z_i + \hat{z}_i),$$

$$(2.11.8)$$

$$z_{i+1} = z_i + \frac{h}{2}[G(y_i, z_i, t_i) + G(\hat{y}_i, \hat{z}_i, t_i + h)],$$

$$t_{i+1} = t_i + h, \quad i = 0, 1, 2, \ldots, n-1,$$

where h, t_0, y_0, and z_0 are given.

Example 2 Use the improved Euler algorithm with $h = 0.1$ to find an approximate solution to the initial value problem of Example 1.

With $G(y, z, t) = -(y + 2z + 2t)$, algorithm (2.11.8) becomes

$$\hat{y}_i = y_i + hz_i, \quad \hat{z}_i = z_i - h(y_i + 2z_i + 2t_i),$$

$$y_{i+1} = y_i + \frac{h}{2}(z_i + \hat{z}_i),$$

$$z_{i+1} = z_i - \frac{h}{2}[(y_i + 2z_i + 2t_i) + (\hat{y}_i + 2\hat{z}_i + 2t_i + 2h)],$$

$$t_{t+1} = t_i + h.$$

Table 2.3 is obtained by letting $h = 0.1$ and starting with $y_0 = 4$, $z_0 = 0$. □

TABLE 2.3

t_i	y_i	z_i
0.	4.0000	0.
0.1	3.9800	-0.3700
0.2	3.9258	-0.7062
0.3	3.8406	-1.0102
0.4	3.7275	-1.2835
0.5	3.5893	-1.5277
0.6	3.4289	-1.7445
0.7	3.2487	-1.9356
0.8	3.0513	-2.1027
0.9	2.8388	-2.2474
1.0	2.6133	-2.3715

The sample program given in Chapter 1 for the Euler and the improved Euler algorithms for first order equations can be easily modified to compute solutions of second order linear equations. See Exercise 2 below for a Fortran program to do the improved Euler method.

EXERCISES 2.11

1. Apply the Euler algorithm to find an approximate solution to the following initial value problems at the points $t = 0.25, 0.50, 0.75, 1.00$ with $h = \frac{1}{4}$. Do your calculations by hand and carry four significant digits in all calculations. In each case find the exact solution and compare it with your numerical result.

a) $y'' + 3y' + 2y = 1$, $y(0) = 0$, $y'(0) = 1$

b) $y'' - y' - 2y = t^2$, $y(0) = 1$, $y'(0) = 0$

c) $y'' - y' - 2y = t + 1$, $y(0) = 0$, $y'(0) = 0$

d) $y'' + 4y = e^{-t}$, $y(0) = 0$, $y'(0) = 0$

2. Repeat Exercise 1 using the improved Euler algorithm.

The Fortran program given in Fig. 2.12 implements the improved Euler algorithm in the text to approximate the solution of the initial value problem:

$$y'' - ty = 0, \qquad y(0) = 0.35503, \qquad y'(0) = 1,$$

at $t = 0.1, 0.2, \ldots, 1.0$ by using the step size $h = 0.1$. The output follows the program given in Fig. 2.12.

3. Using the program of Fig. 2.12 find approximate solutions of the following initial value problems at $t = 0.1, 0.2, \ldots, 1.0$ with a step size of $h = 0.1$.

a) $y'' - y' = \sin t$; $y(0) = 1$, $y'(0) = 0$

b) $y'' + y = 4 \cos t$; $y(0) = 1$, $y'(0) = 1$

c) $y'' + 3y' - 4y = e^{-4t} + te^{-t}$; $y(0) = 0$, $y'(0) = 0$

d) $y'' + 3y' + 2y = \dfrac{1}{e^t + 1}$; $y(0) = 0$, $y'(0) = 1$

```
C    PROGRAM USES THE IMPROVED EULER METHOD TO SOLVE THE INITIAL
C    VALUE PROBLEM
C
C       A(T)*Y"(T) + B(T)*Y'(T) + C(T)*Y(T) = Q(T)
C       Y(T0) = Y0, Y'(TO) = ZO
C
C    DEFINE FUNCTIONS A(T), B(T), C(T), Q(T), G(Y,Z,T)
C    AND INITIAL CONDITIONS
C
        A(T) = 1.0
        B(T) = 0.0
        C(T) = − T
        Q(T) = 0.0
        G(Y,Z,T) = (−C(T)*Y−B(T)*Z+Q(T))/A(T)
        T0 = 0.0
        Y0 = 0.35503
        Z0 = 1.0
C
C    SET FINAL TIME AND NUMBER OF STEPS TO BE TAKEN
C
        TF = 1.0
        N = 10
C
C    BASIC LOOP
C
        T = T0
        Y = Y0
        Z = Z0
        H = (TF−T0)/FLOAT(N)
        WRITE(6,1)T,Y,Z
   1    FORMAT(1X,F6.2,2F10.5)
        DO 2  I = 1,N
           YBAR = Y+H*Z
           GYZ = G(Y,Z,T)
           ZBAR = Z+H*GYZ
           Y = Y+H*(Z+ZBAR)/2.0
           Z = Z+H*(GYZ+G(YBAR,ZBAR,T+H))/2.0
           T = T+H
           WRITE(6,1)T,Y,Z
   2    CONTINUE
        STOP
        END
```

Figure 2.12

0.	0.35503	1.00000
0.10	0.45503	1.00227
0.20	0.55548	1.01010
0.30	0.65705	1.02550
0.40	0.76059	1.05055
0.50	0.86716	1.08740
0.60	0.97807	1.13836
0.70	1.09484	1.20592
0.80	1.21926	1.29285
0.90	1.35343	1.40231
1.00	1.49975	1.53789

Figure 2.12 (continued)

e) $y'' + 4ty' + (4t^2 + 2)y = 0;$ $y(0) = 0,$ $y'(0) = 1$ (the solution here is $y(t) = te^{-t^2}$)

4. Modify the program of Fig. 2.12 to print out every pth step, where p is a positive integer. In applications one may not be interested in knowing the approximate solution at every step h; furthermore, if h is small and $t_F - t_0$ is large, the amount of output could be voluminous.

5. Use the program of Exercise 4 to tabulate the approximate solution of the initial value problem

$$y'' + ty = 0, \qquad y(0) = 0.35503, \qquad y'(0) = 1,$$

at $t = 0.1, 0.2, \ldots, 3.0$. Estimate the first zero of $y(t)$ (i.e., the first value of t, where $y(t) = 0$) from your output. Discuss what procedure you would follow if you wanted to estimate the first zero to an accuracy of two or three decimal places.

MISCELLANEOUS EXERCISES

2.1 Show that the general solution of

$$\epsilon y'' + y' + y = 0, \quad 0 < \epsilon \ll 1, \quad t > 0,$$

will approach the general solution of $y' + y = 0$ as $\epsilon \to 0$. Will this occur for $t < 0$? *Hint:* In your analysis you may find the expansion

$$(1 - u)^{1/2} = 1 - \tfrac{1}{2}u - \tfrac{1}{8}u^2 + \cdots, \quad |u| < 1$$

useful.

2.2 The general solution of $y'' + 3y' + 2y = f(t)$ is of the form

$$y(t) = Ae^{-t} + Be^{-2t} + y_p(t),$$

where $y_p(t)$ is obtained from the variation of parameters formula. If $f(t)$ is bounded for $t \geq 0$ (say, $|f(t)| \leq M$), show that

$$|y(t)| \leq |A| + |B| + \tfrac{3}{2}M, \quad t \geq 0.$$

2.3 Generalize the result of the previous problem to the case

$$y'' + ay' + by = f(t),$$

where the associated characteristic polynomial has roots with negative real parts. Specifically, show that if $f(t)$ is bounded for $t \geq 0$, then so is the general solution. This is an example of the engineering dictum "bounded input gives bounded output."

2.4 Given the homogeneous equation

$$y'' + a(t)y = 0$$

and one nontrivial solution $y_1(t)$, show that

$$y_2(t) = y_1(t) \int \frac{1}{[y_1(t)]^2} \, dt$$

is also a solution. Furthermore, show that $y_1(t)$ and $y_2(t)$ are a pair of linearly independent solutions. (This is a special case of the method of *reduction of order* to be discussed in Chapter 6.)

2.5 Use the result of the previous problem and the one given solution to find the general solution of the following differential equations:

a) $y'' + 9y = 0$, $\quad y_1(t) = \cos 3t$

b) $y'' + \frac{1}{4}t^{-2} y = 0$, $\quad y_1(t) = t^{1/2}$

c) $y'' - (2 \sec^2 t)y = 0$, $\quad y_1(t) = \tan t$

2.6 a) Show that the general second order equation $y'' + a(t)y' + b(t)y = 0$ becomes the first order Riccati equation $du/dt = u^2 - a(t)u + b(t)$ under the transformation $y(t) = \exp[-\int u(t) \, dt]$.

b) Conversely, show that the Riccati equation $du/dt = u^2 + p(t)u + q(t)$ becomes the second order equation $y'' - p(t)y' + q(t)y = 0$ under the transformation $u(t) = -y'(t)/y(t)$.

2.7 Using the results of the previous exercise, solve the following Riccati equations by converting them to second order linear equations:

a) $\dfrac{du}{dt} = -u^2 + 2u - 2$ $\qquad$ b) $\dfrac{du}{dt} = u^2 + 1$ $\qquad$ c) $\dfrac{du}{dt} = u^2 - \dfrac{3}{t}u + \dfrac{1}{t^2}$

2.8 Using the results of Exercise 2.6, find a solution of the following second order linear equations by converting them to Riccati equations and solving:

a) $y'' + (\cos^2 t + \cos t - 1)y = 0$

b) $y'' + \left(\dfrac{3}{t} - 1\right)y' - \dfrac{2}{t}y = 0$

2.9 Given a system governed by the equation $y'' + 2by' + y = \sin 2t$, where $t > 0$, $b > 0$, determine the smallest value of b which insures that the maximum amplitude of the steady-state solution is less than $\frac{1}{4}$.

2.10 Using the idea developed in Exercise 4 of Section 2.8 show that

a) The general solution of $t^2 y'' + t y' + y = 4e^t$ is

$$y(t) = A \sin(\ln t) + B \cos(\ln t) + 4 \sum_{n=0}^{\infty} \frac{t^n}{n!(n^2 + 1)};$$

b) The general solution of $t^2y'' - 5ty' + 9y = 3 \sin t$ is

$$y(t) = At^3 + Bt^3 \ln t + \frac{3}{4} t - \frac{1}{2} t^3 (\ln t)^2 + \frac{3}{4} \sum_{n=2}^{\infty} \frac{(-1)^n t^{2n+1}}{(2n+1)!(n-1)^2}.$$

2.11 The following method uses second order linear equations to solve systems of first order linear equations. Given

$$\frac{dx}{dt} = ax + by, \qquad x(0) = A,$$

$$\frac{dy}{dt} = cx + dy, \qquad y(0) = B,$$

differentiate the first equation and then successively substitute for y and dy/dt:

$$\frac{d^2x}{dt^2} = a\frac{dx}{dt} + b\frac{dy}{dt} = a\frac{dx}{dt} + b(cx + dy)$$

$$= a\frac{dx}{dt} + bcx + bd\left(\frac{1}{b}\frac{dx}{dt} - \frac{a}{b}x\right).$$

We obtain a second order equation in x whose initial conditions are

$$x(0) = A, \qquad x'(0) = ax(0) + by(0) = aA + bB.$$

Find $x(t)$, then substitute in the first differential equation to find $y(t)$. Use this procedure to find solutions $x(t)$, $y(t)$ of the following:

a) $\dfrac{dx}{dt} = 2x + 3y, \quad x(0) = 0; \quad \dfrac{dy}{dt} = 3x + 4y, \quad y(0) = 4$

b) $\dfrac{dx}{dt} = x - 3y, \quad x(0) = 2; \quad \dfrac{dy}{dt} = x - y, \quad y(0) = -1$

Higher order constant coefficient equations of the form

$$y^{(n)} + a_1y^{(n-1)} + \cdots + a_{n-1}y' + a_ny = 0$$

can be solved by following the same procedure used to solve second order equations. Substitution of $y(t) = e^{\lambda t}$ gives the characteristic equation

$$\lambda^n + a_1\lambda^{n-1} + \cdots + a_{n-1}\lambda + a_n = 0,$$

and its roots will determine the values of λ. A simple root $\lambda = \lambda_i$ will give a solution $y_i(t) = e^{\lambda_i t}$, a root $\lambda = \lambda_i$ of multiplicity k will give the k solutions $y_j(t) = t^j e^{\lambda_i t}$, $j = 0, 1, \ldots, k - 1$, and a pair of complex conjugate roots $\lambda = u \pm iv$ will give the two real solutions $e^{ut} \cos vt$ and $e^{ut} \sin vt$. If the complex root has multiplicity k, each of the above solutions will be multiplied by t^j, $j = 0, 1, \ldots, k - 1$.

Since a polynomial of degree n has n roots (real or complex) counting multiplicities, the above procedure will give n linearly independent solutions, and any solution will be a linear combination of these. A general solution will therefore contain n arbitrary constants and these can be determined if n initial conditions are given.

2.12 Find the general solution of the following equations:

a) $y^{(4)} - y'' = 0$ b) $y^{(3)} + y' = 0$

c) $y^{(4)} + 8y'' + 16y = 0$ d) $y^{(3)} - 2y'' - y' + y = 0$

e) $y^{(3)} + y'' + \frac{1}{3}y' + \frac{1}{27}y = 0$ f) $y^{(3)} - 8y'' + 21y' - 18y = 0$

2.13 Find the solution of the following initial value problems:

a) $y^{(3)} - y' = 0$, $\quad y(0) = 4$, $\quad y'(0) = 0$, $\quad y''(0) = 2$

b) $y^{(4)} - y = 0$, $\quad y(0) = y'(0) = 0$, $\quad y''(0) = 2$, $\quad y^{(3)}(0) = -6$

c) $y^{(3)} + y'' + 4y' + 4y = 0$, $\quad y(0) = y'(0) = 1$, $\quad y''(0) = 2$

d) $y^{(4)} + 2y'' + y = 0$, $\quad y(0) = y''(0) = 1$, $\quad y'(0) = y^{(3)}(0) = 2$

e) $y^{(4)} + 5y'' + 4y = 0$, $\quad y(0) = y^{(3)}(0) = 0$, $\quad y'(0) = y''(0) = 3$

The method of comparison of coefficients can be used to find a particular solution of

$$y^{(n)} + a_1 y^{(n-1)} + \cdots + a_{n-1} y' + a_n y = P(t),$$

where $P(t)$ is a polynomial, an exponential, a sine or cosine, or a product of these. The trial solution with undetermined coefficients is selected in the same manner as for the second order equation, and if it is a solution of the homogeneous equation, it must be multiplied by a high enough power of t, so this is no longer the case.

2.14 Find a particular solution of the following equations:

a) $y^{(4)} + y = 3e^{2t}$ b) $y^{(3)} + 2y' = 2 \sin t$

c) $y^{(4)} - y'' = 6t + 1$ d) $y^{(3)} + y'' - y' - y = 8e^{-t}$

e) $y^{(3)} - 4y' = \cos^2 t$ f) $y^{(3)} + y = 4 \cosh t$

REFERENCES

In Chapter 1 we gave some references to introductory texts in ordinary differential equations. Additional references with an emphasis on applications and/or some historical background are

1. P. D. Ritger and N. D. Rose, *Differential Equations with Applications,* McGraw Hill, New York, 1968.

2. G. F. Simmons, *Differential Equations with Applications and Historical Notes,* McGraw Hill, New York, 1972.

A more theoretically inclined treatment is given in

3. O. Plaat, *Ordinary Differential Equations,* Holden Day, San Francisco, 1971.

For a discussion of the broader aspects of mathematics applied to the physical world, the reader might enjoy

4. R. Haberman, *Mathematical Models,* Prentice-Hall, Englewood Cliffs, 1977.

5. B. Noble, *Applications of Undergraduate Mathematics in Engineering,* Macmillan, New York, 1967.

6. G. Polya, *Mathematical Methods in Science,* Mathematical Association of America, 1977.

Elementary Numerical Methods

3.1 INTRODUCTION

Why do we need numerical methods? It should be clear from what has been done so far that only a few differential equations can be solved completely in terms of elementary functions. For example, in Chapter 1 we found that first order equations are usually solvable if they are linear and sometimes solvable if they are not (e.g., separable or exact equations). Equations of order two and higher are generally solvable only if they are linear with constant coefficients. In fact, very few differential equations that arise in applications can be solved exactly, and so one must resort to numerical or approximate methods to find solutions.

3.2 A GENERAL ONE-STEP METHOD

In Chapter 1 two simple numerical methods were introduced: the Euler and the improved Euler algorithms. These methods are adequate only for very simple problems. In this chapter methods will be developed that apply to a wider class of problems and that lay the groundwork for the highly efficient computer program RKF45, which will be discussed in Chapter 8. Before we begin, however, let us briefly review the notion of a numerical solution and introduce some of the notation and definitions that will be used.

To start with, consider the first order initial value problem

$$y' = f(t, y), \quad \alpha \le t \le \beta, \tag{3.2.1}$$

$$y(a) = A, \quad \alpha < a < \beta. \tag{3.2.2}$$

The methods to be developed for solving the first order problem (3.2.1, 3.2.2) can easily be extended to systems of first order differential equations and to higher order differential equations. The reader has already seen an example of this in Chapter 2, where the Euler and improved Euler methods were extended to second order differential equations. These methods are called step-by-step methods and they generate a sequence of t-points, $t_1, t_2, \ldots$, at which the solution to the differential equation is approximated. The *approximate* solution at the point $t = t_i$, denoted by y_i, is computed from approximations at prior t-points.

If k previous values $y_{i-1}, \ldots, y_{i-k}$ are used to compute y_i, a numerical method is called a *k-step method*. If $k = 1$, it is called a *single-step, or one-step, method*; if $k > 1$, a *multistep method*.

In this book we shall concentrate on one-step methods. As in previous sections, the quantity $y(t_i)$ will always be used to denote the *exact* solution to the problem (3.2.1, 3.2.2) at the point $t = t_i$, and y_i will denote an approximation to $y(t_i)$.

Assume that the t-points are equally spaced in the interval $a \le t \le b$, where $\alpha \le a$, $b \le \beta$. Therefore, $t_i = a + ih$, $i = 0, 1, \ldots, n$ for some positive integer n, and $h = (b - a)/n$ is a positive number. For some problems h could be negative ($h = (a - b)/n$), as in the case when one is solving for the initial point of an integral curve given the terminal point, i.e., integrating backwards. The general one-step method for solving (3.2.1, 3.2.2) can be written in the form

$$y_{k+1} = y_k + h\theta(t_k, y_k) \tag{3.2.3}$$

with

$$y_0 = A. \tag{3.2.4}$$

Referring to Section 1.6, one sees that Euler's method has

$$\theta(t_k, y_k) = f(t_k, y_k),$$

while for the improved Euler method

$$\theta(t_k, y_k) = \frac{f(t_k, y_k) + f(t_k + h, y_k + hf(t_k, y_k))}{2}.$$

In what follows, two different θ functions are developed, giving rise to the Taylor series methods and the Runge–Kutta methods.

We seek *accurate* algorithms (3.2.3, 3.2.4). By this we mean algorithms for which the *true* solution, $y(t)$, almost satisfies (3.2.3), i.e.,

$$y(t_{k+1}) = y(t_k) + h\theta(t_k, y(t_k)) + h\tau_k \qquad (3.2.5)$$

with τ_k small. The method (3.2.3) is said to be of *order p* if for all t_k, $a \le t_k \le b$, and for all sufficiently small h, there are constants C and p such that

$$|\tau_k| \le Ch^p.$$

This can be interpreted as meaning that $|\tau_k|$ goes to zero no slower than Ch^p when h approaches zero. The order p condition can be abbreviated by using the big-oh notation[1]

$$\tau_k = O(h^p)$$

as h approaches zero. The constant C, in general, depends on the solution $y(t)$, its derivatives and the length of the interval over which the solution is to be found, but is independent of h. The quantity $h\tau_k$ is called the *local (truncation) error*.

The importance of local error and of the order of a method are easily explained. Suppose the exact value $y(t_k)$ of the solution is known (as it is, for instance, at the initial point $t = a$). Then $y_k = y(t_k)$ and the next step of the method gives

$$y_{k+1} = y_k + h\theta(t_k, y_k) = y(t_k) + h\theta(t_k, y(t_k))$$

as an approximation to $y(t_{k+1})$. From relation (3.2.5) it is seen that if $h\tau_k$ is very small, then y_{k+1} will be very close to $y(t_{k+1})$. For if the method is of order p, where p is a positive integer, then $|h\tau_k| \le Ch^{p+1}$, which will be very small for small enough step size h. The size of p is also important. For instance, if $h = 0.1$ and $p = 1$, then the local error is of order $(0.1)^{1+1} = 0.01$. Whereas if $p = 3$, then the same step size will give an error of order $(0.1)^{1+3} = 0.0001$.

---

[1] The big-oh notation is a useful one for error estimating. We say $g(h) = O(h^k)$ as $h \to 0$ if $\lim_{h\to 0} |g(h)|/|h|^k$ is bounded. For instance, $\sin(h) = O(h)$ and $\cos(h^2) - 1 = O(h^4)$ as $h \to 0$.

Example 1   Find the order p of Euler's method.

Suppose $y(t)$ is a solution of $y' = f(t, y)$ and $y''(t)$ is continuous for $a \leq t \leq b$. Then Taylor's formula with remainder implies

$$y(t_{k+1}) = y(t_k + h) = y(t_k) + hy'(t_k) + \frac{h^2}{2} y''(\zeta_k)$$

$$= y(t_k) + hf(t_k, y(t_k)) + h \left[\frac{h}{2} y''(\zeta_k) \right],$$

where $t_k \leq \zeta_k \leq t_{k+1}$, and we have used the fact that $y' = f(t, y)$. Letting

$$\tau_k = \frac{h}{2} y''(\zeta_k)$$

and observing that the continuity of $y''(t)$ on $[a, b]$ implies that it is bounded there, we see that

$$|\tau_k| = \left| \frac{h}{2} y''(\zeta_k) \right| \leq Ch,$$

where C could be chosen to be the maximum value of $|y''(x)/2|$ on $a \leq x \leq b$. Hence Euler's method has order $p = 1$. $\square$

The order of a one-step method may also be viewed as a measure of how fast the error in the computed solution goes to zero at a fixed point t as more and more steps are taken, i.e., as $h \to 0$. Recall that in the examples in Chapter 1 the error in the Euler algorithm was proportional to h, which is consistent with the method being of order $p = 1$. It was also indicated that the error in the improved Euler algorithm was proportional to h^2. As will be seen shortly, this algorithm has order $p = 2$. Clearly, the larger the value of p, the faster the error goes to zero as $h \to 0$. In the next section we shall describe a class of methods known as Taylor series methods, which can be used to develop algorithms of very high orders.

3.3   TAYLOR SERIES METHODS

Perhaps the simplest methods of order p are based on the Taylor series expansion of the solution $y(t)$ of the differential equation (3.2.1). If $y^{(p+1)}(t)$ is continuous on $[a, b]$, then Taylor's formula gives

$$y(t_{k+1}) = y(t_k + h)$$

$$= y(t_k) + y'(t_k)h + \cdots + y^{(p)}(t_k) \frac{h^p}{p!} + y^{(p+1)}(\zeta_k) \frac{h^{p+1}}{(p + 1)!}$$

$$= y(t_k) + h \left[y'(t_k) + \cdots + y^{(p)}(t_k) \frac{h^{p-1}}{p!} \right] + y^{(p+1)}(\zeta_k) \frac{h^{p+1}}{(p + 1)!},$$

$$(3.3.1)$$

where $t_k \leq \zeta_k \leq t_{k+1}$. The continuity of $y^{(p+1)}(t)$ implies that it is bounded on $a \leq t \leq b$ and therefore

$$y^{(p+1)}(\zeta_k) \frac{h^{p+1}}{(p+1)!} = O(h^{p+1}) = hO(h^p).$$

Since $y' = f(t, y)$, Eq. (3.3.1) can be simplified somewhat. To evaluate the higher derivatives of $y(t)$, we use the chain rule and the fact that $dy/dt = f(t, y)$ and proceed as follows:

$$y''(t) \overset{\text{d}}{=} f^{(1)}(t, y) = \frac{d}{dt} f(t, y) = \frac{\partial}{\partial t} f(t, y) \frac{dt}{dt} + \frac{\partial}{\partial y} f(t, y) \frac{dy}{dt}$$

$$= f_t(t, y) + f_y(t, y) f(t, y) = f_t + f_y f,$$

$$y'''(t) \overset{\text{d}}{=} f^{(2)} = \frac{d}{dt} f^{(1)} = \frac{\partial}{\partial t} f^{(1)} \frac{dt}{dt} + \frac{\partial}{\partial y} f^{(1)} \frac{dy}{dt}$$

$$= f_t^{(1)} + f_y^{(1)} f = f_{tt} + f_{yt} f + f_y f_t + [f_{ty} + f_{yy} f + f_y f_y] f$$

$$= f^2 f_{yy} + 2f f_{ty} + f_{tt} + f_t f_y + f f_y^2,$$

etc. The symbol $\overset{\text{d}}{=}$ means "by definition." In general, all the higher derivatives can be found from the recurrence relation

$$y^{(n+1)}(t) \overset{\text{d}}{=} f^{(n)} = \frac{d}{dt} f^{(n-1)}(t)$$

$$= f_t^{(n-1)} + f_y^{(n-1)} f, \quad n = 1, 2, \ldots, \tag{3.3.2}$$

where $f^{(0)}(t, y) = f(t, y)$. The quantity $f^{(n)}(t, y)$ is called the nth *total derivative* of $f(t, y)$. By using (3.3.2), the expansion (3.3.1) can now be put into the form

$$y(t_{k+1}) = y(t_k) + h \left\{ f(t_k, y(t_k)) + \cdots + f^{(p-1)}(t_k, y(t_k)) \frac{h^{p-1}}{p!} \right\}$$

$$+ hO(h^p), \tag{3.3.3}$$

and comparison of (3.2.5) and (3.3.3) shows that if a method of order p is desired, we can let

$$\theta(t_k, y(t_k)) = f(t_k, y(t_k)) + \cdots + f^{(p-1)}(t_k, y(t_k)) \frac{h^{p-1}}{p!}. \tag{3.3.4}$$

One therefore obtains a family of one-step methods of order p for each integer $p \geq 1$ called *Taylor series methods*. These are given in the following algorithm.

The Taylor series algorithm. To obtain an approximate solution of order p to the initial value problem (3.2.1, 3.2.2) on $[a, b]$, we let $h =$

$(b - a)/n$ for some positive integer n and generate the sequences

$$y_{k+1} = y_k + h\left[f(t_k, y_k) + \cdots + \frac{h^{p-1}}{p!}f^{(p-1)}(t_k, y_k)\right],$$

$$t_{k+1} = t_0 + (k + 1)h, \quad k = 0, 1, \ldots, n - 1,$$

where $t_0 = a$, $y_0 = A$, and y_{k+1} is an approximation to $y(t_{k+1})$. (Note that Euler's method is the Taylor series method of order $p = 1$.)

Example 1   Approximate the solution to $dy/dt = y$, $y(0) = 1.00$ at $t = 1.00$ by using the Taylor series method of order $p = 2$ with $h = 0.25$.

The recursion relation for y_k is

$$y_{k+1} = y_k + h[f(t_k, y_k) + \frac{h}{2}f^{(1)}(t_k, y_k)].$$

Since

$$f(t, y) = y,$$
$$f^{(1)}(t, y) = f_t(t, y) + f_y(t, y)f(t, y) = 0 + 1 \cdot y = y,$$

the algorithm becomes

$$y_{k+1} = y_k + h\left[y_k + \frac{h}{2}y_k\right] = y_k\left(1 + h + \frac{h^2}{2}\right),$$

$$t_{k+1} = (k + 1)\,h, \, t_0 = 0, \, y_0 = 1.00.$$

Now using the fact that $h = 0.25$, we have

$$1 + h + \frac{h^2}{2} = 1 + 0.25 + \frac{0.25^2}{2} = 1.28125,$$

and therefore the sequences generated are

$$y_1 = y_0\left(1 + h + \frac{h^2}{2}\right) = 1.00 \cdot 1.28125 = 1.28125, \quad t_1 = 0.25;$$

$$y_2 = 1.28125y_1 = 1.28125 \cdot 1.28125 = 1.64160, \quad t_2 = 0.50;$$

$$y_3 = 1.28125y_2 = 2.10330, \quad t_3 = 0.75;$$

$$y_4 = 1.28125y_3 = 2.69486, \quad t_4 = 1.00.$$

The actual solution is $y(t) = e^t$ and $y(1) = 2.71828$, which gives an absolute error of 2.3×10^{-2}. If one is willing to do more work, the following table can be constructed.

n	$h = 1/n$	y_n	$\vert \text{Error} \vert$
8	1.25×10^{-1}	2.71184	6.4×10^{-3}
16	6.25×10^{-2}	2.71659	1.7×10^{-3}
32	3.125×10^{-2}	2.71785	4.3×10^{-4}
64	1.5625×10^{-2}	2.71817	1.1×10^{-4}
128	7.8125×10^{-3}	2.71825	2.7×10^{-5}

A little calculation shows that the absolute error at $t = 1.00$ is proportional to h^2; in fact, for $n \geq 16$ we have

$$\vert \text{Error} \vert_{t=1.00} \approx 0.5 h^2,$$

as the theory predicted.

As a matter of interest, the same rather large value $h = 0.25$ could be used and the order increased to attain better accuracy instead of keeping the order fixed and decreasing h. For example, for $p = 3, 4$ the formulas become

$$p = 3: \quad y_{k+1} = y_k \left(1 + h + \frac{h^2}{2} + \frac{h^3}{6} \right),$$

$$p = 4: \quad y_{k+1} = y_k \left(1 + h + \frac{h^2}{2} + \frac{h^3}{6} + \frac{h^4}{24} \right),$$

and we get

Order p	y_4	$\vert \text{Error} \vert$
3	2.71683	1.5×10^{-3}
4	2.71821	7.2×10^{-5}

Therefore the same order of accuracy is obtained for $p = 4$ and $h = 0.25$ as for $p = 2$ and $h = 7.8125 \times 10^{-3}$. Of course, the higher total derivatives $f^{(n)}$ are easy to compute for $f(t, y) = y$. $\square$

Example 2   Approximate the solution to $y' = 2ty^2$, $y(0) = 0.5$ at $t = 1.0$ by using the Taylor series algorithm of order $p = 2$ with $h = 0.1$.

Here $f(t, y) = 2ty^2$, so

$$f_t = 2y^2, \qquad f_y = 4ty.$$

TABLE 3.1

k	t_k	y_k	k	t_k	y_k
0	0.0	0.50000	6	0.6	0.60828
1	0.1	0.50250	7	0.7	0.65962
2	0.2	0.51013	8	0.8	0.73051
3	0.3	0.52335	9	0.9	0.83120
4	0.4	0.54304	10	1.0	0.98108
5	0.5	0.57060			

Therefore

$$f^{(1)} = f_t + f_y f = 2y^2 + 8t^2 y^3$$

and the algorithm for the sequence y_k is

$$y_{k+1} = y_k + h \left[2t_k y_k^2 + \frac{h}{2} (2y_k^2 + 8t_k^2 y_k^3) \right],$$

$$t_{k+1} = (k + 1)h, \qquad t_0 = 0, \qquad y_0 = 0.5.$$

The results are presented in Table 3.1. The actual solution (obtained by separation of variables) is

$$y(t) = \frac{1}{2 - t^2},$$

so $y(1) = 1$ and the absolute error in the approximate solution at $t = 1$ is 1.9×10^{-2}. Using 100 steps ($h = 0.01$) instead of 10 steps ($h = 0.1$) would have produced an error of 2.6×10^{-4}. Again it looks as if the error is proportional to h^2. □

In the above example, if we had wanted to use the Taylor series method of order $p = 3$, we would have needed to compute $f^{(2)}$. Since

$$f^{(1)} = 2y^2 + 8t^2 y^3,$$

then

$$f_t^{(1)} = 16ty^3, \qquad f_y^{(1)} = 4y + 24t^2 y^2,$$

and therefore

$$f^{(2)} = f_t^{(1)} + f_y^{(1)} f = 16ty^3 + (4y + 24t^2 y^2)(2ty^2).$$

The algorithm for the sequence y_k is

$$y_{k+1} = y_k + h \left[2t_k y_k^2 + \frac{h}{2} (2y_k^2 + 8t_k^2 y_k^3) + \frac{h^2}{6} (24t_k y_k^3 + 48t_k^3 y_k^4) \right],$$

$$t_{k+1} = (k + 1) h, \qquad t_0 = 0, \qquad y_0 = 0.5.$$

With $h = 0.1$ an error of order $10^{-3} = 0.1^3$ for the approximation of $y(1)$ would be expected.

Taylor series methods can be quite effective if the total derivatives $f^{(n)}$ of $f(t, y)$ are not too difficult to evaluate. In Example 1 above, they were easy to evaluate and in Example 2, not much more difficult. If, on the other hand, the function is something like

$$f(t, y) = \cos t \sin y,$$

then

$$f^{(1)} = -\sin t \sin y + (\cos t \cos y) \cos t \sin y,$$
$$f^{(2)} = -\cos t \sin y - 2\cos t \sin t \cos y \sin y$$
$$+ (-\sin t \cos y + \cos^2 t \cos^2 y - \cos^2 t \sin^2 y)\cos t \sin y,$$

etc., and we see that the total derivatives of f can become quite messy.

EXERCISES 3.3

1. Use the Taylor series method of order $p = 2$ with $h = 0.1$ to compute an approximate solution at $t = 0.5$ to the following initial value problems. In each case find the exact solution and compare it with your numerical result.

a) $y' = 2y - 1, \quad y(0) = 1$

b) $y' = 1 + y^2, \quad y(0) = 0$

c) $y' = -y, \quad y(0) = 1$

d) $y' = y - t, \quad y(0) = 2$

e) $y' = -ty^2, \quad y(0) = 2$

f) $y' = t/y, \quad y(0) = 1$

2. For the linear equation $y' = P_1(t)y + Q_1(t)$ show that the derivatives needed in the Taylor series method can be obtained from the recursion

$$y^{(r)} = P_r(t)y + Q_r(t),$$

where

$$P_r(t) = P'_{r-1}(t) + P_1(t)P_{r-1}(t),$$

$$Q_r(t) = Q'_{r-1}(t) + Q_1(t)P_{r-1}(t), \quad r = 2, 3, \ldots$$

3. Apply the method of Exercise 2 to develop a 3rd order Taylor series formula to approximate the solution of the initial value problem

$$y' = ty + 1, \qquad y(0) = 0.$$

Implement your formula by using a step size of $h = 0.25$ to approximate $y(1)$.

4. Consider the initial value problem

$$y' = -2y, \qquad y(0) = 1.$$

a) Solve the problem analytically to find $y(1)$.

b) Write a computer program which implements the Taylor series algorithm of order $p = 2$ for this problem. Let your program begin with a fixed step size $h = 1/N$ and repeatedly halve it until a given accuracy, say $10^{-\alpha}$ is achieved, i.e., continue to halve until

$$|y(1) - y_N| < 10^{-\alpha}.$$

Let $\alpha = 3$ and start with $N = 2$.

c) Repeat part (b) using Taylor algorithms of order $p = 3$ and 4 and note the differences in the final value as p increases for fixed N. The example points out that Taylor algorithms can be quite effective when the derivatives are easy to evaluate.

3.4 RUNGE–KUTTA METHODS

We now introduce another class of one-step methods, the Runge–Kutta methods, for solving the initial value problem (3.2.1, 3.2.2). These methods are designed to approximate Taylor series methods but have the advantage of not requiring explicit evaluations of the derivatives of $f(t, y)$. The basic idea is to use a linear combination of values of $f(t, y)$ to approximate $y(t)$. This linear combination is matched up as well as possible with the Taylor series for $y(t)$ to obtain methods of the highest possible order p. An example with one evaluation of f is Euler's method; an example with two evaluations of f is the improved Euler method.

We proceed now to derive a family of Runge–Kutta formulas by using two evaluations of $f(t, y)$ per step. It will turn out that the improved Euler method is a special case of these formulas. The techniques employed in the derivation extend easily to the development of all Runge–Kutta type formulas. To begin with, we are given the values t_k and y_k and the object is to choose the points $\hat{t}_k$ and $\hat{y}_k$ and the constants a_1 and a_2 so as to match

$$y_{k+1} = y_k + h[a_1 f(t_k, y_k) + a_2 f(\hat{t}_k, \hat{y}_k)] \tag{3.4.1}$$

with the Taylor expansion

$$
\begin{aligned}
y(t_{k+1}) &= y_k + h\left[y'(t_k) + y''(t_k)\frac{h}{2} + y'''(t_k)\frac{h^2}{6} + \cdots \right] \\
&= y_k + h\left[f(t_k, y_k) + f^{(1)}(t_k, y_k)\frac{h}{2} + f^{(2)}(t_k, y_k)\frac{h^2}{6} + \cdots \right] \\
&= y_k + h\left[f + (f_t + f_y f)\frac{h}{2} + (f^2 f_{yy} + 2ff_{ty} + f_{tt} + f_t f_y + ff_y^2)\frac{h^2}{6} + \cdots \right].
\end{aligned}
$$

In the last expression and in what follows, all arguments of f and its derivatives will be suppressed when they are evaluated at the point (t_k, y_k). It will be convenient to express $\hat{t}_k$ and $\hat{y}_k$ as

$$\hat{t}_k = t_k + b_1 h, \qquad \hat{y}_k = y_k + b_2 hf,$$

and so the object is to match

$$\theta = a_1 f + a_2 f(t_k + b_1 h, y_k + b_2 hf) \tag{3.4.2}$$

with

$$f + \frac{h}{2}(f_t + f_y f) + \frac{h^2}{6}(f^2 f_{yy} + 2ff_{ty} + f_{tt} + f_t f_y + ff_y^2) + \cdots . \tag{3.4.3}$$

Expanding (3.4.2) in a Taylor series about (t_k, y_k) (see Exercise 1 on p. 168 for a hint on how to do this if you don't know how), we have

$$\theta = (a_1 + a_2)f + a_2 h(b_1 f_t + b_2 ff_y)$$

$$+ \frac{a_2 h^2}{2}(b_2^2 f^2 f_{yy} + 2b_1 b_2 ff_{ty} + b_1^2 f_{tt}) + O(h^3). \tag{3.4.4}$$

Equating coefficients of like powers of h in (3.4.4) and in the previous Taylor expansion (3.4.3) gives

$$h^0: \quad a_1 + a_2 = 1;$$
$$h^1: \quad a_2 b_1 = \tfrac{1}{2}, \quad a_2 b_2 = \tfrac{1}{2};$$
$$h^2: \quad a_2 b_2^2 = \tfrac{1}{3}, \quad a_2 b_1 b_2 = \tfrac{1}{3}, \quad a_2 b_1^2 = \tfrac{1}{3},$$

$$? = \frac{f_t f_y}{6}, \quad ? = \frac{ff_y^2}{6},$$

where the question marks indicate that there are no terms in (3.4.4) that match up with the given terms in the Taylor expansion. Therefore a match up can only be obtained through first powers of h. This means that the two-point evaluation will agree *exactly* with the result obtained from the Taylor series method of order $p = 2$. The relations for h^0 and h^1 give three equations in the four unknowns a_1, a_2, b_1, b_2, so we can choose $a_2 = c$, an arbitrary parameter. The system of equations

$$a_1 + a_2 = 1, \qquad a_2 b_1 = \tfrac{1}{2}, \qquad a_2 b_2 = \tfrac{1}{2}$$

can then be solved exactly to give

$$a_2 = c, \quad a_1 = 1 - c, \quad b_1 = b_2 = \frac{1}{2c}, \quad c \neq 0.$$

Therefore,

$$\theta = (1 - c)f(t, y) + cf\left(t + \frac{h}{2c}, y + \frac{h}{2c}f(t, y)\right)$$

gives a one-step method of order $p = 2$ if $c \neq 0$ and f is sufficiently smooth. This brings us to the following method.

Runge–Kutta method of order $p = 2$. To obtain an approximate solution of order $p = 2$ to the initial value problem (3.2.1, 3.2.2) on $[a, b]$, let $h = (b - a)/n$ and generate the sequences

$$y_{k+1} = y_k + h \left[(1 - c)f(t_k, y_k) + cf\left(t_k + \frac{h}{2c}, y_k + \frac{h}{2c} f(t_k, y_k) \right) \right],$$

$$t_{k+1} = a + (k + 1)h, \quad k = 0, 1, \ldots, n - 1,$$

where $c \neq 0$, $t_0 = a$, $y_0 = A$, and y_{k+1} is an approximation to $y(t_{k+1})$.

If $c = \frac{1}{2}$, we recognize this as the improved Euler method discussed in Chapter 1. The choice $c = 1$ gives another method called the *Euler–Cauchy method*.

Example 1   Approximate the solution to $y' = y^2 + 1$, $y(0) = 0.0$ at $t = 0.1, 0.2, \ldots, 1.0$ by using the Euler–Cauchy method with $h = 0.1$.

The recursion relation for y_k is

$$y_{k+1} = y_k + hf\left(t_k + \frac{h}{2}, y_k + \frac{h}{2} f(t_k, y_k) \right)$$

$$= y_k + h\left\{ 1 + \left[y_k + \frac{h}{2}\left(1 + y_k^2 \right) \right]^2 \right\}, \qquad y_0 = 0.0, \quad (3.4.5)$$

and the resulting approximation is given in Table 3.2. How does this compare with the actual solution, $y(t) = \tan t$? □

Higher order Runge–Kutta formulas can be derived in a similar manner by considering linear combinations of 3, 4, or more evaluations of f and trying to match them up with the appropriate Taylor series expansion of $y(t)$. Notice that so far one evaluation of f produced a method of order $p = 1$ and two evaluations produced a method of order $p = 2$. In fact, k evaluations per step produce methods of order $p = k$ for $k = 1, 2, 3$, and 4 but not for $k = 5$. For this reason fourth order formulas are quite common, and since the local truncation is $O(h^4)$, they are quite accurate. As in the second order case, where the parameter c was arbitrary, there is a family of fourth order formulas that depend on several arbitrary parameters. One choice of these parameters leads to the so-called *classical* formulas.

TABLE 3.2

t_k	y_k	t_k	y_k	t_k	y_k
0.0	0.00000	0.4	0.42223	0.8	1.02534
0.1	0.10025	0.5	0.54539	0.9	1.25256
0.2	0.20252	0.6	0.68263	1.0	1.54327
0.3	0.30900	0.7	0.83977		

Classical Runge-Kutta Formulas.

$$y_0 = A, \ t_0 = a,$$

$$k_1 = f(t_k, y_k),$$

$$k_2 = f\left(t_k + \frac{h}{2}, y_k + \frac{h}{2} k_1\right),$$

$$k_3 = f\left(t_k + \frac{h}{2}, y_k + \frac{h}{2} k_2\right), \qquad (3.4.6)$$

$$k_4 = f(t_k + h, y_k + h k_3),$$

$$y_{k+1} = y_k + \frac{h}{6} (k_1 + 2k_2 + 2k_3 + k_4),$$

$$t_{k+1} = t_0 (k + 1) h.$$

Example 2 Solve the problem of Example 2 on p. 155:

$$y' = 2ty^2, \qquad y(0) = 0.5, \qquad h = 0.1, \qquad n = 10,$$

by using the classical fourth order Runge–Kutta formula.

For the reader's benefit, the first two steps are calculated below. The output from the algorithm is given step by step in Table 3.3.

$$k_1 = f(0, 0.5) = 2(0)(0.5)^2 = 0.0000,$$

$$k_2 = f\left(0 + \frac{0.1}{2}, 0.5 + \frac{0.1}{2} k_1\right) = 2(0.05)(0.5)^2 = 0.0250,$$

$$k_3 = f\left(0 + \frac{0.1}{2}, 0.5 + \frac{0.1}{2} k_2\right) = 2(0.05)(0.5013)^2 = 0.0251,$$

$$k_4 = f(0.1, 0.5 + 0.1 k_3) = 2(0.1)(0.5025)^2 = 0.0505,$$

$$y_1 = 0.5 + \frac{0.1}{6} (k_1 + 2k_2 + 2k_3 + k_4) = 0.5025, \ t_1 = 0.1,$$

$$k_1 = f(0.1, 0.5025) = 2(0.1)(0.5025)^2 = 0.0505,$$

$$k_2 = f\left(0.1 + \frac{0.1}{2}, 0.5025 + \frac{0.1}{2} k_1\right) = 2(0.15)(0.5050)^2 = 0.0765,$$

$$k_3 = f\left(0.1 + \frac{0.1}{2}, 0.5025 + \frac{0.1}{2} k_2\right) = 2(0.15)(0.5063)^2 = 0.0769,$$

$$k_4 = f(0.2, 0.5025 + 0.1 k_3) = 2(0.2)(0.5102)^2 = 0.1041,$$

$$y_2 = 0.5025 + \frac{0.1}{6} (k_1 + 2k_2 + 2k_3 + k_4) = 0.5102, \ t_2 = 0.2, \text{ etc.}$$

TABLE 3.3	t	y	k_1	k_2	k_3	k_4	$\dfrac{h}{6}(k_1 + 2k_2 + 2k_3 + k_4)$
	0.0000	0.5000	0.0000	0.0250	0.0251	0.0505	0.0025
	0.1000	0.5025	0.0505	0.0765	0.0769	0.1041	0.0077
	0.2000	0.5102	0.1041	0.1328	0.1336	0.1645	0.0134
	0.3000	0.5236	0.1645	0.1980	0.1992	0.2363	0.0199
	0.4000	0.5435	0.2363	0.2775	0.2796	0.3265	0.0280
	0.5000	0.5714	0.3265	0.3800	0.3835	0.4462	0.0383
	0.6000	0.6098	0.4462	0.5194	0.5254	0.6141	0.0525
	0.7000	0.6623	0.6140	0.7203	0.7314	0.8653	0.0730
	0.8000	0.7353	0.8651	1.0304	1.0524	1.2717	0.1050
	0.9000	0.8403	1.2711	1.5523	1.6010	2.0017	0.1597
	1.0000	0.9999					

As one would expect, the result is considerably more accurate than that obtained by using the Taylor algorithm of order $p = 2$. In fact, the absolute error here is 5.5×10^{-6} (the numerical result correct to seven significant digits is 0.9999945). □

As was pointed out in the beginning of this chapter, the methods developed extend readily to systems of first order differential equations. For example, if we want to solve the system

$$y' = \mathbf{f}(t, \mathbf{y}), \qquad (3.4.7)$$

$$\mathbf{y}(a) = \mathbf{A} \qquad (3.4.8)$$

where $\mathbf{y}$, $\mathbf{f}$, and $\mathbf{A}$ are N dimensional vectors, the classical Runge–Kutta formulas (3.4.6) become

$$\mathbf{y}_0 = \mathbf{A}, \ t_0 = a,$$

$$\mathbf{k}_1 = \mathbf{f}(t_k, \mathbf{y}_k),$$

$$\mathbf{k}_2 = \mathbf{f}\left(t_k + \frac{h}{2}, \mathbf{y}_k + \frac{h}{2}\mathbf{k}_1\right),$$

$$\mathbf{k}_3 = \mathbf{f}\left(t_k + \frac{h}{2}, \mathbf{y}_k + \frac{h}{2}\mathbf{k}_2\right), \qquad (3.4.9)$$

$$\mathbf{k}_4 = \mathbf{f}(t_k + h, \mathbf{y}_k + h\mathbf{k}_3),$$

$$\mathbf{y}_{k+1} = \mathbf{y}_k + \frac{h}{6}(\mathbf{k}_1 + 2\mathbf{k}_2 + 2\mathbf{k}_3 + \mathbf{k}_4),$$

$$t_{k+1} = t_0 + (k + 1)h.$$

Example 3   Write out Eq. (3.4.9) for the system of first order differential equations

$$y_1' = y_2, \qquad y_2' = y_1 + t,$$

$$y_1(0) = 1, \qquad y_2(0) = 1.$$

First of all, using the notation of (3.4.7, 3.4.8) we can write

$$\mathbf{y} = \begin{bmatrix} y_1 \\ y_2 \end{bmatrix}, \qquad \mathbf{f} = \begin{bmatrix} f_1 \\ f_2 \end{bmatrix} = \begin{bmatrix} y_2 \\ y_1 + t \end{bmatrix}, \qquad \mathbf{A} = \begin{bmatrix} 1 \\ 1 \end{bmatrix}.$$

So

$$\mathbf{y}_0 = \begin{bmatrix} 1 \\ 1 \end{bmatrix}, \qquad \mathbf{k}_1 = \mathbf{f}(t, \mathbf{y}) = \begin{bmatrix} y_2 \\ y_1 + t \end{bmatrix},$$

which implies

$$\mathbf{y} + \frac{h}{2}\mathbf{k}_1 = \begin{bmatrix} y_1 \\ y_2 \end{bmatrix} + \frac{h}{2}\begin{bmatrix} y_2 \\ y_1 + t \end{bmatrix} = \begin{bmatrix} y_1 + \dfrac{h}{2}y_2 \\[2mm] y_2 + \dfrac{h}{2}(y_1 + t) \end{bmatrix},$$

and therefore

$$\mathbf{k}_2 = \mathbf{f}\left(t + \frac{h}{2}, \mathbf{y} + \frac{h}{2}\mathbf{k}_1\right) = \begin{bmatrix} y_2 + \dfrac{h}{2}(y_1 + t) \\[2mm] y_1 + \dfrac{h}{2}y_2 + t + \dfrac{h}{2} \end{bmatrix},$$

etc. At the kth step, the values of t, y_1, and y_2 are assumed to be evaluated at (t_k, y_k).   □

The methods given also apply to higher order differential equations in which the highest order derivative can be solved for explicitly since equations of this type can always be written in the form (3.4.7, 3.4.8).

Example 4   Write the initial value problem

$$y'' + ty' + y = \sin t, \qquad y(0) = 1, \qquad y'(0) = 2$$

in the form (3.4.7, 3.4.8).

Let

$$y_1 = y, \qquad y_2 = y';$$

```
      SUBROUTINE RK4(F,T,Y,H,N)
C
C  SUBROUTINE RK4 IMPLEMENTS THE CLASSICAL 4TH ORDER RUNGE-KUTTA
C  FORMULAS OVER ONE STEP OF LENGTH H.
C
C
C  INPUT:
C
C     F — NAME OF A SUBROUTINE SUBPROGRAM DEFINING THE DIFFER-
C          ENTIAL EQUATIONS. F MUST HAVE THE FORM
C
C          SUBROUTINE F(T,Y,YP,N)
C          REAL Y(N),YP(N)
C          YP(1)=...
C             .
C             .
C             .
C          YP(N)=...
C          RETURN
C          END
C
C     T — VALUE OF THE INDEPENDENT VARIABLE
C     Y — SOLUTION VECTOR OF LENGTH N EVALUATED AT T
C     H — STEP SIZE TO BE USED (H#0.0)
C     N — NUMBER OF COMPONENTS IN SOLUTION VECTOR
C
C  OUTPUT:
```

Figure 3.1 A listing of the subroutine RK4.

then

$$y_1' = y' = y_2,$$
$$y_2' = y'' = -y_1 - ty_2 + \sin t$$

with $y_1(0) = 1$ and $y_2(0) = 2$.

Using the notation of (3.4.7, 3.4.8), we get

$$\mathbf{y} = \begin{bmatrix} y_1 \\ y_2 \end{bmatrix}, \quad \mathbf{f} = \begin{bmatrix} f_1 \\ f_2 \end{bmatrix} = \begin{bmatrix} y_2 \\ -y - ty_2 + \sin t \end{bmatrix}, \quad \mathbf{A} = \begin{bmatrix} 1 \\ 2 \end{bmatrix}. \quad \square$$

The Fortran subroutine RK4, which implements the 4th order Runge–Kutta algorithm defined by the formula (3.4.9), is given in Fig. 3.1. It is followed by a sample program in Example 5, which illustrates its use. For convenience we have included the name of the subroutine F (defining the dif-

```
C
C     T — NEW VALUE OF INDEPENDENT VARIABLE (INPUT VALUE PLUS H)
C     Y — SOLUTION VECTOR EVALUATED AT T
C
C  NOTE: Y AND T MUST BE VARIABLES IN THE CALLING PROGRAM AND F MUST
C  APPEAR IN AN EXTERNAL STATEMENT IN THE CALLING PROGRAM.
C
C
C
      REAL Y(N),K1(10),K2(10),K3(10),K4(10),YTEMP(10)
      CALL F(T,Y,K1,N)
      DO 1 I=1,N
   1     YTEMP(I)=Y(I)+0.5*H*K1(I)
      CALL F(T+0.5*H,YTEMP,K2,N)
      DO 2  I=1,N
   2     YTEMP(I)=Y(I)+0.5*H*K2(I)
      CALL F(T+0.5*H,YTEMP,K3,N)
      DO 3  I=1,N
   3     YTEMP(I)=Y(I)+H*K3(I)
      CALL F(T+H,YTEMP,K4,N)
C
      DO 4  I=1,N
   4     Y(I)=Y(I)+H*(K1(I)+2.0*(K2(I)+K3(I))+K4(I))/6.0
      T=T+H
C
      RETURN
      END
```

Figure 3.1 (*continued*)

ferential equations to be solved) in the argument list of RK4. This allows the user to choose the name for the subroutine. This could be useful, for example, if two different differential equations were to be solved in the same program. To illustrate this capability, we have used the name DERIV for the differential equation subroutine F in Example 5. A programming point worth noting here is that whenever the name of a subroutine appears in the argument list of another subroutine in a Fortran program, that name must appear in an EXTERNAL statement in the calling program.

Example 5   Using the subroutine RK4, set up a program to solve the initial value problem

$$y_1' = y_2, \qquad y_2' = y_1, \qquad y_1(0) = 1.0, \qquad y_2(0) = 1.0.$$

Use a step size of $h = 0.1$ and print out the solution at $t = 0.0, 0.5$, and 1.0 (see Fig. 3.2). $\square$

```
C   SAMPLE PROGRAM TO ILLUSTRATE SUBROUTINE RK4
C
      EXTERNAL F
      REAL Y(2)
      Y(1)=1.0
      Y(2)=1.0
      T=0.0
      WRITE(6,10)T,Y(1),Y(2)
      H=0.1
      NEQN=2
      NP=0
      NPRINT=5
      DO 1   I=1,10
          NP=NP+1
          CALL RK4(F,T,Y,H,NEQN)
          IF(NP.LT.NPRINT)GO TO 1
          WRITE(6,10)T,Y(1),Y(2)
          NP=0
    1 CONTINUE
      STOP
   10 FORMAT(1X,F10.2,2E15.4)
      END
C
      SUBROUTINE F(T,Y,YP,N)
      REAL Y(N),YP(N)
      YP(1)=Y(2)
      YP(2)=Y(1)
      RETURN
      END
```

T	Y(1)	Y(2)
0.	0.1000E+01	0.1000E+01
0.50	0.1649E+01	0.1649E+01
1.00	0.2718E+01	0.2718E+01

Figure 3.2 Sample program and output using RK4 to solve the problem of Example 5.

An issue of practical importance arises here. How does one intelligently pick a step size h for any of the numerical methods, and for a given h what can be said about the accuracy of the computed result? It can be shown that for any of the methods given, the error at any fixed point has the qualitative behavior illustrated in Fig. 3.3.

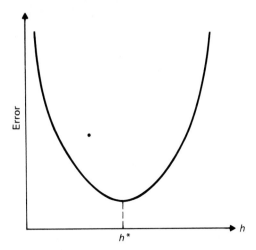

Figure 3.3 Qualitative behavior of absolute error at a fixed point as a function of the step size h.

Values of $h < h^*$ tend to produce poor results due to round-off error (error induced because the arithmetic is not exact), while values of $h > h^*$ produce poor results due to discretization error (error induced because the algorithm is not exact).

In practice, the following simple approach is useful: Choose a reasonable step size h, for instance 0.1, and perform the integration. Reduce h by a factor of 2 and integrate again. Compare the two solutions at all common values of t for which the solution is desired. If they agree to the required accuracy, accept the second result. If not, halve again, integrate and compare. Continue until two consecutive results agree to the required accuracy.

Of course, if the choice $h = 0.1$ produces a solution that agrees with the solution for $h = 0.05$ to more accuracy than you need, a larger h value should be considered to start with. In general, this procedure should keep the step size h as large as possible, so as to avoid round-off error problems. It should be pointed out, however, that this process gives an indication of the accuracy of the computed results, *but not a proof.*

Another point of some practical importance should be made. What if a formula of order $p = 2$ is used to solve an initial value problem whose solution has two continuous derivatives, but not three? Examination of the local truncation error will show that the formula is then of order $p = 1$ and therefore the rate of convergence is of order h and not h^2. The point here is that numerical procedures can be used on problems whose solutions are not smooth, but the rate of convergence may be reduced.

A common example of problems whose solutions will not be smooth are those where the coefficients have a jump discontinuity at some point. In numerically solving such a problem one should not integrate across the point. For example, suppose we are solving the problem

$$y'' + y = f(t), \quad 0 \le t < 2, \quad y(0) = 0, \quad y'(0) = 1,$$

$$f(t) = \begin{cases} 1 \text{ for } 0 \le t < 1, \\ 0 \text{ for } 1 \le t \le 2. \end{cases}$$

A good procedure would be to integrate from $t = 0$ to $t = 1$ and then from $t = 1$ to $t = 2$, and not to integrate across the discontinuity in $f(t)$ at $t = 1$. On each subinterval ($0 \le t \le 1$ and $1 \le t \le 2$), the differential equation has smooth solutions and convergence rates will be as advertised.

In Chapter 8 we will discuss numerical methods for solving initial value problems in more detail. There we will analyze the types of errors that are present in numerical solutions and will discuss techniques for estimating and controlling some of them.

EXERCISES 3.4

1. Derive the expansion (3.4.4) in the text. *Hint:* Proceed by a succession of one-variable expansions, e.g.,

$$f(t + \Delta, y + \delta) = f(t, y + \delta)$$
$$+ f_t(t, y + \delta)\,\Delta + \cdots$$
$$= f(t, y) + f_y(t, y)\,\delta$$
$$+ f_t(t, y)\,\Delta + f_{ty}(t, y)\,\Delta\delta$$
$$+ \cdots.$$

2. Write a program using the Euler–Cauchy method with step size $h = 0.1$ to obtain an approximate solution to each of the following initial value problems. Tabulate each solution at $t = 0.0, 0.1, \ldots, 1.0$.

a) $y' = y, \quad y(0) = 1$

b) $y' = t - y, \quad y(0) = 0$

c) $y' = 1 + y^2, \quad y(0) = 0$

d) $y' = t^2 + y^2, \quad y(0) = 0$

3. Approximate the solution to the following initial value problems at $t = 0.3$ by using the classical fourth order Runge–Kutta formulas with step size $h = 0.1$. Do your computations by hand and carry four significant digits in all

results. Organize your calculations as in Example 2.

a) $y' = -y + t + 4, \quad y(0) = 2$

b) $y' = -e^t y^3, \quad y(0) = 1$

c) $y' = 4t/y - ty, \quad y(0) = 3$

4. Write a program using the subroutine RK4 to obtain an approximate solution to each of the following initial value problems. Use $h = 0.1$ and tabulate each solution at $t = 0.0, 0.5$ and 1.0.

a) $y' = -4y + 1, \quad y(0) = 2$

b) $y' = y - t, \quad y(0) = 2$

c) $y' = 1 + y^2, \quad y(0) = 0$ (Compare with Exercise 2(c) and with the actual solution.)

d) $y' = 1 - 2ty, \quad y(0) = 0$

e) $y' = ty^{1/3}, \quad y(0) = 1$

f) $y' = -ty^2, \quad y(0) = 2$

5. Use RK4 with a step-size $h = 0.1$ to solve each of the following initial value problems. Tabulate the solution at the indicated t-values.

a) $y'' - 3y' + 2y = e^{-t}, y(1) = 0, y'(1) = 0$; $t = 1.0, 1.5, 2.0$.

b) $y'' + 4y' + 8y = \sin t$, $y(0) = 1$,
   $y'(0) = 0$; $t = 0.0, 0.2, 0.4, \ldots, 1.0$.

c) $y'' + y = \sin t$, $y(0) = 0$, $y'(0) = 2$;
   $t = 0.0, 1.0, 2.0, 3.0$.

d) $y'' + y' = \ln t$, $y(1) = 0$, $y'(1) = -1$;
   $t = 1.0, 2.0, \ldots, 10.0$.

e) $y'' + 6y' + 2y = t^2 + 1$, $y(0) = 1$,
   $y'(0) = 1$; $t = 0.0, 0.5, 1.0$.

6. Use RK4 with a step-size $h = 0.1$ to solve each of the following initial value problems. Tabulate the solution at $t = 0.0, 0.5, 1.0$.

a) $x' = -4x - y$, $x(0) = 0$,
   $y' = x - 2y$, $y(0) = -1$

b) $x' = 2x + 6y$, $x(0) = 1$,
   $y' = -2x - 5y$, $y(0) = 1$

c) $x' = -3x - 4y$, $x(0) = 0$,
   $y' = 3x + 2y$, $y(0) = 1$

d) $x' = y$, $x(0) = 0$
   $y' = 3x + 2y$, $y(0) = 1$

e) $x' = 3x - 2y$, $x(0) = 2$,
   $y' = -2x + 6y$, $y(0) = 0$

7. Consider the problem

$$y' = 2|t|y, \qquad y(-1) = 1/e$$

on the interval $[-1, 1]$. Find the analytical solution $y(t)$ and show that $y'(t)$ is continuous on $[-1, 1]$ but $y''(t)$ is not. Study the behavior of the error of Euler's method at some fixed points in $[-1, 1]$ as $h \to 0$.

8. When $f(t, y)$ depends only on t, show that the classical fourth order Runge–Kutta formula reduces to Simpson's rule

$$\int_a^{a+h} f(t)\, dt$$

$$= \frac{h}{6}\left[f(a) + 4f\left(a + \frac{h}{2}\right) + f(a + h) \right].$$

What is the order of Simpson's rule, i.e., what is the highest degree polynomial $P(t)$ which the rule integrates exactly? To what quadrature rule does the Runge–Kutta method of order 2 correspond when $c = \frac{1}{2}$? When $c = 1$?

9. The differential equation $y' = y - 0.5y^2$ represents a simple population growth model. Let $y(0) = 1$.

a) Write a program using RK4 to find an approximate solution on $[0, 2]$. Obtain four-figure accuracy.

b) Solve the problem analytically and compare with your results in (a).

10. Consider a pendulum consisting of a mass m at the end of a weightless rod of length l. Assume that there is no air resistance and no friction at the pivot point. Physical arguments similar to those in Chapter 2 can be used to show that the equation of motion is

$$\frac{d^2\theta}{dt^2} + \frac{g}{l} \sin \theta = 0, \qquad (3.4.10)$$

where g is the acceleration of gravity and $\theta = \theta(t)$ is the angular displacement (in radians) of the pendulum from the vertical equilibrium position at time t. (Notice that the mass m does not occur in the equation.) Henceforth take $g/l = 4.0 \text{ sec}^{-2}$.

a) Obtain numerical solutions to (3.4.10) for $0 \le t \le 10$ by using RK4 with $h = 0.1$ and the initial conditions $\theta(0) = \theta_0$ and $\theta'(0) = 0$, where $\theta_0 = \pi/20, \pi/10, \pi/5, \pi/2$. You will discover (as suggested by physics) that the motion is periodic. Find an approximate value for the period and construct phase plane plots in the (θ, θ') plane when $\theta_0 = \pi/10$ and $\theta_0 = \pi/2$.

b) If $\theta = \theta(t)$ is small, $(g/l) \sin \theta$ can be replaced by $g\theta/l$ in (3.4.10). For the cases $\theta_0 = \pi/20$ and $\theta_0 = \pi/10$ solve this problem analytically and compare your results with the numerical results you obtained in (a).

c) If viscous damping due to air resistance or friction at the pivot is added, Eq. (3.4.10) becomes

$$\frac{d^2\theta}{dt^2} + k\frac{d\theta}{dt} + \frac{g}{l} \sin \theta = 0.$$

Obtain numerical solutions with the same initial conditions as in (a), where $\theta_0 = \pi/10$, $\pi/5$, and $\pi/2$, and $k = 1.0 \text{ sec}^{-1}$. Construct a phase plane plot in the case $\theta_0 = \pi/2$.

11. A very simple system of differential equations, which has been used to model the growth of two populations, where one is a predator and the other its prey, is

$$\frac{dR}{dt} = R(a - bP), \qquad \frac{dP}{dt} = P(-c + dR),$$

where a, b, c, and d are all positive constants. Here $P = P(t)$ is the number of predators (measured in thousands) at time t (measured in years); $R = R(t)$ is the prey population similarly measured. For $a = 5.0$, $b = 1.0$, $c = 4.0$, $d = 0.8$ use RK4 with $h = 0.1$ for $0 \le t \le 2.0$ to approximate $R(t)$ and $P(t)$ numerically when

a) $R(0) = 10$, $P(0) = 1$

b) $R(0) = 5$, $P(0) = 5$

c) $R(0) = 1$, $P(0) = 10$

In each case construct a phase plane plot in the (R, P) plane. (The biological assumptions leading to such a model will be discussed in Chapter 7.)

12. Approximate the solution to the second order differential equation

$$y'' + t^2 y' + (\sin t)y = 3e^t,$$
$$y(0) = y'(0) = 1,$$

using RK4 with $h = 0.1$. Tabulate your results for $t = 0.0, 0.2, \ldots, 1.0$.

13. An interesting fact about Runge–Kutta methods is that the error depends on the form of the equation as well as on the solution itself. Verify that $y(t) = (t - 1)^2$ is a solution to both initial value problems

$$y' = 2(t - 1), \qquad y(0) = 1$$

and

$$y' = \frac{2y}{t - 1}, \qquad y(0) = 1.$$

Show that the Euler–Cauchy method is exact for the first equation but not for the second.

14. Consider the initial value problem for the damped pendulum

$$\frac{d^2\theta}{dt^2} + \frac{d\theta}{dt} + 4 \sin \theta = 0,$$

$$\theta(0) = \frac{\pi}{10}, \qquad \theta'(0) = 0.$$

Write a program to use the subroutine RK4 with step size $h = 0.05$ to approximate the first time $\hat{t}$ when the pendulum reaches the rest position $\theta(\hat{t}) = 0$.

REFERENCES

A classical introductory test on numerical analysis, which has an excellent discussion of numerical methods for ordinary differential equations in Chapter 14, is

1. P. Henrici, *Elements of Numerical Analysis,* Wiley, New York, 1968.

Another good discussion is Chapter 4 of

2. R. C. Buck and E. F. Buck, *Introduction to Differential Equations,* Houghton Mifflin, Boston, 1976.

More advanced discussions can be found in

3. L. F. Shampine and R. C. Allen, *Numerical Computing: An Introduction,* W. B. Saunders, Philadelphia, 1973.

4. G. E. Forsythe, M. A. Malcolm, and C. B. Moler, *Computer Methods for Mathematical Computations,* Prentice-Hall, Englewood Cliffs, 1977.

The Laplace Transform

4.1 INTRODUCTION

In this chapter we introduce another technique, which comes under the general classification of *transform methods,* for solving the initial value problem for constant coefficient linear differential equations. Transform methods are used extensively in the analysis of linear systems; the common features of these methods are:

1. The solution of the system $y(t)$, which exists in the t-domain, is transformed into a function $Y(s)$ of another independent variable s; call this the s-domain. The initial conditions are usually incorporated in the transformation.

2. The system of differential or integral equations, which $y(t)$ satisfies in the t-domain, is in turn transformed into a system of algebraic equations in the s-domain. If one can solve these equations for $Y(s)$, one obtains a relation of the form $Y(s) = \Phi(s)$.

3. If one can find a function $\phi(t)$ whose transform is $\Phi(s)$, then one can apply the inverse transformation that takes us from the s-domain to the t-domain and assert that $y(t) = \phi(t)$ is the required solution.

171

Different transform methods are used, depending on the nature of the t-domain (e.g., the whole line or a half-line) and on the class of functions admissible as solutions (e.g., continuous, differentiable, or analytic). The two most commonly used transforms are the Fourier transform and the Laplace transform and we will study the latter. It was first introduced by the great French mathematician Pierre Laplace (1749–1827), but its application and the techniques associated with it were not developed until about a hundred years later.

The Laplace transform, used as a technique for solving constant coefficient linear ordinary differential equations, is particularly effective for problems where the forcing function is discontinuous or has corners. For systems of linear differential equations its use obviates some of the algebraic difficulties associated with such systems. But in addition to this, the Laplace transform is used extensively to study input–output relations in systems analysis, to analyze feedback control systems, and to solve certain classes of partial differential equations of mathematical physics. It is one of the important tools of applied mathematics.

4.2 THE LAPLACE TRANSFORM OF e^{at}

Given a function $f(t)$, defined for $0 \le t < \infty$, define the *Laplace transform* of $f(t)$ by

$$L\{f\} = \int_0^\infty e^{-st}f(t)\ dt = F(s), \qquad (4.2.1)$$

where s is a complex variable, whenever the improper integral exists. Recall that the existence of the integral means that the limit

$$\lim_{R \to \infty} \int_0^R e^{-st}f(t)\ dt$$

exists. The fact that the variable s is complex is not significant in many applications, but the reader should be aware that the Laplace transform takes functions $f(t)$ defined on the half line $0 \le t < \infty$ into functions $F(s)$ defined on the complex plane.

The Laplace transform of $f(t) = e^{at}$ will now be worked out in detail, first of all because of its importance and frequent occurrence, and second, because it easily allows us to point out the interpretation of the variable s. Proceeding formally from the definition (4.2.1) above, we get

$$L\{e^{at}\} = \int_0^\infty e^{-st}e^{at}\ dt = \int_0^\infty e^{-(s-a)t}\ dt.$$

For the transform to exist, the improper integral must exist, so we must determine whether

$$\lim_{R \to \infty} \int_0^R e^{-(s-a)t}\, dt = \lim_{R \to \infty} \left[\frac{1}{s-a} (1 - e^{-(s-a)R}) \right] \qquad (4.2.2)$$

exists. Since a is real and s is complex, let $s = x + iy$ and therefore $s - a = (x - a) + iy$ and

$$e^{-(s-a)R} = e^{-(x-a)R}e^{-iyR} = e^{-(x-a)R}[\cos(yR) - i\sin(yR)].$$

Also note that the modulus of $e^{-(s-a)R}$ is

$$|e^{-(s-a)R}| = e^{-(x-a)R}|\cos(yR) - i\sin(yR)| = e^{-(x-a)R}$$

since $|u + iv| = \sqrt{u^2 + v^2}$. One can now immediately see that the limit in (4.2.2)

a) does not exist if $x - a < 0$, since $\lim\limits_{R \to \infty} e^{-(x-a)R} = \infty$.

b) does not exist if $x - a = 0$, since

$$\lim_{R \to \infty} \left\{ \frac{1}{iy} [1 - \cos(yR) + i\sin(yR)] \right\}$$

does not exist for any value of y.

c) does exist if $x - a > 0$, since $\lim\limits_{R \to \infty} e^{-(x-a)R} = 0$.

We conclude that if $x = \text{Re}(s) > a$, then

$$\lim_{R \to \infty} \left\{ \frac{1}{s-a} [1 - e^{-(s-a)R}] \right\} = \frac{1}{s-a},$$

and our first Laplace transform is

$$L\{e^{at}\} = f(s) = \frac{1}{s-a}, \qquad \text{Re}(s) > a. \qquad (4.2.3)$$

Therefore, L takes the real function e^{at} into the complex rational function $1/(s-a)$.

The reader should note the following: if it had been merely assumed that s was real, we would have easily derived the formula

$$L\{e^{at}\} = F(s) = \frac{1}{s-a}, \qquad s > a,$$

since the limit in (4.2.2) exists only if $s - a > 0$. A general rule of thumb is that in computing transforms of elementary functions, one can proceed as if s were real, and in any inequalities defining an s-region replace s by $\text{Re}(s)$ (see Fig. 4.1).

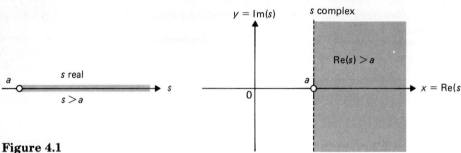

Figure 4.1

EXERCISES 4.2

1. Find the Laplace transform of each of the following functions. Determine and sketch the region in the complex plane in which it is valid.

a) e^{2t} b) e^{-7t} c) e^{10t}

d) $e^{-0.5t}$ e) $4e^t$ f) e^{2-t}

2. By following the analysis given to find $L\{e^{at}\}$ for a real, prove that for k real we have

$$L\{e^{ikt}\} = \frac{1}{s - ik}, \qquad \text{Re}(s) > 0.$$

3. Prove the formula

$$L\{e^{at}\} = \frac{1}{s - a}, \qquad \text{Re}(s) > \text{Re}(a),$$

where a is a complex number.

4. What is $L\{1\}$ and for what region in the complex plane is it valid? What is $L\{6\}$?

5. Using properties of the integral and the formula for $L\{e^{at}\}$, establish the formula

$$L\{Ae^{\alpha t} + Be^{\beta t}\} = \frac{A}{s - \alpha} + \frac{B}{s - \beta},$$

valid for $\text{Re}(s) > \max(\alpha, \beta)$, where α, β are real and A, B are arbitrary constants.

4.3 AN APPLICATION TO FIRST ORDER EQUATIONS

At this point, rather than developing a collection of formulas for the Laplace transform of various functions, we will state two of its important properties and then proceed directly to use it to solve a first order differential equation.

Property of linearity. If $f(t) = Ag(t) + Bh(t)$, where A and B are constants and $g(t)$ and $h(t)$ have Laplace transforms $G(s)$ and $H(s)$, respectively, valid for $\text{Re}(s) > a$, then

$$L\{f(t)\} = F(s) = AL\{g(t)\} + BL\{h(t)\} = AG(s) + BH(s). \qquad (4.3.1)$$
$$\text{Re}(s) > a.$$

The reader can see that this follows directly from the definition and the linearity of the integral.

Since the Laplace transform is to be used to study ordinary differential equations, we ask the question: Given $L\{f(t)\} = F(s)$ defined for $\text{Re}(s) > a$, can it be used to obtain $L\{f'(t)\}$, the Laplace transform of its derivative? If it is assumed that $f'(t)$ is continuous or piecewise continuous for $0 \leq t < 0$, then the answer is yes and it depends on the following integration by parts:

$$\int_0^R e^{-st} f'(t) \, dt = e^{-st} f(t) \big|_0^R + s \int_0^R e^{-st} f(t) \, dt$$

$$= e^{-sR} f(R) - f(0) + s \int_0^R e^{-st} f(t) \, dt.$$

The limit of the last integral as $R \to \infty$ is $sF(s)$, valid for $\text{Re}(s) > a$. If it is further assumed that $f(t)$ satisfies $\lim_{R \to \infty} e^{-sR} f(R) = 0$ for $\text{Re}(s) > a$, we obtain the second important property.

The differentiation formula.

$$L\{f'(t)\} = sL\{f(t)\} - f(0) = sF(s) - f(0). \tag{4.3.2}$$

Note that if $F(s)$ exists for $\text{Re}(s) > a$, then so does $L\{f'(t)\}$, and the above result was obtained under the assumptions that $f'(t)$ was continuous or piecewise continuous and $\lim_{t \to \infty} e^{st} f(t) = 0$ for $\text{Re}(s) > a$. The class of functions that satisfy the last relation are called *functions of exponential order*, and they represent a large class of functions whose Laplace transforms exist. This will be discussed in the next section.

With our meager collection of tools, consisting of the formula (4.2.3) for $L\{e^{at}\}$, of the Property of Linearity (4.3.1), and of the Differentiation Formula (4.3.2), we will show how to use the Laplace transform to solve the initial value problem

$$y' + ky = be^{at}, \quad y(0) = A, \quad a \neq -k. \tag{4.3.3}$$

Although the problem is relatively simple, the reader should carefully follow each step in the solution process, since the problem will be a prototype for all that follows. Let $y(t)$ be the solution of (4.3.3) and $Y(s) = L\{y(t)\}$ be its Laplace transform. Apply the transform to both sides of the equation, use the Property of Linearity and the formula for $L\{e^{at}\}$ to obtain

$$L\{y' + ky\} = L\{y'\} + kL\{y\} = L\{be^{at}\} = \frac{b}{s - a}.$$

Now applying the Differentiation Formula to the first term on the left, we get

$$sL\{y\} - y(0) + kL\{y\} = \frac{b}{s - a},$$

or, since $L\{y\} = Y(s)$ and $y(0) = A$,

$$sY(s) - A + kY(s) = \frac{b}{s-a}.$$

Solving the last expression for $Y(s)$ gives

$$Y(s) = \frac{A}{s+k} + \frac{b}{(s+k)(s-a)}. \qquad (4.3.4)$$

Thus, the first step is done, namely, by using the properties of the transform and algebraic operations an expression for $Y(s)$, the transform of the solution $y(t)$ has been obtained.

It is important to note in (4.3.4) that the initial value $y(0) = A$ is incorporated into the expression for $Y(s)$. The fact that the initial data are a part of the transform is an attractive feature of the method since one does not have to perform additional operations at the end to find values of arbitrary constants. Of course the initial data enter through the Differentiation Formula.

As it stands, the expression for $Y(s)$ is not in a form we recognize because of the second term. But if this term is decomposed by using partial fractions, then

$$Y(s) = \frac{A}{s+k} + \frac{b}{k+a}\left[\frac{1}{s-a} - \frac{1}{s+k}\right]$$

or

$$Y(s) = \left(A - \frac{b}{k+a}\right)\frac{1}{s+k} + \frac{b}{k+a}\frac{1}{s-a}. \qquad (4.3.5)$$

Now, using the formula for $L\{e^{at}\}$ and linearity, we see that

$$y(t) = \left(A - \frac{b}{k+a}\right)e^{-kt} + \frac{b}{k+a}e^{at} \qquad (4.3.6)$$

is a function whose Laplace transform is $Y(s)$ and that it is the solution of the initial value problem.

The astute reader may well ask how did we get from the expression (4.3.5) to the solution (4.3.6) since the Laplace transform only tells us how to get from $y(t)$ to $Y(s) = L\{y(t)\}$. The answer is that associated with the operator L is an operator L^{-1}, called the *inverse Laplace transform*, and it has the following properties:

Uniqueness property. If $f(t)$ is continuous and has a transform $F(s) = L\{t\}$, then

$$L^{-1}\{F(s)\} = f(t).$$

Linearity property. If $g(t)$ and $h(t)$ are continuous and have Laplace transforms $G(s)$ and $H(s)$, respectively, and A and B are constants, then

$$L^{-1}\{AG(s) + BH(s)\} = Ag(t) + Bh(t).$$

One sees now that applying L^{-1} to both sides of (4.3.5) gives us the solution $y(t)$.

We finally remark that, given a certain general class of functions $F(s)$ in the complex plane, there is an integral that permits us to calculate $f(t) = L^{-1}\{F(s)\}$. It is a rather complicated line integral in the complex plane, and if it had to be used each time to get back to the t-domain, then the Laplace transform method would not be worth using. The Uniqueness Property above is what saves matters, for it means that for continuous functions $f(t)$ one can go back and forth via a table once $F(s)$ has been calculated. Our table now has only one entry

$f(t)$	$F(s) = L\{f(t)\}$
e^{at}	$\dfrac{1}{s - a}$

or equivalently

$F(s)$	$f(t) = L^{-1}\{F(s)\}$
$\dfrac{1}{s - a}$	e^{at}

In the next section the table will be expanded to include most functions that will arise in the analysis of linear constant coefficient differential equations.

EXERCISES 4.3

1. Use the Laplace transform to solve the following initial value problems:

a) $y' + 3y = 4e^{-2t}$, $y(0) = 1$

b) $y' - y = e^{-t}$, $y(0) = -2$

c) $y' = 2e^{3t}$, $y(0) = 1$

d) $y' + 3y = 6$, $y(0) = 3$

e) $y' - \frac{3}{2}y = e^{t/2}$, $y(0) = 0$

2. a) Find the solution of the initial value problem

$$y' + y = 2e^{-t}, \qquad y(0) = 0$$

by using the variation of parameters method or otherwise.

b) Use the Laplace transform in (a) and show that

$$Y(s) = \frac{2}{(s+1)^2}.$$

From this infer that $L\{te^{-t}\} = 1/(s+1)^2$.

c) Generalize the above to find the solution of

$$y' + ky = be^{kt}, \qquad y(0) = 0$$

and from this infer that $L\{te^{kt}\} = 1/(s-k)^2$.

3. Use the Laplace transform and the Property of Linearity to solve the following initial value problems:

a) $y' + 2y = 6e^t - 3e^{4t}, \quad y(0) = 1$

b) $y' - y = 1 + 4e^{3t}, \quad y(0) = 5$

c) $y' + y = \cosh t, \quad y(0) = 1$ (see Exercise 2 above).

4. Follow the proof of the Differentiation Formula and integrate by parts twice to prove that

$$L\{f''(t)\} = s^2 F(s) - sf(0) - f'(0),$$

where $L\{f(t)\} = F(s)$.

4.4 FURTHER PROPERTIES AND TRANSFORM FORMULAS

In the previous section, when the Differentiation Formula was derived, we mentioned the class of functions of exponential order. It is defined as follows:

> The class of functions E_α of exponential order α is the set of all continuous functions on $0 \le t < \infty$ having the property that, if $f(t)$ is in E_α, there exists a constant α and a positive constant M such that
>
> $$|f(t)| \le Me^{\alpha t}, \quad 0 \le t < \infty.$$

In a later section the requirement that members of E_α be continuous will be dropped, but for now such generality is not needed.

In a nutshell, the functions of exponential order α are functions which grow no faster than the exponential $e^{\alpha t}$. For instance,

a) $f(t) = 3e^{2t}$ belongs to E_2 since $|f(t)| \le 3e^{2t}$, so $M = 3$ and $\alpha = 2$.

b) $f(t) = t^n$ belongs to E_α for any $\alpha > 0$ since

$$\frac{\alpha^n t^n}{n!} \le 1 + \alpha t + \cdots + \frac{\alpha^n t^n}{n!} + \cdots = e^{\alpha t},$$

which implies that $|t^n| \le (n!/\alpha^n)e^{\alpha t}$ and so $M = n!/\alpha^n$.

c) $f(t) = 4e^{-t} \cos 3t$ belongs to E_{-1} since

$$|4e^{-t} \cos 3t| \le 4e^{-t}.$$

d) $f(t) = e^{t^2}$ does not belong to E_α for any α, since $e^{t^2} \le Me^{\alpha t}$ implies that $e^{t^2 - \alpha t} \le M$. But if $\alpha > 0$, then for $t > \alpha$ the exponent $t^2 - \alpha t$ is positive and grows without bound, so the exponential cannot be bounded. If $\alpha \leq 0$, then $t^2 - \alpha t$ is positive for $t > 0$ and, again, the exponential is unbounded.

A little thought shows that the set of all possible solutions of homogeneous constant coefficient linear differential equations is of exponential order. These solutions are linear combinations of functions like e^{at}, $e^{at} \cos bt$, $e^{at} \sin bt$, and integral powers of t times them.

The important property of functions of exponential order is that they are Laplace transformable.

> If $f(t)$ belongs to E_α, then $L\{f(t)\} = F(s)$ exists for $\mathrm{Re}(s) > \alpha$.

The proof of this statement follows from these simple estimates:

$$\left| \int_0^\infty f(t)e^{-st} \, dt \right| \leq \int_0^\infty |f(t)| \, |e^{-st}| \, dt$$

$$\leq M \int_0^\infty e^{at} e^{-\mathrm{Re}(s)t} \, dt = M \int_0^\infty e^{-[\mathrm{Re}(s)-\alpha]t} \, dt.$$

The last improper integral converges if $\mathrm{Re}(s) - \alpha > 0$, which implies that

$$L\{f\} = \int_0^\infty f(t)e^{-st} \, dt$$

converges also. Since the last integral equals $M/[\mathrm{Re}(s) - \alpha]$, this also shows that

> If $f(t)$ belongs to E_α, than $F(s)$ approaches zero as $\mathrm{Re}(s)$ approaches infinity.

Functions like 1, s^n, $n > 0$, $s^2/(s^2 - 3)$ or e^{2s} cannot be Laplace transforms of functions of exponential order.

We now proceed to compute the Laplace transforms of a few elementary functions, then give a few other properties of the transform which allow the table of transforms to be extended.

Example 1   Show that $L\{\sin bt\} = \dfrac{b}{s^2 + b^2}$, $\mathrm{Re}(s) > 0$.

One can find this directly by computing

$$L\{\sin bt\} = \int_0^\infty e^{-st} \sin bt \, dt$$

through integration by parts. However, this integral and the transform of cos bt can be found by using complex exponentials as well. We remind the reader from our discussion in Chapter 2 that the differentiation (and hence the integration) formula for the exponential e^{zt} is the same regardless of whether z is real or complex. Therefore, from the previous formula for $L\{e^{at}\}$ it follows that

$$L\{e^{ibt}\} = \frac{1}{s - ib}, \quad \mathrm{Re}(s) > 0,$$

and furthermore

$$\frac{1}{s - ib} = \frac{1}{s - ib}\frac{s + ib}{s + ib} = \frac{s}{s^2 + b^2} + i\frac{b}{s^2 + b^2}.$$

Now use linearity and the fact that

$$L\{e^{ibt}\} = L\{\cos bt + i \sin bt\} = L\{\cos bt\} + i\,L\{\sin bt\},$$

then equate real and imaginary parts of the two expressions. □

Example 2 Show that $L\{\cos bt\} = \dfrac{s}{s^2 + b^2}$, $\mathrm{Re}(s) > 0$.

This follows from the previous example. □

Example 3 Show that $L\{\cosh bt\} = \dfrac{s}{s^2 - b^2}$, $\mathrm{Re}(s) > |b|$.

Recall that $\cosh bt = \frac{1}{2}(e^{bt} + e^{-bt})$; then, by the linearity property,

$$L\{\cosh bt\} = \frac{1}{2}\,[L\{e^{bt}\} + L\{e^{-bt}\}] = \frac{1}{2}\left[\frac{1}{s - b} + \frac{1}{s + b}\right] = \frac{s}{s^2 - b^2}. \quad □$$

Example 4 Show that $L\{\sinh bt\} = \dfrac{b}{s^2 - b^2}$, $\mathrm{Re}(s) > |b|$.

This follows from the identity $\sinh bt = \frac{1}{2}(e^{bt} - e^{-bt})$. □

Example 5 Show that $L\{t\} = \dfrac{1}{s^2}$, $\mathrm{Re}(s) > 0$.

A direct way to prove this is to use the definition of the transform and integrate by parts:

$$L\{t\} = \int_0^\infty e^{-st}t\, dt = -\frac{1}{s}\,te^{-st}\Big|_0^\infty + \frac{1}{s}\int_0^\infty e^{-st}\, dt.$$

Since t belongs to E_α for any $\alpha > 0$, then the first term equals zero when $\text{Re}(s) > 0$. The second term equals

$$\frac{1}{s} L\{1\} = \frac{1}{s} L\{e^{0t}\} = \frac{1}{s^2}.$$

A more general result left to the reader to prove is the formula

$$L\{t^n\} = \frac{n!}{s^{n+1}}, \quad \text{Re}(s) > 0.$$

It is interesting to note that all of the formulas above essentially depend on the transform of e^{at}.   □

We next state a property that allows the table of transforms to be extended considerably. It is called the *shift property*.

If $L\{f(t)\} = F(s)$, then $L\{e^{at} f(t)\} = F(s - a)$.

Therefore the computation of the Laplace transform of $e^{at} f(t)$ is simply accomplished by shifting the argument of $F(s)$ by α units. If $F(s)$ is valid for $\text{Re}(s) > a$, then $F(s - \alpha)$ is valid for $\text{Re}(s) > a + \alpha$. A proof of the shift property is immediate since

$$L\{e^{at} f(t)\} = \int_0^\infty e^{-st} e^{at} f(t) \, dt = \int_0^\infty e^{-(s-a)t} f(t) \, dt,$$

and the last expression is merely the Laplace transform of $f(t)$ evaluated at $s - \alpha$.

Example 6   Show that $L\{e^{2t} \cos 3t\} = \dfrac{s - 2}{(s - 2)^2 + 9}$, $\text{Re}(s) > 2$.

Since $L\{\cos 3t\} = s/(s^2 + 3^2)$, from Example 2 above, the shift from s to $(s - 2)$ gives the transform.   □

Example 7   Show that $L\{e^{at} t\} = \dfrac{1}{(s - \alpha)^2}$, $\text{Re}(s) > \alpha$.

From Example 5, $L\{t\} = 1/s^2$, and the required transform is obtained by the shift from s to $(s - \alpha)$.   □

Example 8   Find $L^{-1} \left\{ \dfrac{3s - 2}{s^2 + 6s + 25} \right\}$.

Completing the square in the denominator and applying a little algebra gives

$$\frac{3s - 2}{s^2 + 6s + 25} = \frac{3s - 2}{(s + 3)^2 + 16} = \frac{3(s + 3)}{(s + 3)^2 + 16} - \frac{11}{(s + 3)^2 + 16}$$

$$= 3\frac{s + 3}{(s + 3)^2 + 16} - \frac{11}{4}\frac{4}{(s + 3)^2 + 16}.$$

Therefore,

$$L^{-1}\left\{\frac{3s - 2}{s^2 + 6s + 2s}\right\} = 3e^{-3t}\cos 4t - \frac{11}{4}e^{-3t}\sin 4t.$$

The reader should follow the algebraic steps carefully in the last example since such manipulations will be performed frequently in the coming sections. □

Two other properties of the Laplace transform, which are occasionally useful, are:

1. If $f(t)$ belongs to E_α and its Laplace transform is $F(s)$, then

$$L\left\{\int_0^t f(r) \, dr\right\} = \frac{1}{s} F(s), \quad \text{Re}(s) > \alpha.$$

2. If $f(t)$ belongs to E_α, then its Laplace transform $F(s)$ has derivatives of all orders given by the formula

$$\frac{d^n}{ds^n} F(s) = L\{(-t)^n f(t)\}.$$

These two properties provide alternative ways of computing other transforms from the given ones; examples of this are left for the exercises.

Finally, we state a more general form of the differentiation property of Laplace transforms. Its proof is a repeated application of integration by parts combined with the argument given in the previous derivation.

The differentiation formulas. If $f(t), f'(t), \ldots, f^{(n)}(t)$ belong to E_α and $L\{f(t)\} = F(s)$, then, for $\text{Re}(s) > \alpha$, the following is valid:

$$L\{f'(t)\} = sF(s) - f(0),$$
$$L\{f''(t)\} = s^2F(s) - sf(0) - f'(0),$$
$$L\{f^{(n)}(t)\} = s^nF(s) - s^{n-1}f(0) - \cdots - sf^{(n-2)}(0) - f^{(n-1)}(0).$$

One sees, therefore, that the transform of the nth derivative depends only on the transform of the function and n initial data values.

Example 9 If $y(t)$ is the solution of a third order differential equation with initial conditions

$$y(0) = 1, \qquad y'(0) = -3, \qquad y''(0) = 2,$$

and $Y(s)$ is its Laplace transform, then

$$L\{y'(t)\} = sY(s) - 1,$$
$$L\{y''(t)\} = s^2 Y(s) - s + 3,$$
$$L\{y'''(t)\} = s^3 Y(s) - s^2 + 3s - 2. \quad \square$$

We conclude this section with a short table of Laplace transforms (Table 4.1) and then proceed to solve some differential equations. The last three entries in the table have not been previously discussed, but will be derived later. They are added at this point for the reader's convenience.

EXERCISES 4.4

1. Use Table 4.1 to find the Laplace transforms of the following functions. Give the region of validity.

a) t^3

b) $5 \cos 2t$

c) $\sin 3t$

d) $e^{-t} \cos 4t$

e) $4e^{2t} \sin t$

f) $t \sin 2t$

g) $-2t \sin t$

h) $3 \sinh 7t$

i) $e^{-3t} \cosh t$

j) $e^{-t} t^2$

k) $e^{2t} t \cos t$

l) $\int_0^t r \sin r \, dr$

m) $4t^2 \sin 3t$

n) $5 + 4t$

o) $3e^{-t} + 4 \sin 2t$

p) $e^{2t} \cos (t + \pi/4)$

q) $9 \sinh t \sin 3t$

r) $t^4 \cosh 2t$

2. Use Table 4.1 to find the inverse Laplace transform of the following functions:

a) $\dfrac{4}{s^3}$

b) $\dfrac{3}{s + 1}$

c) $\dfrac{5}{s - 2}$

d) $\dfrac{3}{s^2 + 3}$

e) $\dfrac{4s}{s^2 + 3}$

f) $\dfrac{8}{s^2 - 16}$

g) $\dfrac{1}{s^2 - 2s + 5}$

h) $\dfrac{s + 4}{s^2 + 8s + 25}$

i) $\dfrac{6}{s^2 + 2s + 4}$

j) $\dfrac{4s - 9}{s^2 + 9}$

k) $\dfrac{3s + 1}{s^2 - 6s + 18}$

l) $\dfrac{6s}{(s^2 + 4)^2}$

m) $\dfrac{-9}{(s - 2)^2}$

n) $\dfrac{12}{(s + \pi)^4}$

o) $\dfrac{6}{s - 3} + \dfrac{s + 4}{s^2 + 25}$

p) $\dfrac{4}{(s + 1)^2} - \dfrac{2s + 3}{s^2 + s + 5/2}$

q) $\dfrac{2}{(s^2 + 1)^2}$

r) $\dfrac{-(s^2 + 9)}{(s^2 - 9)^2}$

TABLE 4.1 **Short Table of Laplace Transforms**

	$f(t)$	$F(s)$	Region of validity
1.	1	$\dfrac{1}{s}$	$\text{Re}(s) > 0$
2.	$t^n, n = 0, 1, 2, \ldots$	$\dfrac{n!}{s^{n+1}}$	$\text{Re}(s) > 0$
3.	e^{at}	$\dfrac{1}{s - a}$	$\text{Re}(s) > a$
4.	$\sin bt$	$\dfrac{b}{s^2 + b^2}$	$\text{Re}(s) > 0$
5.	$\cos bt$	$\dfrac{s}{s^2 + b^2}$	$\text{Re}(s) > 0$
6.	$\sinh bt$	$\dfrac{b}{s^2 - b^2}$	$\text{Re}(s) > \lvert b \rvert$
7.	$\cosh bt$	$\dfrac{s}{s^2 - b^2}$	$\text{Re}(s) > \lvert b \rvert$
8.	$e^{at}\sin bt$	$\dfrac{b}{(s - a)^2 + b^2}$	$\text{Re}(s) > a$
9.	$e^{at}\cos bt$	$\dfrac{s - a}{(s - a)^2 + b^2}$	$\text{Re}(s) > a$
10.	$t \sin bt$	$\dfrac{2bs}{(s^2 + b^2)^2}$	$\text{Re}(s) > 0$
11.	$t \cos bt$	$\dfrac{s^2 - b^2}{(s^2 + b^2)^2}$	$\text{Re}(s) > 0$
12.	$e^{at}f(t)$	$F(s - a)$	$\text{Re}(s) > \alpha + a$ if $F(s)$ valid for $\text{Re}(s) > \alpha$
13.	$\displaystyle\int_0^t f(r)\, dr$	$\dfrac{1}{s} F(s)$	$\text{Re}(s) > \max(a, 0)$ if $F(s)$ valid for $\text{Re}(s) > a$
14.	$(-t)^n f(t)$	$\dfrac{d^n}{ds^n} F(s)$	valid for same region as $F(s)$
15.	$\dfrac{1}{b}\sin bt - t \cos bt$	$\dfrac{2b^2}{(s^2 + b^2)^2}$	$\text{Re}(s) > 0$
16.	$f(t - c)u(t - c)$	$e^{-sc}F(s)$	valid for same region as $F(s)$
17.	$f(t + T) = f(t)$	$\dfrac{\displaystyle\int_0^T e^{-sr}f(r)\, dr}{1 - e^{-sT}}$	$\text{Re}(s) > 0$

3. Establish entry (2) of Table 4.1 by using entries (14) and (1) or induction.

4. Establish entry (11) of Table 4.1 by using entries (14) and (5).

5. Using entries (13) and (10) of Table 4.1 and integration by parts, derive the formula

$$L^{-1}\left\{\frac{2b^2}{(s^2 + b)^2}\right\} = \frac{1}{b}\sin bt - t\cos bt.$$

6. Establish entry (13) of Table 4.1.

$$\textit{Hint: } \frac{d}{dt}\int_0^t f(r)\,dr = f(t).$$

7. Given that $y(t)$ is the solution of a second order differential equation and satisfies the following initial conditions, find $L\{y'(t)\}$ and $L\{y''(t)\}$ in terms of $Y(s) = L\{y(t)\}$.

a) $y(0) = 0,\quad y'(0) = 1$

b) $y(0) = 3,\quad y'(0) = -2$

c) $y(0) = 1,\quad y'(0) = 0$

d) $y(0) = -1,\quad y'(0) = 1$

e) $y(0) = 0,\quad y'(0) = 0$

f) $y(0) = -2,\quad y'(0) = -1$

8. Find $L\{y^{(n)}(t)\}$, $n = 1, 2, 3, 4$, in terms of

$Y(s) = L\{y(t)\}$ if $y(t)$ satisfies the following initial conditions:

a) $y(0) = 0, y'(0) = -1, y''(0) = 0, y^{(3)}(0) = 1$

b) $y(0) = y'(0) = 1, y''(0) = y^{(3)}(0) = -1$

c) $y(0) = 1, y'(0) = y''(0) = 0, y^{(3)}(0) = 2$

9. The following formula for the integration of the transform:

$$L\left\{\frac{f(t)}{t}\right\} = \int_s^\infty F(r)\,dr$$

where $F(s) = L\{f(t)\}$ is similar to entry (13) of Table 4.1. The formula is valid if $f(t)$ is of class E_α and $\lim_{t\to 0+}[f(t)/t]$ exists. Use the formula to show that

a) $L\left\{\dfrac{\sin kt}{t}\right\} = \text{arc cot }\dfrac{s}{k}$

b) $L\left\{\displaystyle\int_0^t \dfrac{\sin \tau}{\tau}\,d\tau\right\} = \dfrac{1}{s}\text{ arc cot }s$

c) $L\left\{\dfrac{1 - e^{-t}}{t}\right\} = \ln\left(1 + \dfrac{1}{s}\right)$

10. Verify that the following functions are of class E_α and find a suitable value of α and positive constant M:

a) $t^2 \sin t$ b) $3\cosh 2t$

c) 2^{3t-1} d) $\exp[t^{2/3}]$

e) $6(t^2 + 1)^{200}$ f) $\tanh t$

4.5 SOLVING CONSTANT COEFFICIENT LINEAR EQUATIONS

We begin with a fairly simple example which partially avoids one of the difficulties of the Laplace transform technique, namely the need to compute lengthy partial fraction decompositions. The idea here is to clearly demonstrate the method free from a lot of computational obfuscation. The examples following will contain more partial fraction decompositions, but in the next section will be given a set of formulas, called *Heaviside's formulas*, which will make the task much easier.

Example 1   Solve the initial value problem

$$y'' + 9y = \cos 3t, \qquad y(0) = 1, \qquad y'(0) = -1.$$

If $Y(s) = L\{y(t)\}$, then from the initial data and the differentiation formulas it follows that

$$L\{y''(t)\} = s^2 Y(s) - s + 1.$$

Applying the Laplace transform to both sides of the differential equation and using entry (5) from Table 4.1 gives

$$s^2 Y(s) - s + 1 + 9 Y(s) = L\{\cos 3t\} = \frac{s}{s^2 + 9}.$$

Now solve for $Y(s)$ to obtain

$$Y(s) = \frac{s - 1}{s^2 + 9} + \frac{s}{(s^2 + 9)^2}.$$

The denominator is $s^2 + 9 = s^2 + 3^2$, and an examination of the last term tells us that the key entries in Table 4.1 are (4), (5), and (10). A little rewriting gives

$$Y(s) = \frac{s}{s^2 + 9} - \frac{1}{3} \frac{3}{s^2 + 9} + \frac{1}{6} \frac{6s}{(s^2 + 9)^2},$$

and now the inverse transform can be applied (by using Table 4.1 directly!) to obtain

$$y(t) = \cos 3t - \tfrac{1}{3} \sin 3t + \tfrac{1}{6} t \sin 3t,$$

the desired solution. □

Example 2 Solve the initial value problem

$$y'' + 3y' + 2y = 6, \qquad y(0) = 0, \qquad y'(0) = 2.$$

Proceeding as above, first use the initial data to compute

$$L\{y''(t)\} = s^2 Y(s) - 2, \qquad L\{y'(t)\} = s Y(s);$$

applying the Laplace transform to the equation gives

$$s^2 Y(s) - 2 + 3s Y(s) + 2 Y(s) = L\{6\} = \frac{6}{s}.$$

Now solve for $Y(s)$ to obtain

$$Y(s) = \frac{2}{s^2 + 3s + 2} + \frac{6}{s(s^2 + 3s + 2)}$$

or

$$Y(s) = \frac{2}{(s + 1)(s + 2)} + \frac{6}{s(s + 1)(s + 2)} = \frac{2s + 6}{s(s + 1)(s + 2)}.$$

From the theory of partial fractions we know that

$$\frac{2s + 6}{s(s + 1)(s + 2)} = \frac{A}{s} + \frac{B}{s + 1} + \frac{C}{s + 2};$$

combining terms on the right side and equating the numerators gives the relation

$$2s + 6 = A(s + 1)(s + 2) + Bs(s + 2) + Cs(s + 1).$$

Equating like powers of s leads to the equations

$$A + B + C = 0, \qquad 3A + 2B + C = 2, \qquad 2A = 6.$$

These have the solution $A = 3$, $B = -4$, $C = 1$, which could have also been obtained by letting $s = 0$, $s = -1$, and $s = -2$, successively, in the previous equation. Therefore,

$$Y(s) = \frac{3}{s} - \frac{4}{s + 1} + \frac{1}{s + 2},$$

and now Table 4.1 gives

$$y(t) = 3 - 4e^{-t} + e^{-2t},$$

the required solution. $\square$

At this point the reader may well ask what is the advantage of using Laplace transforms to solve problems such as Example 1 and Example 2? Example 1 is a case of resonance, so we know that the solution is of the form

$$y(t) = A \cos 3t + B \sin 3t + Ct \cos 3t + Dt \sin 3t$$

and substitution and comparison of coefficients would give the answer. From the standpoint of algebra it looks like the transform method has a slight edge, especially since the initial conditions are built in.

In Example 2 this advantage is certainly not evident since we had to factor the characteristic polynomial $s^2 + 3s + 2 = (s + 1)(s + 2)$ anyway, and this would indicate that the trial solution is of the form

$$y(t) = Ae^{-t} + Be^{-2t} + C.$$

In view of the time spent on the partial fraction decomposition, direct substitution would probably win the day. In the next section some formulas will be introduced, which reduce the computational time but, at best, the matter will be a toss up.

In the authors' opinion, the real advantage of the Laplace transform method occurs when one is faced with a constant coefficient differential equation in which the forcing term (the right-hand side) is a discontinuous function or one with corners. For instance, the function

$$f(t) = \begin{cases} 0 \text{ for } t < 1, \\ I \text{ for } t \geq 1, \end{cases}$$

would ideally model, say, a current or force which is zero for $t < 1$, and then is switched on to full power I at $t = 1$. Finding the solution of the initial value problem

$$y'' + ay' + by = f(t), \qquad y(0) = A, \qquad y'(0) = B$$

cannot be done by comparison of coefficients, since the right-hand side is not a polynomial, or an exponential, or a sine, or a cosine function. Variation of parameters will work if used with care, but the Laplace transform method is the sure winner. This class of problems will be considered shortly.

Another class of problems where the Laplace transform has a clear advantage is the solving of systems of linear differential equations. The reason for this is that some theoretical algebraic considerations are bypassed and, again, that the initial conditions are contained in the transformation. This is demonstrated in the next example.

Example 3 Find the solution $x(t)$, $y(t)$ of the initial value problem

$$\dot{x} = 2x + y, \qquad x(0) = 1,$$
$$\dot{y} = 3x + 4y, \qquad y(0) = 0.$$

This is a constant coefficient system of two linear differential equations, and so we must apply the Laplace transform to each equation. Letting $X(s) = L\{x(t)\}$ and $Y(s) = L\{y(t)\}$, we get by the differentiation formula and the initial data that

$$L\{\dot{x}(t)\} = sX(s) - 1,$$
$$L\{\dot{y}(t)\} = sY(s).$$

Now apply the transform to each differential equation to get

$$sX(s) - 1 = 2X(s) + Y(s),$$
$$sY(s) = 3X(s) + 4Y(s),$$

or, equivalently,

$$(s - 2)X(s) - Y(s) = 1,$$
$$-3X(s) + (s - 4)Y(s) = 0.$$

This is a system of linear algebraic equations which must be solved for $X(s)$

and $Y(s)$; using Cramer's rule we obtain

$$X(s) = \frac{\begin{vmatrix} 1 & -1 \\ 0 & s-4 \end{vmatrix}}{\begin{vmatrix} s-2 & -1 \\ -3 & s-4 \end{vmatrix}}$$

$$= \frac{s-4}{s^2 - 6s + 5} = \frac{s-4}{(s-1)(s-5)},$$

$$Y(s) = \frac{\begin{vmatrix} s-2 & 1 \\ -3 & 0 \end{vmatrix}}{\begin{vmatrix} s-2 & -1 \\ -3 & s-4 \end{vmatrix}}$$

$$= \frac{3}{(s-1)(s-5)}.$$

Finally, a partial fraction decomposition gives

$$X(s) = \frac{s-4}{(s-1)(s-5)}$$

$$= \frac{3/4}{s-1} + \frac{1/4}{s-5},$$

$$Y(s) = \frac{3}{(s-1)(s-5)}$$

$$= \frac{-3/4}{s-1} + \frac{3/4}{s-5},$$

and therefore

$$x(t) = \frac{3}{4}e^t + \frac{1}{4}e^{5t}, \qquad y(t) = -\frac{3}{4}e^t + \frac{3}{4}e^{5t}. \quad \square$$

After studying linear systems in Chapter 5, the reader will see how efficient the Laplace transform is as a solution technique and how many of the algebraic difficulties are circumvented.

As a simple application of linear systems of differential equations consider the following:

A mixing problem

Two 100-gallon tanks are connected in the manner depicted in Fig. 4.2, where the arrows indicate the flow of liquid from and to the outside and between the two tanks. Fresh water is flowing into tank A from the outside,

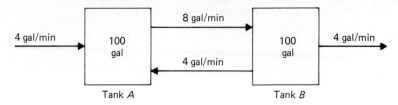

Figure 4.2

and at $t = 0$ tank A contains 100 gallons of fresh water, whereas tank B contains 100 gallons of brine consisting of 10 pounds of dissolved salt (that is, 0.1 pound of salt per gallon). Find the amounts of salt $x(t)$, $y(t)$ (in pounds) in tanks A and B, respectively, as functions of time assuming instantaneous stirring. The solution will lead to a simple linear system of differential equations.

Since $x(t)$ is the number of pounds of salt in tank A at time t, the rate of change of $x(t)$ is

$$\dot{x}(t) = (\text{Rate in}) - (\text{Rate out}),$$

in units of lb/min. To determine how many pounds of salt per minute are going into A, we notice that the only salt contribution to A is from tank B since the 4 gal/min coming from the outside is fresh water. At time t, tank B contains $y(t)$ pounds of salt dissolved in 100 gallons of water, and this flows into tank A at the rate of 4 gal/min. Hence for tank A we can write:

$$(\text{Rate in}) = \frac{y(t)}{100} \text{ (lb/gal)} \cdot 4 \text{ (gal/min)} = \frac{4}{100} y(t) \text{ (lb/min)},$$

whereas

$$(\text{Rate out}) = \frac{x(t)}{100} \text{ (lb/gal)} \cdot 8 \text{ (gal/min)} = \frac{8}{100} x(t) \text{ (lb/min)}.$$

Also, at $t = 0$ tank A contains no salt, so $x(0) = 0$. This gives the first differential equation and initial condition

$$\dot{x} = \frac{4}{100} y - \frac{8}{100} x, \qquad x(0) = 0$$

or

$$\dot{x} = -\frac{2}{25} x + \frac{1}{25} y, \qquad x(0) = 0. \qquad (4.5.1)$$

The same analysis of tank B leads to the second differential equation and initial condition

$$\dot{y} = \frac{8}{100} x - \frac{8}{100} y, \qquad y(0) = 10$$

or

$$\dot{y} = \frac{2}{25} x - \frac{2}{25} y, \qquad y(0) = 10. \tag{4.5.2}$$

Equations (4.5.1) and (4.5.2) are first order linear systems of differential equations with initial conditions at $t = 0$. Letting $X(s)$ and $Y(s)$ be the Laplace transforms of $x(t)$ and $y(t)$, respectively, and applying the differential formula, we get

$$sX(s) = -\frac{2}{25} X(s) + \frac{1}{25} Y(s),$$

$$sY(s) - 10 = \frac{2}{25} X(s) - \frac{2}{25} Y(s),$$

which must be solved for $X(s)$ and $Y(s)$. From Cramer's rule we obtain

$$X(s) = \frac{10}{25} \frac{1}{(s + 2/25)^2 - (\sqrt{2}/25)^2} = \frac{10}{\sqrt{2}} \frac{\sqrt{2}/25}{(s + 2/25)^2 - (\sqrt{2}/25)^2},$$

$$Y(s) = 10 \frac{s + 2/25}{(s + 2/25)^2 - (\sqrt{2}/25)^2},$$

and now the shift formula and entries (6) and (7) of Table 4.1 give the solution

$$x(t) = \frac{10}{\sqrt{2}} e^{-2t/25} \sinh \frac{\sqrt{2}}{25} t, \qquad y(t) = 10e^{-2t/25} \cosh \frac{\sqrt{2}}{25} t$$

for the amounts of salt in pounds at time t in tanks A and B, respectively. By converting the hyperbolic sine and cosine into their exponential form, the reader can further analyze the system and easily see that both $x(t)$ and $y(t)$ approach zero as t approaches infinity. After a long time both tanks will essentially contain fresh water.

EXERCISES 4.5

1. Use the Laplace transform to solve the following initial value problems:

   a) $y'' + 9y = 0$, $y(0) = 1$, $y'(0) = 2$

   b) $y'' - y' - 6y = 0$, $y(0) = 0$, $y'(0) = 3$

   c) $y'' + y = \sin t$, $y(0) = 2$, $y'(0) = -1$

   d) $y'' + 5y' + 6y = 4e^t$, $y(0) = y'(0) = 0$

   e) $y'' - 4y = 8$, $y(0) = 2$, $y'(0) = 0$

   f) $y'' + 2y' + 5y = 0$, $y(0) = 6$,
      $y'(0) = -3$

2. Use the Laplace transform to find the solutions $x(t)$, $y(t)$ of the following initial value problems:

   a) $\dot{x} = x + y$, $x(0) = 3$,
      $\dot{y} = 9x + y$, $y(0) = -3$

   b) $\dot{x} = 4x + 5y$, $x(0) = -1$,
      $\dot{y} = -4x - 4y$, $y(0) = 2$

   c) $\dot{x} = x + 3y$, $x(0) = 1$,
      $\dot{y} = 3x + y$, $y(0) = 0$

   d) $\dot{x} = 2x + 4y$, $x(0) = 1$,
      $\dot{y} = -4x + 2y$, $y(0) = -1$

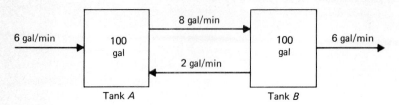

Figure 4.3

e) $\dot{x} = 3x - 3y, \quad x(0) = 4,$
$\quad \dot{y} = 6x - 3y, \quad y(0) = 3$

f) $\dot{x} = 2x - y, \quad x(0) = 1,$
$\quad \dot{y} = x + 4y, \quad y(0) = 0$

3. Given the inhomogeneous system of linear differential equations

$$\dot{x} = ax + by + f(t), \qquad x(0) = 0,$$
$$\dot{y} = cx + dy + g(t), \qquad y(0) = 0,$$

show that the Laplace transforms of the solutions $x(t)$, $y(t)$ are

$$X(s) = \frac{(s-d)F(s)}{P(s)} + \frac{bG(s)}{P(s)},$$

$$Y(s) = \frac{(s-a)G(s)}{P(s)} + \frac{cF(s)}{P(s)},$$

where $F(s)$ and $G(s)$ are the Laplace transforms of $f(t)$ and $g(t)$, respectively, and

$$P(s) = s^2 - (a+d)s + (ad - bc).$$

(The above initial value problem would represent a system at rest being excited by driving forces $f(t)$ and $g(t)$.)

4. Use the results of Exercise 3 or any other method to find the solutions $x(t)$, $y(t)$ of the following initial value problems:

a) $\dot{x} = x + y, \quad x(0) = 0$
$\quad \dot{y} = 9x + y + e^t, \quad y(0) = 0$

b) $\dot{x} = -2x + y + 1, \quad x(0) = 0$
$\quad \dot{y} = x - 2y + e^{-t}, \quad y(0) = 0$

5. Consider a mixing problem with flow configuration given by Fig. 4.3 and with the following initial data: Tank A contains 100 gal of brine consisting of 10 lb of dissolved salt. Fresh water is flowing in at the rate of 6 gal/min. Tank B contains 100 gal of fresh water.

a) Find the amounts of salt $x(t)$, $y(t)$ (in pounds) in tanks A and B. Show that both approach zero as t approaches infinity.

b) Instead of salt, let the dissolved quantity be 10 lb of a toxic substance. Then one would be interested in finding the time T at which the system has been flushed sufficiently, so that the outflow from tank B is at a "safe" level. Using the result from (a), show that the maximum amount of toxic substance in tank B is approximately 3.85 lb (a concentration of 0.0385 lb/gal) and that this maximum occurs after approximately 13.7 minutes. Also show that for $T \simeq 138$ min the concentration in tank B is less than 0.0004 lb/gal.

c) Suppose that to reduce the toxicity in tank B faster, we pump fresh water directly into it from the outside at a rate of 4 gal/min. To maintain the volume at 100 gal we must therefore pump water out at the rate of 10 gal/min instead of 6 gal/min. Find the amount of toxic substance $y(t)$ in tank B. Show that for $T \simeq 98$ min the concentration in tank B is less than 0.0004 lb/gal.

d) Show that the strategy in (c) is equivalent to draining tank A to the outside at a rate of 4 gal/min and pumping fresh water into it at a rate of 10 gal/min instead of 6 gal/min. Do this by finding $L\{y(t)\}$.

6. A closed two-tank system has the configuration depicted in Fig. 4.4. The initial data for each tank are: Tank A contains 8 lb of chemical X dissolved in 100 gal of fresh water. Tank B contains 4 lb of chemical X dissolved in 100 gal of fresh water.

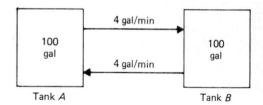

Tank A Tank B

Figure 4.4

a) Find the amounts $x(t)$, $y(t)$ of chemical X in tanks A and B.

b) Show that $x(t)$ is a decreasing function and $y(t)$ is an increasing function of time, and both approach 6 lb as t approaches infinity.

c) If we require only that the amount $y(T)$ of chemical X equal $(6 - \epsilon)$ lb, where $0 < \epsilon < 2$, show that $T = (25/2) \ln (2/\epsilon)$. Graph T as a function of ϵ.

4.6 THE HEAVISIDE FORMULAS

A better title for this section would be "Partial Fractions Made Partially Easier" and the material is based on some formulas developed by a reclusive English engineer, Oliver Heaviside (1850–1925). He developed heuristically many of the techniques described in this chapter under the title of "the operational calculus," and at that time his work was derided by the scientists of the day. Later researchers justified the correctness of his calculations and pointed out their relation to the Laplace transform introduced a hundred years earlier.

As the reader has seen in the last few sections, the Laplace transforms we have obtained are expressions of the form

$$F(s) = P(s)/Q(s),$$

where P and Q are polynomials in s, the degree of P is less than the degree of Q, and P and Q have no common factors.

This is a rational function of s, and the object of a partial fraction decomposition is to reduce it to a sum of simpler rational functions whose inverse Laplace transform can be easily looked up in Table 4.1.

Since $Q(s)$ is a polynomial, we know that over the real number field its only possible factors are

1. $(s - c)$, where c is a real simple root; hence $(s - c)^2$ is not a factor of $Q(s)$;

2. $(s - c)^k$, $k = 1, 2, \ldots, r$, so c is a real root of multiplicity r; hence $(s - c)^{r+1}$ is not a factor of $Q(s)$;

3. $(s - a)^2 + b^2$, $b \neq 0$, which is a simple irreducible quadratic factor; hence $[(s - a)^2 + b^2]^2$ is not a factor of $Q(s)$;

4. $[(s - a)^2 + b^2]^k$, $k = 1, 2, \ldots, q$, which is an irreducible quadratic factor of order q; hence $[(s - a)^2 + b^2]^{q+1}$ is not a factor of $Q(s)$.

By irreducible is meant having only complex roots, and the reader can see clearly that $(s - a)^2 + b^2$ cannot have real roots if a and b are real, $b \neq 0$. By completing the square, we can write any quadratic polynomial with complex roots in this form.

The Heaviside formulas, which will be developed, deal specifically with Cases 1 and 2 above; afterwards something will be said about Cases 3 and 4 in the common situation where $q = 2$.

Case 1. If $Q(s)$ has simple roots $c_1, c_2, \ldots, c_n$, then the partial fraction expansion of $P(s)/Q(s)$ will be of the form

$$\frac{P(s)}{Q(s)} = \frac{A_1}{s - c_1} + \frac{A_2}{s - c_2} + \cdots + \frac{A_n}{s - c_n} + N(s),$$

where $N(s)$ is a collection of terms (if any) not involving $s - c_i$, $i = 1, 2, \ldots, n$.

Heaviside formula 1.

$$A_i = \frac{P(c_i)}{Q'(c_i)} \text{ or, equivalently, } A_i = \frac{P(c_i)}{Q_i(c_i)},$$

where $Q_i(s) = Q(s)/(s - c_i)$, $i = 1, 2, \ldots, n$. This is an easy result to verify. Multiply $P(s)/Q(s)$ by $s - c_i$ to obtain

$$\frac{P(s)}{Q(s)} (s - c_i) = \frac{(s - c_i)A_1}{s - c_1} + \cdots + A_i + \cdots$$

$$+ \frac{(s - c_i)A_n}{s - c_n} + (s - c_i)N(s).$$

Now let s approach c_i. On the right, all terms go to zero leaving A_i, whereas on the left $P(s)$ approaches $P(c_i)$ and the denominator approaches $Q'(c_i)$ because $Q(c_i) = 0$. Since c_i is a simple root, $Q(s) = (s - c_i)Q_i(s)$, where c_i is not a root of $Q_i(s)$.

Example 1 Decompose

$$\frac{P(s)}{Q(s)} = \frac{3s + 2}{s^2 + 2s - 3} = \frac{A_1}{s - 1} + \frac{A_2}{s + 3}.$$

Here $P(s) = 3s + 2$, $Q'(s) = 2s + 2$, and $c_1 = 1$, $c_2 = -3$; hence $A_1 = 5/4$ and $A_2 = -7/(-4) = 7/4$. ☐

Example 2 Decompose

$$\frac{P(s)}{Q(s)} = \frac{s^2 + 1}{(s - 1)(s^2 + 4)} = \frac{A_1}{s - 1} + \frac{B_1 s + C}{s^2 + 4}.$$

Here $P(s) = s^2 + 1$, $c_1 = 1$, and $Q_1(s) = s^2 + 4$; hence $A_1 = \frac{2}{5}$. □

Example 3 Decompose

$$\frac{P(s)}{Q(s)} = \frac{1}{s^4 - 1} = \frac{A_1}{s - 1} + \frac{A_2}{s + 1} + \frac{B_1 s + c}{s^2 + 1}.$$

Here $P(s) = 1$, $Q'(s) = 4s^3$, and $c_1 = 1$, $c_2 = -1$; hence $A_1 = \frac{1}{4}$ and $A_2 = -\frac{1}{4}$. □

A simple summary of Case 1 is: cancel $s - c_j$ from the denominator of $P(s)/Q(s)$ and evaluate the remainder at $s = c_j$ to get A_j.

Case 2. If $Q(s)$ has a root c of multiplicity r, then the partial fraction expansion of $P(s)/Q(s)$ will be of the form

$$\frac{P(s)}{Q(s)} = \frac{A_1}{s - c} + \frac{A_2}{(s - c)^2} + \cdots + \frac{A_r}{(s - c)^r} + N(s),$$

where, again, $N(s)$ is a collection of terms (if any) not involving $s - c$.

Heaviside formula 2. Let $\phi(s) = (s - c)^r \dfrac{P(s)}{Q(s)}$. Then

$$A_r = \phi(c), \quad A_{r-1} = \frac{\phi'(c)}{1!}, \quad A_{r-2} = \frac{\phi''(c)}{2!}, \quad \cdots, \quad A_1 = \frac{\phi^{(r-1)}(c)}{(r - 1)!}.$$

Rather than doing the general case, perhaps it is easier to give the idea of the proof by considering the case when $r = 3$. Multiply the general expression for the expansion by $(s - c)^3$ to get

$$\phi(s) = A_1(s - c)^2 + A_2(s - c) + A_3 + (s - c)^3 N(s);$$

when s approaches c, we find that $A_3 = \phi(c)$. Now, differentiate both sides to obtain

$$\phi'(s) = 2A_1(s - c) + A_2 + 3(s - c)^2 N(s) + (s - c)^3 N'(s),$$

from which it follows that $A_2 = \phi'(c)$. Differentiating again gives

$$\phi''(s) = 2A_1 + 6(s - c)N(s) + 6(s - c)^2 N'(s) + (s - c)^3 N''(s),$$

and so $\phi''(c) = 2A_1$ or $A_1 = \phi''(c)/2!$.

Example 4   Decompose

$$\frac{P(s)}{Q(s)} = \frac{s^2 + 2}{(s - 4)^3} = \frac{A_1}{s - 4} + \frac{A_2}{(s - 4)^2} + \frac{A_3}{(s - 4)^3}.$$

Here $\phi(s) = s^2 + 2$ and $c = 4$; so $A_3 = \phi(4) = 18$, $A_2 = \phi'(4) = 8$, and $A_1 = \phi''(4)/2 = 1$. ☐

Example 5   Decompose

$$\frac{P(s)}{Q(s)} = \frac{4s + 1}{s^2(s^2 - 2s + 4)} = \frac{A_1}{s} + \frac{A_2}{s^2} + \frac{Bs + C}{s^2 - 2s + 4}.$$

Since $c = 0$ and the quadratic factor $s^2 - 2s + 4 = (s - 1)^2 + 3$ is irreducible,

$$\phi(s) = \frac{4s + 1}{s^2 - 2s + 4}, \qquad \phi'(s) = \frac{-4s^2 - 2s + 18}{(s^2 - 2s + 4)^2};$$

hence $A_2 = \phi(0) = \frac{1}{4}$ and $A_1 = \phi'(0) = \frac{18}{16} = \frac{9}{8}$. ☐

Case 3.   If $Q(s)$ has a simple irreducible quadratic factor $(s - a)^2 + b^2$, then in the partial fraction expansion there will be a term of the form

$$\frac{Bs + C}{(s - a)^2 + b^2},$$

where B and C are constants to be determined.

There is a Heaviside formula for this case, but it is considerably more complicated than the previous ones. Presumably in applications one will not be faced with more than one or two such terms, so it is easier just to compute them directly. First, calculate all the coefficients corresponding to simple or multiple roots (if any) to reduce the bookkeeping, then either combine terms and compare like powers of s, or just substitute various selected values for s. In either case, linear equations for the unknown coefficients will result.

Example 6   Find $L^{-1} \left\{ \dfrac{4s + 1}{s^2(s^2 - 2s + 4)} \right\}$.

This was Example 5 of the previous case, and we found that

$$\frac{4s + 1}{s^2(s^2 - 2s + 4)} = \frac{9/8}{s} + \frac{1/4}{s^2} + \frac{Bs + C}{s^2 - 2s + 4},$$

which we can rewrite as

$$Bs + C = \frac{4s + 1}{s^2} - \left(\frac{9/8}{s} + \frac{1/4}{s^2} \right) (s^2 - 2s + 4).$$

Now let $s = 1$ and $s = -1$ (any two nonzero values would do) to obtain the pair of linear equations

$$B + C = \frac{7}{8} \quad \text{and} \quad -B + C = \frac{25}{8},$$

which gives $B = -\%$ and $C = 2$. Expressing $s^2 - 2s + 4$ as $(s - 1)^2 + (\sqrt{3})^2$, we now have to find

$$L^{-1}\left\{\frac{9}{8}\frac{1}{s} + \frac{1}{4}\frac{1}{s^2} + \frac{-(9/8)s + 2}{(s - 1)^2 + (\sqrt{3})^2}\right\}.$$

The first two terms are no problem, but the numerator of the last term must be expressed in terms of $(s - 1)$, so we can go directly to Table 4.1. An easy calculation yields

$$-\frac{9}{8}s + 2 = -\frac{9}{8}(s - 1) + \frac{7}{8} = -\frac{9}{8}(s - 1) + \frac{7\sqrt{3}}{8\sqrt{3}},$$

which finally requires us to find

$$f(t) = L^{-1}\left\{\frac{9}{8}\frac{1}{s} + \frac{1}{4}\frac{1}{s^2} - \frac{9}{8}\frac{s - 1}{(s - 1)^2 + (\sqrt{3})^2} + \frac{7}{8\sqrt{3}}\frac{\sqrt{3}}{(s - 1)^2 + (\sqrt{3})^2}\right\}.$$

The answer can now be read directly from Table 4.1, and

$$f(t) = \frac{9}{8} + \frac{1}{4}t - \frac{9}{8}e^t \cos \sqrt{3}\, t + \frac{7}{8\sqrt{3}}e^t \sin \sqrt{3}\, t. \quad \Box$$

Case 4 (a few remarks). If $Q(s)$ has an irreducible quadratic factor $(s - a)^2 + b^2$ of order $q \geq 1$, it can be shown that this corresponds to transforming a differential equation of order $2q$ or more. In practice, $q = 1$ (Case 3) and $q = 2$ are the only values that frequently occur.

Corresponding to a factor $[(s - a)^2 + b^2]^2$ in $Q(s)$, the partial fraction expansion of $P(s)/Q(s)$ contains a term of the form

$$\frac{As + B}{(s - a)^2 + b^2} + \frac{Cs + D}{[(s - a)^2 + b^2]^2},$$

and the values of A, B, C, and D have to be found by "brute force." Unbelievably, Heaviside devised a formula for this case, but it is even more laborious than the already complicated one he devised for Case 3!

The first term above presents no problem since its inverse transform will be a sum of $e^{at} \cos bt$ and $e^{at} \sin bt$. If the second term is written as

$$\frac{C}{2b}\frac{2b(s - a)}{[(s - a)^2 + b^2]^2} + \frac{D + Ca}{[(s - a)^2 + b^2]^2},$$

then the shift formula and entry (10) of Table 4.1 show that the first part has an inverse transform

$$\frac{C}{2b}e^{at}t \sin bt.$$

For the second part we need the identity

$$\frac{1}{[(s-a)^2 + b^2]^2} = \frac{1}{2b^3}\frac{b}{(s-a)^2 + b^2} - \frac{1}{2b^2}\frac{(s-a)^2 - b^2}{[(s-a)^2 + b^2]^2},$$

and now the shift formula and entries (8) and (11) give the inverse transform

$$(D + Ca)\left[\frac{1}{2b^3}e^{at}\sin bt - \frac{1}{2b^2}e^{at}t\cos bt\right].$$

To help the reader avoid these calculations in the future, the following entry has been added to Table 4.1 as entry (15):

	$f(t)$	$F(s)$	Region of validity
15.	$\dfrac{1}{b}\sin bt - t\cos bt$	$\dfrac{2b^2}{(s^2 + b^2)^2}$	$\mathrm{Re}(s) > 0$

Example 7 Find $L^{-1}\left\{\dfrac{3s + 2}{[(s+1)^2 + 1]^2}\right\}$.

A little arithmetic juggling gives

$$\frac{3s + 2}{[(s+1)^2 + 1]^2} = \frac{3}{2}\frac{2(s+1)}{[(s+1)^2 + 1]^2} - \frac{1}{2}\frac{2}{[(s+1)^2 + 1]^2}$$

and the shift formula and entries (10) and (15) give

$$L^{-1}\left\{\frac{3s + 2}{[(s+1)^2 + 1]^2}\right\} = \frac{3}{2}e^{-t}t\sin t - \frac{1}{2}e^{-t}(\sin t - t\cos t). \quad \Box$$

EXERCISES 4.6

1. Using the Heaviside formulas and/or algebraic manipulation, find the unknown constants in the following partial fraction expansions:

a) $\dfrac{3s - 1}{(s - 2)(s + 1)} = \dfrac{A_1}{s - 2} + \dfrac{A_2}{s + 1}$

b) $\dfrac{4s^2 + 2}{s(s - 1)(s - 3)} = \dfrac{A_1}{s} + \dfrac{A_2}{s - 1} + \dfrac{A_3}{s - 3}$

c) $\dfrac{2s + 5}{(s + 2)^2} = \dfrac{A_1}{s + 2} + \dfrac{A_2}{(s + 2)^2}$

d) $\dfrac{s^2 - 3s + 1}{s^3} = \dfrac{A_1}{s} + \dfrac{A_2}{s^2} + \dfrac{A_3}{s^3}$

e) $\dfrac{4s^2 + 9}{(s - 1)(s^2 + 2s + 10)} = \dfrac{A_1}{s - 1} +$

   $A_2\dfrac{s + 1}{(s + 1)^2 + 3^2} + A_3\dfrac{3}{(s + 1)^2 + 3^2}$

   Hint: First write the expansion in the form

   $$\dfrac{A_1}{s - 1} + \dfrac{Bs + C}{(s + 1)^2 + 3^2}$$

   and find A_1, B, and C.

f) $\dfrac{s + 4}{s^2(s^2 + 4)} = \dfrac{A_1}{s} + \dfrac{A_2}{s^2} + A_3\dfrac{s}{s^2 + 4}$

   $+ A_4\dfrac{2}{s^2 + 4}$

g) $\dfrac{6s}{(s-2)(s^2-2s+3)} = \dfrac{A_1}{s-2}$

$+\, A_2 \dfrac{s-1}{(s-1)^2+(\sqrt{2})^2} + A_3 \dfrac{\sqrt{2}}{(s-1)^2+(\sqrt{2})^2}$

h) $\dfrac{s^3+2s^2-5s+14}{(s^2+9)^2} = A_1 \dfrac{s}{s^2+9}$

$+\, A_2 \dfrac{3}{s^2+9} + A_3 \dfrac{6s}{(s^2+9)^2} + A_4 \dfrac{18}{(s^2+9)^2}$

Hint: First write the expansion in the form

$$\frac{As+B}{s^2+9} + \frac{Cs+D}{(s^2+9)^2}$$

and clear out the denominators to get

$(As+B)(s^2+9) + Cs + D$

$\qquad\qquad = s^3 + 2s^2 - 5s + 14.$

Now compare like powers of s to find A, B, C, and D.

2. Write down the inverse Laplace transforms of whichever expansions you computed in Exercise 1.

3. Find the following inverse Laplace transforms:

a) $L^{-1}\left\{\dfrac{2s+1}{s^2+5s+6}\right\}$

b) $L^{-1}\left\{\dfrac{s+10}{s^3-2s^2-15s}\right\}$

c) $L^{-1}\left\{\dfrac{s-5}{s^2+2s+26}\right\}$

d) $L^{-1}\left\{\dfrac{s^2-5}{s^3+s^2+9s+9}\right\}$

e) $L^{-1}\left\{\dfrac{4s^2+2}{s^4-4s^2}\right\}$

f) $L^{-1}\left\{\dfrac{21}{(s-2)(s^2+4s+9)}\right\}$

g) $L^{-1}\left\{\dfrac{3s-1}{(s^2-2s+5)^2}\right\}$

h) $L^{-1}\left\{\dfrac{3s^2+8s+3}{(s^2+1)(s^2+9)}\right\}$

i) $L^{-1}\left\{\dfrac{1}{s^2+4} + \dfrac{5}{(s^2+4)^2}\right\}$

j) $L^{-1}\left\{\dfrac{3s+51}{(s+17)^5}\right\}$ (Look before you leap!)

4. Using the factorization in the complex numbers

$$(s-a)^2 + b^2 = [s-(a-bi)][s-(a+bi)],$$

one can apply Heaviside formula 1 to treat the single irreducible quadratic in Case 3. After the partial fraction expansion is accomplished, one uses the fact that

$$L^{-1}\left\{\frac{1}{s-(a\pm bi)}\right\} = e^{(a\pm bi)t}$$

$$= e^{at}(\cos bt \pm i \sin bt)$$

valid for $\mathrm{Re}(s) > a$. Use these facts to derive the formulas:

a) $L^{-1}\left\{\dfrac{b}{s^2+b^2}\right\} = \sin bt$

b) $L^{-1}\left\{\dfrac{s}{s^2+b^2}\right\} = \cos bt$

c) $L^{-1}\left\{\dfrac{b}{(s-a)^2+b^2}\right\} = e^{at}\sin bt$

d) $L^{-1}\left\{\dfrac{s-a}{(s-a)^2+b^2}\right\} = e^{at}\cos bt$

5. Using the Laplace transform solve the following inhomogeneous initial value problems:

a) $y'' + 3y' + 2y = 4e^{2t}$, $y(0) = 2$, $y'(0) = 0$

b) $y'' + 5y' + 6y = 3e^{-2t}$, $y(0) = 0$, $y'(0) = 1$

c) $y'' + 9y = 2\cos 3t$, $y(0) = 1$, $y'(0) = 1$

d) $y'' + 25y = 6e^{-t}$, $y(0) = 1$, $y'(0) = 0$

e) $y'' + 2y' + 5y = 4t$, $y(0) = 0$, $y'(0) = 0$

f) $y'' - 4y' + 13y = 5\sin t$, $y(0) = 1$, $y'(0) = 4$

g) $y'' + 4y' + 4y = -e^{-2t}$, $y(0) = 2$, $y'(0) = 0$

6. Use the Laplace transform to solve the following inhomogeneous systems of differential equations:

a) $\dot{x} = x + 3y + t$, $x(0) = 0$
   $\dot{y} = 3x + y + e^{-3t}$, $y(0) = 0$

b) $\dot{x} = 2x - 2y + e^{-t}$, $x(0) = 1$
   $\dot{y} = 4x - 2y - 2t$, $y(0) = 1$

c) $\dot{x} = x - 2y - 4$, $x(0) = 0$
   $\dot{y} = 2x + y + \sin t$, $y(0) = 0$

d) $\dot{x} = 3x - 4y$, $x(0) = 1$
   $\dot{y} = 4x - 5y + 2e^{-t}$, $y(0) = 0$.

4.7 THE CONVOLUTION INTEGRAL: WEIGHTING FUNCTIONS

A natural question to ask is the following one: If $f(t)$ has a Laplace transform $F(s)$ and $g(t)$ has a Laplace transform $G(s)$, what function $h(t)$ has the Laplace transform $H(s) = F(s) G(s)$? A moment's reflection clearly shows the answer is not $h(t) = f(t) g(t)$. For instance, if $f(t) = g(t) = 1$, then $f(t)g(t) = 1$, but

$$H(s) = \frac{1}{s}\frac{1}{s} = \frac{1}{s^2},$$

whose inverse Laplace transform is $t \neq f(t) g(t)$.

The answer to the question is that $h(t)$ is an expression involving the integral of $f(t)$ and $g(t)$, called the *convolution integral,* or *convolution of f and g*:

$$f*g = \int_0^t f(t - \tau) g(\tau) \, d\tau.$$

Note that in the standard notation on the left the dependence on the variable t is understood and is omitted. To emphasize the dependence, it is sometimes useful to write $(f*g)(t)$.

Example 1   Compute $t*\cos t$.

$$t*\cos t = \int_0^t (t - \tau) \cos \tau \, d\tau = t \int_0^t \cos \tau \, d\tau - \int_0^t \tau \cos \tau \, d\tau$$

$$= t \sin t - \cos t - t \sin t + 1 = 1 - \cos t. \quad \square$$

The simple change of variables $\tau = t - s$ in the convolution integral gives

$$f*g = -\int_t^0 f(s) g(t - s) \, ds = \int_0^t g(t - s) f(s) \, ds = g*f,$$

so we see that the order of the functions is immaterial.

The convolution integral, or, equivalently, the operation $(*)$, can be thought of as a generalized product between functions, and simple calculations show that $f*g$ satisfies

$$f*g = g*f \qquad \text{(commutativity)},$$

$$f*(g + h) = f*g + f*h \quad \text{(distributivity)},$$

$$f*(g*h) = (f*g)*h \qquad \text{(associativity)}.$$

However, the important property is the one mentioned at the beginning of this section:

Convolution property. If $L\{f(t)\} = F(s)$ and $L\{g(t)\} = G(s)$ exist for $\text{Re}(s) > a$, then

$$L\{f*g\} = F(s)\ G(s), \qquad \text{Re}(s) > a,$$

or, equivalently,

$$f*g = L^{-1}\{F(s)\ G(s)\}.$$

The proof of this very important property is left until the end of this section, since we prefer to give some applications of the property and a physical motivation of the convolution integral itself first.

A very useful application of the convolution property is in the calculation of inverse Laplace transforms $L\{H(s)\}$, where $H(s)$ is recognized as the product of two simple functions $F(s)$ and $G(s)$. An integration will have to be performed at the end, but that may be easier than working out a complicated partial fraction expansion.

Example 2   Compute $L^{-1}\left\{\dfrac{1}{s^3(s^2 + 1)}\right\}$.

This example, which can be easily done by using Heaviside's formulas, is mainly intended to demonstrate the convolution integral approach. One can write

$$\frac{1}{s^3(s^2 + 1)} = \left(\frac{1}{s^3}\right)\left(\frac{1}{s^2 + 1}\right),$$

and so

$$F(s) = \frac{1}{s^3} = L^{-1}\left\{\frac{t^2}{2}\right\}$$

and

$$G(s) = \frac{1}{s^2 + 1} = L^{-1}\{\sin t\}.$$

Therefore,

$$L^{-1}\left\{\frac{1}{s^3(s^2 + 1)}\right\} = \frac{t^2}{2} * \sin t = \frac{1}{2}\int_0^t (t - \tau)^2 \sin \tau\ d\tau$$

$$= \frac{1}{2}t^2 \int_0^t \sin \tau\ d\tau - t\int_0^t \tau \sin \tau\ d\tau + \frac{1}{2}\int_0^t \tau^2 \sin \tau\ d\tau$$

$$= -1 + \frac{t^2}{2} + \cos t. \quad \square$$

Example 3   Compute $L^{-1}\left\{\dfrac{2b^2}{[s^2 + b^2]^2}\right\}$.

This example, which was discussed in the previous section, is ideally set up for using the convolution property. Just write

$$\frac{2b^2}{[s^2 + b^2]^2} = 2 \left(\frac{b}{s^2 + b^2} \right) \left(\frac{b}{s^2 + b^2} \right),$$

and since each of the terms in parentheses is the transform of sin bt, we get

$$L^{-1} \left\{ \frac{2b^2}{[s^2 + b^2]^2} \right\} = 2[\sin bt * \sin bt]$$

$$= 2 \int_0^t \sin b(t - \tau) \sin b\tau \, d\tau$$

$$= 2 \sin bt \int_0^t \cos b\tau \sin b\tau \, d\tau - 2 \cos bt \int_0^t \sin^2 b\tau \, d\tau$$

$$= \frac{1}{b} \sin^3 bt - t \cos bt + \frac{1}{2b} \sin 2bt \cos bt.$$

Since sin $2bt = 2 \sin bt \cos bt$, the answer previously given will be obtained after a little trigonometric manipulation. □

The convolution integral approach is especially efficient when the denominator of $F(s)$ is a product of irreducible quadratic factors.

The real importance of the convolution integral is that it is one of the building blocks in the analysis of linear systems. This is best illustrated by an example. Consider the general initial value problem

$$y'' + ay' + by = f(t), \qquad y(0) = 0, \qquad y'(0) = 0,$$

which can be regarded as a mechanical or electrical system at rest excited by an input or driving term $f(t)$. First apply the Laplace transform to obtain

$$s^2 Y(s) + as Y(s) + b Y(s) = L\{f(t)\} = F(s)$$

and hence

$$Y(s) = \frac{F(s)}{s^2 + as + b} = P(s) F(s).$$

The function $P(s) = (s^2 + as + b)^{-1}$ is called the *transfer function* of the system described by the differential equation above.

In the s-domain the system could be described by the diagram in Fig. 4.5, where the box represents the action of the system at rest on the input, namely, multiplication by the transfer function. Now apply the inverse Laplace transform to get

$$y(t) = L^{-1}\{Y(s)\} = L^{-1}\{P(s) F(s)\},$$

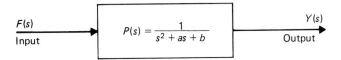

Figure 4.5

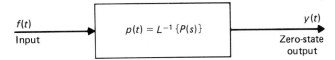

Figure 4.6

and by the convolution property

$$y(t) = p(t)*f(t),$$

where $p(t) = L^{-1}\{P(s)\}$ is called the *weighting function* of the system, and $y(t)$ is called the *zero-state output*.

In the t-domain the system could then be described by the diagram in Fig. 4.6, where the box represents the action of the system at rest on the input, namely convolution with the weighting function. In the last section of this chapter this analysis will be extended further, but for now we wish to study the effect of the weighting function on the input by examining the convolution integral a little closer.

If the equation $y(t) = p(t)*f(t)$, which describes the zero-state output, is written in the integral form, then

$$y(t) = \int_0^t p(\tau) f(t - \tau) \, d\tau$$

since $p*f = f*p$. Since $0 \leq \tau \leq t$, the integral can be interpreted by saying that the input τ units in the past, namely $f(t - \tau)$, is weighted by the value of the weighting function evaluated at time τ. The idea of regarding the zero-state output as a weighted integral in turn allows us to discuss the *memory* of the system.

Suppose, for example, that the input $f(t)$ is a bounded function and that the weighting function $p(t)$ is a function that goes to zero as τ approaches ∞ and is negligible for $\tau \geq T$. Its graph might look like that in Fig. 4.7. Then $y(t)$ for $t > T$ can be expressed as

$$y(t) = \int_0^T p(\tau) f(t - \tau) \, d\tau + \int_T^t p(\tau) f(t - \tau) \, d\tau,$$

and by the assumptions on $p(\tau)$ the second integral can be considered negligible. A way of saying this is that values of the input for more than T units

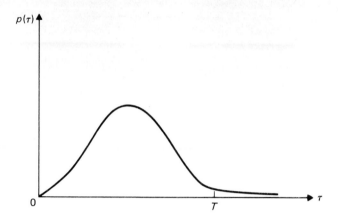

Figure 4.7

in the past have very little effect on the output, i.e., for any t, the contributions of $f(t - \tau)$ to the output are negligible for $\tau \geq T$.

On the other hand, if $p(\tau)$ is a function which is, say, strictly increasing, then the second integral in the above expression for $y(t)$ will not be negligible for any $T > 0$. In this case, the system is said to have *infinite memory*, and for any t and any τ, where $0 \leq \tau \leq t$, the contribution of $f(t - \tau)$ will strongly influence the zero-state output. For practical "real-world" systems, one would not expect this kind of behavior.

We conclude this section with a proof of the convolution property which depends on the following very useful formula for the change of order of integration in a double integral:

$$\int_a^b \left[\int_a^t F(t, \tau)\, d\tau \right] dt = \int_a^b \left[\int_\tau^b F(t, \tau)\, dt \right] d\tau.$$

The justification for this formula follows easily from a diagram of the region of integration shown in Fig. 4.8. In the first integral the region of integration is described by $a \leq \tau \leq t$ and $a \leq t \leq b$, whereas in the second integral it is described by $\tau \leq t \leq b$ and $a \leq \tau \leq b$. These clearly describe the same region.

One can now proceed directly to prove the convolution property. By the definition of the convolution integral and of the Laplace transform, we have

$$L\{f*g\} = \int_0^\infty e^{-st} \left[\int_0^t f(t - \tau)\, g(\tau)\, d\tau \right] dt$$

$$= \int_0^\infty \left[\int_0^t e^{-st} f(t - \tau)\, g(\tau)\, d\tau \right] dt.$$

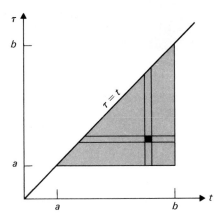

Figure 4.8

Apply the formula for the interchange of the order of integration with $a = 0$, $b = \infty$ to obtain

$$L\{f*g\} = \int_0^\infty \left[\int_\tau^\infty e^{-st} f(t - \tau) \, g(\tau) \, dt \right] d\tau.$$

In the bracketed integral make the change of variables $t = \tau + r$ to obtain finally

$$L\{f*g\} = \int_0^\infty \left[\int_0^\infty e^{-s(\tau+r)} f(r) \, g(\tau) \, dr \right] d\tau$$

$$= \int_0^\infty e^{-s\tau} g(\tau) \left[\int_0^\infty e^{-sr} f(r) \, dr \right] d\tau$$

$$= \left[\int_0^\infty e^{-sr} f(r) \, dr \right] \left[\int_0^\infty e^{-s\tau} g(\tau) \, d\tau \right]$$

$$= L\{f\} \, L\{g\} = F(s) \, G(s),$$

which is the desired result.

EXERCISES 4.7

1. Compute the following convolution integrals $f*g$ directly, then verify your result by computing $L^{-1}\{F(s) \, G(s)\}$:

a) $t*t$ b) $t*e^{-t}$ c) e^t*e^{-t} d) $e^{-t}*\cos t$

2. Use the convolution property to compute the following inverse Laplace transforms:

a) $L^{-1}\left\{ \dfrac{1}{s^2} \cdot \dfrac{1}{s + 4} \right\}$ b) $L^{-1}\left\{ \dfrac{1}{s^2} \cdot \dfrac{1}{s^2 + 9} \right\}$

c) $L^{-1}\left\{\dfrac{1}{s^2+1} \cdot \dfrac{1}{s^2+9}\right\}$

d) $L^{-1}\left\{\dfrac{s^2}{(s^2+4)^2}\right\}$

3. Find the transfer function and weighting function of the systems described by the following differential equations:

a) $y'' - 3y' + 10y = f(t)$

b) $y'' + 16y = f(t)$

c) $y'' - 9y = f(t)$

d) $y'' + 4y' + 10y = f(t)$

4. Assume that the weighting function of a system is

$$p(t) = \begin{cases} t & \text{if } 0 \le t \le \frac{1}{2}, \\ 1 - t & \text{if } \frac{1}{2} \le t \le 1, \\ 0 & \text{if } t > 1, \end{cases}$$

and the input is $f(t) = e^t$.

a) Show that the zero-state output for $t > 1$ is $Kf(t - 1) = Ke^{t-1}$, where K is some constant, and hence the system has memory $T = 1$.

b) Using directly the definition of the Laplace transform, find the transfer function $P(s)$.

5. a) Using the formula for the change of order of integration in a double integral, show that

$$\int_0^t \int_0^r f(\tau)\, d\tau\, dr = \int_0^t (t - \tau)\, f(\tau)\, d\tau.$$

b) Using the convolution property and the above result, verify the following formulas.

$$L^{-1}\left\{\dfrac{1}{s} F(s)\right\} = \int_0^t f(\tau\, d\tau),$$

$$L^{-1}\left\{\dfrac{1}{s^2} F(s)\right\} = \int_0^t \int_0^r f(\tau)\, d\tau\, dr.$$

6. A Volterra integral equation of convolution or renewal type is of the form

$$y(t) = f(t) + \int_0^t k(t - \tau)\, y(\tau)\, d\tau,$$

where $f(t)$ and $k(t)$ (called the *kernel of the equation*) are given functions, and $y(t)$ is the solution sought. It is ideally suited for analysis using the Laplace transform. Assuming that $f(t)$ and $k(t)$ have Laplace transforms $F(s)$ and $K(s)$, respectively, show that the Laplace transform $Y(s)$ satisfies the relation

$$Y(s) = \dfrac{F(s)}{1 - K(s)}.$$

7. Use the result of Exercise 6 to find the solution $y(t)$ of the following integral equations:

a) $y(t) = 4t + \displaystyle\int_0^t \sin(t - \tau)\, y(\tau)\, d\tau$

b) $y(t) = \cos t - \displaystyle\int_0^t e^{t-\tau} y(\tau)\, d\tau$

c) $y(t) = a \sin t + c \displaystyle\int_0^t \sin(t - \tau)\, y(\tau)\, d\tau$;

consider the three cases $c < 1$, $c = 1$, and $c > 1$.

d) $y(t) = \sin t - \displaystyle\int_0^t (t - \tau)\, y(\tau)\, d\tau$

4.8 THE UNIT STEP FUNCTION: TRANSFORMS OF NONSMOOTH FUNCTIONS

It was stated in Section 4.5 that the Laplace transform is a very effective tool for solving differential equations where the forcing term is a nonsmooth function. Such a function could have one or more jump discontinuities (i.e., the graph of the function could have breaks in it) or it could be a continuous function whose first derivative is discontinuous. In the last case the graph of the function would have "corners" in it—a simple example of such a function

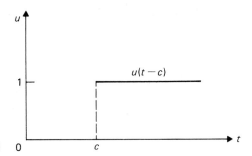

Figure 4.9

would be $f(t) = |t|$. We see that $f'(t) = -1$ if $t < 0$ and $f'(t) = 1$ if $t > 0$, whereas $f(t)$ is continuous for all t.

One of the basic tools for describing such nonsmooth functions and for computing their Laplace transforms is the *unit step, or Heaviside, function*. It is described as follows:

$$u(t) = \begin{cases} 0 & \text{if} \quad t < 0, \\ 1 & \text{if} \quad t \geq 0. \end{cases}$$

The reader should note that in some texts $u(0)$ may be defined to be 0 instead of 1, or the equality sign at $t = 0$ may be dropped completely, so that $u(0)$ is undefined. The value of $u(0)$ is immaterial to the discussion. It follows that for $c > 0$

$$u(t - c) = \begin{cases} 0 & \text{if} \quad t < c, \\ 1 & \text{if} \quad t \geq c, \end{cases}$$

so that $u(t - c)$ has a simple jump discontinuity of value 1 at $t = c$ (see Fig. 4.9).

To use the unit step function as a tool, first observe the following important property: If $f(t)$ is a function defined for $t \geq 0$ and if $c > 0$, then

$$f(t - c) u(t - c) = \begin{cases} 0 & \text{if} \quad t < c, \\ f(t - c) & \text{if} \quad t \geq c, \end{cases}$$

so the effect of multiplying $f(t - c)$ by $u(t - c)$ is to shift the graph of $f(t)$ by c units to the right (see Fig. 4.10). This basic property will be used to construct some nonsmooth functions in the examples below.

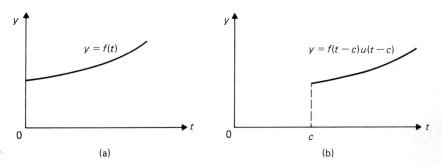

Figure 4.10   (a)   (b)

Figure 4.11   **Figure 4.12**

Example 1   A pulse of magnitude 4 and duration from $t = 1$ to $t = 2$ could be described by the function

$$f(t) = \begin{cases} 0 & \text{if } t < 1, \\ 4 & \text{if } 1 \le t < 2, \\ 0 & \text{if } t \ge 2, \end{cases}$$

as shown in Fig. 4.11.

From the property above with $f(t) = 4$ it follows that the function $4u(t - 1)$ describes a function which is zero for $t < 1$ and equals 4 for $t \ge 1$ (see Fig. 4.12).

To "lop off" the contribution for $t \ge 2$, a function must be subtracted that is zero for $t < 2$ and equals 4 for $t \ge 2$. But that function by the same reasoning is $4u(t - 2)$, hence

$$f(t) = 4u(t - 1) - 4u(t - 2). \quad \square$$

Example 2   The preceding example leads to another useful fact which the reader can easily verify: if $a < b$, then

$$u(t - a) - u(t - b) = \begin{cases} 0 & \text{if } t < a, \\ 1 & \text{if } a \le t < b, \\ 0 & \text{if } t \ge b. \end{cases}$$

This fact is used to construct the triangular pulse (Fig. 4.13) given by

$$f(t) = \begin{cases} t & \text{if } 0 \le t \le 1, \\ 2 - t & \text{if } 1 < t \le 2; \\ 0 & \text{if } t > 2. \end{cases}$$

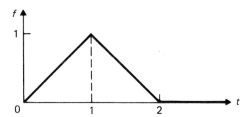

Figure 4.13

This is a continuous function but its derivative is discontinuous at $t = 1$ and $t = 2$, where the graph has corners.

Using the fact above we see that

$$t[1 - u(t - 1)] = \begin{cases} t & \text{if } 0 \le t < 1, \\ 0 & \text{if } t \ge 1, \end{cases}$$

and

$$(2 - t)[u(t - 1) - u(t - 2)] = \begin{cases} 0 & \text{if } t < 1, \\ 2 - t & \text{if } 1 \le t < 2, \\ 0 & \text{if } t \ge 2. \end{cases}$$

Thus we conclude that

$$\begin{aligned} f(t) &= t[1 - u(t - 1)] + (2 - t)[u(t - 1) - u(t - 2)] \\ &= t + (2 - 2t)\, u(t - 1) - (2 - t)\, u(t - 2), \quad t \ge 0, \end{aligned}$$

is the required pulse. □

Example 3 Using the facts above and proceeding interval by interval, it is easy to construct the representation of the following square wave $S(t)$ with period $2a$ shown in Fig. 4.14.

Its representation is given by the series

$$\begin{aligned} S(t) &= (1)[1 - u(t - a)] + (-1)[u(t - a) - u(t - 2a)] \\ &\quad + (1)[u(t - 2a) - u(t - 3a)] \\ &\quad + (-1)[u(t - 3a) - u(t - 4a)] \\ &\quad + (1)[u(t - 4a) - u(t - 5a)] \\ &\quad + \cdots, \quad t \ge 0. \end{aligned}$$

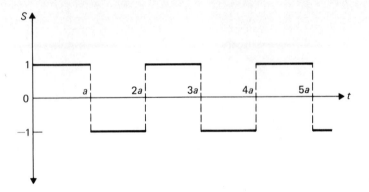

Figure 4.14 A square wave $S(t)$.

Combining the terms gives

$$S(t) = 1 - 2u(t - a) + 2u(t - 2a) - 2u(t - 3a) + 2u(t - 4a) + \cdots$$

$$= 1 + 2 \sum_{n=1}^{\infty} (-1)^n u(t - na), \quad t \geq 0.$$

A simpler representation is obtained by noting the periodicity of $f(t)$ and just describing $S(t)$ over the first period:

$$S(t) = 1 - 2u(t - a) + u(t - 2a), \quad 0 \leq t \leq 2a,$$
$$S(t + 2a) = S(t), \quad t \geq 0. \quad \Box$$

The above examples should indicate to the reader the usefulness of the unit step function in representing nonsmooth functions.

We now come to the business of computing the Laplace transform of nonsmooth functions, such as those in the above examples. The first task is to compute the Laplace transform of $u(t - c)$, but this follows directly from the definition. Since $u(t - c)$ is zero for $t < c$ and equals 1 for $t \geq c$, then

$$L\{u(t - c)\} = \int_0^\infty e^{-st} u(t - c) \, dt$$

$$= \int_c^\infty e^{-st} \, dt = \frac{e^{-sc}}{s},$$

valid for $\mathrm{Re}(s) > 0$. Notice that the Laplace transform of $u(t - c)$ is simply the Laplace transform of $f(t) = 1$ multiplied by e^{-sc}.

It is an equally simple matter to compute the Laplace transform of $f(t - c) u(t - c)$. Suppose $f(t)$ has a Laplace transform $F(s)$ valid for

$\text{Re}(s) > a$. Then,

$$L\{f(t - c) \, u(t - c)\} = \int_0^\infty e^{-st} f(t - c) \, u(t - c) \, dt$$

$$= \int_c^\infty e^{-st} f(t - c) \, dt.$$

Making the change of variables $r = t - c$ gives

$$L\{f(t - c) \, u(t - c)\} = \int_0^\infty e^{-s(r+c)} f(r) \, dr$$

$$= e^{-sc} \int_0^\infty e^{-sr} f(r) \, dr = e^{-sc} L\{f\} = e^{-sc} F(s),$$

also valid for $\text{Re}(s) > a$. This useful result is entry (16) of Table 4.1:

	f(t)	F(s)	Region of validity
16.	$f(t - c) \, u(t - c)$	$e^{-sc} F(s)$	Valid for same region as $F(s)$

The reader should notice that this entry is the counterpart in the s-domain of the shift property (entry 12) in the t-domain.

We are now ready to consider some differential equations with non-smooth forcing terms. For the first example consider the problem

$$y'' + y = f(t), \qquad y(0) = 1, \qquad y'(0) = 0,$$

where $f(t)$ is the rectangular pulse given in Example 1 above. Therefore we must solve

$$y'' + y = 4u(t - 1) - 4u(t - 2), \qquad y(0) = 1, \qquad y'(0) = 0.$$

If $Y(s)$ is the Laplace transform of $y(t)$, use the differentiation formulas and entry (16) above to obtain

$$s^2 Y(s) - s + Y(s) = 4 \frac{e^{-s}}{s} - 4 \frac{e^{-2s}}{s}.$$

Solving for $Y(s)$ and computing the required partial fraction expansions gives

$$Y(s) = \frac{s}{s^2 + 1} + 4e^{-s} \left[\frac{1}{s} - \frac{s}{s^2 + 1} \right] - 4e^{-2s} \left[\frac{1}{s} - \frac{s}{s^2 + 1} \right].$$

From Table 4.1,

$$L^{-1}\left\{\frac{1}{s} - \frac{s}{s^2 + 1}\right\} = 1 - \cos t,$$

and from entry (16) it follows again that the solution is

$$y(t) = \cos t + 4[1 - \cos(t - 1)]\,u(t - 1)$$
$$- 4[1 - \cos(t - 2)]\,u(t - 2).$$

The reader can easily verify from the above expression that the solution can also be written as

$$y(t) = \begin{cases} \cos t & \text{if } 0 \le t \le 1, \\ \cos t + 4 - 4\cos(t - 1) & \text{if } 1 \le t \le 2, \\ \cos t - 4\cos(t - 1) + 4\cos(t - 2) & \text{if } t \ge 2. \end{cases}$$

The second example is a slight variant of the previous example. We wish to find the zero-state output for the equation

$$y'' + y = \phi(t),$$

where $\phi(t)$ is the function (see Fig. 4.15)

$$\phi(t) = t[1 - u(t - 1)].$$

As it stands, $\phi(t)$ is not in a form suitable for application of entry (16), but it can be rewritten as

$$\phi(t) = t - tu(t - 1) = t - [(t - 1) + 1]\,u(t - 1).$$

The second term in the last expression is in the form $f(t - 1)\,u(t - 1)$, where $f(t) = t + 1$, and now the Laplace transform can be applied. Since zero-state output implies that the initial conditions are $y(0) = y'(0) = 0$, applying the Laplace transform gives

$$s^2 Y(s) + Y(s) = L\{\phi(t)\} = \frac{1}{s^2} - e^{-s}\left[\frac{1}{s^2} + \frac{1}{s}\right].$$

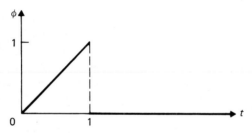

Figure 4.15

It is left to the reader to show that

$$y(t) = t - \sin t - [t - \sin(t - 1) - \cos(t - 1)] \, u(t - 1)$$

is the zero-state output corresponding to the input $\phi(t)$.

The last example, which really shows the effectiveness of the Laplace transform, is to find the zero-state output for the equation

$$y'' + y = S(t),$$

where $S(t)$ is the square wave of period $2a$ given in Example 3 above. If one uses its series representation

$$S(t) = 1 + 2 \sum_{n=1}^{\infty} (-1)^n u(t - na),$$

then $L\{S(t)\}$ can be computed by transforming term by term and by using the linearity of the transform. Therefore

$$L\{S(t)\} = \frac{1}{s} + 2 \sum_{n-1}^{\infty} (-1)^n \frac{e^{-sna}}{s};$$

now applying the transform to the differential equation gives

$$s^2 Y(s) + Y(s) = \frac{1}{s} + 2 \sum_{n=1}^{\infty} (-1)^n \frac{e^{-sna}}{s}$$

or

$$Y(s) = \frac{1}{s(s^2 + 1)} + 2 \sum_{n=1}^{\infty} (-1)^n \frac{e^{-sna}}{s(s^2 + 1)}.$$

But the inverse Laplace transform can also be applied term by term and hence

$$y(t) = (1 - \cos t) + 2 \sum_{n=1}^{\infty} (-1)^n [1 - \cos(t - na)] \, u(t - na)$$

is the desired output. Notice that for any given value of t the sum will be finite since $(t - na) < 0$ for all n sufficiently large, and the corresponding $u(t - na)$ will be zero.

In the computation of the Laplace transform of the periodic square wave its given series representation was used. More often one is given $f(t)$ explicitly defined for $0 \le t \le T$, and the periodicity condition $f(t + T) = f(t)$. A general formula will now be developed for the transform of a periodic function and added to the Table of Laplace transforms.

If $f(t + T) = f(t)$, where T is the period, then it follows easily that

$f(t + nT) = f(t)$ for any positive (or negative) integer n. Now write the integral for $L\{f\}$ as a sum of integrals over intervals of length T as follows:

$$L\{f\} = \int_0^\infty e^{-st} f(t) \, dt = \sum_{n=0}^\infty \int_{nT}^{(n+1)T} e^{-st} f(t) \, dt = \sum_{n=0}^\infty e^{-snT} \int_0^T e^{-sr} f(r) \, dr$$

since $f(t)$ is periodic with period T, or

$$L\{f\} = \left[\int_0^T e^{-sr} f(r) \, dr \right] \sum_{n=0}^\infty (e^{-sT})^n.$$

Now make use of the fact that

$$\sum_{n=0}^\infty z^n = 1 + z + z^2 + \cdots = \frac{1}{1 - z} \quad \text{if } \mathrm{Re}(z) < 1,$$

and let $z = e^{-sT}$. Then $|z| < 1$ if $\mathrm{Re}(s) > 0$, and the formula gives

$$\sum_{n=0}^\infty (e^{-sT})^n = \frac{1}{1 - e^{-sT}}.$$

Thus another useful entry can be added to Table 4.1:

	$f(t)$	$F(s)$	Region of validity
17.	$f(t + T) = f(t)$	$\dfrac{\int_0^T e^{-sr} f(r) \, dr}{1 - e^{-sT}}$	$\mathrm{Re}(s) > 0$

If entry (17) is used instead of the series to compute the transform of the square wave $S(t)$, one obtains

$$L\{S(t)\} = \frac{\int_0^{2a} e^{-sr} S(r) \, dr}{1 - e^{-2sa}} = \frac{\int_0^a e^{-sr}(1) \, dr + \int_a^{2a} e^{-sr}(-1) \, dr}{1 - e^{-2sa}}$$

$$= \frac{1}{s} \frac{(1 - e^{-sa})^2}{1 - e^{-2sa}} = \frac{1}{s} \left[\frac{1 - e^{-sa}}{1 + e^{-sa}} \right] = \frac{1}{s} \tanh \frac{sa}{2}.$$

The Laplace transform of the zero-state solution $y'' + y = S(t)$ is then

$$Y(s) = \left(\frac{1}{s^2 + 1} \right) \left(\frac{1}{s} \tanh \frac{sa}{2} \right) = L\{\sin t\} \, L\{S(t)\},$$

and now the convolution property gives another representation of the solution

$$y(t) = (\sin t) * (S(t)) = \int_0^t \sin(t - \tau)\, S(\tau)\, d\tau.$$

The series representation obtained above for $y(t)$ is therefore an explicit representation of the convolution integral.

This section is concluded with a comment that modifies our previous discussion of E_α, of the functions of exponential order, and of the Laplace transform. As has been clearly demonstrated in this section, one can drop the requirement of continuity for functions whose Laplace transform exists. Instead, it can be replaced with the weaker requirement that it be sectionally continuous, defined as follows:

A function is *sectionally continuous* on $0 \le t < \infty$ if it is continuous on every finite interval with exception of at most a finite number of points, where it has a simple jump discontinuity.

A little contemplation will convince the reader that every sectionally continous function can be represented by a sum of continuous functions, each multiplied by the difference of two unit step functions.

Given the above, we next modify the definition of functions of exponential order.

A function $f(t)$, $0 \le t < \infty$, is said to be of *exponential order for t sufficiently large* if there exist positive constants M and T and a constant α such that

$$|f(t)| \le Me^{\alpha t}, \quad t \ge T.$$

Finally, putting the two definitions together provides us with a more general class of functions whose Laplace transforms exist.

If $f(t)$, $0 \le t < \infty$, is sectionally continuous and is of exponential order for t sufficiently large, then its Laplace transform exists.

The proof of this statement follows from the estimate:

$$\left| \int_0^\infty f(t)e^{-st} \, dt \right| \leq \int_0^\infty |f(t)| \, |e^{-st}| \, dt$$

$$\leq \int_0^T |f(t)| \, |e^{-st}| \, dt + \int_T^\infty |f(t)| \, |e^{-st}| \, dt.$$

In the last expression, the first integral is bounded since $f(t)$ is sectionally continuous, whereas for T sufficiently large, the second integral converges by an argument exactly like the one given previously in Section 4.4.

EXERCISES 4.8

1. For each of the following functions, sketch its graph, find an expression for it in terms of the unit step function $u(t)$, and find its Laplace transform. All functions are defined only for $t \geq 0$.

a) $f(t) = \begin{cases} 1 & \text{if } t < 2, \\ -1 & \text{if } 2 \leq t < 4, \\ 0 & \text{if } t \geq 4 \end{cases}$

b) $f(t) = \begin{cases} 0 & \text{if } t < 1, \\ 2 & \text{if } 1 \leq t < 2, \\ -2 & \text{if } 2 \leq t < 3, \\ 0 & \text{if } t \geq 3 \end{cases}$

c) $f(t) = \begin{cases} t & \text{if } t < 1, \\ 2 - t & \text{if } 1 \leq t < 3, \\ t - 4 & \text{if } 3 \leq t < 4, \\ 0 & \text{if } t \geq 4 \end{cases}$

d) $f_n(t) = \begin{cases} nt & \text{if } t < 1/n, \\ 1 & \text{if } t \geq 1/n. \end{cases}$

Discuss what happens when $n \to \infty$.

e) $f(t) = \begin{cases} 0 & \text{if } t < 2\pi, \\ \sin t & \text{if } t \geq 2\pi \end{cases}$

f) $f(t) = \begin{cases} 4e^{1-t} & \text{if } t < 1, \\ 4 & \text{if } t \geq 1 \end{cases}$

2. Sketch the graphs of the following functions and find their Laplace transforms:

a) $u(t - 1) + 4u(t - 2)$

b) $2tu(t - \frac{1}{2})$

c) $e^{t-2}u(t - 3)$

d) $u(t - a) \sin t$; discuss the cases $a = \pi$, $a = 2\pi$.

e) $t + (t - 1)u(t - 1)$

3. Find the solution of the following initial value problems:

a) $y'' + 3y' + 2y = f(t)$, $y(0) = 0$, $y'(0) = 0$, where $f(t)$ is given in Exercise 1(a) above. Also, describe the solution without using unit step functions.

b) $y'' + 9y = f(t)$, $y(0) = 1$, $y'(0) = 0$, where $f(t)$ is given in Exercise 1(b) above. Also, describe the solution without using unit step functions.

c) $y'' + y = f(t)$, $y(0) = 0$, $y'(0) = 0$, where $f(t)$ is given in Exercise 1(e) above. Are $y(t)$, $y'(t)$, and $y''(t)$ continuous at $t = 2\pi$?

d) $y'' + 4y' + 3y = f(t)$, $y(0) = 2$, $y'(0) = 0$, where $f(t)$ is given in Exercise 1(f) above.

e) $y'' + y = f_n(t)$, $y(0) = 0$, $y'(0) = 0$, where $f_n(t)$ is given in Exercise 1(d) above. Does the solution approach the solution of $y'' + y = 1$, $y(0) = 0$, $y'(0) = 0$, as $n \to \infty$?

4. Sketch the graphs of the following periodic functions and find their Laplace transforms. Do this in two ways: by expressing the function as a series of terms involving unit step functions then calculating the transform term by term, and by using entry (17).

a) $f(t) = \begin{cases} 2 & \text{if } 0 \le t < 1, \\ 0 & \text{if } 1 \le t < 2; \end{cases}$ $f(t + 2) = f(t)$

b) $f(t) = 2 - t$ if $0 \le t < 4$; $f(t + 4) = f(t)$

c) $f(t) = \begin{cases} 0 & \text{if } 0 \le t < \frac{1}{2}, \\ 1 & \text{if } \frac{1}{2} \le t < \frac{3}{2}, \\ 0 & \text{if } \frac{3}{2} \le t < 2; \end{cases}$ $f(t + 2) = f(t)$

d) $f(t) = |\cos t|$ if $t \ge 0$

5. Find the solutions of the following initial value problems:

a) $y'' - 9y = \frac{3}{2}f(t)$, $y(0) = 0$, $y'(0) = 2$, where $f(t)$ is given in Exercise 4(a) above.

b) $y'' + 4y = f(t)$, $y(0) = 0$, $y'(0) = 0$, where $f(t)$ is given in Exercise 4(b) above.

c) $y'' + y = |\sin t|$, $y(0) = 4$, $y'(0) = 0$

4.9 THE UNIT IMPULSE FUNCTION: TRANSFER FUNCTIONS

Given a mechanical or electrical system, we wish to model a force or voltage of large magnitude which occurs over a very short period of time. A function that describes this behavior is

$$\delta_\epsilon(t - t_0) = \begin{cases} 0 & \text{if } t < t_0 - \epsilon, \\ \dfrac{1}{2\epsilon} & \text{if } t_0 - \epsilon \le t < t_0 + \epsilon, \\ 0 & \text{if } t \ge t_0 + \epsilon, \end{cases}$$

where $\epsilon > 0$ is a small positive number and $t_0 > 0$. Using the unit step function one can also write

$$\delta_\epsilon(t - t_0) = \frac{1}{2\epsilon} [u(t - t_0 + \epsilon) - u(t - t_0 - \epsilon)],$$

and the graph of $\delta_\epsilon(t - t_0)$ is shown in Fig. 4.16. Clearly, as ϵ approaches zero, the rectangular pulse $\delta_\epsilon(t - t_0)$ gets taller and thinner but the limit does not exist.

In spite of the limit $\lim_{\epsilon \to 0} \delta_\epsilon(t - t_0)$ not existing, one can derive some interesting properties of the function. First of all, since $\delta_\epsilon(t - t_0) = 0$ for all t outside the interval $t_0 - \epsilon < t \le t_0 + \epsilon$, it follows that

$$\int_{-\infty}^{\infty} \delta_\epsilon(t - t_0) \, dt = \int_{t_0 - \epsilon}^{t_0 + \epsilon} \frac{1}{2\epsilon} \, dt = 1. \tag{4.9.1}$$

If $\delta_\epsilon(t - t_0)$ is thought of as a force, then (4.9.1) says the *total impulse* is unity.

Second, given any continuous function $f(t)$ defined on $-\infty < t < \infty$, we have

$$\int_{-\infty}^{\infty} \delta_\epsilon(t - t_0) f(t) \, dt = \int_{t_0 - \epsilon}^{t_0 + \epsilon} \frac{1}{2\epsilon} f(t) \, dt = \frac{1}{2\epsilon} \int_{t_0 - \epsilon}^{t_0 + \epsilon} f(t) \, dt.$$

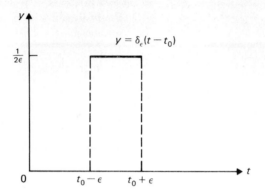

Figure 4.16

Recall that the mean value theorem for integrals states that $\int_a^b f(t)\ dt = (b - a)\ f(\tau)$ for some τ in $a \le t \le b$, and consequently

$$\int_{-\infty}^{\infty} \delta_\epsilon(t - t_0)\ f(t)\ dt = \frac{1}{2\epsilon}\ [2\epsilon f(\tau_\epsilon)] = f(\tau_\epsilon) \qquad (4.9.2)$$

for some τ_ϵ in $t_0 - \epsilon \le t \le t_0 + \epsilon$.

Third, the representation of $\delta_\epsilon(t - t_0)$ in terms of unit step functions implies immediately that for $t_0 > 0$

$$L\{\delta_\epsilon(t - t_0)\} = \frac{1}{2\epsilon} \left[\frac{e^{-s(t_0 - \epsilon)}}{s} - \frac{e^{-s(t_0 + \epsilon)}}{s} \right] = e^{-st_0} \left[\frac{\sinh s\epsilon}{s\epsilon} \right]. \qquad (4.9.3)$$

For the last property, $\epsilon > 0$ must be sufficiently small so that $t_0 - \epsilon > 0$, in order for the Laplace transform to be defined.

Now suppose that ϵ is allowed to approach zero in each of the properties (4.9.1), (4.9.2), and (4.9.3). For the first two properties some not-so-trivial mathematical justification of the limiting process is required but it can be accomplished. If $\delta(t - t_0)$ denotes the result of letting ϵ go to zero, then (4.9.1) becomes

$$\int_{-\infty}^{\infty} \delta(t - t_0)\ dt = 1. \qquad (4.9.4)$$

For the property (4.9.2), τ_ϵ must remain in the interval $t_0 - \epsilon \le t \le t_0 + \epsilon$ as ϵ goes to zero. Consequently $\lim_{\epsilon \to 0} \tau_\epsilon = t_0$ and

$$\int_{-\infty}^{\infty} \delta(t - t_0)\ f(t)\ dt = f(t_0) \qquad (4.9.5)$$

for any continuous function defined for $-\infty < t < \infty$. Actually, $f(t)$ has to be continuous only in some neighborhood of $t = t_0$, and it can then be defined to be zero outside this interval. Property (4.9.5) is sometimes called the *sifting property*.

In the last property, (4.9.3), which defines the Laplace transform of $\delta_\epsilon(t - t_0)$, observe by L'Hospital's rule that

$$\lim_{\epsilon \to 0} \frac{\sinh s\epsilon}{s\epsilon} = \lim_{\epsilon \to 0} \frac{\cosh s\epsilon}{1} \doteq 1.$$

Therefore

$$L\{\delta(t - t_0)\} = e^{-st_0} \qquad (4.9.6)$$

is valid for $t_0 > 0$ since the transform of $\delta_\epsilon(t - t_0)$ can be calculated only when $t_0 - \epsilon > 0$ or $t_0 > \epsilon > 0$. But nothing prevents us from letting $t_0 \to 0$, and this gives

$$L\{\delta(t)\} = 1. \qquad (4.9.7)$$

This clearly points out that $\delta(t)$ is not at all a "nice" function since, if it were, its Laplace transform would be a function that goes to zero as s approaches infinity.

The "function" $\delta(t)$ defined by the relations (4.9.4), (4.9.5), (4.9.6), and (4.9.7) is called the *unit impulse function* or *Dirac delta function*, named after the Nobel Laureate physicist Paul Dirac (born 1902). It is not a function in any traditional sense but is an example of what modern-day mathematicians call a "generalized function." The theory of such functions was only recently developed, but physicists, engineers, and applied mathematicians have been using them for years.

The users of such functions have developed such definitions as:

$\delta(t)$ is a function that is zero for all $t \neq 0$ and satisfies

$$\int_{-\infty}^{\infty} \delta(t)\ dt = 1$$

and

$$\int_{-\infty}^{\infty} \delta(t - t_0)\ f(t)\ dt = f(t_0).$$

One can see why this would trouble a mathematician, but a physicist would be content to think of $\delta(t)$ as a unit impulse or a point mass. Another definition comes from the representation of $\delta_\epsilon(t - t_0)$ in terms of unit step functions. Recall that

$$\delta_\epsilon(t - t_0) = \frac{1}{2\epsilon} [u(t - t_0 + \epsilon) - u(t - t_0 - \epsilon)];$$

letting $t - t_0 = \tau + \epsilon$, we obtain

$$\delta_\epsilon(\tau + \epsilon) = \frac{u(\tau + 2\epsilon) - u(\tau)}{2\epsilon}.$$

The right-hand side is the difference quotient for the unit step function, so letting ϵ approach zero leads to the definition

$$\delta(t) \text{ is the derivative of } u(t).$$

This of course makes no sense at $t = 0$, where $u(t)$ is not even continuous, much less differentiable. But engineers and control systems analysts frequently use this definition; mathematically it turns out that $\delta(t)$ is a "generalized derivative" of $u(t)$.

The unit impulse function is a useful tool in the analysis of linear systems. Recall that in our discussion in Section 4.7, where the zero-state initial value problem

$$y'' + ay' + by = f(t), \qquad y(0) = 0, \qquad y'(0) = 0,$$

was studied, the relation

$$Y(s) = \frac{F(s)}{s^2 + as + b} = P(s)\,F(s)$$

was obtained. The function $P(s) = (s^2 + as + b)^{-1}$ was called the *transfer function*. But if $f(t) = \delta(t)$, then $F(s) = 1$, so if the *solution* of the differential equation with $f(t) = \delta(t)$ is defined as the *impulse response*, this leads to the statement

---

The transfer function $P(s)$ is the Laplace transform of the impulse response.

Another definition of the impulse response would be the zero-state output corresponding to the input $\delta(t)$.

Recall also that the weighting function $p(t)$ was defined by the relation $p(t) = L^{-1}\{P(s)\}$ and the zero-state output was

$$y(t) = L^{-1}\{Y(s)\} = L^{-1}\{P(s)\,F(s)\}.$$

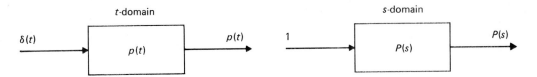

t-domain

s-domain

$\delta(t)$

$p(t)$

$p(t)$

1

$P(s)$

$P(s)$

$P(s)$

Figure 4.17

But if $F(s) = 1$, then $y(t) = p(t)$, and this leads to the statement

> The weighting function $p(t)$ is the zero-state output corresponding to the unit impulse function input.

Using the terminology developed in the last paragraph, we can say that the weighting function is simply the impulse response. In the t-domain and s-domain the system diagrams would look like those in Fig. 4.17. In the first diagram, the "black box" performs a convolution; in the second it multiplies.

EXERCISES 4.9

1. An initial value problem of the form

$$y'' + ay' + by = \delta(t - t_0),$$

$$t_0 > 0, \qquad y(0) = A, \qquad y'(0) = B,$$

can be interpreted as a mechanical or electrical system subjected to a unit impulse input at $t = t_0$. Using the formula (4.9.6), we can easily solve it by Laplace transform methods, and the solution will be a displacement of the solution with zero input. Solve the following:

a) $y'' + 4y' = \delta(t - \pi)$, $y(0) = 1$, $y'(0) = 0$

b) $y'' + 6y' + 9y = 3 + \delta(t - 1)$, $y(0) = 0$, $y'(0) = 0$

c) $y'' + 2y' + 5y = 8e^{-t} + \delta(t - 2)$, $y(0) = 2$, $y'(0) = 0$

d) $y'' - 16y = 4\delta(t - 1) + 8t$, $y(0) = 0$, $y'(0) = 0$

e) $y'' + 9y = \delta(t - \pi/3)$, $y(0) = 1$, $y'(0) = 0$

2. Using the property (4.9.5) of the unit impulse function and the definition of the Laplace transform, show that

$$L\{\delta(t - t_0) f(t)\} = e^{-st_0}f(t_0),$$

where $t_0 > 0$ and $f(t)$ is defined to be zero for $t < 0$. Use this result to solve the following initial value problems:

a) $y'' + 3y' + 3y = \delta(t - 1)t^2$, $y(0) = 0$, $y'(0) = 0$

b) $y'' + 16y = \delta(t - \pi/8) \sin 4t$, $y(0) = 2$, $y'(0) = 0$

c) $y'' + y = \delta(t - \alpha) \cos t$, $\alpha > 0$, $y(0) = 0$, $y'(0) = 1$

As α approaches $\pi/2$, does the solution approach the solution of

$$y'' + y = 0, y(0) = 0, y'(0) = 1?$$

MISCELLANEOUS EXERCISES

4.1 Noting that $d^{n+1}(t^n)/dt^{n+1} = 0$, use the Differentiation Formula directly to show that $L\{t^n\} = n!/s^{n+1}$.

4.2 The transform formula

$$L\{t^n y(t)\} = (-1)^n \frac{d^n}{ds^n} Y(s)$$

where $L\{y(t)\} = Y(s)$, allows one to "solve" linear differential equations whose coefficients are polynomials in t. Applying the transform, one obtains a linear differential equation in $Y(s)$ whose coefficients are polynomials in s (if it is easier to solve than the original, one is in luck!).

a) Use the above formula to show that

$$L\{t^2 y'(t)\} = s Y''(s) + 2 Y'(s)$$

and

$$L\{t y''(t)\} = -s^2 Y'(s) - 2s Y(s) + y(0).$$

b) Use the Laplace transform to solve

$$ty'' - ty' - y = 0, \qquad y(0) = 0, \qquad y'(0) = 4.$$

c) Use the Laplace transform to show that the solution of

$$ty'' + y' + ty = 0, \qquad y(0) = 1$$

is

$$y(t) = J_0(t)$$

$$= \sum_{n=0}^{\infty} \frac{(-1)^n}{(2^n n!)^2} t^{2n},$$

the Bessel function of order zero of the first kind. Do this by showing first that

$$L\{y(t)\} = Y(s) = \frac{c}{\sqrt{s^2 + 1}} = \frac{c}{s} \left(1 + \frac{1}{s^2}\right)^{-1/2},$$

where c is an arbitrary constant. Expand the second expression by using the binomial formula, then invert and use the initial condition.

4.3 A transform that is used to analyze discrete or sampled data systems is the z-transform. It is defined as follows: if T is a fixed positive number and $f(t)$ a function defined for $t \geq 0$, then its z-transform $F(z)$ is given by

$$F(z) = \sum_{n=0}^{\infty} f(nT) z^{-n},$$

where z is a complex number. One may regard the fixed number T as the sampling interval. Hence, functions in the t-domain, $0 \leq t < \infty$, are transformed into functions in the z-domain which is the complex plane.

a) If $f(t) = 1$, $t \geq 0$, show that $F(z) = z/(z - 1)$, valid for $|z| > 1$. *Hint:* Recall that

$$1 + u + u^2 + \cdots + u^n + \cdots = \frac{1}{1 - u},$$

valid for $|u| < 1$.

b) If $f(t) = t$, $t \geq 0$, show that $F(z) = zT/(z - 1)^2$. *Hint:* Convergent power series can be differentiated term by term.

c) If $f(t) = e^{2t}$, $t \geq 0$, show that $F(z) = z/(z - e^{2T})$, valid for $|z| > e^{2T}$.

4.4 When the z-transform is used, translation through integer multiples of the sampling interval T corresponds to differentiation. If $f(t) = 0$ for $t < 0$ and $Z[f(t)] = F(z)$ is the z-transform of $f(t)$, establish the formulas

$$Z[f(t - kT)] = z^{-k}F(z)$$

and

$$Z[f(t + kT)] = z^k F(z) - [z^k f(0) + z^{k-1} f(T) + \cdots + z f((k - 1)T)],$$

where k is a positive integer.

4.5 Use the z-transform to verify that the given function $Y(z)$ is the z-transform of the solution $y(t)$ of the following finite difference equations where $T = 1$:

a) $y(t + 1) + 2y(t) = 0$, $\quad y(0) = 4$; $\quad Y(z) = \dfrac{4z}{z + 2} = 4 - \dfrac{8}{z + 2}$

b) $y(t + 1) + 2y(t) = 1$, $\quad y(0) = 0$;

$$Y(z) = \frac{z}{(z + 2)(z - 1)} = \frac{2}{3}\frac{1}{z + 2} + \frac{1}{3}\frac{1}{z - 1}$$

c) $y(t + 2) + 3y(t + 1) - 10y(t) = 0$, $\quad y(0) = 0$, $\quad y(1) = 1$;

$$Y(z) = \frac{z}{z^2 + 3z - 10} = \frac{2}{7}\frac{1}{z - 2} + \frac{5}{7}\frac{1}{z + 5}$$

Remark: The last two problems show that the z-transform corresponds to the Laplace transform when solving finite difference equations.

4.6 Using the z-transform show that:

a) If $f(t) = a[u(t) - u(t - T)]$, then its z-transform is

$$F(z) = a \sum_{n=0}^{\infty} [u(nT) - u(nT - T)]z^{-n} = a.$$

b) If $f(t) = a^{(t-T)/T}u(t - T)$, then its z-transform is

$$F(z) = \sum_{n=0}^{\infty} a^{n-1}u(nT - T)z^{-n}$$

$$= z^{-1} \sum_{n=0}^{\infty} \left(\frac{a}{z}\right)^n = \frac{1}{z - a},$$

valid for $|z| > |a|$.

4.7 Use the results above to find the solutions of the following difference equations by employing the inverse z-transform:

a) In Problem 4.5(a) show that the solution at the sample values $t = 0, 1, 2, \ldots$ is given by

$$y(t) = \begin{cases} 4 & \text{if } t = 0, \\ -8(-2)^{t-1} = 4(-2)^t & \text{if } t = 1, 2, \ldots \end{cases}$$

b) In Problem 4.5(b) show that

$$y(t) = \begin{cases} 0 & \text{if } t = 0, \\ \tfrac{2}{3}(2)^{t-1} + \tfrac{1}{3} & \text{if } t = 1, 2, \ldots \end{cases}$$

c) In Problem 4.5(c) show that

$$y(t) = \begin{cases} 0 & \text{if } t = 0, \\ \tfrac{2}{7}(2)^{t-1} + \tfrac{5}{7}(-5)^{t-1} & \text{if } t = 1, 2, \ldots \end{cases}$$

REFERENCES

A standard text on the Laplace transform with an extensive discussion of its application to partial differential equations is

1. R. V. Churchill, *Operational Mathematics,* 2nd ed., McGraw-Hill, New York, 1958.

A book that discusses a variety of transforms as well as the Laplace transform is

2. W. Kaplan, *Operational Methods for Linear Systems,* Addison-Wesley, Reading, 1962.

For a discussion of systems theory including transform methods see

3. J. J. DiStefano III, A. R. Stubberud, and I. J. Williams, *Feedback and Control Systems,* Schaum, New York, 1967.

4. C. A. Desoer and L. A. Zadeh, *Linear Systems Theory,* McGraw-Hill, New York, 1963.

5. E. V. Bohn, *The Transform Analysis of Linear Systems,* Addison-Wesley, Reading, 1963.

Linear Systems of Differential Equations

5.1 INTRODUCTION

A physical system with several interdependent quantities can often be modeled by a system of simultaneous ordinary differential equations. If the independent variable is denoted by t and the dependent variables by $y_1, y_2, \ldots, y_n$, a typical system is

$$\frac{dy_1}{dt} = g_1(t, y_1, y_2, \ldots, y_n),$$

$$\frac{dy_2}{dt} = g_2(t, y_1, y_2, \ldots, y_n),$$

$$\vdots \qquad\qquad\qquad \vdots \qquad\qquad\qquad (5.1.1)$$

$$\frac{dy_n}{dt} = g_n(t, y_1, y_2, \ldots, y_n).$$

The system of differential equations is said to be *linear* if each of the functions $g_j(t, y_1, y_2, \ldots, y_n)$ is a linear function of the dependent variables, hence,

$$\frac{dy_1}{dt} = a_{11}(t)y_1 + a_{12}(t)y_2 + \cdots + a_{1n}(t)y_n + f_1(t),$$

$$\frac{dy_2}{dt} = a_{21}(t)y_1 + a_{22}(t)y_2 + \cdots + a_{2n}(t)y_n + f_2(t),$$

$$\vdots \qquad\qquad\qquad\qquad\qquad \vdots \qquad\qquad (5.1.2)$$

$$\frac{dy_n}{dt} = a_{n1}(t)y_1 + a_{n2}(t)y_2 + \cdots + a_{nn}(t)y_n + f_n(t).$$

The coefficients $a_{ij}(t)$ and $f_j(t)$, $i, j = 1, 2, \ldots, n$, are assumed to be continuous functions of t on some nonempty interval. If all the $f_j(t)$ are identically zero, the system is called *homogeneous;* otherwise, the system is called *nonhomogeneous.* For convenience and simplicity we shall consider only linear systems with constant coefficients in this chapter.

Numerous physical problems can be modeled by linear systems with constant coefficients. Three such problems are considered in the following examples.

Example 1 Consider the consecutive unimolecular reactions

$$A \xrightarrow{k_1} B \xrightarrow{k_2} C,$$

where A, B, and C are compounds and k_1 and k_2 are reaction rates. Let $y_1 = [A]$, $y_2 = [B]$, and $y_3 = [C]$ denote the concentrations of the respective compounds in grams per cubic centimeter. If the rate of each reaction is proportional to the concentration of reactant present, the reaction can be modeled by the linear system of differential equations

$$\frac{dy_1}{dt} = -k_1 y_1, \qquad\qquad (5.1.3)$$

$$\frac{dy_2}{dt} = -k_2 y_2 + k_1 y_1, \qquad\qquad (5.1.4)$$

$$\frac{dy_3}{dt} = k_2 y_2. \qquad\qquad (5.1.5)$$

By adding the above equations, we get

$$\frac{d}{dt}(y_1 + y_2 + y_3) = 0,$$

and, after integrating, the mass conservation law

$$y_1 + y_2 + y_3 = \text{const.}$$

In this example the equations are fairly easy to solve. Suppose the concentrations of the reactants are specified at time $t = 0$, say,

$$y_1(0) = a, \qquad y_2(0) = b, \qquad y_3(0) = c.$$

Then from Eq. (5.1.3) and the initial condition $y_1(0) = a$, we obtain, using the methods of Chapter 1,

$$y_1(t) = ae^{-k_1 t}.$$

Substituting this value for $y_1(t)$ into Eq. (5.1.4) gives

$$\frac{dy_2}{dt} = -k_2 y_2(t) + k_1 a e^{-k_1 t}, \qquad y_2(0) = b.$$

If $k_1 \neq k_2$, this initial value problem has the solution

$$y_2(t) = \frac{k_1 a}{k_2 - k_1} e^{-k_1 t} + \left(b - \frac{k_1 a}{k_2 - k_1} \right) e^{-k_2 t}. \qquad (5.1.6)$$

Finally, from the mass conservation law we get

$$y_3(t) = a + b + c - y_1(t) - y_2(t)$$

$$= a + b + c - \frac{ak_2}{k_2 - k_1} e^{-k_1 t} - \left(b - \frac{k_1 a}{k_2 - k_1} \right) e^{-k_2 t}.$$

Of course, $y_3(t)$ could have also been obtained by solving Eq. (5.1.5), using (5.1.6) and the initial condition $y_3(0) = c$. $\square$

Example 2 Consider the simple LRC circuit shown in Fig. 5.1. If the current in the circuit is denoted by I and the charge on the capacitor by Q, an application of Kirchhoff's voltage law gives

$$L\frac{dI}{dt} + RI + \frac{Q}{C} = 0,$$

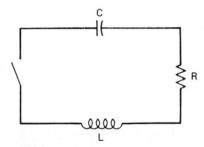

Figure 5.1

where

$$\frac{dQ}{dt} = I.$$

This can be written as the linear system

$$\frac{dQ}{dt} = I,$$

$$\frac{dI}{dt} = -\frac{1}{LC}Q - \frac{R}{L}I.$$

Suppose that the switch is closed at time $t = 0$ and that $I(0) = 0$, $Q(0) = Q_0$. If $R^2 < 4L/C$, the solution to this initial value problem is given by

$$Q = Q_0 e^{-Rt/2L}\left[\cos \omega t + \frac{R}{2L\omega}\sin \omega t\right],$$

$$I = \frac{-Q_0}{\omega LC} e^{-Rt/2L}\sin \omega t,$$

where

$$\omega = \left(\frac{4L}{C} - R^2\right)^{1/2}\bigg/ 2L.$$

We shall not go into the details here of how to construct the solution since it was shown earlier that the system is equivalent to the second order equation

$$L\frac{d^2Q}{dt} + R\frac{dQ}{dt} + \frac{Q}{C} = 0,$$

which was studied in Chapter 2. The point to be stressed here is that the circuit problem is quite naturally formulated as a linear system of first order differential equations. □

Example 3   Consider a mechanical system consisting of two equal masses and three identical springs. The masses are on a smooth level surface and are connected to each other and to two walls as shown in Fig. 5.2. It is assumed that the

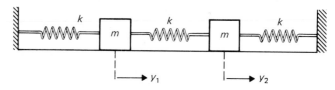

Figure 5.2

motion is along a straight line and that the only force on the masses are spring forces. Let y_1 and y_2 denote the displacements of the masses from their equilibrium positions. By Hooke's law, the spring forces are proportional to the displacements from the equilibrium positions, and so the forces acting on the masses are those shown in the figure. By Newton's second law, the time rate of change of the momentum of each mass is equal to the sum of the forces acting on it. This leads to the pair of differential equations

$$m\,\frac{d^2y_1}{dt^2} = -ky_1 + k(y_2 - y_1),$$

$$m\,\frac{d^2y_2}{dt^2} = -ky_2 - k(y_2 - y_1).$$

This system of two coupled second order differential equations can be analyzed as formulated, but we shall rewrite it to illustrate a method of representing higher order equations as a system of first order equations. To do this, introduce the new dependent variables, $y_3 = dy_1/dt$, $y_4 = dy_2/dt$, to get

$$\frac{dy_1}{dt} = y_3,$$

$$\frac{dy_2}{dt} = y_4,$$

$$\frac{dy_3}{dt} = -\frac{2k}{m}\,y_1 + \frac{k}{m}\,y_2,$$

$$\frac{dy_4}{dt} = \frac{k}{m}\,y_1 - 2\,\frac{k}{m}\,y_2,$$

(5.1.7)

where $d^2y_1/dt^2 = dy_3/dt$ and $d^2y_2/dt^2 = dy_4/dt$.

From symmetry one can expect two simple motions called *normal modes of vibrations*. The first motion has the two masses moving together as a rigid body, the second—as mirror images. Let us examine them in turn. To induce the masses to move together, they are displaced in the same direction by equal amounts and then released with zero initial velocity. Hence,

$$y_1(0) = y_2(0) = a, \qquad y_3(0) = y_4(0) = 0.$$

The solution to the system (5.1.7) subject to these initial conditions is

$$y_1(t) = y_2(t) = a\cos\omega t,$$

$$y_3(t) = y_4(t) = -a\omega\sin\omega t,$$

where the frequency $\omega = (k/m)^{1/2}$ is a *natural frequency* of the system. To induce the masses to move as mirror images they are displaced in opposite directions by the same amount and then released with zero initial velocity. Hence,

$$y_1(0) = -y_2(0) = b, \qquad y_3(0) = y_4(0) = 0,$$

and the corresponding solution is

$$y_1(t) = -y_2(t) = b \cos \sigma t,$$

$$y_3(t) = -y_4(t) = -b\sigma \sin \sigma t,$$

where σ, the second natural frequency, will be found by substitution in (5.1.7). The first two equations are automatically satisfied and from the second two we have

$$-b\sigma^2 \cos \sigma t = -2 \frac{k}{m} b \cos \sigma t - \frac{k}{m} b \cos \sigma t,$$

$$b\sigma^2 \cos \sigma t = \frac{k}{m} b \cos \sigma t + 2 \frac{k}{m} \cos \sigma t.$$

Either of these equations can be used to find $\sigma = (3k/m)^{1/2} = 3^{1/2}\omega$. Therefore, the frequency of the second normal mode is higher than that of the first normal mode. This could have been predicted since in the second mode the middle spring is active. □

EXERCISES 5.1

1. Solve the problem of Example 1 when $k_1 = k_2$.

2. Verify by substitution that

$$y_1(t) = a \cos \omega t + a' \sin \omega t$$
$$+ b \cos \sigma t + b' \sin \sigma t,$$

$$y_2(t) = a \cos \omega t + a' \sin \omega t$$
$$- b \cos \sigma t - b' \sin \sigma t,$$

is a solution to the mass–spring system of Example 3, where a, a', b, b' are arbitrary constants, $\omega = (k/m)^{1/2}$, $\sigma = (3k/m)^{1/2}$.

3. Use Laplace transforms to solve (5.1.7) for the two normal mode solutions.

5.2 VECTORS AND MATRICES

This section contains a résumé of some topics of linear algebra needed for the study of linear systems of differential equations. Some results will be proved, others stated without proof; the omitted proofs are assigned as exercises or as reading projects.

We start with n-vectors, which are ordered n-tuples

$$\mathbf{y} = \begin{bmatrix} y_1 \\ y_2 \\ \cdot \\ \cdot \\ \cdot \\ y_n \end{bmatrix} = \text{col}(y_1, y_2, \ldots, y_n)$$

of real or complex numbers. The number y_i is called the ith component of $\mathbf{y}$. The algebra of n-vectors is the same as the algebra of the two- or three-dimensional vectors used in mechanics and calculus courses. For example, two vectors are equal if and only if all their components are pairwise equal. Vector addition and subtraction and the product of a vector and a number (also called a *scalar*) are defined by

$$\mathbf{y} \pm \mathbf{x} = \begin{bmatrix} y_1 \pm x_1 \\ y_2 \pm x_2 \\ \cdot \\ \cdot \\ \cdot \\ y_n \pm x_n \end{bmatrix}, \qquad c\mathbf{y} = \begin{bmatrix} cy_1 \\ cy_2 \\ \cdot \\ \cdot \\ \cdot \\ cy_n \end{bmatrix}.$$

The *scalar* or *dot product* of two vectors $\mathbf{x}$ and $\mathbf{y}$ is

$$\mathbf{x} \cdot \mathbf{y} = x_1 y_1 + x_2 y_2 + \cdots + x_n y_n.$$

Vectors $\mathbf{y}^1, \mathbf{y}^2, \ldots, \mathbf{y}^k$ are *linearly dependent* if there are numbers c_1, $c_2, \ldots, c_k$, *not all zero*, such that

$$c_1 \mathbf{y}^1 + c_2 \mathbf{y}^2 + \cdots + c_k \mathbf{y}^k = \mathbf{0},$$

where $\mathbf{0}$ denotes the n-vector whose components are all 0. If the vectors $\mathbf{y}^1$, $\mathbf{y}^2, \ldots, \mathbf{y}^k$ are not linearly dependent, they are *linearly independent;* hence,

$$c_1 \mathbf{y}^1 + c_2 \mathbf{y}^2 + \cdots + c_k \mathbf{y}^k = \mathbf{0}$$

is satisfied only if $c_1 = c_2 = \cdots = c_k = 0$.

Example 1 Show that the vectors $\mathbf{y}^1, \mathbf{y}^2, \mathbf{y}^3$ are linearly independent while the vectors $\mathbf{y}^1, \mathbf{y}^2, \mathbf{y}^3, \mathbf{y}^4$ are linearly dependent, where

$$\mathbf{y}^1 = \begin{bmatrix} 1 \\ 1 \\ 1 \\ 1 \end{bmatrix}, \qquad \mathbf{y}^2 = \begin{bmatrix} 1 \\ 0 \\ 1 \\ 0 \end{bmatrix}, \qquad \mathbf{y}^3 = \begin{bmatrix} 1 \\ 1 \\ 0 \\ 1 \end{bmatrix}, \qquad \mathbf{y}^4 = \begin{bmatrix} 0 \\ 2 \\ -1 \\ 2 \end{bmatrix}.$$

First note that

$$c_1\mathbf{y}^1 + c_2\mathbf{y}^2 + c_3\mathbf{y}^3 = c_1 \begin{bmatrix} 1 \\ 1 \\ 1 \\ 1 \end{bmatrix} + c_2 \begin{bmatrix} 1 \\ 0 \\ 1 \\ 0 \end{bmatrix} + c_3 \begin{bmatrix} 1 \\ 1 \\ 0 \\ 1 \end{bmatrix}$$

$$= \begin{bmatrix} c_1 + c_2 + c_3 \\ c_1 + c_3 \\ c_1 + c_2 \\ c_1 + c_3 \end{bmatrix} = \begin{bmatrix} 0 \\ 0 \\ 0 \\ 0 \end{bmatrix}$$

implies that

$$\begin{aligned} c_1 + c_2 + c_3 &= 0, \\ c_1 + c_3 &= 0, \\ c_1 + c_2 &= 0. \end{aligned}$$

Clearly, $c_1 = c_2 = c_3 = 0$ and therefore $\mathbf{y}^1$, $\mathbf{y}^2$, and $\mathbf{y}^3$ are linearly independent. On the other hand, it is easy to check that $\mathbf{y}^4 = \mathbf{y}^1 - 2\mathbf{y}^2 + \mathbf{y}^3$. Equivalently, if $c_1 = 1$, $c_2 = -2$, $c_3 = 1$, $c_4 = -1$, then

$$\mathbf{y}^1 - 2\mathbf{y}^2 + \mathbf{y}^3 - \mathbf{y}^4 = \begin{bmatrix} 1 \\ 1 \\ 1 \\ 1 \end{bmatrix} - 2 \begin{bmatrix} 1 \\ 0 \\ 1 \\ 0 \end{bmatrix} + \begin{bmatrix} 1 \\ 1 \\ 0 \\ 1 \end{bmatrix} - \begin{bmatrix} 0 \\ 2 \\ -1 \\ 2 \end{bmatrix}$$

$$= \begin{bmatrix} 1 - 2 + 1 - 0 \\ 1 - 0 + 1 - 2 \\ 1 - 2 + 0 + 1 \\ 1 + 0 + 1 - 2 \end{bmatrix} = \begin{bmatrix} 0 \\ 0 \\ 0 \\ 0 \end{bmatrix}.$$

Therefore, $\mathbf{y}^1$, $\mathbf{y}^2$, $\mathbf{y}^3$, $\mathbf{y}^4$ are linearly dependent. □

The solutions to systems of linear differential equations can be represented as *vector-valued functions*. For example,

$$\mathbf{y}(t) = \begin{bmatrix} y_1(t) \\ y_2(t) \\ y_3(t) \\ y_4(t) \end{bmatrix} = \begin{bmatrix} a\cos t \\ a\cos t \\ -a\sin t \\ -a\sin t \end{bmatrix}$$

is a normal mode solution to the mass–spring system of Example 3 if $k/m = 1$. Vector-valued functions are integrated and differentiated by components. For example,

$$\int_0^t \begin{bmatrix} a \cos s \\ a \cos s \\ -a \sin s \\ -a \sin s \end{bmatrix} ds = \begin{bmatrix} a \sin t \\ a \sin t \\ a (\cos t - 1) \\ a (\cos t - 1) \end{bmatrix}$$

and

$$\frac{d}{dt} \begin{bmatrix} a \sin t \\ a \sin t \\ a \cos(t - 1) \\ a \cos(t - 1) \end{bmatrix} = \begin{bmatrix} a \cos t \\ a \cos t \\ -a \sin t \\ -a \sin t \end{bmatrix}.$$

Vector-valued functions satisfy the same algebraic rules as constant vectors, but the definitions of linear dependence and independence must be modified.

The functions $\mathbf{y}^1(t), \mathbf{y}^2(t), \ldots, \mathbf{y}^k(t)$ are said to be *linearly dependent on the interval* $t_1 \le t \le t_2$ if there exist *constants* $c_1, c_2, \ldots, c_k$, not all zero, such that

$$c_1\mathbf{y}^1(t) + c_2\mathbf{y}^2(t) + \cdots + c_k\mathbf{y}^k(t) \equiv \mathbf{0}, \quad t_1 \le t \le t_2.$$

Otherwise they are said to be *linearly independent on the interval*.

Therefore if the relation

$$c_1\mathbf{y}^1(t) + c_2\mathbf{y}^2(t) + \cdots + c_k\mathbf{y}^k(t) \equiv 0, \quad t_1 \le t \le t_2,$$

implies that the constants $c_1, c_2, \ldots, c_n$ are all zero, then $\mathbf{y}^1(t), \mathbf{y}^2(t), \ldots, \mathbf{y}^k(t)$ are linearly independent on $t_1 \le t \le t_2$.

Example 2 Show that

$$\mathbf{y}^1(t) = \begin{bmatrix} t \\ t \end{bmatrix}, \quad \mathbf{y}^2(t) = \begin{bmatrix} t^2 \\ t^2 \end{bmatrix}$$

are linearly independent vector-valued functions on any nonempty interval, but for each fixed t they are dependent.

If c_1 and c_2 are constants satisfying

$$c_1 \begin{bmatrix} t \\ t \end{bmatrix} + c_2 \begin{bmatrix} t^2 \\ t^2 \end{bmatrix} = \begin{bmatrix} c_1 t + c_2 t^2 \\ c_1 t + c_2 t^2 \end{bmatrix} \equiv \begin{bmatrix} 0 \\ 0 \end{bmatrix}$$

for $t_1 \le t \le t_2$, with $t_1 < t_2$, it follows that $c_1 = c_2 = 0$. This is a consequence of the fact that a polynomial of degree two can vanish at most twice on any interval. On the other hand, for any *fixed* t,

$$t \begin{bmatrix} t \\ t \end{bmatrix} - \begin{bmatrix} t^2 \\ t^2 \end{bmatrix} = \begin{bmatrix} 0 \\ 0 \end{bmatrix},$$

but the coefficients t, -1 are not both constants as required. $\square$

An $n \times n$ matrix is an array of n^2 numbers a_{ij}, where $i, j = 1, 2, \ldots, n$. It will be denoted by a capital letter and displayed as follows:

$$A = \begin{bmatrix} a_{11} & a_{12} & \cdots & a_{1n} \\ a_{21} & a_{22} & \cdots & a_{2n} \\ \vdots & \vdots & & \vdots \\ a_{n1} & a_{n2} & \cdots & a_{nn} \end{bmatrix}.$$

It is also written as $A = (a_{ij})$. The entry a_{ij} can be real or complex, with the first subscript referring to its row, the second to its column. Since the number of columns equals the number of rows, an $n \times n$ matrix is called a *square* matrix.

Equality of matrices means pairwise equality of entries. Addition, subtraction, and scalar multiplication are defined by

$$A \pm B = \begin{bmatrix} a_{11} \pm b_{11} & a_{12} \pm b_{12} & \cdots & a_{1n} \pm b_{1n} \\ a_{21} \pm b_{21} & a_{22} \pm b_{22} & \cdots & a_{2n} \pm b_{2n} \\ \vdots & \vdots & & \vdots \\ a_{n1} \pm b_{n1} & a_{n2} \pm b_{n2} & \cdots & a_{nn} \pm b_{nn} \end{bmatrix}$$

and

$$cA = \begin{bmatrix} ca_{11} & ca_{12} & \cdots & ca_{1n} \\ ca_{21} & ca_{22} & \cdots & ca_{2n} \\ \vdots & \vdots & & \vdots \\ ca_{n1} & ca_{n2} & \cdots & ca_{nn} \end{bmatrix},$$

or, more concisely, $(a_{ij}) \pm (b_{ij}) = (a_{ij} \pm b_{ij})$ and $c(a_{ij}) = (ca_{ij})$.

The product of two $n \times n$ matrices $A = (a_{ij})$ and $B = (b_{ij})$ is the $n \times n$ matrix AB defined by

$$AB = \sum_{k=1}^{n} a_{ik}b_{kj}.$$

Unlike real or complex numbers, AB is not in general equal to BA. The rule for multiplying matrices can be easily remembered by noting that the (i, j)th entry of AB is the scalar, or dot, product of the ith row of A and the jth column of B. The vectors formed from the rows of A are denoted $\mathbf{A}_j$, those formed from its columns are denoted $\mathbf{A}^j$. Therefore if c_{ij} is the (i, j)th entry of $C = AB$, then $c_{ij} = \mathbf{A}_i \cdot \mathbf{B}^j$.

Example 3   Let

$$A = \begin{bmatrix} 2 & 3 & 0 \\ 0 & -2 & 0 \\ 1 & 0 & 4 \end{bmatrix}, \qquad B = \begin{bmatrix} -1 & 2 & 1 \\ 1 & -2 & -1 \\ 1 & 0 & 1 \end{bmatrix},$$

and form $3A - 4B$, AB, and BA.

Since $3A - 4B = (3a_{ij} - 4b_{ij})$, we can write

$$3A - 4B = \begin{bmatrix} 6+4 & 9-8 & 0-4 \\ 0-4 & -6+8 & 0+4 \\ 3-4 & 0+0 & 12-4 \end{bmatrix} = \begin{bmatrix} 10 & 1 & -4 \\ -4 & 2 & 4 \\ -1 & 0 & 8 \end{bmatrix}.$$

From the rule for matrix multiplication it follows that

$$AB = \begin{bmatrix} 1 & -2 & -1 \\ -2 & 4 & 2 \\ 3 & 2 & 5 \end{bmatrix}, \qquad BA = \begin{bmatrix} -1 & -7 & 4 \\ 1 & 7 & -4 \\ 3 & 2 & 4 \end{bmatrix}.$$

A typical step in the computation of AB is the scalar, or dot, product of the third row of A by the first column of B to obtain the $(3, 1)$th entry of AB:

$$\sum_{k=1}^{3} a_{3k}b_{k1} = \begin{bmatrix} 1 \\ 0 \\ 4 \end{bmatrix} \cdot \begin{bmatrix} -1 \\ 1 \\ 1 \end{bmatrix} = -1 + 0 + 4 = 3.$$

Note that $AB \neq BA$. $\square$

The $n \times n$ matrix whose (i, j)th entry is 1 if $i = j$ and 0 if $i \neq j$ is called the *identity* matrix and is denoted by I. It is simply the square matrix with ones on the main diagonal and zeros elsewhere. It has the easily verified property of

$$IA = AI = A$$

for any $n \times n$ matrix A. For example,

$$\begin{bmatrix} 1 & 0 \\ 0 & 1 \end{bmatrix}\begin{bmatrix} 2 & 3 \\ 3 & 1 \end{bmatrix} = \begin{bmatrix} 2 & 3 \\ 3 & 1 \end{bmatrix}\begin{bmatrix} 1 & 0 \\ 0 & 1 \end{bmatrix} = \begin{bmatrix} 2 & 3 \\ 3 & 1 \end{bmatrix}.$$

The following algebraic rules for vectors and matrices are used extensively:

$$y + z = z + y, \qquad A + B = B + A;$$
$$(y + w) + z = y + (w + z), \qquad (A + B) + C = A + (B + C);$$
$$c(y + z) = cy + cz, \qquad c(A + B) = cA + cB, \quad c \text{ a scalar;}$$
$$A(BC) = (AB)C;$$
$$A(B + C) = AB + AC.$$

The multiplication of a vector y by a matrix A is a vector Ay whose ith component is the scalar, or dot, product of the ith row of A with the vector y. Thus $(Ay)_i = A_i \cdot y = \sum_{k=1}^{n} a_{ik} y_k$.

Example 4 Let

$$A = \begin{bmatrix} 2 & 1 & 4 \\ 4 & 0 & -2 \\ 5 & 0 & 1 \end{bmatrix}, \qquad y = \begin{bmatrix} 0 \\ 1 \\ -3 \end{bmatrix}, \qquad I = \begin{bmatrix} 1 & 0 & 0 \\ 0 & 1 & 0 \\ 0 & 0 & 1 \end{bmatrix}.$$

Then

$$Ay = \begin{bmatrix} 2 & 1 & 4 \\ 4 & 0 & -2 \\ 5 & 0 & 1 \end{bmatrix} \begin{bmatrix} 0 \\ 1 \\ -3 \end{bmatrix} = \begin{bmatrix} 0 + 1 - 12 \\ 0 + 0 + 6 \\ 0 + 0 - 3 \end{bmatrix} = \begin{bmatrix} -11 \\ 6 \\ -3 \end{bmatrix},$$

$$Iy = \begin{bmatrix} 1 & 0 & 0 \\ 0 & 1 & 0 \\ 0 & 0 & 1 \end{bmatrix} \begin{bmatrix} 0 \\ 1 \\ -3 \end{bmatrix} = \begin{bmatrix} 0 \\ 1 \\ -3 \end{bmatrix}.$$

The last computation illustrates a simple but useful fact, namely that $y = Iy$ for all n-vectors y. □

The vectors formed from the columns of a matrix A can be either linearly dependent or linearly independent. To examine this, write

$$c_1 A^1 + c_2 A^2 + \cdots + c_n A^n = 0$$

or, more concisely, $Ac = 0$. Clearly, $Ac = 0$ for some $c \neq 0$ implies linear dependence of the columns of A, while $Ac = 0$ only if $c = 0$ implies linear independence.

Example 5 Use the matrices of Example 3 and show that the column vectors of A are linearly independent and that the column vectors of B are linearly dependent.

To show that the column vectors of A are linearly independent one writes

$$c_1 A^1 + c_2 A^2 + c_3 A^3 = c_1 \begin{bmatrix} 2 \\ 0 \\ 1 \end{bmatrix} + c_2 \begin{bmatrix} 3 \\ -2 \\ 0 \end{bmatrix} + c_3 \begin{bmatrix} 0 \\ 0 \\ 4 \end{bmatrix} = \begin{bmatrix} 0 \\ 0 \\ 0 \end{bmatrix}$$

or

$$2c_1 + 3c_2 + 0c_3 = 0,$$

$$0c_1 - 2c_2 + 0c_3 = 0,$$

$$c_1 + 0c_2 + 4c_3 = 0.$$

Clearly, $c_1 = c_2 = c_3 = 0$. On the other hand, the column vectors of B are dependent since

$$\mathbf{B}^1 + \mathbf{B}^2 - \mathbf{B}^3 = \begin{bmatrix} -1 \\ 1 \\ 1 \end{bmatrix} + \begin{bmatrix} 2 \\ -2 \\ 0 \end{bmatrix} - \begin{bmatrix} 1 \\ -1 \\ 1 \end{bmatrix} = \begin{bmatrix} 0 \\ 0 \\ 0 \end{bmatrix}. \quad \square$$

It is a theorem of linear algebra that the n column vectors of an $n \times n$ matrix A are linearly independent if and only if the determinant, $\det(A)$, is nonzero. This theorem gives a test for checking the independence of n vectors of dimension n: form an $n \times n$ matrix whose columns are the vectors and compute its determinant. Unfortunately, the determinant of a square matrix of arbitrary dimension is both difficult to define and to evaluate. For simplicity, only determinants of 2×2 and 3×3 matrices will be considered. They are given for the 2×2 case by

$$\det(A) = \begin{vmatrix} a_{11} & a_{12} \\ a_{21} & a_{22} \end{vmatrix} = a_{11}a_{22} - a_{21}a_{12}$$

and for the 3×3 case by

$$\det(B) = \begin{vmatrix} b_{11} & b_{12} & b_{13} \\ b_{21} & b_{22} & b_{23} \\ b_{31} & b_{32} & b_{33} \end{vmatrix} = b_{11} \begin{vmatrix} b_{22} & b_{23} \\ b_{32} & b_{33} \end{vmatrix} - b_{12} \begin{vmatrix} b_{21} & b_{23} \\ b_{31} & b_{33} \end{vmatrix} + b_{13} \begin{vmatrix} b_{21} & b_{22} \\ b_{31} & b_{32} \end{vmatrix}$$

$$= b_{11}(b_{22}b_{33} - b_{32}b_{23}) - b_{12}(b_{21}b_{33} - b_{31}b_{23}) + b_{13}(b_{21}b_{32} - b_{31}b_{22}).$$

Example 6   Evaluate the determinants of the matrices A and B of Example 3:

$$\det(A) = \begin{vmatrix} 2 & 3 & 0 \\ 0 & -2 & 0 \\ 1 & 0 & 4 \end{vmatrix} = 2 \begin{vmatrix} -2 & 0 \\ 0 & 4 \end{vmatrix} - 3 \begin{vmatrix} 0 & 0 \\ 1 & 4 \end{vmatrix} + 0 \begin{vmatrix} 0 & -2 \\ 1 & 0 \end{vmatrix}$$

$$= 2(-8 - 0) - 3(0 - 0) + 0(0 + 2) = -16,$$

$$\det(B) = \begin{vmatrix} -1 & 2 & 1 \\ 1 & -2 & -1 \\ 1 & 0 & 1 \end{vmatrix}$$

$$= -1 \begin{vmatrix} -2 & -1 \\ 0 & 1 \end{vmatrix} - 2 \begin{vmatrix} 1 & -1 \\ 1 & 1 \end{vmatrix} + 1 \begin{vmatrix} 1 & -2 \\ 1 & 0 \end{vmatrix}$$

$$= -1(-2 - 0) - 2(1 + 1) + 1(0 + 2) = 0. \quad \square$$

Example 7   Use the determinant test to determine the dependence or independence of the following vectors:

$$\mathbf{x} = \begin{bmatrix} 2 \\ 2 \\ 0 \end{bmatrix}, \qquad \mathbf{y} = \begin{bmatrix} 0 \\ -1 \\ 2 \end{bmatrix}, \qquad \mathbf{z} = \begin{bmatrix} 3 \\ 0 \\ 2 \end{bmatrix}.$$

Form a matrix C with columns $\mathbf{C}^1 = \mathbf{x}$, $\mathbf{C}^2 = \mathbf{y}$, $\mathbf{C}^3 = \mathbf{z}$ and evaluate

$$
\begin{aligned}
\det(C) &= \begin{vmatrix} 2 & 0 & 3 \\ 2 & -1 & 0 \\ 0 & 2 & 2 \end{vmatrix} \\
&= 2 \begin{vmatrix} -1 & 0 \\ 2 & 2 \end{vmatrix} - 0 \begin{vmatrix} 2 & 0 \\ 0 & 2 \end{vmatrix} + 3 \begin{vmatrix} 2 & -1 \\ 0 & 2 \end{vmatrix} \\
&= (2)(-2) - 0 + (3)(4) = 8.
\end{aligned}
$$

Since $\det(C) \neq 0$, the vectors are linearly independent. $\square$

The relationship between the linear independence of the columns of a square matrix and the nonvanishing of its determinant plays a key role in the theory of linear algebraic equations. The algebraic system of equations

$$a_{11}x_1 + a_{12}x_2 + a_{13}x_3 = b_1,$$

$$a_{21}x_1 + a_{22}x_2 + a_{23}x_3 = b_2, \qquad (5.2.1)$$

$$a_{31}x_1 + a_{32}x_2 + a_{33}x_3 = b_3$$

can be written in two ways as

$$A\mathbf{x} = \mathbf{b} \qquad \text{and} \qquad x_1\mathbf{A}^1 + x_2\mathbf{A}^2 + x_3\mathbf{A}^3 = \mathbf{b},$$

where the coefficient matrix is $A = (a_{ij})$. By Cramer's rule, if $\det(A) \neq 0$, there is a unique solution vector $\mathbf{x}$ whose components are

$$x_1 = \frac{\begin{vmatrix} b_1 & a_{12} & a_{13} \\ b_2 & a_{22} & a_{23} \\ b_3 & a_{32} & a_{33} \end{vmatrix}}{\det(A)}, \qquad x_2 = \frac{\begin{vmatrix} a_{11} & b_1 & a_{13} \\ a_{21} & b_2 & a_{23} \\ a_{31} & b_3 & a_{33} \end{vmatrix}}{\det(A)},$$

$$x_3 = \frac{\begin{vmatrix} a_{11} & a_{12} & b_1 \\ a_{21} & a_{22} & b_2 \\ a_{31} & a_{32} & b_3 \end{vmatrix}}{\det(A)}. \qquad (5.2.2)$$

It follows from (5.2.2) that $\mathbf{x}$ can be written as

$$\mathbf{x} = b_1\mathbf{C}^1 + b_2\mathbf{C}^2 + b_3\mathbf{C}^3 \qquad \text{or} \qquad \mathbf{x} = C\mathbf{b}, \qquad (5.2.3)$$

where C is a 3×3 matrix. For example, the components of the first column of C are the factors of b_1 obtained from computing each of the determinants in (5.2.2):

$$\mathbf{C}^1 = \frac{1}{\det(A)} \begin{bmatrix} a_{22}a_{33} - a_{32}a_{23} \\ -(a_{21}a_{33} - a_{31}a_{23}) \\ a_{21}a_{32} - a_{31}a_{22} \end{bmatrix}.$$

This is illustrated in the following example.

Example 8   Write the solution to the following system as $\mathbf{x} = C\mathbf{b}$.

$$\begin{array}{rl} -x_1 + 5x_2 - 3x_3 &= b_1, \\ x_1 - 3x_2 + 2x_3 &= b_2, \\ -x_2 + x_3 &= b_3, \end{array} \quad \text{or} \quad \begin{bmatrix} -1 & 5 & -3 \\ 1 & -3 & 2 \\ 0 & -1 & 1 \end{bmatrix} \begin{bmatrix} x_1 \\ x_2 \\ x_3 \end{bmatrix} = \begin{bmatrix} b_1 \\ b_2 \\ b_3 \end{bmatrix}.$$

This could be solved by systematic elimination, but we notice that when the three equations are added, the coefficients of x_1 and x_3 sum to 0. Hence,

$$x_2 = b_1 + b_2 + b_3.$$

From the third equation, we get

$$x_3 = x_2 + b_3 = b_1 + b_2 + 2b_3,$$

and from the second equation

$$x_1 = 3x_2 - 2x_3 + b_2 = b_1 + 2b_2 - b_3.$$

Thus

$$\mathbf{x} = \begin{bmatrix} x_1 \\ x_2 \\ x_3 \end{bmatrix} = \begin{bmatrix} 1 & 2 & -1 \\ 1 & 1 & 1 \\ 1 & 1 & 2 \end{bmatrix} \begin{bmatrix} b_1 \\ b_2 \\ b_3 \end{bmatrix} = C\mathbf{b}.$$

Another route to the same solution is via Cramer's rule:

$$x_1 = \frac{\begin{vmatrix} b_1 & 5 & -3 \\ b_2 & -3 & 2 \\ b_3 & -1 & 1 \end{vmatrix}}{\det(A)}, \qquad x_2 = \frac{\begin{vmatrix} -1 & b_1 & -3 \\ 1 & b_2 & 2 \\ 0 & b_3 & 1 \end{vmatrix}}{\det(A)},$$

$$x_3 = \frac{\begin{vmatrix} -1 & 5 & b_1 \\ 1 & -3 & b_2 \\ 0 & -1 & b_3 \end{vmatrix}}{\det(A)},$$

where the ith column of C has as its components the factors of b_i in each of the above expressions. We leave these computations to the reader.   □

We now generalize the previous discussion and introduce the notion of the inverse of a square matrix. Two n-dimensional linear systems of equations

$$A\mathbf{x} = \mathbf{b} \quad \text{and} \quad C\mathbf{b} = \mathbf{x}$$

with $\det(A) \neq 0$, $\det(C) \neq 0$ are dual in the sense that either $\mathbf{b}$ or $\mathbf{x}$ can be considered as being given and then either $\mathbf{x}$ or $\mathbf{b}$ is the corresponding solution. By substitution we have (recall that $\mathbf{b} = I\mathbf{b}$ and $\mathbf{x} = I\mathbf{x}$) that

$$A\mathbf{x} = AC\mathbf{b} = \mathbf{b} \quad \text{or} \quad (AC - I)\mathbf{b} = \mathbf{0},$$
$$C\mathbf{b} = CA\mathbf{x} = \mathbf{x} \quad \text{or} \quad (CA - I)\mathbf{x} = \mathbf{0}.$$

Because of the arbitrariness of $\mathbf{b}$ (or $\mathbf{x}$), these equations imply that

$$AC = I \quad \text{and} \quad CA = I.$$

If these equations are satisfied, then C is called the *inverse* of A and is denoted by A^{-1}. The last pair of equations implies that

$$AA^{-1} = A^{-1}A = I, \tag{5.2.4}$$

and it is not hard to show that a matrix A^{-1} satisfying (5.2.4) is unique.

In general, an algebraic formula for the inverse of a matrix is rather complicated. For $n = 3$, one can be extracted from (5.2.2), as was done in Example 7. For $n = 2$, the inverse is especially easy to determine:

$$\text{If } A = \begin{bmatrix} a & b \\ c & d \end{bmatrix}, \quad \text{then } A^{-1} = \frac{1}{ad - bc} \begin{bmatrix} d & -b \\ -c & a \end{bmatrix}.$$

The verification of this formula is left to the reader.

The next topics in our summary of linear algebra are eigenvalues and eigenvectors, strange terms coming from the German words *Eigenwert* and *Eigenvektor* meaning *characteristic value* and *characteristic vector*, respectively. They occur for instance in the study of normal modes of vibrations.

If A is a square matrix and $\mathbf{x}$ is a nonzero vector, such that $A\mathbf{x} = \lambda\mathbf{x}$ for some real or complex number λ, then $\mathbf{x}$ is called an *eigenvector* of A corresponding to the *eigenvalue* λ. The defining equation can also be written as

$$(A - \lambda I)\mathbf{x} = \mathbf{0} \tag{5.2.5}$$

with the aid of the identity $\mathbf{x} = I\mathbf{x}$. Equation (5.2.5) will have a nonzero solution if and only if the column vectors of $A - \lambda I$ are linearly dependent. But, as we have seen, this is equivalent to the requirement that

$$\det(A - \lambda I) = 0. \tag{5.2.6}$$

If A is an $n \times n$ matrix, then Eq. (5.2.6) is a polynomial of degree n in λ called the *characteristic equation* of A. It must be satisfied by any eigenvalue of A, and since a polynomial of degree n has at most n distinct roots, an $n \times n$ matrix has at most n distinct eigenvalues. The process of finding eigenvalues and eigenvectors is illustrated in the following example.

Example 9 Find the eigenvalues and eigenvectors of

$$A = \begin{bmatrix} 0 & 1 & 0 \\ 0 & 0 & 1 \\ -2 & 1 & 2 \end{bmatrix}.$$

The characteristic equation is

$$0 = \det(A - \lambda I) = \det \left[\begin{bmatrix} 0 & 1 & 0 \\ 0 & 0 & 1 \\ -2 & 1 & 2 \end{bmatrix} - \begin{bmatrix} \lambda & 0 & 0 \\ 0 & \lambda & 0 \\ 0 & 0 & \lambda \end{bmatrix} \right]$$

$$= \begin{vmatrix} -\lambda & 1 & 0 \\ 0 & -\lambda & 1 \\ -2 & 1 & 2 - \lambda \end{vmatrix} = -\lambda^3 + 2\lambda^2 + \lambda - 2.$$

Since

$$-\lambda^3 + 2\lambda^2 + \lambda - 2 = -(\lambda - 1)(\lambda + 1)(\lambda - 2),$$

the roots $\lambda = 1$, $\lambda = -1$, and $\lambda = 2$ are the eigenvalues of A. To find the corresponding eigenvectors, we select nontrivial solutions to the homogeneous equations $(A - \lambda I)\mathbf{x} = 0$, where λ is an eigenvalue. The equations are

$$(A - I)\mathbf{x} = \mathbf{0}, \quad (A + I)\mathbf{x} = \mathbf{0}, \quad (A - 2I)\mathbf{x} = \mathbf{0},$$

or

$$(\lambda = 1) \qquad\qquad (\lambda = -1) \qquad\qquad (\lambda = 2)$$

$$
\begin{aligned}
-x_1 + x_2 &= 0, & x_1 + x_2 &= 0, & -2x_1 + x_2 &= 0, \\
-x_2 + x_3 &= 0, & x_2 + x_3 &= 0, & -2x_2 + x_3 &= 0, \\
-2x_1 + x_2 + x_3 &= 0; & -2x_1 + x_2 + 3x_3 &= 0; & -2x_1 + x_2 &= 0.
\end{aligned}
$$

One set of nontrivial solutions can be constructed by taking $x_1 = 1$ and solving for x_2 and x_3. The respective eigenvectors are

$$(\lambda = 1) \qquad\qquad (\lambda = -1) \qquad\qquad (\lambda = 2)$$

$$\mathbf{x}^1 = \begin{bmatrix} 1 \\ 1 \\ 1 \end{bmatrix}, \quad \mathbf{x}^2 = \begin{bmatrix} 1 \\ -1 \\ 1 \end{bmatrix}, \quad \mathbf{x}^3 = \begin{bmatrix} 1 \\ 2 \\ 4 \end{bmatrix}.$$

Any other set of nontrivial solutions can be obtained by another choice of x_1 and, clearly, this is equivalent to multiplying the above solutions by a constant. This is a general result: an eigenvector remains an eigenvector upon multiplication by any nonzero scalar. ☐

If an $n \times n$ matrix has n distinct eigenvalues, it can be shown that the corresponding n eigenvectors are linearly independent. But if there are multiple eigenvalues (corresponding to multiple roots of the characteristic equation), there may or may not be n linearly independent eigenvectors. This matter will not be discussed further here, but it will come up in subsequent sections.

Our final topic is matrix valued functions. Their algebra is the same as that of constant matrices and, just as with vector valued functions, they are differentiated and integrated entry by entry. For example, if $\Theta(t)$ is the matrix

$$\Theta(t) = \begin{bmatrix} \cos t & \sin t \\ -\sin t & \cos t \end{bmatrix},$$

then

$$\frac{d}{dt}\Theta(t) = \begin{bmatrix} -\sin t & \cos t \\ -\cos t & -\sin t \end{bmatrix}.$$

Note that the matrix $\Theta(t)$ is a solution to a *matrix differential equation,*

$$\frac{d\Theta}{dt} = J\Theta, \quad J = \begin{bmatrix} 0 & 1 \\ -1 & 0 \end{bmatrix},$$

since

$$J\Theta = \begin{bmatrix} 0 & 1 \\ -1 & 0 \end{bmatrix} \begin{bmatrix} \cos t & \sin t \\ -\sin t & \cos t \end{bmatrix} = \begin{bmatrix} -\sin t & \cos t \\ -\cos t & -\sin t \end{bmatrix} = \frac{d\Theta}{dt}.$$

Matrix differential equations will be encountered in Section 5.3 of this chapter.

The theorem that the n columns of an $n \times n$ matrix A are linearly independent if and only if the determinant $\det(A)$ is nonzero is not valid for general matrix valued functions. However, if the columns of the matrix are solutions to a homogeneous linear system of differential equations, then the theorem is valid (see [1, p. 69]) and the determinant test for linear dependence or linear independence can be employed.

This completes our résumé of linear algebra. We have presented a few concepts that are needed in an introductory study of linear systems of ordinary differential equations. Many important topics were not mentioned; they can be found in more advanced differential equation texts such as [1].

EXERCISES 5.2

1. For each of the following pairs of vectors $\mathbf{x}$, $\mathbf{y}$ below find $\mathbf{x} + \mathbf{y}$, $\mathbf{x} - \mathbf{y}$, $\mathbf{x} \cdot \mathbf{y}$, and determine whether they are linearly independent or linearly dependent.

a) $\mathbf{x} = \begin{bmatrix} 1 \\ 0 \end{bmatrix}$, $\mathbf{y} = \begin{bmatrix} 2 \\ 0 \end{bmatrix}$

b) $\mathbf{x} = \begin{bmatrix} 1 \\ 2 \end{bmatrix}$, $\mathbf{y} = \begin{bmatrix} 3 \\ 4 \end{bmatrix}$

c) $\mathbf{x} = \begin{bmatrix} 1 \\ 2 \end{bmatrix}$, $\mathbf{y} = \begin{bmatrix} 2 \\ 4 \end{bmatrix}$

d) $\mathbf{x} = \begin{bmatrix} 1 \\ 2 \\ 3 \end{bmatrix}$, $\mathbf{y} = \begin{bmatrix} 4 \\ 5 \\ 6 \end{bmatrix}$

e) $\mathbf{x} = \begin{bmatrix} 1 \\ 0 \\ 1 \end{bmatrix}$, $\mathbf{y} = \begin{bmatrix} 2 \\ 3 \\ -1 \end{bmatrix}$

f) $\mathbf{x} = \begin{bmatrix} 1 \\ -1 \\ 2 \end{bmatrix}$, $\mathbf{y} = \begin{bmatrix} 1 \\ -1 \\ -1 \end{bmatrix}$

2. Determine which of the following sets of vectors are linearly independent and which are linearly dependent:

a) $\begin{bmatrix} 1 \\ 0 \\ 0 \end{bmatrix}$, $\begin{bmatrix} 0 \\ 1 \\ 0 \end{bmatrix}$, $\begin{bmatrix} 0 \\ 0 \\ 1 \end{bmatrix}$

b) $\begin{bmatrix} 3 \\ 2 \\ 1 \end{bmatrix}$, $\begin{bmatrix} 2 \\ -1 \\ 2 \end{bmatrix}$, $\begin{bmatrix} -1 \\ 2 \\ 1 \end{bmatrix}$

c) $\begin{bmatrix} 2 \\ -2 \\ 1 \end{bmatrix}$, $\begin{bmatrix} 1 \\ 3 \\ 0 \end{bmatrix}$, $\begin{bmatrix} 0 \\ 0 \\ 1 \end{bmatrix}$

d) $\begin{bmatrix} 1 \\ 2 \end{bmatrix}$, $\begin{bmatrix} 3 \\ 4 \end{bmatrix}$, $\begin{bmatrix} 5 \\ 6 \end{bmatrix}$

e) $\begin{bmatrix} 1 \\ 0 \\ 0 \end{bmatrix}$, $\begin{bmatrix} -2 \\ 1 \\ 0 \end{bmatrix}$, $\begin{bmatrix} 3 \\ -1 \\ 1 \end{bmatrix}$, $\begin{bmatrix} 7 \\ -2 \\ 1 \end{bmatrix}$

f) $\begin{bmatrix} 1 \\ 3 \\ -2 \end{bmatrix}$, $\begin{bmatrix} 0 \\ 0 \\ 1 \end{bmatrix}$, $\begin{bmatrix} 1 \\ 3 \\ -1 \end{bmatrix}$

3. For each of the following vector functions $\mathbf{x}(t)$, find $d\mathbf{x}(t)/dt$ and $\int_0^t \mathbf{x}(s)\, ds$.

a) $\mathbf{x}(t) = \begin{bmatrix} t \\ t^2 \end{bmatrix}$

b) $\mathbf{x}(t) = \begin{bmatrix} t \\ \sin t \\ e^t \end{bmatrix}$

c) $\mathbf{x}(t) = \begin{bmatrix} te^t \\ \sin 2t \\ 2t + 1 \end{bmatrix}$

d) $\mathbf{x}(t) = \begin{bmatrix} e^t \\ 2e^t \\ -e^t \end{bmatrix}$

e) $\mathbf{x}(t) = \begin{bmatrix} \sin t + \cos t \\ \sin t - \cos t \end{bmatrix}$

f) $\mathbf{x}(t) = \begin{bmatrix} 1 \\ e^{2t} \end{bmatrix}$

4. Determine which of the following sets of vector functions are linearly independent and which are linearly dependent for $0 \le t \le 1$:

a) $\begin{bmatrix} 1 \\ t \end{bmatrix}$, $\begin{bmatrix} t \\ t^2 \end{bmatrix}$

b) $\begin{bmatrix} e^t \\ -e^t \end{bmatrix}$, $\begin{bmatrix} e^{2t} \\ 3e^{2t} \end{bmatrix}$

c) $\begin{bmatrix} 2e^t \\ -e^t \\ 3e^t \end{bmatrix}$, $\begin{bmatrix} e^{-t} \\ 4e^{-t} \\ -e^{-t} \end{bmatrix}$, $\begin{bmatrix} 3e^{-2t} \\ e^{-2t} \\ -e^{-2t} \end{bmatrix}$

d) $\begin{bmatrix} e^t \\ 1 \\ e^t \end{bmatrix}$, $\begin{bmatrix} e^{-t} \\ 1 \\ -e^{-t} \end{bmatrix}$, $\begin{bmatrix} \cosh t \\ 1 \\ \sinh t \end{bmatrix}$

5. Find the Laplace transform $\mathbf{X}(s)$ for each of the vector functions $\mathbf{x}(t)$ in Exercise 3.

6. Let

$$A = \begin{bmatrix} 1 & 2 & 1 \\ 2 & 1 & 0 \\ 0 & 1 & 1 \end{bmatrix}, \quad B = \begin{bmatrix} 2 & 1 & 2 \\ 1 & 0 & 1 \\ 0 & 1 & 0 \end{bmatrix}.$$

Find

a) $A + B$ b) $A - B$ c) $A - 2B$

d) $2A - B$ e) AB f) BA

7. If

$$A = \begin{bmatrix} 2 & 1 & 1 \\ 3 & 1 & 0 \\ 0 & 2 & 1 \end{bmatrix}, \quad B = \begin{bmatrix} 1 & 1 & 1 \\ 2 & 3 & 1 \\ 1 & 0 & 1 \end{bmatrix},$$

find

a) $3A + B$ b) $2A - 3B$ c) AB d) BA

8. Let

$$A = \begin{bmatrix} 1 & 2 & 3 \\ 1 & 2 & 1 \\ 0 & 1 & 0 \end{bmatrix}, \quad B = \begin{bmatrix} 1 & 1 & 1 \\ 6 & 1 & 4 \\ 1 & 4 & 2 \end{bmatrix},$$

$$C = \begin{bmatrix} 0 & 1 & 3 \\ 0 & 4 & 1 \\ 2 & 0 & 4 \end{bmatrix}.$$

Verify that

a) $A + (B + C) = (A + B) + C$

b) $A(B + C) = AB + AC$

c) $(A + B)C = AC + BC$

d) $A(BC) = (AB)C$

e) $2(A + B) = 2A + 2B$

9. The transpose of a matrix $A = (a_{ij})$ denoted by A^T is defined by $A^T = (a_{ji})$. It is the matrix obtained from A by interchanging the rows and columns of A. If

$$A = \begin{bmatrix} 1 & 2 \\ 3 & 4 \end{bmatrix}, \quad B = \begin{bmatrix} 5 & 6 \\ 7 & 8 \end{bmatrix},$$

find

a) A^T b) B^T c) $(A + B)^T$

d) $A^T + B^T$ e) $(AB)^T$ f) $B^T A^T$

10. Verify that three columns of the identity matrix

$$I = \begin{bmatrix} 1 & 0 & 0 \\ 0 & 1 & 0 \\ 0 & 0 & 1 \end{bmatrix}$$

form a linearly independent set. Verify that

$IA = AI = A$ if

$$A = \begin{bmatrix} 1 & 2 & 3 \\ 4 & 5 & 6 \\ 7 & 8 & 9 \end{bmatrix}.$$

11. Find the determinant of each of the following matrices:

a) $\begin{bmatrix} 1 & 2 \\ 3 & 4 \end{bmatrix}$ b) $\begin{bmatrix} 2 & 1 \\ 4 & 2 \end{bmatrix}$

c) $\begin{bmatrix} 1 & 2 & 3 \\ 2 & 4 & 5 \\ 3 & 5 & 6 \end{bmatrix}$ d) $\begin{bmatrix} 1 & 2 & 1 \\ -2 & 1 & 8 \\ 1 & -2 & -7 \end{bmatrix}$

e) $\begin{bmatrix} 1 & 0 & 0 \\ 0 & 1 & 0 \\ 0 & 0 & 1 \end{bmatrix}$ f) $\begin{bmatrix} 1 & 2 & 1 \\ 3 & 1 & 4 \\ 1 & 0 & 1 \end{bmatrix}$

In which of the above matrices do the columns form a linearly independent set of vectors?

12. Complete the details of Example 8 by finding C via Cramer's rule.

13. Solve each of the following systems of algebraic equations $A\mathbf{x} = \mathbf{b}$ and find the matrix C so that the solution $\mathbf{x}$ can be written in the form $\mathbf{x} = C\mathbf{b}$.

a) $\begin{bmatrix} 1 & 1 \\ 2 & 1 \end{bmatrix} \begin{bmatrix} x_1 \\ x_2 \end{bmatrix} = \begin{bmatrix} 1 \\ 0 \end{bmatrix}$

b) $\begin{bmatrix} 1 & -3 \\ 2 & 5 \end{bmatrix} \begin{bmatrix} x_1 \\ x_2 \end{bmatrix} = \begin{bmatrix} -5 \\ 12 \end{bmatrix}$

c) $\begin{bmatrix} 7 & 4 \\ 5 & -3 \end{bmatrix} \begin{bmatrix} x_1 \\ x_2 \end{bmatrix} = \begin{bmatrix} 26 \\ 1 \end{bmatrix}$

d) $\begin{bmatrix} 2 & 1 & 3 \\ 3 & 4 & 1 \\ 1 & 0 & -2 \end{bmatrix} \begin{bmatrix} x_1 \\ x_2 \\ x_3 \end{bmatrix} = \begin{bmatrix} 6 \\ 8 \\ -1 \end{bmatrix}$

e) $\begin{bmatrix} 1 & 1 & -1 \\ 3 & -1 & 0 \\ 9 & 1 & 1 \end{bmatrix} \begin{bmatrix} x_1 \\ x_2 \\ x_3 \end{bmatrix} = \begin{bmatrix} 0 \\ 3 \\ 10 \end{bmatrix}$

14. Verify that $\begin{bmatrix} a & b \\ c & d \end{bmatrix}^{-1} = \dfrac{1}{ad - bc} \begin{bmatrix} d & -b \\ -c & a \end{bmatrix}$.

15. Find the inverse of each of the following 2×2 matrices. Verify that $AA^{-1} = A^{-1}A = I$.

a) $\begin{bmatrix} 3 & 5 \\ 1 & 2 \end{bmatrix}$ b) $\begin{bmatrix} 2 & 5 \\ 1 & 3 \end{bmatrix}$

c) $\begin{bmatrix} 3 & 2 \\ 1 & 2 \end{bmatrix}$
d) $\begin{bmatrix} 25 & 36 \\ 9 & 13 \end{bmatrix}$

e) $\begin{bmatrix} 1 & 2 \\ 3 & 4 \end{bmatrix}$
f) $\begin{bmatrix} 3 & -5 \\ -1 & 2 \end{bmatrix}$

16. Verify that the inverse of

$$\begin{bmatrix} 1 & 2 & 3 \\ 2 & 3 & 4 \\ 3 & 4 & 6 \end{bmatrix} \text{ is } \begin{bmatrix} -2 & 0 & 1 \\ 0 & 3 & -2 \\ 1 & -2 & 1 \end{bmatrix}.$$

17. Find the characteristic polynomial, the eigenvalues, and eigenvectors of the following matrices:

a) $\begin{bmatrix} 2 & -1 \\ -1 & 2 \end{bmatrix}$
b) $\begin{bmatrix} 1 & -2 \\ -3 & 2 \end{bmatrix}$

c) $\begin{bmatrix} 1 & 1 \\ 2 & 2 \end{bmatrix}$
d) $\begin{bmatrix} 3 & -1 & -1 \\ -12 & 0 & 5 \\ 4 & -2 & -1 \end{bmatrix}$

e) $\begin{bmatrix} 2 & 2 & 1 \\ 1 & 3 & 1 \\ 1 & 2 & 2 \end{bmatrix}$
f) $\begin{bmatrix} 1 & 1 & -2 \\ -1 & 2 & 1 \\ 0 & 1 & -1 \end{bmatrix}$

18. For each of the following matrix valued functions $A(t)$ find $dA(t)/dt$.

a) $A(t) = \begin{bmatrix} e^t & e^{-t} \\ 2e^t & -e^t \end{bmatrix}$

b) $A(t) = \begin{bmatrix} 1 & e^{2t} \\ 2 & -2e^{2t} \end{bmatrix}$

c) $A(t) = \begin{bmatrix} e^{2t}\cos t & 2e^{2t}\sin t \\ 3e^{2t}\cos t & e^{2t}\sin t \end{bmatrix}$

d) $A(t) = \begin{bmatrix} e^t & e^{2t} & e^{5t} \\ e^t & -2e^{2t} & e^{5t} \\ -e^t & -3e^{2t} & 3e^{5t} \end{bmatrix}$

e) $A(t) = \begin{bmatrix} 0 & 2\cos t & 5\sin t \\ 2e^t & 2\cos t & 5\sin t \\ e^t & 3\cos t & -5\sin t \end{bmatrix}$

f) $A(t) = \begin{bmatrix} 0 & e^{-t} & 3te^{-t} \\ 2e^t & -4e^{-t} & (3+6t)e^{-t} \\ e^t & -2e^{-t} & (3+3t)e^{-t} \end{bmatrix}$

19. In each of the following problems verify that the given matrix satisfies the given matrix differential equation:

a) $\dfrac{d\Theta}{dt} = \begin{bmatrix} 2 & 1 \\ 3 & 4 \end{bmatrix}\Theta, \quad \Theta(t) = \begin{bmatrix} e^t & e^{5t} \\ -e^t & 3e^{5t} \end{bmatrix}$

b) $\dfrac{d\Theta}{dt} = \begin{bmatrix} 1 & 1 \\ -2 & 3 \end{bmatrix}\Theta,$

$\Theta(t) = e^{2t} \begin{bmatrix} \cos t & \sin t \\ \cos t - \sin t & \cos t + \sin t \end{bmatrix}$

c) $\dfrac{d\Theta}{dt} = \begin{bmatrix} 2 & -1 & 2 \\ 1 & 0 & 2 \\ -2 & 1 & -1 \end{bmatrix}\Theta,$

$\Theta(t) = \begin{bmatrix} 0 & \cos t + \sin t & 2\sin t \\ 2e^t & \cos t + \sin t & 2\sin t \\ e^t & -\sin t & \cos t - \sin t \end{bmatrix}$

20. Not every square matrix has an inverse. Show that this is the case for

$$A = \begin{bmatrix} 1 & 2 \\ 1 & 2 \end{bmatrix}.$$

21. Find the Laplace transformation of the matrices in Exercise 18.

5.3 HOMOGENEOUS LINEAR SYSTEMS WITH CONSTANT COEFFICIENTS

We begin with two-dimensional constant coefficient homogeneous linear systems, so that the ideas and techniques will not be obscured by algebraic complexities. However, our choice of topics and notation allows their extension to higher dimensional systems. The differential equations

$$\frac{dy_1}{dt} = a_{11}y_1 + a_{12}y_2, \qquad \frac{dy_2}{dt} = a_{21}y_1 + a_{22}y_2 \tag{5.3.1}$$

in vector–matrix notation are

$$\frac{d\mathbf{y}}{dt} = A\mathbf{y},$$

where

$$A = \begin{bmatrix} a_{11} & a_{12} \\ a_{21} & a_{22} \end{bmatrix}, \qquad \mathbf{y} = \begin{bmatrix} y_1 \\ y_2 \end{bmatrix}.$$

In finding a solution to Eqs. (5.3.1) we are guided by our experience with both first and second order linear differential equations; we set

$$\mathbf{y} = \begin{bmatrix} y_1 \\ y_2 \end{bmatrix} = \begin{bmatrix} \alpha e^{\lambda t} \\ \beta e^{\lambda t} \end{bmatrix} = \begin{bmatrix} \alpha \\ \beta \end{bmatrix} e^{\lambda t} \tag{5.3.2}$$

and try to determine α, β, and λ, so that we have a solution. Before studying the general problem, we shall discuss a specific example in some detail.

Consider the system

$$\frac{dy_1}{dt} = y_1 + y_2, \qquad \frac{dy_2}{dt} = 4y_1 - 2y_2 \tag{5.3.3}$$

or

$$\frac{d}{dt}\begin{bmatrix} y_1 \\ y_2 \end{bmatrix} = \begin{bmatrix} 1 & 1 \\ 4 & -2 \end{bmatrix}\begin{bmatrix} y_1 \\ y_2 \end{bmatrix}.$$

Substituting (5.3.2) into (5.3.3) gives

$$\alpha\lambda e^{\lambda t} = \alpha e^{\lambda t} + \beta e^{\lambda t}, \qquad \beta\lambda e^{\lambda t} = 4\alpha e^{\lambda t} - 2\beta e^{\lambda t},$$

which, upon dividing by $e^{\lambda t}$, simplifies to

$$(1 - \lambda)\alpha + \beta = 0, \qquad 4\alpha - (2 + \lambda)\beta = 0. \tag{5.3.4}$$

The Eqs. (5.3.4) can be thought of as the equations of two straight lines in the $\alpha\beta$ plane intersecting at the origin (see Fig. 5.3). This intersection corresponds to the trivial solution $y_1 = 0$, $y_2 = 0$. In order to construct a nontrivial solution, λ is chosen so that the lines coincide. Since both lines contain the origin, they will be coincident if and only if their slopes are equal. This is equivalent to the algebraic condition

$$\begin{vmatrix} 1 - \lambda & 1 \\ 4 & -2 - \lambda \end{vmatrix} = 0, \tag{5.3.5}$$

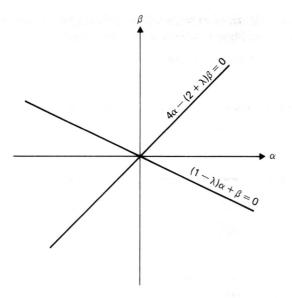

Figure 5.3   Graphs of equations (5.3.4).

or

$$\lambda^2 + \lambda - 6 = (\lambda + 3)(\lambda - 2) = 0.$$

An important observation is that the last equation is just the characteristic equation of the coefficient matrix

$$A = \begin{bmatrix} 1 & 1 \\ 4 & -2 \end{bmatrix}.$$

Hence we can take $\lambda = -3$ or $\lambda = 2$. If $\lambda = -3$, α and β must satisfy $4\alpha + \beta = 0$; if $\lambda = 2$, α and β must satisfy $-\alpha + \beta = 0$ (see Fig. 5.4). Any point on either line (other than the origin) will determine a nontrivial solution to the system of differential equations (5.3.3).

First consider the case where $\lambda = -3$. Geometrically, it is clear that all points on the line $4\alpha + \beta = 0$ are scalar multiples of any nontrivial point, for example, $\alpha = 1$, $\beta = -4$. Therefore, we obtain a one-parameter family of solutions

$$\mathbf{y}^1(t) = c_1 \begin{bmatrix} 1 \\ -4 \end{bmatrix} e^{-3t}, \tag{5.3.6}$$

where c_1 is an arbitrary constant. In contrast to the scalar equation $dy/dt = ry$ with solution $y = ce^{rt}$, the coefficient of the exponential function is a vector whose direction is determined by the coefficient matrix of the system of

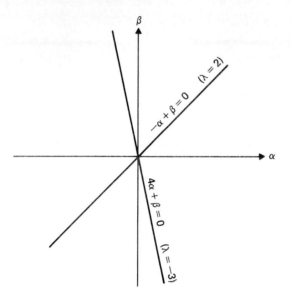

Figure 5.4

differential equations. Repeating the argument for $\lambda = 2$ gives a second one-parameter family of solutions

$$\mathbf{y}^2(t) = c_2 \begin{bmatrix} 1 \\ 1 \end{bmatrix} e^{2t}. \tag{5.3.7}$$

The two values of λ, -3 and 2, are eigenvalues of the coefficient matrix

$$A = \begin{bmatrix} 1 & 1 \\ 4 & -2 \end{bmatrix},$$

and the corresponding vectors

$$\begin{bmatrix} 1 \\ -4 \end{bmatrix} \quad \text{and} \quad \begin{bmatrix} 1 \\ 1 \end{bmatrix}$$

are eigenvectors of A.

The solution vectors $\mathbf{y}^1(t)$ and $\mathbf{y}^2(t)$ are linearly independent, as can be shown by the determinant test (see p. 237):

$$\begin{vmatrix} y_1^1(t) & y_1^2(t) \\ y_2^1(t) & y_2^2(t) \end{vmatrix} = \begin{vmatrix} e^{-3t} & e^{2t} \\ -4e^{-3t} & e^{2t} \end{vmatrix} = 5e^{-t} \neq 0. \tag{5.3.8}$$

A linear combination of these solutions,

$$\mathbf{y}(t) = c_1\mathbf{y}^1(t) + c_2\mathbf{y}^2(t) = c_1 \begin{bmatrix} 1 \\ -4 \end{bmatrix} e^{-3t} + c_2 \begin{bmatrix} 1 \\ 1 \end{bmatrix} e^{2t}, \tag{5.3.9}$$

is easily shown to be a solution to the differential equation (5.3.3). Furthermore, it is a *general solution,* since, as we shall show, the constants c_1 and c_2 can be chosen to match arbitrary initial data.

For example, suppose the solution is required to satisfy

$$\mathbf{y}(0) = \begin{bmatrix} 6 \\ -4 \end{bmatrix}.$$

Substituting $t = 0$ into (5.3.9) implies that c_1 and c_2 must satisfy the vector equation

$$c_1 \begin{bmatrix} 1 \\ -4 \end{bmatrix} + c_2 \begin{bmatrix} 1 \\ 1 \end{bmatrix} = \begin{bmatrix} 6 \\ -4 \end{bmatrix}$$

or

$$c_1 + c_2 = 6,$$
$$-4c_1 + c_2 = -4.$$

Solving this system gives $c_1 = 2$, $c_2 = 4$, and the required solution is

$$\mathbf{y}(t) = 2 \begin{bmatrix} 1 \\ -4 \end{bmatrix} e^{-3t} + 4 \begin{bmatrix} 1 \\ 1 \end{bmatrix} e^{2t}.$$

If $y(t)$ is required to satisfy the general initial conditions

$$\mathbf{y}(t_0) = \mathbf{y}_0 = \begin{bmatrix} y_1^0 \\ y_2^0 \end{bmatrix},$$

then, by using (5.3.8), we obtain the equations

$$\mathbf{y}(t_0) = \begin{bmatrix} y_1^0 \\ y_2^0 \end{bmatrix} = c_1 \begin{bmatrix} 1 \\ -4 \end{bmatrix} e^{-3t_0} + c_2 \begin{bmatrix} 1 \\ 1 \end{bmatrix} e^{2t_0}$$

or

$$y_1^0 = c_1 e^{-3t_0} + c_2 e^{2t_0},$$
$$y_2^0 = -4c_1 e^{-3t_0} + c_2 e^{2t_0}.$$

Using Cramer's rule, we write

$$c_1 = \frac{\begin{vmatrix} y_1^0 & e^{2t_0} \\ y_2^0 & e^{2t_0} \end{vmatrix}}{\begin{vmatrix} e^{-3t_0} & e^{2t_0} \\ -4e^{-3t_0} & e^{2t_0} \end{vmatrix}} = \frac{1}{5}(y_1^0 - y_2^0)e^{3t_0},$$

$$c_2 = \frac{\begin{vmatrix} e^{-3t_0} & y_1^0 \\ -4e^{-3t_0} & y_2^0 \end{vmatrix}}{\begin{vmatrix} e^{-3t_0} & e^{2t_0} \\ -4e^{-3t_0} & e^{2t_0} \end{vmatrix}} = \frac{1}{5}(y_2^0 + 4y_1^0)e^{-2t_0};$$

thus the constants c_1 and c_2 are uniquely determined.

Let us now return to the general system

$$\frac{d}{dt}\begin{bmatrix} y_1 \\ y_2 \end{bmatrix} = \begin{bmatrix} a_{11} & a_{12} \\ a_{21} & a_{22} \end{bmatrix}\begin{bmatrix} y_1 \\ y_2 \end{bmatrix} \quad \text{or} \quad \frac{d\mathbf{y}}{dt} = A\mathbf{y}. \quad (5.3.10)$$

Set

$$\mathbf{y} = \begin{bmatrix} y_1 \\ y_2 \end{bmatrix} = \begin{bmatrix} b_1 \\ b_2 \end{bmatrix} e^{\lambda t} = \mathbf{b}e^{\lambda t} \quad (5.3.11)$$

and substitute into the differential equation to obtain

$$\lambda \mathbf{b}e^{\lambda t} = A\mathbf{b}e^{\lambda t}.$$

Cancelling the scalar quantity $e^{\lambda t}$ and writing $\mathbf{b} = I\mathbf{b}$ gives $\lambda I\mathbf{b} = A\mathbf{b}$ or, equivalently,

$$\begin{aligned} (a_{11} - \lambda)b_1 + a_{12}b_2 &= 0, \\ a_{21}b_1 + (a_{22} - \lambda)b_2 &= 0, \end{aligned} \quad (5.3.12)$$

or $(A - \lambda I)\mathbf{b} = \mathbf{0}$. Hence, λ is an eigenvalue and $\mathbf{b}$ is an eigenvector of the matrix A. Equations (5.3.12) have solutions b_1 and b_2 (not both zero) if and only if λ satisfies the *characteristic equation*

$$\det(A - \lambda I) = \begin{vmatrix} a_{11} - \lambda & a_{12} \\ a_{21} & a_{22} - \lambda \end{vmatrix} = 0. \quad (5.3.13)$$

This quadratic equation can have distinct real, distinct complex, or equal real roots. Just as in the study of second order linear differential equations with constant coefficients, we must examine each of the three types of eigenvalues. Each will provide a different type of solutions to the system of differential equations (5.3.10).

Type 1—Distinct Real Eigenvalues

To construct the solutions (5.3.11), we must find values of b_1 and b_2 for each of λ_1 and λ_2. From (5.3.12) it is clear that the following algebraic systems must be solved:

$$(A - \lambda_1 I)\mathbf{b} = 0 \quad \text{and} \quad (A - \lambda_2 I)\mathbf{b} = 0.$$

Denoting these solutions by $\mathbf{b}^1$ and $\mathbf{b}^2$, respectively, we find the two solutions of (5.3.10):

$$\mathbf{y}^1(t) = \mathbf{b}^1 e^{\lambda_1 t} = \begin{bmatrix} b_1^1 \\ b_2^1 \end{bmatrix} e^{\lambda_1 t}$$

and

$$\mathbf{y}^2(t) = \mathbf{b}^2 e^{\lambda_2 t} = \begin{bmatrix} b_1^2 \\ b_2^2 \end{bmatrix} e^{\lambda_2 t}.$$

As mentioned previously, the corresponding eigenvectors are linearly independent since the eigenvalues are distinct. By the determinant test,

$$\begin{vmatrix} y_1^1(t) & y_1^2(t) \\ y_2^1(t) & y_2^2(t) \end{vmatrix} = \begin{vmatrix} b_1^1 e^{\lambda_1 t} & b_1^2 e^{\lambda_2 t} \\ b_2^1 e^{\lambda_1 t} & b_2^2 e^{\lambda_2 t} \end{vmatrix} = e^{\lambda_1 t} e^{\lambda_2 t} \begin{vmatrix} b_1^1 & b_1^2 \\ b_2^1 & b_2^2 \end{vmatrix} \neq 0$$

(the last determinant is nonzero since $\mathbf{b}^1$ and $\mathbf{b}^2$ are linearly independent), and the solutions $\mathbf{y}^1(t)$, $\mathbf{y}^2(t)$ are linearly independent. Hence, the coefficients c_1 and c_2 in the linear combination

$$\mathbf{y}(t) = c_1 \mathbf{y}^1(t) + c_2 \mathbf{y}^2(t) \tag{5.3.14}$$

can be chosen to satisfy arbitrary initial data. We conclude that (5.3.14) is a *general solution* to the homogeneous differential equation (5.3.10).

All of this can be summarized in the following algorithm.

Algorithm for solving $d\mathbf{y}/dt = A\mathbf{y}$ (distinct real eigenvalue case).

1. Determine the eigenvalues λ_1 and λ_2 which are roots of the characteristic equation

$$\det(A - \lambda I) = 0.$$

2. Find the corresponding eigenvectors $\mathbf{b}^1$ and $\mathbf{b}^2$ which are solutions of

$$(A - \lambda_j I)\mathbf{b} = 0, \quad j = 1, 2.$$

3. Two linearly independent solutions are given by

$$\mathbf{y}^1(t) = \mathbf{b}^1 e^{\lambda_1 t}, \qquad \mathbf{y}^2(t) = \mathbf{b}^2 e^{\lambda_2 t}$$

and the general solution is

$$\mathbf{y}(t) = c_1 \mathbf{y}^1(t) + c_2 \mathbf{y}^2(t).$$

4. If initial data $\mathbf{y}(t_0) = \mathbf{y}^0$ are given, find c_1, c_2 by solving

$$\mathbf{y}^0 = c_1 \mathbf{y}^1(t_0) + c_2 \mathbf{y}^2(t_0).$$

Example 1 Find the general solution of $\dfrac{d\mathbf{y}}{dt} = \begin{bmatrix} 2 & 6 \\ -2 & -5 \end{bmatrix} \mathbf{y}$.

The eigenvalues are roots of the quadratic equation

$$\begin{vmatrix} 2 - \lambda & 6 \\ -2 & -5 - \lambda \end{vmatrix} = (2 - \lambda)(-5 - \lambda) + 12 = \lambda^2 + 3\lambda + 2 = 0,$$

hence $\lambda_1 = -1$, $\lambda_2 = -2$. To obtain the solution corresponding to $\lambda_1 = -1$, solve

$$[2 - (-1)]b_1 + 6b_2 = 0, \qquad -2b_1 + [-5 - (-1)]b_2 = 0$$

to get $b_1 = -2$, $b_2 = 1$, and therefore

$$\mathbf{b}^1 = \begin{bmatrix} -2 \\ 1 \end{bmatrix}.$$

The corresponding solution to the differential equation is

$$\mathbf{y}^1(t) = \begin{bmatrix} -2 \\ 1 \end{bmatrix} e^{-t}.$$

To get the solution corresponding to $\lambda_2 = -2$, solve

$$[2 - (-2)]b_1 + 6b_2 = 0, \qquad -2b_1 + [-5 - (-2)]b_2 = 0$$

to obtain

$$\mathbf{b}^2 = \begin{bmatrix} -3 \\ 2 \end{bmatrix},$$

which gives

$$\mathbf{y}^2(t) = \begin{bmatrix} -3 \\ 2 \end{bmatrix} e^{-2t}.$$

The general solution is

$$\mathbf{y}(t) = c_1\mathbf{y}^1(t) + c_2\mathbf{y}^2(t)$$

$$= c_1 \begin{bmatrix} -2 \\ 1 \end{bmatrix} e^{-t} + c_2 \begin{bmatrix} -3 \\ 2 \end{bmatrix} e^{-2t}. \quad \square$$

Example 2 Find the solution of the initial value problem

$$\frac{d\mathbf{y}}{dt} = \begin{bmatrix} -5 & 4 \\ 1 & -2 \end{bmatrix} \mathbf{y}, \qquad \mathbf{y}(0) = \begin{bmatrix} 3 \\ -2 \end{bmatrix}.$$

The eigenvalues are the roots of

$$\begin{vmatrix} -5 - \lambda & 4 \\ 1 & -2 - \lambda \end{vmatrix} = \lambda^2 + 7\lambda + 6 = 0,$$

so $\lambda_1 = -1$, $\lambda_2 = -6$. A vector $\mathbf{b}^1$, found by solving the system of equations

$$[-5 - (-1)]b_1 + 4b_2 = 0, \qquad b_1 + [-2 - (-6)]b_2 = 0,$$

is

$$\begin{bmatrix} 1 \\ 1 \end{bmatrix}.$$

Similarly,

$$\mathbf{b}^2 = \begin{bmatrix} 4 \\ -1 \end{bmatrix}$$

satisfies

$$[-5 - (-6)]b_1 + 4b_2 = 0, \qquad b_1 + [-2 - (-6)]b_2 = 0.$$

The corresponding solutions are

$$\mathbf{y}^1(t) = \begin{bmatrix} 1 \\ 1 \end{bmatrix} e^{-t}, \qquad \mathbf{y}^2(t) = \begin{bmatrix} 4 \\ -1 \end{bmatrix} e^{-6t},$$

and the general solution is

$$\mathbf{y}(t) = c_1 \mathbf{y}^1(t) + c_2 \mathbf{y}^2(t) = c_1 \begin{bmatrix} 1 \\ 1 \end{bmatrix} e^{-t} + c_2 \begin{bmatrix} 4 \\ -1 \end{bmatrix} e^{-6t}.$$

To obtain the solution satisfying the initial conditions, c_1 and c_2 must satisfy

$$\mathbf{y}(0) = \begin{bmatrix} 3 \\ -2 \end{bmatrix} = c_1 \begin{bmatrix} 1 \\ 1 \end{bmatrix} + c_2 \begin{bmatrix} 4 \\ -1 \end{bmatrix}$$

or

$$3 = c_1 + 4c_2, \qquad -2 = c_1 - c_2.$$

Solving these gives $c_1 = -1$, $c_2 = 1$, and the desired solution is

$$\mathbf{y}(t) = (-1) \begin{bmatrix} 1 \\ 1 \end{bmatrix} e^{-t} + (1) \begin{bmatrix} 4 \\ -1 \end{bmatrix} e^{-6t}$$

$$= \begin{bmatrix} -1 \\ -1 \end{bmatrix} e^{-t} + \begin{bmatrix} 4 \\ -1 \end{bmatrix} e^{-6t}. \quad \square$$

Type 2—Distinct Complex Conjugate Eigenvalues

In this case both λ and its complex conjugate λ^* are eigenvalues of the coefficient matrix A. If b_1 and b_2 are solutions of (5.3.12) corresponding to λ, then b_1^* and b_2^* are also solutions of (5.3.12) corresponding to λ^*. To see this, write

$$(a_{11} - \lambda)b_1 + a_{12}b_2 = 0, \qquad a_{21}b_1 + (a_{22} - \lambda)b_2 = 0,$$

and take complex conjugates of both equations to get

$$(a_{11} - \lambda)^* b_1^* + a_{12}^* b_2^* = 0^*, \qquad a_{21}^* b_1^* + (a_{22} - \lambda)^* b_2^* = 0^*.$$

Since a_{11}, a_{12}, a_{21}, and a_{22} are real, this becomes

$$(a_{11} - \lambda^*)b_1^* + a_{12}b_2^* = 0, \qquad a_{21}b_1^* + (a_{22} - \lambda^*)b_2^* = 0,$$

thus b_1^* and b_2^* are solutions corresponding to λ^*. The general solution in this case can now be written in its complex form as

$$y_1(t) = c_1 b_1 e^{\lambda t} + c_2 b_1^* e^{\lambda^* t},$$

$$y_2(t) = c_1 b_2 e^{\lambda t} + c_2 b_2^* e^{\lambda^* t},$$

or

$$\mathbf{y}(t) = c_1 \begin{bmatrix} b_1 \\ b_2 \end{bmatrix} e^{\lambda t} + c_2 \begin{bmatrix} b_1^* \\ b_2^* \end{bmatrix} e^{\lambda^* t}. \tag{5.3.15}$$

Since the complex eigenvalues are distinct, the eigenvectors are linearly independent.

The real form of the general solution can be obtained from the complex form by some clever choices of the constants c_1 and c_2 in (5.3.15). The procedure will be illustrated by the following example.

Example 3   Find the general real solution of

$$\frac{d\mathbf{y}}{dt} = \begin{bmatrix} 1 & 5 \\ -1 & -3 \end{bmatrix} \mathbf{y}. \tag{5.3.16}$$

The characteristic equation

$$\begin{vmatrix} 1 - \lambda & 5 \\ -1 & -3 - \lambda \end{vmatrix} = \lambda^2 + 2\lambda + 2 = 0$$

has the roots $\lambda = -1 + i$ and $\lambda^* = -1 - i$. Equations (5.3.12) become

$$(2 - i)b_1 + 5b_2 = 0, \qquad -b_1 - (2 + i)b_2 = 0,$$

and the solution $b_1 = 5$, $b_2 = -2 + i$ is chosen. Therefore, $b_1^* = 5$, $b_2^* = -2 - i$, and the general solution has the form

$$y_1(t) = 5c_1 e^{(-1+i)t} + 5c_2 e^{(-1-i)t},$$

$$y_2(t) = (-2 + i)c_1 e^{(-1+i)t} + (-2 - i)c_2 e^{(-1-i)t}.$$

To find the real form of the general solution, a procedure is followed similar to that used in Chapter 2 to find the general real solution from the general complex solution for second order linear equations. Consider the two cases: (1) $c_1 = c_2 = 1/2$ and (2) $c_1 = -c_2 = 1/2i$ and make use of the Euler formula

$$e^{i\theta} = \cos \theta + i \sin \theta, \tag{5.3.17}$$

which implies that

$$\cos \theta = \frac{e^{i\theta} + e^{-i\theta}}{2}, \qquad \sin \theta = \frac{e^{i\theta} - e^{-i\theta}}{2i}.$$

Case 1: $c_1 = c_2 = 1/2.$

$$y_1(t) = \frac{5}{2} e^{-t} e^{it} + \frac{5}{2} e^{-t} e^{-it} = 5e^{-t} \left(\frac{e^{it} + e^{-it}}{2} \right) = 5 \, e^{-t} \cos t = \mathrm{Re}[5e^{(-1+i)t}],$$

$$y_2(t) = \frac{1}{2} (-2 + i) e^{-t} e^{it} + \frac{1}{2} (-2 - i) e^{-t} e^{-it}$$

$$= e^{-t} \left[(-2) \left(\frac{e^{it} + e^{-it}}{2} \right) + i \left(\frac{e^{it} - e^{-it}}{2} \right) \right]$$

$$= e^{-t} \left[(-2) \left(\frac{e^{it} + e^{-it}}{2} \right) - \left(\frac{e^{it} - e^{-it}}{2i} \right) \right]$$

$$= e^{-t}(-2 \cos t - \sin t) = \mathrm{Re}[(-2 + i)e^{(-1+i)t}].$$

The choice for c_1 and c_2 in Case 1 is motivated by the fact that if a is a complex number $a = \alpha + i\beta$, then $a^* = \alpha - i\beta$ and

$$\frac{1}{2} a + \frac{1}{2} a^* = \alpha = \mathrm{Re}(a).$$

Case 2: $c_1 = -c_2 = 1/2i.$

$$y_1(t) = \frac{5}{2i} e^{-t} e^{-it} - \frac{5}{2i} e^{-t} e^{-it} = 5e^{-t} \left(\frac{e^{it} - e^{-it}}{2i} \right)$$

$$= 5e^{-t} \sin t = \mathrm{Im}[5e^{(-1+i)t}],$$

$$y_2(t) = \frac{1}{2i} (-2 + i) e^{-t} e^{it} - \frac{1}{2i} (-2 - i) e^{-t} e^{-it}$$

$$= e^{-t} \left[(-2) \left(\frac{e^{it} - e^{-it}}{2i} \right) + \left(\frac{e^{it} + e^{-it}}{2} \right) \right].$$

$$= e^{-t}(-2 \sin t + \cos t) = \mathrm{Im}[(-2 + i)e^{(-1+i)t}].$$

As in Case 1, the choice of c_1 and c_2 is based on the fact that for $a = \alpha + i\beta$,

$$\frac{1}{2i} a - \frac{1}{2i} a^* = \beta = \mathrm{Im}(a).$$

Clearly, the vector functions

$$\mathbf{y}^1(t) = \begin{bmatrix} 5 \cos t \\ -2 \cos t - \sin t \end{bmatrix} e^{-t}, \qquad \mathbf{y}^2(t) = \begin{bmatrix} 5 \sin t \\ \cos t - 2 \sin t \end{bmatrix} e^{-t}$$

are solutions to (5.3.16) and they are real. The general real solution is then given by

$$\mathbf{y}(t) = c_1 \begin{bmatrix} 5 \cos t \\ -2 \cos t - \sin t \end{bmatrix} e^{-t} + c_2 \begin{bmatrix} 5 \sin t \\ \cos t - 2 \sin t \end{bmatrix} e^{-t},$$

and the ability to solve an arbitrary initial value problem depends on the fact that for any value of t_0

$$\begin{vmatrix} 5e^{-t_0}\cos t_0 & 5e^{-t_0}\sin t_0 \\ -e^{-t_0}(2\cos t_0 + \sin t_0) & e^{-t_0}(\cos t_0 - 2\sin t_0) \end{vmatrix}$$

$$= 5e^{-2t_0}[\cos t_0(\cos t_0 - 2\sin t_0) + (2\cos t_0 + \sin t_0)\sin t_0]$$

$$= 5e^{-2t_0} \neq 0. \quad \square$$

The technique outlined in Example 3 can be applied in general to (5.3.15). Let $\lambda = p + iq$, $b_1 = u_1 + iv_1$, $b_2 = u_2 + iv_2$. Then we can consider two cases.

Case 1: $c_1 = c = 1/2$.

$$y_1^1(t) = \text{Re}(b_1 e^{\lambda t}) = e^{pt}(u_1 \cos qt - v_1 \sin qt),$$

$$y_2^1(t) = \text{Re}(b_2 e^{\lambda t}) = e^{pt}(u_2 \cos qt - v_2 \sin qt).$$

Case 2: $c_2 = -c_2 = 1/2i$.

$$y_1^2(t) = \text{Im}(b_1 e^{\lambda t}) = e^{pt}(u_1 \sin qt + v_1 \cos qt),$$

$$y_2^2(t) = \text{Im}(b_2 e^{\lambda t}) = e^{pt}(u_2 \sin qt + v_2 \cos qt).$$

The general solution can then be written in the form

$$\mathbf{y}(t) = e^{pt}\left[c_1 \begin{bmatrix} u_1 \cos qt - v_1 \sin qt \\ u_2 \cos qt - v_2 \sin qt \end{bmatrix} + c_2 \begin{bmatrix} u_1 \sin qt + v_1 \sin qt \\ u_2 \sin qt + v_2 \cos qt \end{bmatrix} \right],$$

where c_1 and c_2 are arbitrary constants. That this is the general solution follows again from the fact that the determinant

$$e^{2pt}\begin{vmatrix} u_1 \cos qt - v_1 \sin qt & u_1 \sin qt + v_1 \cos qt \\ u_2 \cos qt - v_2 \sin qt & u_2 \sin qt + v_2 \cos qt \end{vmatrix} = e^{2pt}(u_1 v_2 - v_1 u_2) \neq 0,$$

since the vectors $\begin{bmatrix} b_1 \\ b_2 \end{bmatrix}$, $\begin{bmatrix} b_1^* \\ b_2^* \end{bmatrix}$ are linearly independent (see Exercises 24 and 25, p. 265).

All of this can be summarized in the following algorithm.

Algorithm for solving $dy/dt = Ay$ (complex eigenvalue case).

1. Determine the eigenvalues, λ and λ^*, which are roots of the equation

$$\det(A - \lambda I) = 0.$$

2. Find the corresponding eigenvectors

$$\mathbf{b} = \begin{bmatrix} b_1 \\ b_2 \end{bmatrix} \quad \text{and} \quad \mathbf{b}^* = \begin{bmatrix} b_1^* \\ b_2^* \end{bmatrix},$$

which are solutions of

$$(A - \lambda I)b = 0, \qquad (A - \lambda^* I)b^* = 0.$$

3. Two linearly independent real solutions are given by

$$\mathbf{y}^1(t) = \begin{bmatrix} u_1 \cos qt - v_1 \sin qt \\ u_2 \cos qt - v_2 \sin qt \end{bmatrix} e^{pt},$$

$$\mathbf{y}^2(t) = \begin{bmatrix} u_1 \sin qt + v_1 \cos qt \\ u_2 \sin qt + v_2 \cos qt \end{bmatrix} e^{pt},$$

where $\lambda = p + iq$, $b_1 = u_1 + iv_1$, $b_2 = u_2 + iv_2$, and the general solution is given by

$$\mathbf{y}(t) = c_1 \mathbf{y}^1(t) + c_2 \mathbf{y}^2(t).$$

4. If initial data $\mathbf{y}(t_0) = \mathbf{y}^0$ are given, find c_1, c_2 from

$$\begin{bmatrix} y_1^0 \\ y_2^0 \end{bmatrix} = c_1 \mathbf{y}^1(t_0) + c_2 \mathbf{y}^2(t_0).$$

Example 4   Solve the initial value problem $\dfrac{d\mathbf{y}}{dt} = \begin{bmatrix} 3 & -3 \\ 3 & -1 \end{bmatrix} \mathbf{y}, \ \mathbf{y}(0) = \begin{bmatrix} 3 \\ 0 \end{bmatrix}$.

The characteristic equation

$$\begin{vmatrix} 3 - \lambda & -3 \\ 3 & -1 - \lambda \end{vmatrix} = \lambda^2 - 2\lambda + 6 = 0$$

has roots $\lambda = 1 + \sqrt{5}i$ and $\lambda^* = 1 - \sqrt{5}i$, and therefore $p = 1$, $q = \sqrt{5}$. Equations (5.3.12) to determine the eigenvector **b** are

$$(2 - \sqrt{5}i)b_1 - 3b_2 = 0, \qquad 3b_1 + (-2 - \sqrt{5}i)b_2 = 0,$$

and the solutions $b_1 = 3$, $b_2 = 2 - \sqrt{5}i$ are chosen. Therefore, $u_1 = 3$, $v_1 = 0$, $u_2 = 2$, $v_2 = -\sqrt{5}$, and two linearly independent solutions are

$$\mathbf{y}^1(t) = \begin{bmatrix} 3 \cos \sqrt{5}t \\ 2 \cos \sqrt{5}t + \sqrt{5} \sin \sqrt{5}t \end{bmatrix} e^t,$$

$$\mathbf{y}^2(t) = \begin{bmatrix} 3 \sin \sqrt{5}t \\ 2 \sin \sqrt{5}t - \sqrt{5} \cos \sqrt{5}t \end{bmatrix} e^t.$$

The general solution is $\mathbf{y}(t) = c_1 \mathbf{y}^1(t) + c_2 \mathbf{y}^2(t)$. Using the initial data we obtain the equation

$$\mathbf{y}(0) = c_1 \mathbf{y}^1(0) + c_2 \mathbf{y}^2(0) = c_1 \begin{bmatrix} 3 \\ 2 \end{bmatrix} + c_2 \begin{bmatrix} 0 \\ -\sqrt{5} \end{bmatrix} = \begin{bmatrix} 3 \\ 0 \end{bmatrix}.$$

The solution is $c_1 = 1$, $c_2 = 2\sqrt{5}/5$, and

$$\mathbf{y}(t) = \begin{bmatrix} 3\cos\sqrt{5}t + \dfrac{6}{5}\sqrt{5}\sin\sqrt{5}t \\ \dfrac{9}{5}\sqrt{5}\sin\sqrt{5}t \end{bmatrix} e^t$$

is the solution of the initial value problem. □

Type 3: Equal Real Eigenvalues

Here we consider two situations. The first is with equations (5.3.10) uncoupled, i.e., when $a_{11} = a_{22} = p$ and $a_{12} = a_{21} = 0$. Then the equations become

$$\frac{d}{dt}\begin{bmatrix} y_1 \\ y_2 \end{bmatrix} = \begin{bmatrix} p & 0 \\ 0 & p \end{bmatrix}\begin{bmatrix} y_1 \\ y_2 \end{bmatrix}, \qquad (5.3.18)$$

and they can be solved separately to obtain

$$\mathbf{y}(t) = \begin{bmatrix} c_1 e^{pt} \\ c_2 e^{pt} \end{bmatrix} = c_1 \begin{bmatrix} 1 \\ 0 \end{bmatrix} e^{pt} + c_2 \begin{bmatrix} 0 \\ 1 \end{bmatrix} e^{pt}.$$

This is clearly the general solution.

If equations (5.3.10) are coupled, there is more work to do. Denoting the double root by λ, one can obtain an associated eigenvector $\mathbf{b}$ from (5.3.12) and have the one solution

$$\mathbf{y}^1(t) = \mathbf{b}e^{\lambda t}. \qquad (5.3.19)$$

To obtain a second solution, one might proceed as in the case of second order linear equations with constant coefficients and try

$$\mathbf{y}^2 = \alpha t e^{\lambda t} \qquad (5.3.20)$$

for some constant vector α. Substituting into (5.3.10) gives

$$\alpha e^{\lambda t} + \lambda \alpha t e^{\lambda t} = \mathbf{A}\alpha t e^{\lambda t}. \qquad (5.3.21)$$

In order for this equation to be satisfied for all t, it is necessary for the coefficients of $te^{\lambda t}$ and $e^{\lambda t}$ to be the same on both sides of the equation. So

$$\alpha = 0, \qquad \lambda\alpha = \mathbf{A}\alpha;$$

hence $\alpha = 0$, and there is no nonzero solution of the form (5.3.20) to (5.3.10). Since Eq. (5.3.21) contains terms both in $te^{\lambda t}$ and $e^{\lambda t}$, it might be more reasonable to try a solution of the form

$$\mathbf{y}^2(t) = \zeta e^{\lambda t} + \sigma t e^{\lambda t}, \qquad (5.3.22)$$

where ζ and σ are constant vectors. Substituting this expression for $\mathbf{y}$ into (5.3.10), one obtains

$$\lambda\zeta e^{\lambda t} + \sigma e^{\lambda t} + \lambda\sigma t e^{\lambda t} = A\zeta e^{\lambda t} + A\sigma t e^{\lambda t}.$$

Equating coefficients of $te^{\lambda t}$ and $e^{\lambda t}$ gives the equations

$$\lambda\zeta + \sigma = A\zeta \qquad \text{or} \qquad (A - \lambda I)\zeta = \sigma \qquad (5.3.23)$$

and

$$\lambda\sigma = A\sigma \qquad \text{or} \qquad (A - \lambda I)\sigma = \mathbf{0}. \qquad (5.3.24)$$

To solve Eqs. (5.3.23, 5.3.24), first note that the second equation says that σ is an eigenvector of A corresponding to the eigenvalue λ, so we might as well choose $\sigma = \mathbf{b}$, the eigenvector that we have already found. The vector ζ can now be found by solving Eq. (5.3.23) with $\sigma = \mathbf{b}$. However, if $\mathbf{b}$ has not already been found, it is easier to proceed as follows:

Find ζ, so that (5.3.23) determines σ with $\sigma_1^2 + \sigma_2^2 > 0$ to insure $\sigma \neq \mathbf{0}$. This σ will automatically satisfy (5.3.24) (see Exercise 38, p. 265).

Therefore a second solution is of the form

$$\mathbf{y}^2(t) = \zeta e^{\lambda t} + \sigma t e^{\lambda t},$$

and a general solution is (if we let $\mathbf{b} = \sigma$)

$$\mathbf{y}(t) = c_1 \sigma e^{\lambda t} + c_2(\zeta + \sigma t)e^{\lambda t}. \qquad (5.3.25)$$

To show that (5.3.25) is indeed the general solution to (5.3.10), consider the arbitrary initial data

$$\mathbf{y}(t_0) = \mathbf{y}^0 \qquad (5.3.26)$$

and set

$$y_1^0 = \sigma_1 e^{\lambda t_0} c_1 + (\zeta_1 + \sigma_1 t_0)e^{\lambda t_0} c_2,$$

$$y_2^0 = \sigma_2 e^{\lambda t_0} c_1 + (\zeta_2 + \sigma_2 t_0)e^{\lambda t_0} c_2.$$

The equations have a unique solution c_1, c_2 if and only if the determinant

$$\begin{vmatrix} \sigma_1 e^{\lambda t_0} & (\zeta_1 + \sigma_1 t_0)e^{\lambda t_0} \\ \sigma_2 e^{\lambda t_0} & (\zeta_2 + \sigma_2 t_0)e^{\lambda t_0} \end{vmatrix} = e^{2\lambda t_0}(\sigma_1\zeta_2 - \sigma_2\zeta_1) \neq 0.$$

If the determinant were to vanish, then $\sigma_1\zeta_2 - \sigma_2\zeta_1 = 0$ or, equivalently,

$$\begin{vmatrix} \sigma_1 & \zeta_1 \\ \sigma_2 & \zeta_2 \end{vmatrix} = 0,$$

which implies that vectors σ and ζ are linearly dependent. This means there exists a constant $c \neq 0$ such that $\zeta = c\sigma$. Substituting this into Eq. (5.3.23) gives

$$c(A - \lambda I)\sigma = \sigma,$$

which, when compared with Eq. (5.3.24), implies $\sigma = 0$. But this is impossible since σ is an eigenvector with $\sigma_1^2 + \sigma_2^2 > 0$. All of this can be summarized in the following algorithm.

Algorithm for solving $dy/dt = Ay$ (real equal eigenvalues case).

1. Determine the eigenvalue λ, which is a root of the equation

$$\det (A - \lambda I) = 0.$$

2. Determine vectors $\zeta \neq 0$ and $\sigma \neq 0$ such that

$$(A - \lambda I)\zeta = \sigma, \qquad (A - \lambda I)\sigma = 0.$$

3. Two linearly independent solutions are given by

$$\mathbf{y}^1(t) = \sigma e^{\lambda t}, \qquad \mathbf{y}^2(t) = (\zeta + \sigma t)e^{\lambda t},$$

and the general solution is

$$\mathbf{y}(t) = c_1\mathbf{y}^1(t) + c_2\mathbf{y}^2(t).$$

4. If initial data $\mathbf{y}(t_0) = \mathbf{y}^0$ are given, find c_1, c_2 from

$$\mathbf{y}^0 = c_1\mathbf{y}^1(t_0) + c_2\mathbf{y}^2(t_0).$$

Example 5 Obtain the general solution of

$$\mathbf{y}' = \begin{bmatrix} 3 & -4 \\ 1 & -1 \end{bmatrix} \mathbf{y}.$$

From (5.3.13), we get

$$\begin{vmatrix} 3 - \lambda & -4 \\ 1 & -1 - \lambda \end{vmatrix} = \lambda^2 - 2\lambda + 1 = 0,$$

ence $\lambda = 1$ is a double root. From (5.3.23) it follows that

$$\begin{bmatrix} 3 - 1 & -4 \\ 1 & -1 - 1 \end{bmatrix} \begin{bmatrix} \zeta_1 \\ \zeta_2 \end{bmatrix} = \begin{bmatrix} \sigma_1 \\ \sigma_2 \end{bmatrix}$$

or

$$\begin{bmatrix} 2 & -4 \\ 1 & -2 \end{bmatrix} \begin{bmatrix} \zeta_1 \\ \zeta_2 \end{bmatrix} = \begin{bmatrix} \sigma_1 \\ \sigma_2 \end{bmatrix}.$$

Now the task is to choose ζ_1 and ζ_2, not both zero, which will give values σ_1 and σ_2 which are not both zero. The choice $\zeta_1 = 1, \zeta_2 = 0$ gives $\sigma_1 = 2, \sigma_2 =$

1. As a check, note from (5.3.24) that

$$\begin{bmatrix} 2 & -4 \\ 1 & -2 \end{bmatrix} \begin{bmatrix} 2 \\ 1 \end{bmatrix} = \begin{bmatrix} 0 \\ 0 \end{bmatrix}.$$

Therefore $\sigma = \begin{bmatrix} 2 \\ 1 \end{bmatrix}$ is an eigenvector of A corresponding to $\lambda = 1$, and so

$$\mathbf{y}^1(t) = \begin{bmatrix} 2 \\ 1 \end{bmatrix} e^t.$$

A second solution is given by

$$\mathbf{y}^2(t) = \left[\begin{bmatrix} 1 \\ 0 \end{bmatrix} + \begin{bmatrix} 2 \\ 1 \end{bmatrix} t \right] e^t$$

and the general solution is

$$\mathbf{y}(t) = c_1 \begin{bmatrix} 2 \\ 1 \end{bmatrix} e^t + c_2 \left[\begin{bmatrix} 1 \\ 0 \end{bmatrix} + \begin{bmatrix} 2 \\ 1 \end{bmatrix} t \right] e^t. \quad \square \qquad (5.3.27)$$

Example 6   Find the solution in Example 5 that satisfies

$$\mathbf{y}(0) = \begin{bmatrix} 1 \\ 1 \end{bmatrix}.$$

From (5.3.27),

$$\begin{bmatrix} 1 \\ 1 \end{bmatrix} = c_1 \begin{bmatrix} 2 \\ 1 \end{bmatrix} + c_2 \begin{bmatrix} 1 \\ 0 \end{bmatrix}$$

or

$$1 = 2c_1 + c_2, \qquad 1 = c_1.$$

So $c_1 = 1$, $c_2 = -1$, and the solution is

$$\mathbf{y}(t) = \left[\begin{bmatrix} 2 \\ 1 \end{bmatrix} - \begin{bmatrix} 1 \\ 0 \end{bmatrix} - \begin{bmatrix} 2 \\ 1 \end{bmatrix} t \right] e^t = \left[\begin{bmatrix} 1 \\ 1 \end{bmatrix} - \begin{bmatrix} 2 \\ 1 \end{bmatrix} t \right] e^t. \quad \square$$

Although we will not do much with systems of linear differential equations of dimension greater than two, it is useful to work a few examples. In general, for systems of size $n \geq 3$ the algebraic difficulties become formidable, particularly for the repeated eigenvalue case, and quite often the best approach is the Laplace transform or a numerical approximation.

Example 7   Find the general solution of

$$\mathbf{y}' = \begin{bmatrix} 1 & -1 & -1 \\ 0 & 1 & 3 \\ 0 & 3 & 1 \end{bmatrix} \mathbf{y}.$$

First determine the eigenvalues of the coefficient matrix by solving

$$\begin{vmatrix} 1 - \lambda & -1 & -1 \\ 0 & 1 - \lambda & 3 \\ 0 & 3 & 1 - \lambda \end{vmatrix} = -\lambda^3 + 3\lambda^2 + 6\lambda - 8$$

$$= -(\lambda - 1)(\lambda - 4)(\lambda + 2) = 0;$$

so the eigenvalues are $\lambda = 1, 4, -2$. Since they are distinct, the corresponding eigenvectors will be linearly independent. For $\lambda = 1$, an eigenvector $\mathbf{b}^1$ is obtained as a solution to

$$(A - 1I)\mathbf{b} = 0 \quad \text{or} \quad \begin{bmatrix} 0 & -1 & -1 \\ 0 & 0 & 3 \\ 0 & 3 & 0 \end{bmatrix} \begin{bmatrix} b_1 \\ b_2 \\ b_3 \end{bmatrix} = \begin{bmatrix} 0 \\ 0 \\ 0 \end{bmatrix},$$

and a solution is

$$\mathbf{b}^1 = \begin{bmatrix} 1 \\ 0 \\ 0 \end{bmatrix}.$$

For $\lambda = 4$,

$$(A - 4I)\mathbf{b} = 0 \quad \text{or} \quad \begin{bmatrix} -3 & -1 & -1 \\ 0 & -3 & 3 \\ 0 & 3 & -3 \end{bmatrix} \begin{bmatrix} b_1 \\ b_2 \\ b_3 \end{bmatrix} = \begin{bmatrix} 0 \\ 0 \\ 0 \end{bmatrix},$$

and a choice for $\mathbf{b}^2$ is

$$\mathbf{b}^2 = \begin{bmatrix} 2 \\ -3 \\ -3 \end{bmatrix}.$$

Finally, for $\lambda = -2$, solve

$$[A - (-2)I]\mathbf{b} = 0 \quad \text{or} \quad \begin{bmatrix} 3 & -1 & -1 \\ 0 & 3 & 3 \\ 0 & 3 & 3 \end{bmatrix} \begin{bmatrix} b_1 \\ b_2 \\ b_3 \end{bmatrix} = \begin{bmatrix} 0 \\ 0 \\ 0 \end{bmatrix}$$

to get

$$\mathbf{b}^3 = \begin{bmatrix} 0 \\ -1 \\ 1 \end{bmatrix}.$$

Three linearly independent solutions are

$$\mathbf{y}^1(t) = \begin{bmatrix} 1 \\ 0 \\ 0 \end{bmatrix} e^t, \quad \mathbf{y}^2(t) = \begin{bmatrix} 2 \\ -3 \\ -3 \end{bmatrix} e^{4t}, \quad \mathbf{y}^3(t) = \begin{bmatrix} 0 \\ -1 \\ 1 \end{bmatrix} e^{-2t},$$

and the general solution is

$$\mathbf{y}(t) = c_1 \begin{bmatrix} 1 \\ 0 \\ 0 \end{bmatrix} e^t + c_2 \begin{bmatrix} 2 \\ -3 \\ -3 \end{bmatrix} e^{4t} + c_3 \begin{bmatrix} 0 \\ -1 \\ 1 \end{bmatrix} e^{-2t}. \quad \square$$

Example 8   Obtain the general solution of

$$\mathbf{y}' = \begin{bmatrix} 1 & -1 & -1 \\ 1 & 1 & 0 \\ 3 & 0 & 1 \end{bmatrix} \mathbf{y}.$$

The eigenvalues are found as solutions of

$$\begin{vmatrix} 1-\lambda & -1 & -1 \\ 1 & 1-\lambda & 0 \\ 3 & 0 & 1-\lambda \end{vmatrix} = (1-\lambda)^3 + 3(1-\lambda) + (1-\lambda)$$

$$= (1-\lambda)(\lambda^2 - 2\lambda + 5) = 0;$$

hence $\lambda = 1, 1 \pm 2i$. For $\lambda = 1$, solve

$$(A - 1I)\mathbf{b} = 0 \quad \text{or} \quad \begin{bmatrix} 0 & -1 & -1 \\ 1 & 0 & 0 \\ 3 & 0 & 0 \end{bmatrix} \begin{bmatrix} b_1 \\ b_2 \\ b_3 \end{bmatrix} = \begin{bmatrix} 0 \\ 0 \\ 0 \end{bmatrix}$$

to get

$$\mathbf{b}^1 = \begin{bmatrix} 0 \\ 1 \\ -1 \end{bmatrix}.$$

For $\lambda = 1 + 2i$,

$$[A - (1+2i)I]\mathbf{b} = 0 \quad \text{or} \quad \begin{bmatrix} -2i & -1 & -1 \\ 1 & -2i & 0 \\ 3 & 0 & -2i \end{bmatrix} \begin{bmatrix} b_1 \\ b_2 \\ b_3 \end{bmatrix} = \begin{bmatrix} 0 \\ 0 \\ 0 \end{bmatrix},$$

and the solution is

$$\mathbf{b}^2 = \begin{bmatrix} 2i \\ 1 \\ 3 \end{bmatrix},$$

and for $\lambda = 1 - 2i$

$$\mathbf{b}^3 = \mathbf{b}^{2*} = \begin{bmatrix} -2i \\ 1 \\ 3 \end{bmatrix}.$$

Three linearly independent solutions are:

$$\mathbf{y}^1(t) = \begin{bmatrix} 0 \\ 1 \\ -1 \end{bmatrix} e^t, \qquad \mathbf{y}^2(t) = \begin{bmatrix} 2i \\ 1 \\ 3 \end{bmatrix} e^{(1+2i)t},$$

$$\mathbf{y}^3(t) = \begin{bmatrix} -2i \\ 1 \\ 3 \end{bmatrix} e^{(1-2i)t},$$

and the general complex solution is

$$\mathbf{y}(t) = c_1 \begin{bmatrix} 0 \\ 1 \\ -1 \end{bmatrix} e^t + c_2 \begin{bmatrix} 2i \\ 1 \\ 3 \end{bmatrix} e^{(1+2i)t} + c_3 \begin{bmatrix} -2i \\ 1 \\ 3 \end{bmatrix} e^{(1-2i)t}.$$

To get the general real solution, proceed in a manner similar to that in Example 3 (p. 254). The choices $c_2 = c_3 = 1/2$ and $c_2 = -c_3 = 1/2i$ give

$$\frac{1}{2} \mathbf{y}^2(t) + \frac{1}{2} \mathbf{y}^3(t) = \begin{bmatrix} -2 \sin 2t \\ \cos 2t \\ 3 \cos 2t \end{bmatrix} e^t$$

and

$$\frac{1}{2i} \mathbf{y}^2(t) - \frac{1}{2i} \mathbf{y}^3(t) = \begin{bmatrix} 2 \cos 2t \\ \sin 2t \\ 3 \sin 2t \end{bmatrix} e^t.$$

Therefore the general real solution is

$$\mathbf{y}(t) = c_1 \begin{bmatrix} 0 \\ 1 \\ -1 \end{bmatrix} e^t + c_2 \begin{bmatrix} -2 \sin 2t \\ \cos 2t \\ 3 \cos 2t \end{bmatrix} e^t + c_3 \begin{bmatrix} 2 \cos 2t \\ \sin 2t \\ 3 \sin 2t \end{bmatrix} e^t. \quad \square$$

EXERCISES 5.3

Find the general solution of the following Type 1 systems:

1. $\mathbf{y}' = \begin{bmatrix} 3 & -2 \\ 2 & -2 \end{bmatrix} \mathbf{y}$ 　　**2.** $\mathbf{y}' = \begin{bmatrix} 0 & 1 \\ 8 & -2 \end{bmatrix} \mathbf{y}$

3. $\mathbf{y}' = \begin{bmatrix} -3 & 1 \\ 1 & -2 \end{bmatrix} \mathbf{y}$ 　　**4.** $\mathbf{y}' = \begin{bmatrix} -5 & 4 \\ 1 & -2 \end{bmatrix} \mathbf{y}$

5. $\mathbf{y}' = \begin{bmatrix} -3 & 2 \\ 2 & -3 \end{bmatrix} \mathbf{y}$ 　　**6.** $\mathbf{y}' = \begin{bmatrix} 1 & 0 \\ 1 & 2 \end{bmatrix} \mathbf{y}$

In each of the following solve the given Type 1 initial value problem:

7. $\mathbf{y}' = \begin{bmatrix} -3 & 4 \\ -2 & 3 \end{bmatrix} \mathbf{y}, \quad \mathbf{y}(0) = \begin{bmatrix} 1 \\ 0 \end{bmatrix}$

8. $\mathbf{y}' = \begin{bmatrix} 5 & -2 \\ 6 & -2 \end{bmatrix} \mathbf{y}, \quad \mathbf{y}(2) = \begin{bmatrix} 1 \\ 0 \end{bmatrix}$

9. $\mathbf{y}' = \begin{bmatrix} -1 & 6 \\ 1 & -2 \end{bmatrix} \mathbf{y}, \quad \mathbf{y}(0) = \begin{bmatrix} 1 \\ 1 \end{bmatrix}$

10. $\mathbf{y}' = \begin{bmatrix} 7 & 6 \\ 2 & 6 \end{bmatrix} \mathbf{y}, \quad \mathbf{y}(0) = \begin{bmatrix} 0 \\ 1 \end{bmatrix}$

11. $\mathbf{y}' = \begin{bmatrix} 2 & -1 \\ 3 & -2 \end{bmatrix} \mathbf{y}, \quad \mathbf{y}(0) = \begin{bmatrix} 1 \\ 4 \end{bmatrix}$

In each of the following, find the general real solution of the given Type 2 system of differential equations.

12. $\mathbf{y}' = \begin{bmatrix} -1 & -4 \\ 1 & -1 \end{bmatrix} \mathbf{y}$ **13.** $\mathbf{y}' = \begin{bmatrix} 3 & -5 \\ 2 & 1 \end{bmatrix} \mathbf{y}$

14. $\mathbf{y}' = \begin{bmatrix} 4 & -2 \\ 5 & 2 \end{bmatrix} \mathbf{y}$ **15.** $\mathbf{y}' = \begin{bmatrix} 2 & -5 \\ 2 & -4 \end{bmatrix} \mathbf{y}$

16. $\mathbf{y}' = \begin{bmatrix} 3 & 5 \\ -1 & -1 \end{bmatrix} \mathbf{y}$ **17.** $\mathbf{y}' = \begin{bmatrix} 8 & 16 \\ -5 & -8 \end{bmatrix} \mathbf{y}$

In each of the following, solve the given Type 2 initial value problem.

18. $\mathbf{y}' = \begin{bmatrix} 4 & 2 \\ -13 & -6 \end{bmatrix} \mathbf{y}, \quad \mathbf{y}(0) = \begin{bmatrix} 1 \\ 0 \end{bmatrix}$

19. $\mathbf{y}' = \begin{bmatrix} 4 & -4 \\ 5 & -4 \end{bmatrix} \mathbf{y}, \quad \mathbf{y}(\pi) = \begin{bmatrix} 0 \\ 1 \end{bmatrix}$

20. $\mathbf{y}' = \begin{bmatrix} 2 & -4 \\ 1 & 2 \end{bmatrix} \mathbf{y}, \quad \mathbf{y}(\pi) = \begin{bmatrix} 1 \\ 1 \end{bmatrix}$

21. $\mathbf{y}' = \begin{bmatrix} 3 & 2 \\ -5 & 1 \end{bmatrix} \mathbf{y}, \quad \mathbf{y}(\tfrac{3}{2}\pi) = \begin{bmatrix} 1 \\ 0 \end{bmatrix}$

22. $\mathbf{y}' = \begin{bmatrix} 0 & 1 \\ -1 & 0 \end{bmatrix} \mathbf{y}, \quad \mathbf{y}(\pi) = \begin{bmatrix} 1 \\ 1 \end{bmatrix}$

23. $\mathbf{y}' = \begin{bmatrix} 2 & -1 \\ 3 & 4 \end{bmatrix} \mathbf{y}, \quad \mathbf{y}(0) = \begin{bmatrix} 1 \\ 1 \end{bmatrix}$

24. Show that if λ is a complex number and $A\mathbf{b} = \lambda\mathbf{b}$, $A\mathbf{b}^* = \lambda^*\mathbf{b}^*$, then $\mathbf{b}$ and $\mathbf{b}^*$ are linearly independent.

25. If $b_1 = u_1 + iv_1$, $b_2 = u_2 + iv_2$, show that
$$\begin{vmatrix} b_1 & b_1^* \\ b_2 & b_2^* \end{vmatrix} = -2i \begin{vmatrix} u_1 & v_1 \\ u_2 & v_2 \end{vmatrix}.$$

In each of the following, find the general solution of the given Type 3 differential equations:

26. $\mathbf{y}' = \begin{bmatrix} 2 & 1 \\ -1 & 4 \end{bmatrix} \mathbf{y}$ **27.** $\mathbf{y}' = \begin{bmatrix} -3 & 2 \\ -2 & 1 \end{bmatrix} \mathbf{y}$

28. $\mathbf{y}' = \begin{bmatrix} 3 & -4 \\ 1 & -1 \end{bmatrix} \mathbf{y}$ **29.** $\mathbf{y}' = \begin{bmatrix} 4 & 8 \\ -2 & 12 \end{bmatrix} \mathbf{y}$

30. $\mathbf{y}' = \begin{bmatrix} 2 & -1 \\ 4 & 6 \end{bmatrix} \mathbf{y}$ **31.** $\mathbf{y}' = \begin{bmatrix} 4 & 4 \\ -9 & -8 \end{bmatrix} \mathbf{y}$

In each of the following solve the given Type 3 initial value problem:

32. $\mathbf{y}' = \begin{bmatrix} 1 & -4 \\ 4 & -7 \end{bmatrix} \mathbf{y}, \quad \mathbf{y}(0) = \begin{bmatrix} 3 \\ 1 \end{bmatrix}$

33. $\mathbf{y}' = \begin{bmatrix} 0 & 1 \\ -9 & 6 \end{bmatrix} \mathbf{y}, \quad \mathbf{y}(0) = \begin{bmatrix} 1 \\ 2 \end{bmatrix}$

34. $\mathbf{y}' = \begin{bmatrix} -4 & -1 \\ 1 & -2 \end{bmatrix} \mathbf{y}, \quad \mathbf{y}(0) = \begin{bmatrix} 1 \\ 0 \end{bmatrix}$

35. $\mathbf{y}' = \begin{bmatrix} 5 & 0 \\ 1 & 5 \end{bmatrix} \mathbf{y}, \quad \mathbf{y}(0) = \begin{bmatrix} 1 \\ 1 \end{bmatrix}$

36. $\mathbf{y}' = \begin{bmatrix} 5 & -7 \\ 7 & -9 \end{bmatrix} \mathbf{y}, \quad \mathbf{y}(1) = \begin{bmatrix} 0 \\ 1 \end{bmatrix}$

37. $\mathbf{y}' = \begin{bmatrix} 1 & -2 \\ 2 & -3 \end{bmatrix} \mathbf{y}, \quad \mathbf{y}(0) = \begin{bmatrix} 0 \\ 1 \end{bmatrix}$

38. Show that if λ is a double eigenvalue of the 2×2 matrix A and if $(A - \lambda I)\zeta = \sigma$, where $\sigma \neq \mathbf{0}$, then $(A - \lambda I)\sigma = \mathbf{0}$. *Hint:* Show that $(A - \lambda I)^2$ is the zero matrix, where
$$\lambda = (a_{11} + a_{22})/2.$$

39. Use the Laplace transform to solve the following:

a) The system of Example 7 with initial data
$$\mathbf{y}(0) = \begin{bmatrix} 2 \\ 0 \\ -6 \end{bmatrix};$$

b) The system of Example 8 with initial data
$$\mathbf{y}(0) = \begin{bmatrix} 2 \\ 0 \\ 0 \end{bmatrix}.$$

Use the eigenvalue–eigenvector method or the Laplace transform to find the solution of the following initial value problems:

40. $\mathbf{y}' = \begin{bmatrix} 1 & 0 & 1 \\ 0 & 1 & 2 \\ 1 & 2 & 5 \end{bmatrix} \mathbf{y}, \quad \mathbf{y}(0) = \begin{bmatrix} 1 \\ 0 \\ 1 \end{bmatrix}$

41. $\mathbf{y}' = \begin{bmatrix} 2 & 0 & 9 \\ 0 & 3 & 0 \\ 1 & 0 & 2 \end{bmatrix} \mathbf{y}, \quad \mathbf{y}(0) = \begin{bmatrix} 0 \\ 1 \\ 0 \end{bmatrix}$

42. $\mathbf{y}' = \begin{bmatrix} -2 & 0 & 0 \\ 0 & 3 & -1 \\ 0 & 1 & 3 \end{bmatrix} \mathbf{y}, \quad \mathbf{y}(0) = \begin{bmatrix} 2 \\ 0 \\ 0 \end{bmatrix}$

43. $\mathbf{y}' = \begin{bmatrix} 0 & 1 & 0 \\ 0 & -2 & -5 \\ 0 & 1 & 2 \end{bmatrix} \mathbf{y}, \quad \mathbf{y}(0) = \begin{bmatrix} 0 \\ 5 \\ -2 \end{bmatrix}$

44. $\mathbf{y}' = \begin{bmatrix} 2 & 1 & 1 \\ 1 & 1 & 0 \\ 1 & 0 & 1 \end{bmatrix} \mathbf{y}, \quad \mathbf{y}(0) = \begin{bmatrix} 1 \\ 0 \\ -1 \end{bmatrix}$

5.4 THE FUNDAMENTAL MATRIX

In the previous section we developed techniques for solving the first order linear system of differential equations

$$\mathbf{y}' = A\mathbf{y}, \qquad \mathbf{y}(t_0) = \mathbf{y}^0. \tag{5.4.1}$$

The solution could be obtained by analyzing the eigenvalue–eigenvector structure of the matrix A or directly by Laplace transform methods. But suppose that several different initial conditions $\mathbf{y}(t_0) = \mathbf{y}^1, \mathbf{y}(t_0) = \mathbf{y}^2, \ldots$, etc., are given. Does one have to solve each problem separately or can a general solution be developed in which the various initial conditions can be substituted to obtain the answer directly?

Fortunately, such a general solution can be found, and to accomplish this the concept of the *fundamental* or *transition state matrix* of the system (5.4.1) must be developed. With the fundamental matrix in hand, any initial value problem associated with $\mathbf{y}' = A\mathbf{y}$ can be solved as well as the inhomogeneous problem

$$\mathbf{y}' = A\mathbf{y} + \mathbf{f}(t), \qquad \mathbf{y}(t_0) = \mathbf{y}^0.$$

This last topic will be discussed in the next section.

To motivate the discussion, we go back to the one-dimensional version of the system (5.4.1), namely

$$y' = ay, \qquad y(t_0) = y_0, \tag{5.4.2}$$

where $y = y(t)$ is a scalar function and a is a given constant. For the function $\phi(t) = e^{at}$, the following conclusions are true:

1. $\phi' = a\phi$ since $\phi'(t) = ae^{at} = a\phi(t)$;

2. $\phi(0) = 1$;

3. The solution of $y' = ay$, $y(0) = y_0$, is given by

$$y(t) = \phi(t)y_0 = e^{at} y_0;$$

4. The solution of $y' = ay$, $y(t_0) = y_0$ is given by

$$y(t) = \phi(t - t_0)y_0 = e^{a(t-t_0)}y_0.$$

The reader can easily verify points (3) and (4) by differentiation and direct substitution. In view of this, $\phi(t)$ could be called the fundamental solution of (5.4.2). We also note at this point that $\phi(t)$ has the series representation

$$\phi(t) = e^{at} = 1 + at + \frac{1}{2!} a^2 t^2 + \cdots + \frac{1}{m!} a^m t^m + \cdots$$

which converges for any a and for $-\infty < t < \infty$.

Now, given the first order system (5.4.1), consider the matrix function defined formally by

$$\Phi(t) = e^{tA} = I + tA + \frac{1}{2!} t^2 A^2 + \cdots + \frac{1}{m!} t^m A^m + \cdots, \quad (5.4.3)$$

where A is the $n \times n$ coefficient matrix of (5.4.1) and its powers are defined recursively by

$$A^2 = A \cdot A, \quad A^3 = A \cdot A^2, \quad \ldots, \quad A^m = A \cdot A^{m-1}.$$

Recall that

$$I = \begin{bmatrix} 1 & & & 0 \\ & 1 & & \\ & & \ddots & \\ 0 & & & 1 \end{bmatrix}$$

is the $n \times n$ identity matrix with ones on the main diagonal and zeros elsewhere. The fact that the power series defined above converges to a matrix function for any constant $n \times n$ matrix A and $-\infty < t < \infty$ is beyond the scope of this book. Hopefully, the following examples will convince the reader that it does.

Example 1   Let $A = \begin{bmatrix} 1 & 0 \\ 1 & 2 \end{bmatrix}$; then

$$A^2 = A \cdot A = \begin{bmatrix} 1 & 0 \\ 1 & 2 \end{bmatrix} \begin{bmatrix} 1 & 0 \\ 1 & 2 \end{bmatrix} = \begin{bmatrix} 1 & 0 \\ 3 & 4 \end{bmatrix} = \begin{bmatrix} 1 & 0 \\ 2^2 - 1 & 2^2 \end{bmatrix}$$

$$A^3 = A \cdot A^2 = \begin{bmatrix} 1 & 0 \\ 1 & 2 \end{bmatrix} \begin{bmatrix} 1 & 0 \\ 3 & 4 \end{bmatrix} = \begin{bmatrix} 1 & 0 \\ 7 & 8 \end{bmatrix} = \begin{bmatrix} 1 & 0 \\ 2^3 - 1 & 2^3 \end{bmatrix}$$

$$\vdots \qquad\qquad\qquad\qquad \vdots$$

$$A^m = A \cdot A^{m-1} = \begin{bmatrix} 1 & 0 \\ 1 & 2 \end{bmatrix} \begin{bmatrix} 1 & 0 \\ 2^{m-1} - 1 & 2^{m-1} \end{bmatrix} = \begin{bmatrix} 1 & 0 \\ 2^m - 1 & 2^m \end{bmatrix}.$$

Therefore

$$e^{tA} = \begin{bmatrix} 1 & 0 \\ 0 & 1 \end{bmatrix} + t \begin{bmatrix} 1 & 0 \\ 1 & 2 \end{bmatrix} + \frac{t^2}{2!} \begin{bmatrix} 1 & 0 \\ 2^2 - 1 & 2^2 \end{bmatrix} + \frac{t^3}{3!} \begin{bmatrix} 1 & 0 \\ 2^3 - 1 & 2^3 \end{bmatrix}$$

$$+ \cdots + \frac{t^m}{m!} \begin{bmatrix} 1 & 0 \\ 2^m - 1 & 2^m \end{bmatrix} + \cdots$$

$$= \begin{bmatrix} \displaystyle\sum_{m=0}^{\infty} \frac{t^m}{m!} & 0 \\ \displaystyle\sum_{m=0}^{\infty} \frac{(2t)^m}{m!} - \sum_{m=0}^{\infty} \frac{t^m}{m!} & \displaystyle\sum_{m=0}^{\infty} \frac{(2t)^m}{m!} \end{bmatrix} = \begin{bmatrix} e^t & 0 \\ e^{2t} - e^t & e^{2t} \end{bmatrix}. \quad \square$$

Example 2 An even simpler example is the case when A is a diagonal matrix

$$A = \begin{bmatrix} \lambda_1 & & & 0 \\ & \lambda_2 & & \\ & & \ddots & \\ 0 & & & \lambda_n \end{bmatrix}.$$

Then

$$A^m = \begin{bmatrix} \lambda_1^m & & & 0 \\ & \lambda_2^m & & \\ & & \ddots & \\ 0 & & & \lambda_n^m \end{bmatrix}$$

and

$$e^{tA} = \begin{bmatrix} \displaystyle\sum_{m=0}^{\infty} \frac{(\lambda_1 t)^m}{m!} & & & 0 \\ & \displaystyle\sum_{m=0}^{\infty} \frac{(\lambda_2 t)^m}{m!} & & \\ & & \ddots & \\ 0 & & & \displaystyle\sum_{m=0}^{\infty} \frac{(\lambda_n t)^m}{m!} \end{bmatrix} = \begin{bmatrix} e^{\lambda_1 t} & & & 0 \\ & e^{\lambda_2 t} & & \\ & & \ddots & \\ 0 & & & e^{\lambda_n t} \end{bmatrix}. \quad \square$$

If a power series converges, then it can be differentiated termwise, and this is also true for the power series (5.4.3) that defines e^{tA}. Hence

$$\frac{d}{dt} \Phi(t) = \frac{d}{dt} e^{tA} = \frac{d}{dt} \left[I + tA + \frac{t^2}{2!} A^2 + \cdots + \frac{t^m}{m!} A^m + \cdots \right]$$

$$= 0 + A + tA^2 + \frac{t^2}{2!} A^3 + \cdots + \frac{t^{m-1}}{(m-1)!} A^m + \cdots$$

$$= A \left(I + tA + \frac{t^2}{2!} A^2 + \cdots + \frac{t^{m-1}}{(m-1)!} A^{m-1} + \cdots \right)$$

$$= A e^{tA} = A\Phi(t),$$

or

$$\Phi'(t) = A\Phi(t),$$

which is the matrix analogue of conclusion (1) above (p. 266). For instance, in Examples 1 and 2 above,

$$\frac{d}{dt}\begin{bmatrix} e^t & 0 \\ e^{2t} - e^t & e^{2t} \end{bmatrix} = \begin{bmatrix} e^t & 0 \\ 2e^{2t} - e^t & 2e^{2t} \end{bmatrix} = \begin{bmatrix} 1 & 0 \\ 1 & 2 \end{bmatrix}\begin{bmatrix} e^t & 0 \\ e^{2t} - e^t & e^{2t} \end{bmatrix}$$

and

$$\frac{d}{dt}\begin{bmatrix} e^{\lambda_1 t} & & 0 \\ & e^{\lambda_2 t} & \\ & & \ddots & \\ 0 & & & e^{\lambda_n t} \end{bmatrix} = \begin{bmatrix} \lambda_1 e^{\lambda_1 t} & & 0 \\ & \lambda_2 e^{\lambda_2 t} & \\ & & \ddots & \\ 0 & & & \lambda_n e^{\lambda_n t} \end{bmatrix}$$

$$= \begin{bmatrix} \lambda_1 & & 0 \\ & \lambda_2 & \\ & & \ddots & \\ 0 & & & \lambda_n \end{bmatrix}\begin{bmatrix} e^{\lambda_1 t} & & 0 \\ & e^{\lambda_2 t} & \\ & & \ddots & \\ 0 & & & e^{\lambda_n t} \end{bmatrix}.$$

Furthermore

$$\Phi(0) = e^{0A} = I + 0A + \frac{0^2}{2!}A^2 + \cdots + \frac{0^m}{m!}A^m + \cdots = I,$$

the identity matrix which is the matrix analogue of conclusion (2) above.

Finally, given the vector $\mathbf{y}^0$, consider the vector obtained by multiplying it by $\Phi(t)$:

$$\mathbf{y}(t) = \Phi(t)\mathbf{y}^0 = e^{tA}\mathbf{y}^0 = I\mathbf{y}^0 + tA\mathbf{y}^0 + \frac{t^2}{2!}A^2\mathbf{y}^0 + \cdots + \frac{t^m}{m!}A^m\mathbf{y}^0 + \cdots.$$

Since $I\mathbf{y}^0 = \mathbf{y}^0$, we have

$$\mathbf{y}(0) = \Phi(0)\mathbf{y}^0 = I\mathbf{y}^0 = \mathbf{y}^0,$$

and now differentiating $y(t)$ gives

$$\mathbf{y}'(t) = \frac{d}{dt}\left(I\mathbf{y}^0 + tA\mathbf{y}^0 + \frac{t^2}{2!}A^2\mathbf{y}^0 + \cdots + \frac{t^m}{m!}A^m\mathbf{y}^0 + \cdots\right)$$

$$= A\left(I\mathbf{y}^0 + tA\mathbf{y}^0 + \frac{t^2}{2!}A^2\mathbf{y}^0 + \cdots + \frac{t^{m-1}}{(m-1)!}A^{m-1}\mathbf{y}^0 + \cdots\right)$$

$$= A\Phi(t)\mathbf{y}^0 = A\mathbf{y}(t).$$

The last two relations imply that

$$\mathbf{y}(t) = \Phi(t)\mathbf{y}^0$$

is the solution to the initial value problem

$$\mathbf{y}' = A\mathbf{y}, \qquad \mathbf{y}(0) = \mathbf{y}^0.$$

This is the matrix version of conclusion (3). It is left to the reader to verify that $\mathbf{y}(t) = \Phi(t - t_0)\mathbf{y}^0$ is the solution to the initial value problem

$$\mathbf{y}' = A\mathbf{y}, \qquad \mathbf{y}(t_0) = \mathbf{y}^0.$$

We conclude that $\Phi(t) = e^{tA}$ satisfies all four of the relations satisfied by $\phi(t) = e^{at}$ in the scalar case. For instance, in Example 1 above

$$\Phi(t) = \begin{bmatrix} e^t & 0 \\ e^{2t} - e^t & e^{2t} \end{bmatrix} \qquad \text{and} \qquad \Phi(0) = \begin{bmatrix} 1 & 0 \\ 0 & 1 \end{bmatrix} = I.$$

Let

$$\mathbf{y}(t) = \begin{bmatrix} y_1(t) \\ y_2(t) \end{bmatrix} = \begin{bmatrix} e^t & 0 \\ e^{2t} - e^t & e^{2t} \end{bmatrix} \begin{bmatrix} y_1^0 \\ y_2^0 \end{bmatrix} = \Phi(t)\mathbf{y}^0;$$

then

$$\mathbf{y}(0) = \begin{bmatrix} y_1(0) \\ y_2(0) \end{bmatrix} = \begin{bmatrix} 1 & 0 \\ 0 & 1 \end{bmatrix} \begin{bmatrix} y_1^0 \\ y_2^0 \end{bmatrix} = \begin{bmatrix} y_1^0 \\ y_2^0 \end{bmatrix} = \mathbf{y}^0$$

and

$$\mathbf{y}'(t) = \begin{bmatrix} y_1'(t) \\ y_2'(t) \end{bmatrix} = \begin{bmatrix} e^t & 0 \\ 2e^{2t} - e^t & 2e^{2t} \end{bmatrix} \begin{bmatrix} y_1^0 \\ y_2^0 \end{bmatrix}$$

$$= \begin{bmatrix} 1 & 0 \\ 1 & 2 \end{bmatrix} \begin{bmatrix} e^t & 0 \\ e^{2t} - e^t & e^{2t} \end{bmatrix} \begin{bmatrix} y_1^0 \\ y_2^0 \end{bmatrix}$$

$$= \begin{bmatrix} 1 & 0 \\ 1 & 2 \end{bmatrix} \begin{bmatrix} y_1(t) \\ y_2(t) \end{bmatrix} = \begin{bmatrix} 1 & 0 \\ 1 & 2 \end{bmatrix} \mathbf{y}(t).$$

Hence, if the above relations are expressed by components, then

$$\left. \begin{aligned} y_1(t) &= y_1^0 e^t \\ y_2(t) &= y_1^0(e^{2t} - e^t) + y_2^0 e^{2t} \end{aligned} \right\} = \Phi(t) \begin{bmatrix} y_1^0 \\ y_2^0 \end{bmatrix}$$

is the solution to the initial value problem

$$\left. \begin{aligned} y_1' &= y_1 \\ y_2' &= y_1 + 2y_2 \end{aligned} \right\} = \begin{bmatrix} 1 & 0 \\ 1 & 2 \end{bmatrix} \begin{bmatrix} y_1 \\ y_2 \end{bmatrix}$$

$$y_1(0) = y_1^0, \qquad y_2(0) = y_2^0.$$

It is left to the reader to verify for Example 2 that

$$\mathbf{y}(t) = \begin{bmatrix} e^{\lambda_1 t} & & & 0 \\ & e^{\lambda_2 t} & & \\ & & \ddots & \\ 0 & & & e^{\lambda_n t} \end{bmatrix} \begin{bmatrix} y_1^0 \\ y_2^0 \\ \vdots \\ y_n^0 \end{bmatrix}$$

is the solution of the system

$$\frac{d}{dt} \begin{bmatrix} y_1' \\ y_2' \\ \vdots \\ y_n' \end{bmatrix} = \begin{bmatrix} \lambda_1 y_1 \\ \lambda_2 y_2 \\ \vdots \\ \lambda_n y_n \end{bmatrix} = \begin{bmatrix} \lambda_1 & & & 0 \\ & \lambda_2 & & \\ & & \ddots & \\ 0 & & & \lambda_n \end{bmatrix} \begin{bmatrix} y_1 \\ y_2 \\ \vdots \\ y_n \end{bmatrix}$$

with initial conditions

$$y_1(0) = y_1^0, \qquad y_2(0) = y_2^0, \quad \ldots, \quad y_n(0) = y_n^0.$$

Our conclusions are summarized in the following statement:

For the *fundamental matrix*

$$\Phi(t) = e^{tA} = \sum_{m=0}^{\infty} \frac{t^m A^m}{m!}$$

the following is true:

1. $\Phi' = A\Phi$;

2. $\Phi(0) = I$;

and

3. the solution of $\mathbf{y}' = A\mathbf{y}, \mathbf{y}(t_0) = \mathbf{y}^0$ is given by $\mathbf{y}(t) = \Phi(t - t_0)\mathbf{y}^0$.

There is one remaining problem, and that is to find the fundamental matrix $\Phi(t)$. The power series representation is clearly not useful since it involves computing successive powers of A. For all but the simplest matrices, writing down the first m terms of the power series will not give an inkling as to the nature of $\Phi(t)$.

Hence another way to compute $\Phi(t)$ is needed, and the key to the method lies in the following observation:

The jth column $\Phi^j(t)$ of $\Phi(t)$ is the solution of the initial value problem

$$\mathbf{y}' = A\mathbf{y}, \qquad \mathbf{y}(0) = \mathbf{e}^j,$$

where $\mathbf{e}^j$ is the vector with one in the jth place and zeros elsewhere, i.e., the jth unit coordinate vector.

The proof of this assertion follows easily by noting first of all that $\Phi(0) = I$ implies that the jth column, $\Phi^j(t)$, of $\Phi(t)$ satisfies $\Phi^j(0) = \mathbf{e}^j$. Secondly, from the property of the fundamental matrix it follows that solutions to the above problem can be written in the form $\mathbf{y} = \Phi(t)\mathbf{e}^j$. But this is just $\Phi^j(t)$ since if $\Phi(t) = \phi_{ij}(t)$, then

$$y_k(t) = \sum_{i=1}^{n} \phi_{ki}(t)e_i^j = \phi_{kj}(t)$$

because $e_i^j = 0$ unless $i = j$ when $e_j^j = 1$. Put more simply, the kth entry in the vector $\mathbf{y}(t)$ is the dot product of the kth row of $\Phi(t)$ with $\mathbf{e}^j$, and this will give the kth entry in the jth column of $\Phi(t)$.

In Example 1

$$\begin{aligned} y_1' &= y_1 \\ y_2' &= y_1 + 2y_2, \end{aligned} \qquad \Phi(t) = \begin{bmatrix} e^t & 0 \\ e^{2t} - e^t & e^{2t} \end{bmatrix},$$

and so the columns of $\Phi(t)$ are

$$\Phi^1(t) = \begin{bmatrix} e^t \\ e^{2t} - e^t \end{bmatrix}, \qquad \Phi^2(t) = \begin{bmatrix} 0 \\ e^{2t} \end{bmatrix}.$$

It is seen, for instance, that the entries of $\Phi^1(t)$, $y_1(t) = e^t$, $y_2(t) = e^{2t} - e^t$, satisfy

$$y_1'(t) = e^t = y_1(t),$$

$$y_2'(t) = 2e^{2t} - e^t = e^t + 2(e^{2t} - e^t) = y_1(t) + 2y_2(t),$$

and, furthermore,

$$\begin{bmatrix} y_1(0) \\ y_2(0) \end{bmatrix} = \begin{bmatrix} 1 \\ 0 \end{bmatrix} = \Phi^1(0) = \mathbf{e}^1.$$

The reader can see directly that $y_1(t) = 0$, $y_2(t) = e^{2t}$ also satisfy the system of differential equations and

$$\begin{bmatrix} y_1(0) \\ y_2(0) \end{bmatrix} = \begin{bmatrix} 0 \\ 1 \end{bmatrix} = \Phi^2(0) = \mathbf{e}^2.$$

In Example 2, the system $y_k' = \lambda_k y_k$, $k = 1, 2, \ldots, n$, was given and the jth column of $\Phi(t)$ was

$$\Phi^j(t) = \mathrm{col}(0, \ldots, 0, \underset{\substack{j\text{th} \\ \text{place}}}{e^{\lambda_j t}}, 0, \ldots, 0),$$

which satisfies $\Phi^j(0) = \mathbf{e}^j$. It corresponds to the solution $y_j(t) = e^{\lambda_j t}$, $y_k(t) = 0$, $k \neq j$, of the system.

Our findings are summarized in the following algorithm.

Algorithm for finding $\Phi(t) = e^{tA}$, **where A** *is an* $n \times n$ *matrix.* Given the linear system $\mathbf{y}' = A\mathbf{y}$, where A is an $n \times n$ constant matrix, $\Phi^j(t)$, the jth column of the fundamental matrix $\Phi(t)$, is obtained by solving the initial value problem

$$\mathbf{y}' = A\mathbf{y}, \qquad \mathbf{y}(0) = \mathbf{e}^j,$$

where $\mathbf{e}^j$ is the jth unit coordinate vector in R^n, that is, $y_j(0) = 1$, $y_k(0) = 0$, $k \neq j$.

Example 3 Find the fundamental solution matrix for

$$A = \begin{bmatrix} -1 & 6 \\ 1 & -2 \end{bmatrix}$$

and then solve the initial value problem

$$\mathbf{y}' = A\mathbf{y}, \qquad \mathbf{y}(1) = \begin{bmatrix} 2 \\ -3 \end{bmatrix}.$$

Since

$$\det \begin{bmatrix} -1 - \lambda & 6 \\ 1 & -2 - \lambda \end{bmatrix} = \lambda^2 + 3\lambda - 4 = (\lambda + 4)(\lambda - 1),$$

the eigenvalues of A are $\lambda = -4$ and $\lambda = 1$. The eigenvector corresponding to $\lambda = -4$ is $\begin{bmatrix} -2 \\ 1 \end{bmatrix}$, so

$$\mathbf{y}^1(t) = e^{-4t} \begin{bmatrix} -2 \\ 1 \end{bmatrix} = \begin{bmatrix} -2e^{-4t} \\ e^{-4t} \end{bmatrix}$$

is a solution. The eigenvector corresponding to $\lambda = 1$ is $\begin{bmatrix} 3 \\ 1 \end{bmatrix}$, so

$$\mathbf{y}^2(t) = e^t \begin{bmatrix} 3 \\ 1 \end{bmatrix} = \begin{bmatrix} 3e^t \\ e^t \end{bmatrix}$$

is another linearly independent solution, and

$$\mathbf{y}(t) = \begin{bmatrix} y_1(t) \\ y_2(t) \end{bmatrix} = c_1 \begin{bmatrix} -2e^{-4t} \\ e^{-4t} \end{bmatrix} + c_2 \begin{bmatrix} 3e^t \\ e^t \end{bmatrix}$$

$$= \begin{bmatrix} -2c_1 e^{-4t} + 3c_2 e^t \\ c_1 e^{-4t} + c_2 e^t \end{bmatrix}$$

is the general solution of the system

$$y_1' = -y_1 + 6y_2$$

$$y_2' = y_1 - 2y_2.$$

To obtain $\Phi(t)$, first find its first column $\Phi^1(t)$ by letting

$$\mathbf{y}(0) = \begin{bmatrix} y_1(0) \\ y_2(0) \end{bmatrix} = \begin{bmatrix} -2c_1 + 3c_2 \\ c_1 + c_2 \end{bmatrix} = \begin{bmatrix} 1 \\ 0 \end{bmatrix} = \mathbf{e}^1.$$

Solving for c_1 and c_2 gives $c_1 = -\tfrac{1}{5}$, $c_2 = \tfrac{1}{5}$, so

$$\Phi^1(t) = \begin{bmatrix} \tfrac{2}{5}e^{-4t} + \tfrac{3}{5}e^t \\ -\tfrac{1}{5}e^{-4t} + \tfrac{1}{5}e^t \end{bmatrix}.$$

Now, to get the second column $\Phi^2(t)$, solve

$$\mathbf{y}(0) = \begin{bmatrix} y_1(0) \\ y_2(0) \end{bmatrix} = \begin{bmatrix} -2c_1 + 3c_2 \\ c_1 + c_2 \end{bmatrix} = \begin{bmatrix} 0 \\ 1 \end{bmatrix} = \mathbf{e}^2$$

to get

$$\Phi^2(t) = \begin{bmatrix} -\tfrac{3}{5}e^{-4t} + \tfrac{3}{5}e^t \\ \tfrac{2}{5}e^{-4t} + \tfrac{3}{5}e^t \end{bmatrix}.$$

Therefore

$$\Phi(t) = e^{tA} = \begin{bmatrix} \tfrac{2}{5}e^{-4t} + \tfrac{3}{5}e^t & -\tfrac{3}{5}e^{-4t} + \tfrac{3}{5}e^t \\ -\tfrac{1}{5}e^{-4t} + \tfrac{1}{5}e^t & \tfrac{2}{5}e^{-4t} + \tfrac{3}{5}e^t \end{bmatrix}.$$

The solution of the system with initial conditions $y_1(1) = 2$, $y_2(1) = -3$ is

$$\begin{bmatrix} y_1(t) \\ y_2(t) \end{bmatrix} = \Phi(t-1) \begin{bmatrix} 2 \\ -3 \end{bmatrix}$$

$$= \begin{bmatrix} \tfrac{2}{5}e^{-4(t-1)} + \tfrac{3}{5}e^{t-1} & -\tfrac{3}{5}e^{-4(t-1)} + \tfrac{3}{5}e^{t-1} \\ -\tfrac{1}{5}e^{-4(t-1)} + \tfrac{1}{5}e^{t-1} & \tfrac{2}{5}e^{-4(t-1)} + \tfrac{3}{5}e^{t-1} \end{bmatrix} \begin{bmatrix} 2 \\ -3 \end{bmatrix}$$

$$= \frac{1}{5} \begin{bmatrix} 22e^{-4(t-1)} - 12e^{t-1} \\ -11e^{-4(t-1)} - 4e^{t-1} \end{bmatrix}. \quad \square$$

Example 4 Find the fundamental matrix for

$$A = \begin{bmatrix} 2 & 1 \\ -1 & 4 \end{bmatrix}$$

by using the Laplace transform.

We first solve the initial value problem

$$y_1'(t) = 2y_1(t) + y_2(t), \quad y_1(0) = 1,$$
$$y_2'(t) = -y_1(t) + 4y_2(t), \quad y_2(0) = 0.$$

Applying the Laplace transform gives

$$sY_1(s) - 1 = 2Y_1(s) + Y_2(s),$$
$$sY_2(s) - 0 = -Y_1(s) + 4Y_2(s).$$

Solving the above system for $Y_1(s)$ and $Y_2(s)$, one obtains

$$Y_1(s) = \frac{s - 4}{(s - 3)^2} = \frac{1}{s - 3} - \frac{1}{(s - 3)^2},$$

$$Y_2(s) = \frac{-1}{(s - 3)^2}.$$

Applying the inverse transform, one finds that

$$\Phi^1(t) = \begin{bmatrix} y_1(t) \\ y_2(t) \end{bmatrix}$$

$$= \begin{bmatrix} e^{3t} - te^{3t} \\ -te^{3t} \end{bmatrix}.$$

If the order of the initial values 1 and 0 is reversed in the system of equations for $Y_1(s)$ and $Y_2(s)$, then

$$Y_1(s) = \frac{1}{(s - 3)^2},$$

$$Y_2(s) = \frac{s - 2}{(s - 3)^2} = \frac{1}{s - 3} + \frac{1}{(s - 3)^2}.$$

This implies that

$$\Phi^2(t) = \begin{bmatrix} y_1(t) \\ y_2(t) \end{bmatrix}$$

$$= \begin{bmatrix} te^{3t} \\ e^{3t} + te^{3t} \end{bmatrix},$$

and the fundamental matrix of the system is therefore

$$\Phi(t) = \begin{bmatrix} e^{3t} - te^{3t} & te^{3t} \\ -te^{3t} & e^{3t} + te^{3t} \end{bmatrix}. \quad \square$$

EXERCISES 5.4

1. Verify that the function $y(t) = \Phi(t - t_0)y^0$ is the solution to the initial value problem

$$y' = Ay, \qquad y(t_0) = y^0.$$

In each of the following find the fundamental matrix $\Phi(t)$ by first finding a general solution to the system of differential equations and then finding solutions $\Phi^1(t)$ and $\Phi^2(t)$ which satisfy

$$\Phi^1(0) = \begin{bmatrix} 1 \\ 0 \end{bmatrix}, \Phi^2(0) = \begin{bmatrix} 0 \\ 1 \end{bmatrix}.$$

2. $y' = \begin{bmatrix} 4 & -3 \\ 8 & -6 \end{bmatrix} y$

3. $y' = \begin{bmatrix} 0 & 1 \\ 8 & -2 \end{bmatrix} y$

4. $y' = \begin{bmatrix} 2 & 1 \\ 0 & 2 \end{bmatrix} y$

5. $y' = \begin{bmatrix} 5 & 3 \\ -3 & -1 \end{bmatrix} y$

6. $y' = \begin{bmatrix} -1 & -5 \\ 1 & 1 \end{bmatrix} y$

7. $y' = \begin{bmatrix} 1 & -1 \\ 5 & -1 \end{bmatrix} y$

Use the Laplace transform to find the fundamental matrix $\Phi(t)$ in each of the following:

8. $y' = \begin{bmatrix} 2 & 1 \\ 3 & 4 \end{bmatrix} y$

9. $y' = \begin{bmatrix} 3 & 1 \\ -2 & 1 \end{bmatrix} y$

10. $y' = \begin{bmatrix} 1 & -1 \\ -3 & 3 \end{bmatrix} y$

11. $y' = \begin{bmatrix} 5 & -1 \\ 1 & 3 \end{bmatrix} y$

First find the fundamental matrix, then use it to solve the following initial value problems:

12. $\mathbf{y}' = \begin{bmatrix} -1 & -5 \\ 1 & 3 \end{bmatrix} \mathbf{y}, \quad \mathbf{y}(0) = \begin{bmatrix} 5 \\ 2 \end{bmatrix},$

$\mathbf{y}(\pi) = \begin{bmatrix} 3 \\ 1 \end{bmatrix}$

13. $\mathbf{y}' = \begin{bmatrix} -4 & -1 \\ 1 & -2 \end{bmatrix} \mathbf{y}, \quad \mathbf{y}(0) = \begin{bmatrix} 2 \\ 1 \end{bmatrix},$

$\mathbf{y}(1) = \begin{bmatrix} 1 \\ 2 \end{bmatrix}$

14. $\mathbf{y}' = \begin{bmatrix} 1 & 3 \\ 12 & 1 \end{bmatrix} \mathbf{y}, \quad \mathbf{y}(0) = \begin{bmatrix} 3 \\ -2 \end{bmatrix},$

$\mathbf{y}(2) = \begin{bmatrix} 1 \\ 0 \end{bmatrix}$

15. $\mathbf{y}' = \begin{bmatrix} 3 & -2 \\ 5 & -1 \end{bmatrix} \mathbf{y}, \quad \mathbf{y}(0) = \begin{bmatrix} 1 \\ -1 \end{bmatrix},$

$\mathbf{y}(1) = \begin{bmatrix} 1 \\ 1 \end{bmatrix}$

5.5 THE INHOMOGENEOUS LINEAR SYSTEM— VARIATION OF PARAMETERS

The next problem that we consider is finding solutions of the inhomogeneous linear system:

$$y_1' = a_{11}y_1 + a_{12}y_2 + \cdots + a_{1n}y_n + f_1(t), \qquad y_1(t_0) = y_1^0,$$
$$y_2' = a_{21}y_1 + a_{22}y_2 + \cdots + a_{2n}y_n + f_2(t), \qquad y_2(t_0) = y_2^0,$$
$$\vdots \qquad\qquad \vdots \qquad\qquad\qquad \vdots$$
$$y_n' = a_{n1}y_1 + a_{n2}y_2 + \cdots + a_{nn}y_n + f_n(t), \qquad y_n(t_0) = y_n^0,$$

where the functions $f_1(t), f_2(t), \ldots, f_n(t)$ are continuous in some neighborhood of $t = t_0$. If $\mathbf{f}(t)$ is the vector function

$$\mathbf{f}(t) = \text{col}(f_1(t), f_2(t), \ldots, f_n(t))$$

and $\mathbf{y}^0 = \text{col}(y_1^0, y_2^0, \ldots, y_n^0)$, then the above system can be simply written as

$$\mathbf{y}' = A\mathbf{y} + \mathbf{f}(t), \qquad \mathbf{y}(t_0) = \mathbf{y}^0, \tag{5.5.1}$$

where $A = (a_{ij})$ is the $n \times n$ constant coefficient matrix.

As in the case of first order and second order equations the task is to find any particular solution $\mathbf{y}_p(t)$ of the system of differential equations, i.e., one must find a function $\mathbf{y}_p(t)$ that satisfies

$$\mathbf{y}_p'(t) = A\mathbf{y}_p(t) + \mathbf{f}(t).$$

Then to solve the initial value problem the particular solution is added to the solution $\mathbf{y}_h(t)$ of the homogeneous system

$$\mathbf{y}' = A\mathbf{y}, \qquad \mathbf{y}(t_0) = \mathbf{y}^0 - \mathbf{y}_p(t_0).$$

From the previous discussion we have that

$$\mathbf{y}_h(t) = \Phi(t - t_0)[\mathbf{y}^0 - \mathbf{y}_p(t_0)],$$

and so the solution of the initial value problem (5.5.1) is

$$\mathbf{y}(t) = \mathbf{y}_h(t) + \mathbf{y}_p(t) = \Phi(t - t_0)(\mathbf{y}^0 - \mathbf{y}_p(t_0)) + \mathbf{y}_p(t).$$

To verify this, first note that

$$\begin{aligned}
\mathbf{y}(t_0) &= \Phi(0)[\mathbf{y}^0 - \mathbf{y}_p(t_0)] + \mathbf{y}_p(t_0) \\
&= I(\mathbf{y}^0 - \mathbf{y}_p(t_0)) + \mathbf{y}_p(t_0) = \mathbf{y}^0
\end{aligned}$$

and

$$\begin{aligned}
\mathbf{y}'(t) &= \Phi'(t - t_0)[\mathbf{y}^0 - \mathbf{y}_p(t_0)] + \mathbf{y}'_p(t) \\
&= A\Phi(t - t_0)[\mathbf{y}^0 - \mathbf{y}_p(t_0)] + A\mathbf{y}_p(t) + \mathbf{f}(t) \\
&= A\{\Phi(t - t_0)[\mathbf{y}^0 - \mathbf{y}_p(t_0)] + \mathbf{y}_p(t)\} + \mathbf{f}(t) \\
&= A\mathbf{y}(t) + \mathbf{f}(t),
\end{aligned}$$

since $\Phi'(t) = A\Phi(t)$. Thus $\mathbf{y}(t)$ is the desired solution.

How do we find a particular solution $\mathbf{y}_p(t)$? As in the case of first and second order equations, there is a formula called the variation of parameters formula which is sure-fire. One can also use the Laplace transform and convert the problem into an algebraic one in which the initial conditions can be incorporated and thus solve the entire problem in one (sometimes long!) step.

Another method, which is usually not mentioned in most textbooks when discussing systems, is the method of comparison of coefficients, which might be better called the method of "judicious guessing." The reader, who has used the technique to find solutions of inhomogeneous second order equations, will easily see the extension to inhomogeneous systems in the following examples.

Example 1   Find a particular solution of

$$\begin{aligned}
y'_1 &= 3y_1 + 2y_2 - 3e^t, \\
y'_2 &= y_1 - 4y_2.
\end{aligned}$$

In this case,

$$\mathbf{f}(t) = \begin{bmatrix} f_1(t) \\ f_2(t) \end{bmatrix} = \begin{bmatrix} -3e^t \\ 0 \end{bmatrix},$$

and so a trial solution would be

$$\mathbf{y}_p(t) = \begin{bmatrix} y_1(t) \\ y_2(t) \end{bmatrix} = \begin{bmatrix} ae^t \\ be^t \end{bmatrix},$$

corresponding to the fact that for a scalar differential equation one would try a multiple of e^t. Substitution into the system of equations gives

$$ae^t = 3ae^t + 2be^t - 3e^t,$$

$$be^t = ae^t - 4be^t;$$

cancelling the factor of e^t leads to the system of algebraic equations

$$-2a - 2b = -3,$$

$$a - 5b = 0.$$

The solution is $a = \frac{5}{4}$, $b = \frac{1}{4}$, and therefore

$$\mathbf{y}_p(t) = \begin{bmatrix} \frac{5}{4} e^t \\ \frac{1}{4} e^t \end{bmatrix}.$$

Note that since the inhomogeneous term $-3e^t$ appears only in the first equation, one might have used a trial solution

$$\mathbf{y}_p(t) = \begin{bmatrix} ae^t \\ 0 \end{bmatrix}.$$

As the answer shows, this would not have worked.   □

Example 2   Find a particular solution of

$$y_1' = 3y_1 + 2y_2 + 7t,$$

$$y_2' = y_1 - 4y_2 - 3.$$

The inhomogeneous terms are polynomials of degree 1; this suggests

$$\mathbf{y}_p(t) = \begin{bmatrix} y_1(t) \\ y_2(t) \end{bmatrix} = \begin{bmatrix} at + b \\ ct + d \end{bmatrix}.$$

Substitution into the system of equations gives

$$a = 3at + 3b + 2ct + 2d + 7t,$$

$$c = at + b - 4ct - 4d - 3.$$

If the coefficients of t and the constant terms are equated in each equation, four linear equations in the unknowns a, b, c, and d are obtained. These are

$$a - 3b \qquad\quad - 2d = 0,$$

$$3a \qquad\quad + 2c \qquad\quad = -7,$$

$$b - c - 4d = 3,$$

$$a \qquad\quad - 4c \qquad\quad = 0,$$

and the solution is

$$a = -2, \qquad b = -\frac{3}{14}, \qquad c = -\frac{1}{2}, \qquad d = -\frac{19}{28}.$$

Hence

$$\mathbf{y}_p(t) = \begin{bmatrix} -2t - \frac{3}{14} \\ -\frac{1}{2}t - \frac{19}{28} \end{bmatrix}. \quad \square$$

Example 2 points out one of the drawbacks of the method, namely, that for even the simplest inhomogeneous terms the algebraic calculations become formidable. For instance, in the examples above if

$$\mathbf{f}(t) = \begin{bmatrix} f_1(t) \\ f_2(t) \end{bmatrix} = \begin{bmatrix} 2t \\ \sin t \end{bmatrix},$$

the trial solution would have the form

$$\mathbf{y}_p(t) = \begin{bmatrix} y_1(t) \\ y_2(t) \end{bmatrix} = \begin{bmatrix} at + b + c \cos t + d \sin t \\ et + f + g \cos t + h \sin t \end{bmatrix}.$$

After substitution into the differential equation and equating constant terms and the coefficients of t, $\sin t$, and $\cos t$, respectively, one would have eight linear equations in the eight unknowns a, b, c, d, e, f, g, and h.

Also note that if $f(t)$ contains a term $e^{\lambda t}$ and λ is an eigenvalue of A, then this is the case of resonance and the trial solutions will contain terms of the form $(at + b)e^{\lambda t}$ in each component. This again will increase the level of algebraic difficulties, so we must conclude that the method of "judicious guessing" should only be used in the simplest cases. Therefore, a general technique must be developed; such a technique will depend on knowing e^{tA} $= \Phi(t)$, the fundamental matrix of the homogeneous system.

First observe that e^{tA} satisfies the following property:

For any two real numbers t and s,

$$e^{tA}e^{sA} = e^{sA}e^{tA} = e^{(t+s)A}. \qquad (5.4.2)$$

The above relation can be proved by substituting the series representations for e^{tA} and e^{sA} and comparing like powers of A. It is not a completely trivial assertion since, in general, if A and B are $n \times n$ matrices,

$$e^A e^B \neq e^B e^A \neq e^{A+B},$$

in contrast to the scalar case. From relation (5.4.2) it follows by letting $s =$

$-t$ that

$$e^{tA}e^{-tA} = e^{-tA}e^{tA} = e^{(t-t)A} = e^{0A} = I,$$

or equivalently

$$\Phi(t)\,\Phi(-t) = \Phi(-t)\,\Phi(t) = I.$$

In the terminology of linear algebra, these last relations state that $e^{-tA} = \Phi(-t)$ is the inverse matrix of $e^{tA} = \Phi(t)$.

Example 3 For the given A, find $\Phi(t)$ and verify that $\Phi(t)\,\Phi(-t) = I$.

a) If

$$A = \begin{bmatrix} 0 & 1 \\ -1 & 0 \end{bmatrix},$$

then

$$e^{tA} = \Phi(t) = \begin{bmatrix} \cos t & \sin t \\ -\sin t & \cos t \end{bmatrix},$$

and therefore

$$\Phi(-t) = \begin{bmatrix} \cos(-t) & \sin(-t) \\ -\sin(-t) & \cos(-t) \end{bmatrix} = \begin{bmatrix} \cos t & -\sin t \\ \sin t & \cos t \end{bmatrix}.$$

Thus

$$\Phi(t)\,\Phi(-t) = \begin{bmatrix} \cos^2 t + \sin^2 t & -\cos t \sin t + \sin t \cos t \\ -\sin t \cos t + \cos t \sin t & \sin^2 t + \cos^2 t \end{bmatrix}$$

$$= \begin{bmatrix} 1 & 0 \\ 0 & 1 \end{bmatrix} = I.$$

b) If

$$A = \begin{bmatrix} 2 & 0 & 0 \\ 0 & 1 & 0 \\ 0 & 0 & -3 \end{bmatrix},$$

then

$$e^{tA} = \Phi(t) = \begin{bmatrix} e^{2t} & 0 & 0 \\ 0 & e^t & 0 \\ 0 & 0 & e^{-3t} \end{bmatrix}$$

and therefore

$$\Phi(-t) = \begin{bmatrix} e^{-2t} & 0 & 0 \\ 0 & e^{-t} & 0 \\ 0 & 0 & e^{3t} \end{bmatrix}.$$

Then

$$\Phi(t)\,\Phi(-t) = \begin{bmatrix} e^{2t}e^{-2t} & 0 & 0 \\ 0 & e^{t}e^{-t} & 0 \\ 0 & 0 & e^{-3t}e^{3t} \end{bmatrix} = \begin{bmatrix} 1 & 0 & 0 \\ 0 & 1 & 0 \\ 0 & 0 & 1 \end{bmatrix} = I.$$

The fact that the inverse of the matrix $\Phi(t)$ can be obtained by merely changing the sign of t is an extremely convenient property, as anyone who has computed the inverse of a square matrix will know! □

We are now ready to develop the *variation of parameters* formula for finding a particular solution of the initial value problem

$$\mathbf{y}' = A\mathbf{y} + \mathbf{f}(t), \qquad \mathbf{y}(t_0) = \mathbf{y}^0. \tag{5.5.3}$$

From the previous discussion we know that any solution of the homogeneous equation $\mathbf{y}' = A\mathbf{y}$ can be written as $\mathbf{y}(t) = e^{tA}\mathbf{c} = \Phi(t)\mathbf{c}$, where $\mathbf{c}$ is a constant vector. The *basic assumption* of the method of variation of parameters is that a particular solution of the inhomogeneous equation can be written as

$$\mathbf{y}_p(t) = e^{tA}\mathbf{c}(t), \tag{5.5.4}$$

where $\mathbf{c}(t)$ is a *time-dependent vector*.

To see where the assumption leads, substitute the above expression for $\mathbf{y}_p(t)$ into (5.5.3) to obtain

$$\mathbf{y}_p'(t) = Ae^{tA}\mathbf{c}(t) + e^{tA}\mathbf{c}'(t) = A\mathbf{y}_p(t) + \mathbf{f}(t) = Ae^{tA}\mathbf{c}(t) + \mathbf{f}(t).$$

Cancelling the term $Ae^{tA}\mathbf{c}(t)$ and multiplying both sides of the resulting equation by e^{-tA} gives

$$e^{-tA}[e^{tA}\mathbf{c}'(t)] = e^{-tA}e^{tA}\mathbf{c}'(t) = I\mathbf{c}'(t) = \mathbf{c}'(t) = e^{-tA}\mathbf{f}(t).$$

This is the simplest of all first order differential equations and it can be integrated to find $\mathbf{c}(t)$:

$$\mathbf{c}(t) = \int_{t_0}^{t} e^{-sA}\mathbf{f}(s)\,ds.$$

Note that the lower limit of integration is chosen conveniently so that $\mathbf{c}(t_0) = 0$, although this is not necessary; any antiderivative of $\mathbf{c}'(t)$ would suffice.

Finally, if the expression for $\mathbf{c}(t)$ is substituted into (5.5.4), we have

$$\mathbf{y}_p(t) = e^{tA} \int_{t_0}^{t} e^{-sA}\mathbf{f}(s)\,ds. \tag{5.5.5}$$

It can be verified directly that $\mathbf{y}_p(t)$ is the solution of the initial value problem

$$\mathbf{y}'(t) = A\mathbf{y} + \mathbf{f}(t), \qquad \mathbf{y}(t_0) = \mathbf{0}.$$

It is obvious that $\mathbf{y}_p(t_0) = \mathbf{0}$. Differentiating (5.5.5) gives

$$\mathbf{y}_p'(t) = Ae^{tA} \int_{t_0}^{t} e^{-sA}\mathbf{f}(s) \, ds + e^{tA}e^{-tA}\mathbf{f}(t) = A\mathbf{y}_p(t) + \mathbf{f}(t),$$

hence the assertion is verified and $\mathbf{y}_p(t)$ is a particular solution.

The solution of the original initial value problem (5.5.3) will be the sum of a particular solution and a solution of the homogeneous equation. Since $\mathbf{y}_p(t_0) = \mathbf{0}$ and a solution of $\mathbf{y}' = A\mathbf{y}$, $\mathbf{y}(t_0) = \mathbf{y}^0$ is $\mathbf{y}(t) = e^{(t-t_0)A}\mathbf{y}^0 = \Phi(t - t_0)\mathbf{y}^0$, one can now state the following algorithm:

Variation of parameters algorithm. The solution of the inhomogeneous initial value problem

$$\mathbf{y}' = A\mathbf{y} + \mathbf{f}(t), \qquad \mathbf{y}(t_0) = \mathbf{y}^0$$

can be found as follows:

1. Find the fundamental matrix $\Phi(t) = e^{tA}$ of the homogeneous system $\mathbf{y}' = A\mathbf{y}$ by either the eigenvalue–eigenvector method or by using the Laplace transform.

2. Compute, if possible, the integral

$$\int_{t_0}^{t} e^{-sA}\mathbf{f}(s) \, ds = \int_{t_0}^{t} \Phi(-s) \, \mathbf{f}(s) \, ds.$$

3. A particular solution $\mathbf{y}_p(t)$ satisfying $\mathbf{y}_p(t_0) = \mathbf{0}$ is given by the expression

$$\mathbf{y}_p(t) = e^{tA} \int_{t_0}^{t} e^{-sA} \, \mathbf{f}(s) \, ds = \Phi(t) \int_{t_0}^{t} \Phi(-s) \, \mathbf{f}(s) \, ds.$$

4. The solution of the initial value problem is

$$\mathbf{y}(t) = e^{(t-t_0)A}\mathbf{y}^0 + \mathbf{y}_p(t) = \Phi(t - t_0)\mathbf{y}^0 + \mathbf{y}_p(t),$$

where the first term is the solution of the homogeneous equation satisfying the given initial conditions.

From our previous discussion it follows that equivalent expressions for the solution are

$$\mathbf{y}(t) = e^{(t-t_0)A}\mathbf{y}^0 + \int_{t_0}^{t} e^{(t-s)A}\mathbf{f}(s) \, ds$$

or

$$\mathbf{y}(t) = \Phi(t - t_0)\mathbf{y}^0 + \int_{t_0}^{t} \Phi(t - s) \, \mathbf{f}(s) \, ds.$$

The last expression contains a convolution integral whose significance will be pointed out shortly.

Example 4   Solve

$$y_1' = 2y_1 + t, \qquad\qquad y_1(0) = 2,$$
$$y_2' = -3y_2 + \sin t, \qquad y_2(0) = -1.$$

The example can be solved directly, but its simplicity makes it easier to show how to use the variation of parameters formula. Here $t_0 = 0$,

$$A = \begin{bmatrix} 2 & 0 \\ 0 & -3 \end{bmatrix}, \qquad \mathbf{f}(t) = \begin{bmatrix} t \\ \sin t \end{bmatrix}, \qquad \mathbf{y}^0 = \begin{bmatrix} 2 \\ -1 \end{bmatrix},$$

and therefore

$$e^{tA} = \begin{bmatrix} e^{2t} & 0 \\ 0 & e^{-3t} \end{bmatrix} \qquad \text{and} \qquad e^{-sA} = \begin{bmatrix} e^{-2s} & 0 \\ 0 & e^{3s} \end{bmatrix}.$$

Hence

$$\int_0^t e^{-sA}\mathbf{f}(s)\, ds = \int_0^t \begin{bmatrix} e^{-2s} & 0 \\ 0 & e^{3s} \end{bmatrix} \begin{bmatrix} s \\ \sin s \end{bmatrix} ds$$

$$= \int_0^t \begin{bmatrix} se^{-2s} \\ e^{3s}\sin s \end{bmatrix} ds = \begin{bmatrix} -\tfrac{1}{2}te^{-2t} - \tfrac{1}{4}e^{-2t} + \tfrac{1}{4} \\ \tfrac{3}{10}e^{3t}\sin t - \tfrac{1}{10}e^{3t}\cos t + \tfrac{1}{10} \end{bmatrix},$$

where the integration of a vector is accomplished by integrating each component. A particular solution is therefore

$$\mathbf{y}_p(t) = e^{tA} \int_0^t e^{-sA}\mathbf{f}(s) = \begin{bmatrix} y_{1p}(t) \\ y_{2p}(t) \end{bmatrix}$$

$$= \begin{bmatrix} e^{2t} & 0 \\ 0 & e^{-3t} \end{bmatrix} \begin{bmatrix} -\tfrac{1}{2}te^{-2t} - \tfrac{1}{4}e^{-2t} + \tfrac{1}{4} \\ \tfrac{3}{10}e^{3t}\sin t - \tfrac{1}{10}e^{3t}\cos t + \tfrac{1}{10} \end{bmatrix}$$

$$= \begin{bmatrix} -\tfrac{1}{2}t - \tfrac{1}{4} + \tfrac{1}{4}e^{2t} \\ \tfrac{3}{10}\sin t - \tfrac{1}{10}\cos t + \tfrac{1}{10}e^{-3t} \end{bmatrix}.$$

Since

$$e^{tA}\mathbf{y}^0 = \begin{bmatrix} e^{2t} & 0 \\ 0 & e^{-3t} \end{bmatrix} \begin{bmatrix} 2 \\ -1 \end{bmatrix} = \begin{bmatrix} 2e^{2t} \\ -e^{-3t} \end{bmatrix},$$

this vector is added to the previous one to get the required solution

$$\mathbf{y}(t) = \begin{bmatrix} y_1(t) \\ y_2(t) \end{bmatrix} = \begin{bmatrix} \tfrac{9}{4}e^{2t} - \tfrac{1}{2}t - \tfrac{1}{4} \\ -\tfrac{9}{10}e^{-3t} + \tfrac{3}{10}\sin t - \tfrac{1}{10}\cos t \end{bmatrix}. \qquad \square$$

Example 5   Find a particular solution of

$$y_1' = 2y_1 - y_2 + 1,$$
$$y_2' = 4y_1 - 2y_2 + t^{-1}.$$

This example is an excellent one for the method of variation of parameters since, because of the presence of the term t^{-1}, neither judicious guessing nor the use of the Laplace transform (to be discussed shortly) will work. The first task is to find e^{tA}. Since

$$A = \begin{bmatrix} 2 & -1 \\ 4 & -2 \end{bmatrix}$$

has a double eigenvalue $\lambda = 0$, a solution of the homogeneous problem is of the form

$$y(t) = \begin{bmatrix} y_1(t) \\ y_2(t) \end{bmatrix} = \begin{bmatrix} a \\ b \end{bmatrix} + t \begin{bmatrix} c \\ d \end{bmatrix} = \begin{bmatrix} a + ct \\ b + dt \end{bmatrix}.$$

The first column $\Phi^1(t)$ of e^{tA} is found by solving

$$y_1' = 2y_1 - y_2, \qquad y_1(0) = 1,$$

$$y_2' = 4y_1 - 2y_2, \qquad y_2(0) = 0.$$

The initial condition implies that $a = 1$, $b = 0$, and substituting $y_1(t) = 1 + ct$, $y_2(t) = dt$ into the differential equations gives $d = 2c$ and $c = 2$. Therefore,

$$\Phi^1(t) = \begin{bmatrix} 1 + 2t \\ 4t \end{bmatrix}.$$

Now, changing the initial conditions to

$$y_1(0) = 0, \qquad y_2(0) = 1,$$

implies that $a = 0$, $b = 1$, and substitution gives $d = 2c$ and $c = -1$. Therefore the second column of e^{tA} is

$$\Phi^2(t) = \begin{bmatrix} -t \\ 1 - 2t \end{bmatrix}$$

and

$$e^{tA} = \begin{bmatrix} 1 + 2t & -t \\ 4t & 1 - 2t \end{bmatrix}, \qquad e^{-sA} = \begin{bmatrix} 1 - 2s & s \\ -4s & 1 + 2s \end{bmatrix}.$$

Since $\mathbf{f}(t) = \begin{bmatrix} 1 \\ t^{-1} \end{bmatrix}$, any value for t_0 may be selected except $t_0 = 0$ because of the presence of t^{-1}; for convenience we choose $t_0 = 1$. Hence

$$\int_1^t e^{-sA}\mathbf{f}(s)\, ds = \int_1^t \begin{bmatrix} 1 - 2s & s \\ -4s & 1 + 2s \end{bmatrix} \begin{bmatrix} 1 \\ s^{-1} \end{bmatrix} ds$$

$$= \int_1^t \begin{bmatrix} 2 - 2s \\ -4s + s^{-1} + 2 \end{bmatrix} ds = \begin{bmatrix} 2t - t^2 - 1 \\ -2t^2 + \ln t + 2t \end{bmatrix},$$

and a particular solution is

$$\mathbf{y}_p(t) = e^{tA} \int_1^t e^{-sA}\mathbf{f}(s) \, ds = \begin{bmatrix} y_{1p}(t) \\ y_{2p}(t) \end{bmatrix}$$

$$= \begin{bmatrix} 1 + 2t & -t \\ 4t & 1 - 2t \end{bmatrix} \begin{bmatrix} 2t - t^2 - 1 \\ -2t^2 + \ln t + 2t \end{bmatrix}$$

$$= \begin{bmatrix} -1 + t^2 - t \ln t \\ -2t + 2t^2 + (1 - 2t) \ln t \end{bmatrix}.$$

A sigh of relief is not uncommon after one has solved a problem by the method of variation of parameters! □

In the last example it was mentioned that one could also use the Laplace transform to solve an inhomogeneous system when the inhomogeneous term $\mathbf{f}(t)$ is Laplace transformable. It is an extremely effective method and avoids in part the lengthy integrations of the variation of parameters method. The price to be paid is the sometimes tedious algebraic manipulations and partial fractions expansions. We illustrate the method with an example.

Example 6   Solve

$$y_1' = -4y_1 - y_2 + e^{-t}, \qquad y_1(0) = 1$$

$$y_2' = y_1 - 2y_2 + \sin 2t, \qquad y_2(0) = 2.$$

Let $Y_1(s)$ and $Y_2(s)$ be the Laplace transforms of $y_1(t)$ and $y_2(t)$. Apply the transform to each equation to obtain, after using the differentiation formula,

$$sY_1(s) - 1 = -4Y_1(s) - Y_2(s) + \frac{1}{s + 1},$$

$$sY_2(s) - 2 = Y_1(s) - 2Y_2(s) + \frac{2}{s^2 + 4},$$

or, equivalently,

$$(s + 4)Y_1(s) + Y_2(s) = 1 + \frac{1}{s + 1}$$

$$-Y_1(s) + (s + 2)Y_2(s) = 2 + \frac{2}{s^2 + 4}.$$

Solving this system of equations for $Y_1(s)$ and $Y_2(s)$ gives

$$Y_1(s) = \frac{s}{(s + 3)^2} + \frac{s + 2}{(s + 3)^2(s + 1)} - \frac{2}{(s + 3)^2(s^2 + 4)},$$

$$Y_2(s) = \frac{2s + 9}{(s + 3)^2} + \frac{1}{(s + 3)^2(s + 1)} + \frac{2s + 8}{(s + 3)^2(s^2 + 4)}.$$

Each of the above terms must be expanded in a partial fraction expansion by using Heaviside's formulas or otherwise. After a reasonable amount of time and scratch paper, one obtains

$$Y_1(s) = \frac{459}{676} \cdot \frac{1}{s+3} - \frac{69}{26} \cdot \frac{1}{(s+3)^2} + \frac{1}{4} \cdot \frac{1}{s+1}$$

$$+ \frac{12}{169} \cdot \frac{s}{s^2+4} - \frac{5}{169} \cdot \frac{2}{s^2+4},$$

$$Y_2(s) = \frac{1335}{676} \cdot \frac{1}{s+3} + \frac{69}{26} \cdot \frac{1}{(s+3)^2} + \frac{1}{4} \cdot \frac{1}{s+1}$$

$$- \frac{38}{169} \cdot \frac{s}{s^2+4} + \frac{44}{169} \cdot \frac{2}{s^2+4}.$$

Applying the inverse transform gives the desired answer

$$y_1(t) = \frac{459}{676} e^{-3t} - \frac{69}{26} te^{-3t} + \frac{1}{4} e^{-t} + \frac{12}{169} \cos 2t - \frac{5}{169} \sin 2t,$$

$$y_2(t) = \frac{1339}{676} e^{-3t} + \frac{69}{26} te^{-3t} + \frac{1}{4} e^{-t} - \frac{38}{169} \cos 2t + \frac{44}{169} \sin 2t. \ \square$$

The partial fractions expansions and arithmetic in the last example were lengthy, but using the variation of parameters formula one would get the following expression for a particular solution:

$$\begin{bmatrix} e^{-3t} - te^{-3t} & -te^{-3t} \\ te^{-3t} & e^{-3t} + te^{-3t} \end{bmatrix} \int_0^t \begin{bmatrix} e^{2s} + se^{2s} - se^{3s} \sin 2s \\ -se^{2s} - (e^{3s} - se^{3s}) \sin 2s \end{bmatrix} ds.$$

Computationally, the choice of method is a toss-up with the transform method possibly having a slight edge. Note the special advantage in using the Laplace transform of not having to find the fundamental matrix e^{tA}.

In the discussion preceding the examples we gave the following alternative expression for a particular solution when $t_0 = 0$:

$$\mathbf{y}_p(t) = \int_0^t \Phi(t-s) \, \mathbf{f}(s) \, ds,$$

where $\Phi(t) = e^{tA}$ is the fundamental matrix associated with the matrix A. Observe that $\mathbf{y}_p(t)$ satisfies the initial condition $\mathbf{y}_p(0) = \mathbf{0}$ and can be written simply as the convolution integral

$$\mathbf{y}_p(t) = \Phi * \mathbf{f}.$$

To conclude this section we will derive this expression by using transform techniques.

First of all, recall that the fundamental matrix $\Phi(t)$ satisfies the matrix differential equation and initial condition

$$\Phi' = A\Phi, \qquad \Phi(0) = I.$$

If $\hat{\Phi}(s)$ is the Laplace transform of $\Phi(t)$, the differentiation formula of the Laplace transform can be applied to obtain

$$s\hat{\Phi}(s) - I = A\hat{\Phi}(s). \tag{5.5.6}$$

Note that the Laplace transform of the matrix $\Phi(t)$ will just be the matrix whose components are the transforms of the individual entries of $\Phi(t)$. For example: If

$$A = \begin{bmatrix} -3 & 0 \\ 0 & 2 \end{bmatrix},$$

then

$$\Phi(t) = \begin{bmatrix} e^{-3t} & 0 \\ 0 & e^{2t} \end{bmatrix},$$

hence

$$\hat{\Phi}(s) = \begin{bmatrix} \dfrac{1}{s+3} & 0 \\ 0 & \dfrac{1}{s-2} \end{bmatrix}$$

and

$$s\hat{\Phi}(s) - I = \begin{bmatrix} \dfrac{s}{s+3} & 0 \\ 0 & \dfrac{s}{s-2} \end{bmatrix} - \begin{bmatrix} 1 & 0 \\ 0 & 1 \end{bmatrix}$$

$$= \begin{bmatrix} \dfrac{-3}{s+3} & 0 \\ 0 & \dfrac{2}{s-2} \end{bmatrix}$$

$$= \begin{bmatrix} -3 & 0 \\ 0 & 2 \end{bmatrix} \begin{bmatrix} \dfrac{1}{s+3} & 0 \\ 0 & \dfrac{1}{s-2} \end{bmatrix} = A\hat{\Phi}(s).$$

Returning to Eq. (5.5.6), we see that it can be rewritten as $(sI - A)\hat{\Phi}(s) = I$, which implies that

$$\hat{\Phi}(s) = (sI - A)^{-1}. \tag{5.5.7}$$

This says that $\hat{\Phi}(s)$, the Laplace transform of $\Phi(t)$, is the inverse matrix of the matrix $sI - A$. In the above example, for instance,

$$sI - A = s \begin{bmatrix} 1 & 0 \\ 0 & 1 \end{bmatrix} - \begin{bmatrix} -3 & 0 \\ 0 & 2 \end{bmatrix} = \begin{bmatrix} s+3 & 0 \\ 0 & s-2 \end{bmatrix}$$

and

$$(sI - A)^{-1} = \begin{bmatrix} \dfrac{1}{s+3} & 0 \\ 0 & \dfrac{1}{s-2} \end{bmatrix} = \hat{\Phi}(s).$$

The relation (5.5.7) is what is needed to find an expression for $\mathbf{y}_p(t)$, the particular solution.

Recall that $\mathbf{y}_p(t)$ satisfies the initial value problem

$$\mathbf{y}_p' = A\mathbf{y}_p + \mathbf{f}(t), \qquad \mathbf{y}_p(0) = 0,$$

so if $\mathbf{f}(t)$ is Laplace transformable with Laplace transform $\mathbf{F}(s)$, then

$$s\mathbf{Y}_p(s) - \mathbf{0} = A\mathbf{Y}_p(s) + \mathbf{F}(s),$$

where $\mathbf{Y}_p(s)$ is the transform of $\mathbf{y}_p(t)$. The last relation implies that

$$(sI - A)\mathbf{Y}_p(s) = \mathbf{F}(s)$$

or

$$\mathbf{Y}_p(s) = (sI - A)^{-1}\mathbf{F}(s).$$

But, by using the relation (5.5.7), we see that this is equivalent to the expression

$$\mathbf{Y}_p(s) = \hat{\Phi}(s)\,\mathbf{F}(s).$$

The convolution property states that the inverse transform of the product of two transforms is the convolution integral of their inverse transforms, hence

$$\mathbf{y}_p(t) = \Phi * f = \int_0^t \Phi(t - s)\, f(s)\, ds.$$

This is the result we wanted.

In the language of systems analysis given in Chapter 4, the last two relations state respectively that for the system at rest

$$\mathbf{y}' = A\mathbf{y} + \mathbf{f}(t), \qquad \mathbf{y}(0) = \mathbf{0},$$

the following is true.

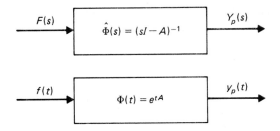

Figure 5.5

In the s-domain, $\Phi(s) = (sI - A)^{-1}$ is the transfer function and $\mathbf{Y}_p(s)$ is the output corresponding to input $\mathbf{F}(s)$.

In the t- domain, $\Phi(t) = e^{tA} = L^{-1}\{\Phi(s)\}$ is the weighting function and $\mathbf{y}_p(t)$ is the zero state output corresponding to the input $\mathbf{f}(t)$.

This can be represented diagrammatically as shown in Fig. 5.5, where the first box represents multiplication of the input by the transfer function and the second box represents the convolution of the input with the weighting function.

Note that the above analysis has given us another way to compute the fundamental matrix $\Phi(t)$. Namely, compute

$$(sI - A)^{-1} = L\{\Phi(t)\} = \hat{\Phi}(s)$$

and then compute

$$\Phi(t) = L^{-1}\{\hat{\Phi}(s)\}.$$

Aside from the algebraic problems in computing the inverse transform, the drawback is in the computing of the matrix inverse $(sI - A)^{-1}$. For dimensions $n = 2$ or 3 the method can be very useful.

Example 7 In Example 5,

$$A = \begin{bmatrix} 2 & -1 \\ 4 & -2 \end{bmatrix},$$

hence

$$(sI - A) = \begin{bmatrix} s - 2 & 1 \\ -4 & s + 2 \end{bmatrix},$$

and

$$\hat{\Phi}(s) = (sI - A)^{-1} = \frac{1}{s^2}\begin{bmatrix} s+2 & -1 \\ 4 & s-2 \end{bmatrix} = \begin{bmatrix} \dfrac{1}{s} + \dfrac{2}{s^2} & -\dfrac{1}{s^2} \\ \dfrac{4}{s^2} & \dfrac{1}{s} - \dfrac{2}{s^2} \end{bmatrix}.$$

Consequently,

$$\Phi(t) = L^{-1}\{\hat{\Phi}(s)\} = \begin{bmatrix} 1+2t & -t \\ 4t & 1-2t \end{bmatrix}. \quad \square$$

Example 8 If $A = \begin{bmatrix} 3 & -4 \\ 1 & -1 \end{bmatrix}$, then $(sI - A) = \begin{bmatrix} s-3 & 4 \\ -1 & s+1 \end{bmatrix}$ and

$$(sI - A)^{-1} = \frac{1}{(s-1)^2}\begin{bmatrix} s+1 & -4 \\ 1 & s-3 \end{bmatrix}$$

$$= \begin{bmatrix} \dfrac{1}{s-1} + \dfrac{2}{(s-1)^2} & \dfrac{-4}{(s-1)^2} \\ \dfrac{1}{(s-1)^2} & \dfrac{1}{s-1} - \dfrac{2}{(s-1)^2} \end{bmatrix}.$$

Hence

$$\Phi(t) = \begin{bmatrix} e^t + 2te^t & -4te^t \\ te^t & e^t - 2te^t \end{bmatrix}. \quad \square$$

As the examples show, the method avoids the need to compute solutions of initial value problems to find each column of $\Phi(t)$.

EXERCISES 5.5

Solve the following initial value problems by using the variation of parameters method. The fundamental matrices have been found in Exercises 5.4.

1. $y' = \begin{bmatrix} 2 & 1 \\ 3 & 4 \end{bmatrix} y + \begin{bmatrix} 2e^{2t} \\ -2e^{2t} \end{bmatrix}$, $y(0) = \begin{bmatrix} 3 \\ -4 \end{bmatrix}$

2. $y' = \begin{bmatrix} 1 & 3 \\ 12 & 1 \end{bmatrix} y + \begin{bmatrix} 3e^{7t} \\ 4 \end{bmatrix}$, $y(0) = \begin{bmatrix} 1 \\ 2 \end{bmatrix}$

3. $y' = \begin{bmatrix} -1 & -5 \\ 1 & 3 \end{bmatrix} y + \begin{bmatrix} 3 \\ 5\cos t \end{bmatrix}$, $y(\pi) = \begin{bmatrix} 0 \\ 0 \end{bmatrix}$

4. $y' = \begin{bmatrix} 3 & 1 \\ -2 & 1 \end{bmatrix} y + \begin{bmatrix} e^{2t} \\ -e^{2t} \end{bmatrix}$, $y(0) = \begin{bmatrix} 2 \\ -1 \end{bmatrix}$

5. $y' = \begin{bmatrix} 5 & -1 \\ 1 & 3 \end{bmatrix} y + \begin{bmatrix} 2e^{-t} \\ e^t \end{bmatrix}$, $y(2) = \begin{bmatrix} -3 \\ 2 \end{bmatrix}$

6. $y' = \begin{bmatrix} -4 & -1 \\ 1 & -2 \end{bmatrix} y + \begin{bmatrix} e^{2t} \\ 4e^{3t} \end{bmatrix}$, $y(-2) = \begin{bmatrix} 3 \\ 4 \end{bmatrix}$

Find the transfer function (the Laplace transform $\hat{\Phi}(s)$ of the fundamental matrix $\Phi(t)$) by computing

$(sI - A)^{-1}$ directly for the following matrices A. Then find $\Phi(t)$ by using the inverse Laplace transform.

7. $A = \begin{bmatrix} 3 & 0 \\ 0 & 1 \end{bmatrix}$

8. $A = \begin{bmatrix} 2 & 1 \\ 4 & 2 \end{bmatrix}$

9. $A = \begin{bmatrix} 1 & 5 \\ -2 & -1 \end{bmatrix}$

10. $A = \begin{bmatrix} 1 & 2 \\ -4 & -3 \end{bmatrix}$

11. $A = \begin{bmatrix} 1 & 0 & 0 \\ 0 & 2 & 1 \\ 0 & 0 & 2 \end{bmatrix}$

12. $A = \begin{bmatrix} 3 & 1 & 0 \\ 0 & 3 & 1 \\ 0 & 0 & 3 \end{bmatrix}$

MISCELLANEOUS EXERCISES

Mass–spring systems without damping, such as those described in Example 3 of Section 5.1, can be expressed by systems of second order differential equations

$$\ddot{\mathbf{y}} + A\mathbf{y} = \mathbf{f}(t),$$

where $\mathbf{y} = \text{col}(y_1, \dots, y_n)$ describes the configuration of the system, $\mathbf{f}(t) = \text{col}(f_1(t), \dots, f_n(t))$ is the vector of forcing functions, A is an $n \times n$ matrix, and n is the number of degrees of freedom.

For the homogeneous equation $\mathbf{f}(t) = \mathbf{0}$, setting $\mathbf{y}(t) = e^{i\lambda t}\mathbf{b}$ leads to the algebraic system

$$(-\lambda^2 I + A)\mathbf{b} = 0,$$

which λ and $\mathbf{b}$ must satisfy. Hence $\mathbf{b}$ is a real eigenvector of A corresponding to the real eigenvalue λ^2, and the condition $\det(-\lambda^2 I + A) = 0$ must be satisfied to have nontrivial solutions. Each value λ^2 that satisfies the last equation will determine two real solutions: $(\cos \lambda t)\mathbf{b}$ and $(\sin \lambda t)\mathbf{b}$, and the general solution will be a linear combination of all such solutions.

5.1 a) Letting $\omega^2 = k/m$, show that for the system described in Example 3 of Section 5.1

$$A = \begin{bmatrix} 2\omega^2 & -\omega^2 \\ -\omega^2 & 2\omega^2 \end{bmatrix}.$$

b) Show that the eigenvalues of A are $\lambda_1 = \omega^2$ and $\lambda_2 = 3\omega^2$ with respective eigenvalues $\mathbf{b} = \text{col}(1,1)$ and $\mathbf{b} = \text{col}(1,-1)$. Write down the general solution of the system.

c) Solve the initial value problems:
   i) $\mathbf{y}(0) = \text{col}(a,a)$, $\dot{\mathbf{y}}(0) = \text{col}(0,0)$
   ii) $\mathbf{y}(0) = \text{col}(a,-a)$, $\dot{\mathbf{y}}(0) = \text{col}(0,0)$

5.2 Modify Example 3 of Section 5.1 so that the spring constants are k_1, k_2, k_3, and write down the corresponding second order system. Letting $k_1 = k_3$, $\omega^2 = k_1/m$, $\eta^2 = k_3/m$, $\sigma^2 = \omega^2 + 2\eta^2$, find the equations for the eigenvalues and eigenvectors of the second order system, and write down the general solution.

The Laplace transform can be used to find a particular solution of the nonhomogeneous second order system

$$\ddot{\mathbf{y}} + A\mathbf{y} = \mathbf{f}(t).$$

If $\mathbf{Y}(s)$ and $\mathbf{F}(s)$ are the transforms of $\mathbf{y}(t)$ and $\mathbf{f}(t)$, respectively, and the initial conditions are chosen as $\mathbf{y}(0) = \dot{\mathbf{y}}(0) = \mathbf{0}$, then the application of the transform gives

$$s^2\mathbf{Y}(s) + A\mathbf{Y}(s) = \mathbf{F}(s).$$

Therefore

$$\mathbf{Y}(s) = (s^2I + A)^{-1}\mathbf{F}(s) = \hat{U}(s)\mathbf{F}(s),$$

and if $U(t) = L^{-1}\{\hat{U}(s)\}$, then by the convolution formula a particular solution is

$$\mathbf{y}_p(t) = \int_0^t U(t - \tau)\mathbf{f}(\tau)\, d\tau.$$

5.3 For the second part of Problem 5.2, find the matrix $U(t)$ and use it to find a particular solution when $\mathbf{f}(t) = \mathrm{col}(\sin \omega t, \sin \sigma t)$.

The differential equation describing a forced mass–spring system

$$m_1\ddot{y} + k_1 y = F \sin \omega t$$

is in resonance if $\omega^2 = k_1/m_1$. The resonance frequency can be shifted by adding a small mass–spring system as a dynamic damper. This system is shown in Fig. 5.6 and the differential equations describing it are

$$m_1\ddot{y}_1 + k_1 y_1 - k_2(y_2 - y_1) = F \sin \omega t,$$

$$m_2\ddot{y}_2 + k_2(y_2 - y_1) = 0.$$

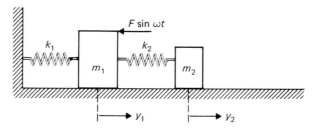

Figure 5.6

5.4 a) Letting $\mathbf{y} = \mathrm{col}(y_1, y_2)$ and $\alpha^2 = k_1/m_1$, $\beta^2 = k_2/m_2$, $\gamma^2 = k_2/m_1$, write the system in the form $\ddot{\mathbf{y}} + A\mathbf{y} = \mathbf{f}(t)$.

b) A particular solution can be found by letting $\mathbf{y}_p(t) = \mathbf{g} \sin \omega t$, substituting it into the system, and finding $\mathbf{g}$. Show that if β^2 is chosen to be equal to ω^2, the first component of the particular solution is zero.

c) The result in part (b) shows that the resonant oscillation is transferred to the dynamic damper by choosing $\alpha^2 = \beta^2$, that is, $k_1/m_1 = k_2/m_2$. This is the defining relation for a *tuned dynamic damper*. Find the eigenvalues λ^2 of A in this case.

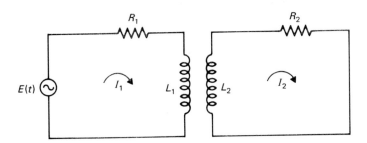

Figure 5.7

d) Let $\omega^2 = \alpha^2 = \beta^2 = 2$, $\gamma^2 = 1$, and $F/m_1 = 4$. Find a particular solution $\mathbf{y}_p(t) = \mathbf{g} \sin \sqrt{2}t$, then find the general solution. Use it to solve the initial value problem $\mathbf{y}(0) = \dot{\mathbf{y}}(0) = \mathbf{0}$.

In the circuit of Fig. 5.7, the currents I_1 and I_2 are coupled by the mutual inductance of the two coils. The system of differential equations describing the system are derived by applying Kirchhoff's voltage law to the two loops:

$$L_1 \frac{dI_1}{dt} - M \frac{dI_2}{dt} + R_1 I_1 = E(t),$$

$$-M \frac{dI_1}{dt} + L_2 \frac{dI_2}{dt} + R_2 I_2 = 0.$$

5.5 a) Assume that $M^2 \neq L_1 L_2$ and by solving the above system for dI_1/dt and dI_2/dt write the system in the normal form

$$\frac{d\mathbf{I}}{dt} = A\mathbf{I} + \mathbf{F}(t),$$

where $\mathbf{I} = \mathrm{col}(I_1, I_2)$. Find the characteristic equation of the matrix A.

b) Set $R_1 = R_2 = 15$ ohms, $L_1 = L_2 = 4$ henries, $M = 1$ henry, and $E(t) = 4 \sin 2t$ volts. Given the initial conditions $I_1(0) = I_2(0) = 0$ amperes, solve the initial value problem by using Laplace transforms.

c) An ideal transformer is described by the equation $M^2 = L_1 L_2$. Let $R_1 = R_2 = 1$ ohm, $L_1 = L_2 = M = 1$ henry, and $E(t) = 4 \sin t$ volts, and find the general solution of the system.

5.6 Time-varying resistors are employed, for instance, in circuits modeling carbon microphones. In the circuit that was described in Exercise 5.5, let $R_1 = 15(1 + p \sin \omega t)$ ohms, $|p| < 1$, $R_2 = 15$ ohms, $L_1 = L_2 = 4$ henries, $M = 1$ henry, $E(t) = 150$ volts. Assume that the initial conditions are $I_1(0) = I_2(0) = 0$ amperes.

a) Write a computer program (or adapt the model program in Chapter 3) using RK4 to compute an approximate solution on $0 \leq t \leq T$. Note that the system must be written in normal form.

b) Let $p = \omega = 0$ and generate an approximate solution on $0 \le t \le 4$ using a step size $h = 0.1$. Plot $I_1(t)$ and $I_2(t)$ versus t and $I_1(t)$ versus $I_2(t)$ on separate graphs.

c) Let $p = 0.5$, $\omega = 2$ and generate an approximate solution on $0 \le t \le 10$ using a step size $h = 0.1$. Plot $I_1(t)$ and $I_2(t)$ versus t and $I_1(t)$ versus $I_2(t)$ on separate graphs. The varying resistance *modulates* the constant ($p = 0$) currents.

5.7 Two tanks with inlet and outlet pipes are connected together as shown in Fig. 5.8. Each tank contains V gallons of brine and the flow rates have been selected to keep the volumes constant. At the start of the process, the brine in tank A contains Q_1 pounds of salt and that in tank B contains Q_2 pounds of salt. The concentration of the incoming mixture is $C_0(t)$ pounds of salt per gallon. Let C_1 and C_2 denote the concentrations (in pounds per gallon) of salt in tanks A and B, respectively. Instantaneous mixing is assumed.

a) Show that the k_i's must satisfy the relations $k_0 = k_2$, $k_1 = k_0 + k_3$, $k_1 = k_2 + k_3$, and that C_1 and C_2 satisfy the initial value problem:

$$\frac{d}{dt}(VC_1) = -k_1C_1 + (k_1 - k_2)C_2 + k_2C_0(t), \qquad C_1(0) = Q_1/V,$$

$$\frac{d}{dt}(VC_2) = k_1C_1 - k_1C_2, \quad C_2(0) = Q_2/V.$$

(The reader may wish to review the previous discussion of tank problems in Chapters 1 and 4.) Subsequent eigenvalue computations can be simplified by introducing a new independent variable $\tau = t/V$. This leads to the system:

$$\frac{dC_1}{d\tau} = -k_1C_1 + (k_1 - k_2)C_2 + k_2C_0(V\tau), \quad C_1(0) = Q_1/V,$$

$$\frac{dC_2}{d\tau} = k_1C_1 - k_1C_2, \quad C_2(0) = Q_2/V.$$

Let $V = 300$, $k_1 = 9$, $k_2 = 4$.

b) Find the fundamental matrix for the corresponding homogeneous system.

c) Write down the variations of parameters formula for a general $C_0(t)$ and general initial conditions.

d) Let $C_0(t) = 3e^{-t/100}$, $Q_1 = 300$ lb, $Q_2 = 1200$ lb, and solve the initial value

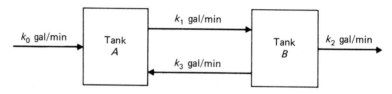

Figure 5.8

problem using the Laplace transform, using the method of judicious guessing, and evaluating the integral in (c).

5.8 Three tanks are connected in a closed system as shown in Fig. 5.9. Each tank contains V gallons of brine, and the flow rates have been selected to keep the volumes constant. Let C_i denote the concentration (in pounds per gallon) of salt in the ith tank, $i = 1, 2, 3$.

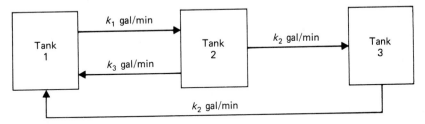

Figure 5.9

a) Derive the system of differential equations describing the system.

b) Show that $C_1 = C_2 = C_3 = a > 0$ is a solution for any positive number a, and therefore $\lambda = 0$ is an eigenvalue of the coefficient matrix.

c) Let $k_2 = 1$, $k_3 = b \geq 0$, and $k_1 = 1 + b$. Derive formulas for the nonzero eigenvalues and characterize them (real distinct, real equal, complex) as functions of the parameter b.

d) Show that $C_1(t) + C_2(t) + C_3(t) = $ constant for all t, and use this constraint to derive a two-dimensional system for $C_2(t)$ and $C_3(t)$.

e) Find the general solutions with

$$C_1(0) + C_2(0) + C_3(0) = 6$$

when
   i) $k_1 = 2$, $k_2 = k_3 = 1$
   ii) $k_1 = 3/2$, $k_2 = 1$, $k_3 = 1/2$
   iii) $k_1 = 1 + \sqrt{3}/2$, $k_2 = 1$, $k_3 = \sqrt{3}/2$.

f) If one of the pumps that runs at a constant rate is replaced by a variable speed pump, then it is not possible to solve the systems of differential equations exactly. Set $C_1(0) = 5$, $C_2(0) = C_3(0) = 0$, $k_1 = 2$ gal/min, $k_3 = 1$ gal/min, and $k_2 = 1 + 0.5 \cos 2t$ gal/min. Construct an approximate solution using RK4 with $h = 0.1$. Plot C_1 versus t. A computer-generated plot is given in Fig. 5.10. Compare your result to the solution corresponding to $k_2 = 1$ gal/min with the other data unchanged. Does a pulsating pump produce better mixing?

5.9 The biological processes of living organisms can often be modeled by splitting them into one or more distinct stages, or *compartments*. Each compartment is characterized by its constituent material or by its spatial assignment. It is assumed that the material in each compartment is homogeneous and that each compartment has a steady-state flux of material into and out of it.

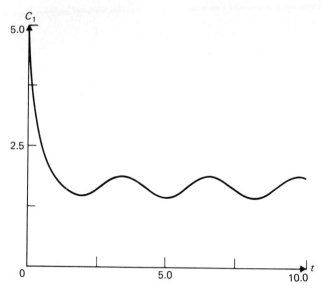

Figure 5.10   A plot of $C_1(t)$.

A compartment system is usually studied experimentally by injecting a pulse or tracer of labeled material into one or more of the compartments. This introduces a transient into the system, which can be used to study its steady-state properties. To accomplish this, the interactions between individual compartments must be determined experimentally.

As an example of this approach, suppose one wanted to analyze the elimination or clearance of creatinine in the dog. Creatinine is a nitrogenous waste product and is a fairly constant component of urine. The compartment system of Fig. 5.11 models this process and the experiment designed to study it: Here l_{21} is the fraction per unit time of creatinine in the plasma compartment transferred to the tissue compartment; l_{12} is the fraction per unit time of creatinine in the tissue compartment transferred to the plasma compartment; l_{01} is the fraction per unit time of creatinine eliminated from the system, and ρ is a bolus input (e.g., a simple injection into a vein) of creatinine introduced at time $t = 0$. It is assumed that the level of creatinine in the plasma at $t = 0$ is null.

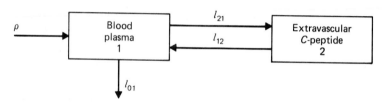

Figure 5.11

If x_1 is the concentration of creatinine in the plasma and x_2 is its concentration in the tissue, the compartment model is described by the following system:

$$\frac{dx_1}{dt} = -(l_{01} + l_{21})x_1 + l_{12}x_2, \qquad x_1(0) = \rho,$$

$$\frac{dx_2}{dt} = l_{21}x_1 - l_{12}x_2, \qquad x_2(0) = 0.$$

To determine the rates l_{ij}, an experiment was performed in which 2 gms of creatinine were injected into the veins of ten dogs, and blood samples were drawn and analyzed at specified times. The constants were found to be

$$l_{01} = 0.0395 \text{ min}^{-1}, \qquad l_{12} = 0.0479 \text{ min}^{-1}, \qquad l_{21} = 0.0391 \text{ min}^{-1}.$$

a) Find the fundamental solution matrix of the system.

b) Evaluate the solution at enough points to enable you to sketch a graph of the concentration of creatinine in the plasma and in the tissue for 10 minutes. (For a discussion of this experiment and compartment models see S. I. Rubinow, *Introduction to Mathematical Biology*, Wiley, New York, 1975, Chap. 3.)

5.10 Published data on the disappearance of connecting peptide, denoted by C-peptide, in the plasma of man indicate that its decay can be modeled by a two-compartment system consisting of an intravascular and an extravascular pool, as shown in Fig. 5.12.

a) Denoting the concentration of intravascular C-peptide by x_1 and that of extravascular C-peptide by x_2, write down the system of differential equations describing the model.

b) Find the fundamental matrix of the system.

c) Assuming $x_1(0) = 4.6$ pmol/mL, $x_2(0) = 0$, evaluate the solution at enough points to plot $x_1(t)$ for $0 \le t \le 60$ min.

(See R. P. Eaton, R. C. Allen, D. S. Schade, K. M. Erickson, and J. Standefer, *Prehepatic Insulin Production in Man: Kinetic Analysis Using Peripheral Connecting Peptide Behavior*, Jour. of Endocrinology and Metabolism, 51 (1980), pp. 520–528.)

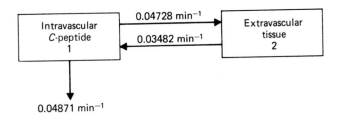

Figure 5.12

5.11 Show that $\mathbf{y}(t) = \sum_{n=0}^{\infty} t^{2n+1}\mathbf{a}^{(2n)}$ is a solution to

$$\frac{d\mathbf{y}}{dt} = \frac{1}{t}\begin{bmatrix} t^2 & 1 \\ 1 & -t^2 \end{bmatrix}\mathbf{y}, \quad (t > 0),$$

where $\mathbf{a}^{(0)} = A_0\mathbf{a}^{(0)}$, $[(2n + 1)I - A_0]\mathbf{a}^{(2n)} = A_2\mathbf{a}^{(2n-2)}$, $n = 1, 2, 3, \ldots$, and

$$A_0 = \begin{bmatrix} 0 & 1 \\ 1 & 0 \end{bmatrix}, \qquad A_2 = \begin{bmatrix} 1 & 0 \\ 0 & -1 \end{bmatrix}.$$

Then show that

$$\mathbf{a}^{(2n)} = \frac{1}{4n(n + 1)}\begin{bmatrix} (2n + 1) & 1 \\ 1 & -(2n + 1) \end{bmatrix}\mathbf{a}^{(2n-2)}, \quad n = 1, 2, 3, \ldots$$

5.12 Write a computer program to compute the vector

$$\mathbf{y}^{(M)}(t) = \sum_{n=0}^{M} t^{2n+1}\mathbf{a}^{(2n)},$$

where the $\mathbf{a}^{2n}$ are defined in Exercise 5.11. Take $\mathbf{a}^{(0)} = \text{col}(1, 1)$, $t = 2$, and find an M, so that convergence to four significant figures *seems* to be obtained.

5.13 Integrate with RK4:

$$\frac{d\mathbf{y}}{dt} = \frac{1}{t}\begin{bmatrix} t^2 & 1 \\ 1 & -t^2 \end{bmatrix}y$$

on $h \leq t \leq 2$ with

a) $\mathbf{y}(0) = \begin{bmatrix} h \\ h \end{bmatrix}$, $h = 0.1$, and $h = 0.2$;

b) $\mathbf{y}(0) = \begin{bmatrix} h \\ h \end{bmatrix} + \frac{1}{4}\begin{bmatrix} h^3 \\ -h^3 \end{bmatrix}$, $h = 0.1$, and $h = 0.2$.

c) Compare your answers with the result of Exercise 5.12. Which initial condition more closely matches the series solution? Why?

REFERENCES

1. E. Coddington and N. Levinson, *Theory of Ordinary Differential Equations*, McGraw-Hill, New York, 1975.

2. B. Noble, J. W. Daniel, *Applied Linear Algebra*, Prentice-Hall, Englewood Cliffs, N. J., 1979.

3. P. C. Shields, *Elementary Linear Algebra*, Worth Publishers, New York, 1980.

4. G. Strang, *Linear Algebra and Its Applications*, Academic Press, New York, 1980.

Nonconstant Coefficient Second Order Linear Equations and Series Solutions

6.1 INTRODUCTION

In Chapter 2, the general theory for the initial value problem

$$a(t)y'' + b(t)y' + c(t)y = 0,$$
$$y(t_0) = r, \qquad y'(t_0) = s,$$

was discussed. There it was assumed that $a(t)$, $b(t)$, and $c(t)$ were continuous functions on some nonempty interval containing t_0 and that $a(t)$ did not vanish at t_0. The theory stated that the solution $y(t)$ could be expressed uniquely as

$$y(t) = c_1 y_1(t) + c_2 y_2(t),$$

where $y_1(t)$ and $y_2(t)$ were a linearly independent pair of solutions of the differential equation, and the constants c_1 and c_2 depended on the initial values r and s. Furthermore, the existence of such a pair of linearly independent solutions $y_1(t)$ and $y_2(t)$ was also guaranteed by the theory.

Following this general discussion, the remainder of Chapter 2 was devoted to a discussion of the constant coefficient case

$$ay'' + by' + cy = 0.$$

In this case finding a pair of linearly independent solutions depended solely on solving the characteristic equation

$$a\lambda^2 + b\lambda + c = 0.$$

The type of solutions depended on the nature of the roots of the characteristic equation (real and unequal, real and equal, or complex) but the solutions could be *explicitly found*.

The reader may well ask whether this simplicity of the solution procedure carries over in some analogous fashion to the nonconstant coefficient case. The answer is a definite NO! With the exception of the special case of the Euler differential equation,

$$t^2 y'' + bty' + cy = 0$$

(see Example 1 and Exercise 4 of Section 2.6), there are no general techniques to reduce the solving of nonconstant coefficient linear differential equations to an algebraic process.

However, if we are willing to extend our notion of the solution of a differential equation to allow for the solution to be expressed as an infinite series, then there is a solution procedure which is applicable to a large class of nonconstant coefficient linear differential equations. The infinite series will be a power series in the independent variable, possibly multiplied by a known function, and the coefficients of the power series can be determined recursively. Given the generality of the problem, one could hardly wish for a more satisfactory outcome.

For instance, a solution of

$$y'' + \frac{1}{t} y' + y = 0$$

is given by the power series

$$y(t) = 1 - \frac{t^2}{2^2} + \frac{t^4}{2^4(2!)^2} - \frac{t^6}{2^6(3!)^2} + \cdots = \sum_{n=0}^{\infty} (-1)^n \frac{t^{2n}}{2^{2n}(n!)^2},$$

as we shall see later in this chapter. The function given by the power series, denoted by $J_0(t)$, is an oscillatory function with an infinite number of zeros on the positive axis, and there are tables and numerical procedures to estimate it for any value of its argument. This is not such a marked contrast from the familiar function $\cos t$, which is a solution of $y'' + y = 0$ and which can also be represented by a power series.

The function $J_0(t)$, called the *Bessel function of the first kind of order zero*, occurs in a number of problems in heat conduction in solids and vibration of membranes. It is an example of what are called *special functions*, many of which arise as series solutions of second order linear differential equations. A few of these special functions will be discussed at the end of this chapter—they are useful tools for the kitbag of any applied scientist.

6.2 SERIES SOLUTIONS—PART 1

In the remainder of this chapter we will denote the independent variable by x rather than t. Therefore the second order linear equation will be written as

$$a(x)y'' + b(x)y' + c(x)y = 0,$$

with solutions $y = y(x)$. The use of x rather than t is somewhat traditional, since many of the problems associated with series solutions or special functions arise from physical problems where x is a spatial rather than a temporal variable.

To motivate the use of infinite series, we start with a simple example that can be solved explicitly. Consider the first order differential equation

$$y' = 2xy,$$

whose general solution is $y(x) = Ae^{x^2}$, where A is an arbitrary constant. Assume that the solution can be represented by a convergent power series, hence

$$y(x) = \sum_{n=0}^{\infty} a_n x^n$$

and the task is to find the a_n.

Since a convergent power series can be differentiated term by term and also multiplied by $2x$, the series can be substituted into the differential equation to obtain

$$y'(x) = \sum_{n=0}^{\infty} na_n x^{n-1} = 2xy(x) = \sum_{n=0}^{\infty} 2a_n x^{n+1}.$$

Expanding both sides gives

$$[0 \cdot a_0 + 1 \cdot a_1 + 2a_2 x + 3a_3 x^2 + 4a_4 x^3 + 5a_5 x^4 + 6a_6 x^5 + \cdots]$$
$$= [2a_0 x + 2a_1 x^2 + 2a_2 x^3 + 2a_3 x^4 + 2a_4 x^5 + 2a_5 x^6 + \cdots],$$

and by comparing the coefficients of like powers of x we have

$$0 \cdot a_0 = 0, \quad 1 \cdot a_1 = 0, \quad 2a_2 = 2a_0, \quad 3a_3 = 2a_1,$$
$$4a_4 = 2a_2, \quad 5a_5 = 2a_3, \quad 6a_6 = 2a_4, \quad \cdots.$$

This implies that a_0 is arbitrary and

$$a_1 = 0, \quad a_2 = a_0, \quad a_3 = \frac{2}{3}a_1 = 0, \quad a_4 = \frac{a_2}{2} = \frac{a_0}{2},$$
$$a_5 = \frac{2}{5}a_3 = 0, \quad a_6 = \frac{a_4}{3} = \frac{a_0}{3 \cdot 2}, \quad \cdots;$$

hence

$$y(x) = a_0 \left[1 + x^2 + \frac{x^4}{2!} + \frac{x^6}{3!} + \cdots \right],$$

which is the first four terms of the Taylor series for $a_0 e^{x^2}$.

The reader can easily see that the general recursion relation for the a_n is $na_n = 2a_{n-2}$, $n = 2, 3, \ldots$. Since a_0 is arbitrary and a_1 is zero, this implies that

$$a_{2n+1} = 0, \quad n = 0, 1, 2, \ldots,$$

$$a_{2n} = \frac{2a_{2n-2}}{2n} = \frac{1}{n} \cdot \frac{2a_{2n-4}}{2n-2} = \frac{1}{n(n-1)} \cdot \frac{2a_{2n-6}}{2n-4} = \cdots = \frac{1}{n!} a_0.$$

Therefore

$$y(x) = \sum_{n=0}^{\infty} \frac{a_0}{n!} x^{2n} = a_0 \sum_{n=0}^{\infty} \frac{x^{2n}}{n!} = a_0 e^{x^2},$$

and the power series solution is exactly the Taylor series of the general solution.

We now return to the second order equation

$$y'' + p(x)y' + q(x)y = 0 \qquad (6.2.1)$$

and consider first the simplest case in which the coefficient functions $p(x)$ and $q(x)$ are smooth, well behaved functions in some neighborhood of a given point, which is assumed to be $x = 0$ for simplicity of notation. Consequently, we can assume that $p(x)$ and $q(x)$ have infinite power series representations

$$p(x) = \sum_{n=0}^{\infty} p_n x^n, \qquad q(x) = \sum_{n=0}^{\infty} q_n x^n,$$

where these series converge in some neighborhood of $x = 0$. If this is the case, $x = 0$ is said to be an *ordinary point* of the differential equation (6.2.1).

With the above assumptions it is natural to suppose that the solution $y(x)$ will also have a power series representation convergent in some neighborhood of $x = 0$, and therefore

$$y(x) = \sum_{n=0}^{\infty} a_n x^n. \qquad (6.2.2)$$

Since a power series can be differentiated term by term, the solution procedure is to substitute the series for $y(x)$ into (6.2.1), with $p(x)$ and $q(x)$ replaced by their power series representation, and compare coefficients of like powers of x. This will lead to a recursion relation by which the a_n's can be determined step by step, and possibly a general expression can be derived which will give all the a_n's. (In the case where the ordinary point is $x_0 \neq 0$, one would use instead $y(x) = \sum_{n=0}^{\infty} a_n(x - x_0)^n$, and the algebraic procedure is the same.)

This procedure is best illustrated by an example known as *Airy's equation:*

$$y'' - xy = 0. \tag{6.2.3}$$

From (6.2.2) we obtain the relations

$$y'(x) = \sum_{n=1}^{\infty} na_n x^{n-1}, \qquad y''(x) = \sum_{n=2}^{\infty} n(n-1)a_n x^{n-2}$$

through term-by-term differentiation. After multiplying the series for $y(x)$ by x, substitution into (6.2.3) gives

$$\sum_{n=2}^{\infty} n(n-1)a_n x^{n-2} - \sum_{n=0}^{\infty} a_n x^{n+1} = 0. \tag{6.2.4}$$

If just the first few terms of the solution are desired, one can expand the above sums to get

$$(2a_2 + 6a_3 x + 12a_4 x^2 + 20a_5 x^3 + 30a_6 x^4 + 42a_7 x^5 + \cdots)$$
$$- (a_0 x + a_1 x^2 + a_2 x^3 + a_3 x^4 + a_4 x^5 + \cdots) = 0.$$

Now, combining the coefficients of like powers of x and setting them equal to zero gives the relations

$$2a_2 = 0, \quad 6a_3 - a_0 = 0, \quad 12a_4 - a_1 = 0, \quad 20a_5 - a_2 = 0,$$
$$30a_6 - a_3 = 0, \quad 42a_7 - a_4 = 0, \quad \ldots,$$

and consequently

$$a_2 = 0, \quad a_3 = \frac{a_0}{6}, \quad a_4 = \frac{a_1}{12}, \quad a_5 = \frac{a_2}{20} = 0,$$

$$a_6 = \frac{a_3}{30} = \frac{a_0}{180}, \quad a_7 = \frac{a_4}{42} = \frac{a_1}{504}, \ldots.$$

Therefore,

$$y(x) = a_0 + a_1 x + 0 + \frac{a_0}{6} x^3 + \frac{a_1}{12} x^4 + 0 + \frac{a_0}{180} x^6 + \frac{a_1}{504} x^7 + \cdots$$

$$= a_0 \left[1 + \frac{1}{6} x^3 + \frac{1}{180} x^6 + \cdots \right] + a_1 \left[x + \frac{1}{12} x^4 + \frac{1}{504} x^7 + \cdots \right],$$

where a_0 and a_1 are arbitrary constants.

Examining the last expression we see that it can be written in the form

$$y(x) = a_0 y_1(x) + a_1 y_2(x),$$

where $y_1(x)$ will have a power series representation in $3n$th powers of x, where $n = 0, 1, 2, \ldots$, and $y_2(x)$ in $(3n + 1)$st powers of x, where $n = 0, 1, 2, \ldots$. Clearly, the two expressions are linearly independent and they represent two linearly independent solutions of Airy's equation. The standard notation for them is

$$\mathrm{Ai}(x) = 1 + \frac{1}{6} x^3 + \frac{1}{180} x^6 + \cdots,$$

$$\mathrm{Bi}(x) = x + \frac{1}{12} x^4 + \frac{1}{504} x^7 + \cdots;$$

$\mathrm{Ai}(x)$ and $\mathrm{Bi}(x)$ are called the Airy functions, and the reader is referred to [10], where they are extensively tabulated and many of their properties are given. The two arbitrary constants a_0 and a_1 can be determined only if initial or boundary conditions are given.

If a general expression for the series representations of $\mathrm{Ai}(x)$ and $\mathrm{Bi}(x)$ is desired, we must return to relation (6.2.4) and try to derive a general recursion relation from it. The procedure used is to shift indices in the two series until they can be combined. Note that the first series leads off with x^0, whereas the second leads off with x, so write

$$2a_2 + \sum_{n=3}^{\infty} n(n - 1)a_n x^{n-2} - \sum_{n=0}^{\infty} a_n x^{n+1} = 0.$$

Now both series lead off with x, but the first starts with $n = 3$, whereas the second starts with $n = 0$, so reindex the second as

$$2a_2 + \sum_{n=3}^{\infty} n(n - 1)a_n x^{n-2} - \sum_{n=3}^{\infty} a_{n-3} x^{n-2} = 0.$$

Both sums can now be combined to obtain

$$2a_2 + \sum_{n=3}^{\infty} [n(n - 1)a_n - a_{n-3}] x^{n-2} = 0.$$

From the above follow the relations

$$a_2 = 0, \qquad a_n = \frac{a_{n-3}}{n(n - 1)}, \qquad n = 3, 4, 5, \ldots, \tag{6.2.5}$$

which clearly imply that

$$a_2, a_5, a_8, \ldots, a_{3n+2}, \ldots \text{ are all zero,}$$

$$a_3, a_6, a_9, \ldots, a_{3n}, \ldots \text{ all depend on } a_0,$$

$$a_4, a_7, a_{10}, \ldots, a_{3n+1}, \ldots \text{ all depend on } a_1.$$

The relations (6.2.5) are the desired general recursion relations.

The matter of reindexing the series, so as to be able to combine terms, is somewhat arbitrary. For instance, in the expression

$$2a_2 + \sum_{n=3}^{\infty} n(n-1)a_n x^{n-2} - \sum_{n=0}^{\infty} a_n x^{n+1} = 0$$

one could reindex the first series instead, so that it starts with $n = 0$. This would give

$$2a_2 + \sum_{n=0}^{\infty} (n+3)(n+2)a_{n+3}x^{n+1} - \sum_{n=0}^{\infty} a_n x^{n+1} = 0,$$

and the recursion relation would be

$$a_2 = 0, \qquad a_{n+3} = \frac{a_n}{(n+3)(n+2)}, \qquad n = 0, 1, 2, \ldots .$$

It is easily seen that this would give exactly the same result.

The reader should now study the analysis below to see how to use (6.2.5) to get the general series expressions for $Ai(x)$ and $Bi(x)$. First,

$$a_n = \frac{a_{n-3}}{n(n-1)}$$

implies that

$$a_{3n} = \frac{a_{3n-3}}{3n(3n-1)} = \frac{a_{3(n-1)}}{3^2 n(n - \frac{1}{3})},$$

and now apply the recursion relation to a_{3n-3}:

$$a_{3n-3} = \frac{a_{3n-6}}{(3n-3)(3n-4)}$$

or

$$a_{3(n-1)} = \frac{a_{3(n-2)}}{3(n-1)(3n-4)} = \frac{a_{3(n-2)}}{3^2(n-1)(n - \frac{4}{3})}.$$

Substitute the last expression for $a_{3(n-1)}$ in the expression for a_{3n} to get

$$a_{3n} = \frac{a_{3(n-2)}}{3^4 n(n-1)(n - \frac{1}{3})(n - \frac{4}{3})}.$$

Proceeding backwards in this fashion, one finally obtains

$$a_{3n} = \frac{a_0}{3^{2n}\, n!(n - \frac{1}{3})(n - \frac{4}{3}) \cdots (\frac{2}{3})}, \qquad n = 1, 2, \ldots ,$$

and since a_0 is arbitrary, the choice of $a_0 = 1$ gives

$$Ai(x) = 1 + \sum_{n=1}^{\infty} \frac{1}{3^{2n}\, n!(n - \frac{1}{3})(n - \frac{4}{3}) \cdots (\frac{2}{3})} x^{3n}.$$

The reader may wish to work out the second case starting with a_{3n+1} and letting $a_1 = 1$ to get

$$Bi(x) = x + \sum_{n=1}^{\infty} \frac{1}{3^{2n} \, n!(n + \frac{1}{3})(n - \frac{2}{3}) \cdots (\frac{4}{3})} x^{3n+1},$$

and also show, by using the ratio test, that both series converge for $-\infty < x < \infty$.

A brief word should be said about convergence of the power series obtained by the method just described. Suppose one is given the linear differential equation

$$y'' + p(x)y' + q(x)y = f(x),$$

where $p(x)$, $q(x)$, and $f(x)$ have power series representations convergent for $|x - x_0| < r, r > 0$, so that x_0 is an *ordinary point*. Then the power series for the solution, obtained formally by substituting the power series

$$y(x) = \sum_{n=0}^{\infty} a_n(x - x_0)^n$$

and recursively solving for the coefficients, will also converge for $|x - x_0| < r$. This useful result, which is proved in more advanced texts, means that there is no need to test for convergence of the power series obtained for the solution. It will have the same radius of convergence as the smallest radius of convergence of the power series for the coefficients.

EXERCISES 6.2

1. Use the power series method to find the general solutions of

   a) $y'' + y = 0$;   b) $y'' - 4y = 0$.

   Verify that you obtain the series for sin x and cos x in (a), and for e^{2x} and e^{-2x} in (b).

2. Find the recurrence relation for the coefficients and the first six nonzero terms of the series solutions of the following:

   a) $y'' + x^2y = 0$

   b) $y'' - xy' - 2y = 0$

   c) $y'' - (1 + x^2)y = 0, \ y(0) = -2, y'(0) = 2$

   d) $y'' + y' + xy = 0, \ \ y(0) = 1, \ \ y'(0) = 0$

3. By expressing the coefficients in a power series, substituting $y(x) = \sum_{n=0}^{\infty} a_nx^n$, and equating like powers of x, find the first five nonzero terms of the series solutions of the following:

   a) $y'' + e^xy = 0, \ \ y(0) = 1, \ \ y'(0) = -1$

   b) $y'' + 2y' + (\sin x)y = 0$

   c) $y'' + (\alpha + \beta \cos 2x)y = 0$ (Mathieu's equation: α, β are parameters.)

4. Series solutions can be used occasionally to find approximations to solutions of nonlinear equations. Find the first three nonzero terms of the series solutions of the following:

   a) $y' = y^2 + (1 + x^2), \ \ y(0) = 0 \quad$ *Hint:* The initial conditions imply that $y(x) = a_1x + a_2x^2 + \cdots$, and a useful fact is

   $$\left(\sum_{k=1}^{n} c_k \right)^2 = \sum_{k=1}^{n} c_k^2 + 2 \sum_{j \neq k} c_jc_k.$$

b) $y' = 1/y + x^2$, $y(0) = 1$. *Hint:* $y(x) = 1 + a_1x + a_2x^2 \cdots$; recall that

$$(1 + u)^{-1} = 1 - u + u^2 - \cdots$$
$$+ (-1)^n u^n + \cdots, \quad |u| < 1.$$

5. *Method of Taylor Series.* Given the initial value problem

$$y'' + p(x)y' + q(x)y = 0,$$
$$y(0) = a, \quad y'(0) = b,$$

where $p(x)$ and $q(x)$ are smooth functions near $x = 0$, one can find $y''(0)$ by direct substitution:

$$y''(0) = -p(0)y'(0) - q(0)y(0)$$
$$= -bp(0) - aq(0).$$

Differentiating the equation and then substituting again will give $y'''(0)$. For example,

$$y'''(0) + p'(0)y'(0) + p(0)y''(0)$$
$$+ q'(0)y(0) + q(0)y'(0) = 0$$

implies that

$$y'''(0) = -bp'(0) - p(0)y''(0)$$
$$- aq'(0) - bq(0).$$

Proceeding in this manner, one can use the derivatives obtained to develop the Taylor series of the solution:

$$y(x) = y(0) + y'(0)x + \frac{y''(0)}{2!}x^2$$
$$+ \cdots + \frac{y^{(n)}(0)}{n!}x^n + \cdots.$$

Use this method to find the first four nonzero terms of the series solutions of the following:

a) $y'' + x^2y = 0$, $y(0) = 0$, $y'(0) = -1$

b) $y'' + (e^{2x})y' + y = 0$, $y(0) = 0$, $y'(0) = 1$

c) Problem 2(c) above

d) Problem 2(d) above

6.3 SERIES SOLUTIONS—PART 2

In the previous section we considered series solutions for

$$y'' + p(x)y' + q(x)y = 0,$$

where it was assumed that $p(x)$ and $q(x)$ were smooth functions in some neighborhood of a given point x_0. In this case the desired series solutions was obtained by substitution and comparison of coefficients of like powers of x. Does this procedure work when $p(x)$ or $q(x)$ are discontinuous at $x = x_0$? The answer is NO, except when the discontinuity is of a special kind, but fortunately this class of equations contains many of the equations which arise in applied mathematics and mathematical physics.

Suppose as before that the given point in question is $x_0 = 0$, and suppose further that the differential equation can be written in the form

$$y'' + \frac{p(x)}{x}y' + \frac{q(x)}{x^2}y = 0, \tag{6.3.1}$$

where $p(x)$ and $q(x)$ are smooth functions in a neighborhood of $x = 0$; hence

$$p(x) = \sum_{n=0}^{\infty} p_n x^n, \qquad q(x) = \sum_{n=0}^{\infty} q_n x^n.$$

In this case $x = 0$ is said to be a *regular singular point,* and it is this type of discontinuity for which the method of power series, suitably modified, also works. The following are examples of differential equations with a regular singular point at $x = 0$:

a) $x^2y'' + y = 0$ or $y'' + \dfrac{1}{x^2} y = 0$ with $p(x) = 0$, $q(x) = 1$;

b) $4x^2y'' - 3xy' + 2y = 0$ or $y'' + \dfrac{-3/4}{x} y' + \dfrac{1/2}{x^2} y = 0$ with $p(x) = -3/4$, $q(x) = 1/2$;

c) $xy'' + (x + 2)y' + e^x y = 0$ or $y'' + \dfrac{x + 2}{x} y' + \dfrac{xe^x}{x^2} y = 0$ with $p(x) = x + 2$, $q(x) = xe^x$

d) $y'' + 2y' + \dfrac{4}{x} y = 0$ or $y'' + \dfrac{2x}{x} y' + \dfrac{4x}{x^2} y = 0$ with $p(x) = 2x$, $q(x) = 4x$.

The *method of Frobenius* provides a technique for finding a series solution, but the series will be of the form

$$y(x) = x^r \sum_{n=0}^{\infty} a_n x^n = \sum_{n=0}^{\infty} a_n x^{n+r}, \qquad (6.3.2)$$

where r is some number *(not necessarily an integer!)* to be determined from the differential equation. This is the modification incurred by the fact that $x = 0$ is a regular singular point. Term-by-term differentiation of the expression for $y(x)$ gives

$$y'(x) = \sum_{n=0}^{\infty} (n + r)a_n x^{n+r-1}$$

and

$$y''(x) = \sum_{n=0}^{\infty} (n + r)(n + r - 1)a_n x^{n+r-2}.$$

Now write the differential equation (6.3.1) in the form

$$x^2y'' + xp(x)y' + q(x)y = 0$$

and substitute the expansions for $p(x)$, $q(x)$, $y(x)$, $y'(x)$, and $y''(x)$ to obtain

$$\sum_{n=0}^{\infty} (n + r)(n + r - 1)a_n x^{n+r} + \left(\sum_{n=0}^{\infty} p_n x^n \right) \sum_{n=0}^{\infty} (n + r)a_n x^{n+r}$$

$$+ \left(\sum_{n=0}^{\infty} q_n x^n \right) \sum_{n=0}^{\infty} a_n x^{n+r} = 0.$$

The term x^r can be factored out of each term, and combining the coefficients of like powers of x we obtain the expression

$$x^r[r(r-1)a_0 + p_0 ra_0 + q_0 a_0]$$
$$+ x^r[(1+r)ra_1 + p_0(1+r)a_1 + p_1 ra_0 + q_0 a_1 + q_1 a_0]x$$
$$+ x^r[\quad]x^2 + \cdots + x^r[\quad]x^n + \cdots = 0,$$

where we have explicitly written down the coefficients of x^0 and x. Equating to zero the first bracketed term and assuming $a_0 \neq 0$ gives

$$r(r-1) + p_0 r + q_0 = r^2 + (p_0 - 1)r + q_0 = 0. \qquad (6.3.3)$$

This quadratic polynomial in r is called the *indicial equation*, and its two roots, r_1 and r_2, will be the admissible values of the exponent r in the expression (6.3.2) for the solution. Note that p_0 and q_0 are merely the respective values of $p(0)$ and $q(0)$ and can be easily obtained from the differential equation.

The theory of the regular singular point case, which the reader may wish to examine in more detail in some of the references, states the following:

a) Choose as r_1 that root of the indicial equation for which the real part of $r_1 - r_2$ is nonnegative; if r_1 and r_2 are real and unequal, then r_1 would be the largest root, for instance. Then for $r = r_1$, one can always obtain recursively all the a_n's, and the series

$$y_1(x) = x^{r_1} \sum_{n=0}^{\infty} a_n x^n$$

converges in some deleted neighborhood of $x = 0$ and is a solution. By a deleted neighborhood we mean the set of all x satisfying $0 < |x| < \alpha$ for some $\alpha > 0$. In some cases the series may also converge at $x = 0$.

b) If r_2 is the second root and $r_1 - r_2$ is not zero or a positive integer, then one can proceed as in (a). The series

$$y_2(x) = x^{r_2} \sum_{n=0}^{\infty} b_n x^n$$

with the b_n's determined recursively converges in some deleted neighborhood of $x = 0$ and is a solution. The two solutions, $y_1(x)$ and $y_2(x)$, are linearly independent.

c) If $r_1 - r_2$ equals zero or a positive integer, then there are two possibilities: either a solution of the form $y_2(x)$ above exists or the second solution is of the form

$$y_2(x) = y_1(x)\, \beta \ln x + x^{r_2} \sum_{n=0}^{\infty} b_n x^n,$$

with β a constant dependent on $y_1(x)$ and $p(x)$.

The last possibility is the so-called *logarithmic case,* and if $r_1 - r_2$ equals zero, it always occurs since the indicial equation has only one root. If $r_1 - r_2$ is a positive integer, there is a technique using $p(x)$ and $y_1(x)$ to determine whether the logarithmic case occurs; it uses the reduction of order formula to be discussed shortly, and the interested reader is referred to [2] or [3] for details.

The problem with the case where the real part of $r_1 - r_2$ is a positive integer is that if one substitutes the series $y_2(x) = x^{r_2} \sum_{n=0}^{\infty} b_n x^n$ into the differential equation, then one cannot find a recursion relation for all the b_n. Specifically, for some positive integer m one can no longer solve for b_m in terms of the previous b_j and the process comes to a halt. One must resort to other techniques to find a second linearly independent solution.

A simple example of the logarithmic case, which the reader has seen before, is the Euler differential equation

$$x^2 y'' + axy' + by = 0, \text{ where } a, b = \text{const.}$$

Clearly, $x = 0$ is a regular singular point and $p(x) = a$, $q(x) = b$, so the indicial equation is

$$r^2 + (a - 1)r + b = 0.$$

If the roots of the indicial equation are r_1 and r_2 and both are real and equal, then two linearly independent solutions are

$$y_1(x) = x^{r_1} \quad \text{and} \quad y_2(x) = x^{r_1} \ln x = y_1(x) \ln x.$$

This is an example of case (c) above with $\beta = 1$ and $\sum_{n=0}^{\infty} b_n x^n = 0$.

The Euler differential equation can be used to illustrate two observations about regular singular points. The first is that solutions need not be undefined at a regular singular point. For instance, a linearly independent pair of solutions of

$$y'' - \frac{2}{x} y' + \frac{2}{x^2} y = 0$$

are $y_1(x) = x$ and $y_2(x) = x^2$ and both are smooth, well defined functions at the regular singular point $x_0 = 0$. The example also illustrates the second observation, namely that the solutions at a regular singular point need not be infinite series, but may merely be polynomials. Some examples of equations with a regular singular point and their corresponding indicial equations are given below:

Example 1   **a)** $y'' - (1/2x)y' + [(1 + x^2)/2x^2]y = 0$, so $p(x) = -1/2$, $q(x) = 1/2 + (1/2)x^2$; hence $p_0 = -1/2$, $q_0 = 1/2$. The indicial equation is

$$r^2 + \left(-\frac{1}{2} - 1\right)r + \frac{1}{2} = (r - 1)\left(r - \frac{1}{2}\right) = 0$$

with roots $r_1 = 1$ and $r_2 = 1/2$, and $r_1 - r_2 = 1/2$, which is not zero or a positive integer. We conclude that two linearly independent series solutions,

$$y_1(x) = x \sum_{n=0}^{\infty} a_n x^n \quad \text{and} \quad y_2(x) = x^{1/2} \sum_{n=0}^{\infty} b_n x^n,$$

can be found.

b) $y'' + (2/x)y' + xy = 0$; if the last term is written as $x = x^3/x^2$, it is seen that $p(x) = 2$, $q(x) = x^3$, hence $p_0 = 2$, $q_0 = 0$. The indicial equation is

$$r^2 + (2 - 1)r = r(r + 1) = 0$$

with roots $r_1 = 0$ and $r_2 = -1$, so $r_1 - r_2 = 1$ a positive integer. Corresponding to the root $r_1 = 0$ there is a solution $y_1(x) = \sum_{n=0}^{\infty} a_n x^n$, and the second solution is either the logarithmic case or has the form $y_2(x) = x^{-1} \sum_{n=0}^{\infty} b_n x^n$, corresponding to $r_2 = -1$, which turns out to be the case.

c) $y'' + (3/x)y' + [(1 + x)/x^2]y = 0$, so $p(x) = 3$, $q(x) = 1 + x$, and $p_0 = 3$, $q_0 = 1$. The indicial equation is

$$r^2 + (3 - 1)r + 1 = (r + 1)^2 = 0$$

with roots $r_1 = r_2 = -1$ and $r_1 - r_2 = 0$. This is a logarithmic case, and the solutions are of the form

$$y_1(x) = x^{-1} \sum_{n=0}^{\infty} a_n x^n$$

corresponding to $r_1 = -1$, and

$$y_2(x) = y_1(x) \beta \ln x + x^{-1} \sum_{n=0}^{\infty} b_n x^n. \quad \square$$

Once the value of r_1 is determined from the indicial equation, finding the series solution for $y_1(x)$ proceeds as in the case of an ordinary point. The series is substituted and like powers of x are compared after shifting indices if necessary.

Example 2 Find $y_1(x)$ for the differential equation of Example 1(c) above:

$$x^2 y'' + 3xy' + (1 + x)y = 0.$$

Set $y_1(x) = \sum_{n=0}^{\infty} a_n x^{n-1}$, then differentiate the series term by term and substitute to get

$$x^2 \sum_{n=0}^{\infty} (n - 1)(n - 2)a_n x^{n-3}$$

$$+ 3x \sum_{n=0}^{\infty} (n - 1)a_n x^{n-2} + (1 + x) \sum_{n=0}^{\infty} a_n x^{n-1} = 0.$$

Multiply each series by its coefficient and combine terms to obtain

$$\sum_{n=0}^{\infty} [(n-1)(n-2) + 3(n-1) + 1]a_n x^{n-1} + \sum_{n=0}^{\infty} a_n x^n$$

$$= \sum_{n=0}^{\infty} n^2 a_n x^{n-1} + \sum_{n=0}^{\infty} a_n x^n = 0.$$

When $n = 0$ in the first series, the coefficient of x^{-1} is zero, so we can shift its index by one to get

$$\sum_{n=0}^{\infty} (n+1)^2 a_{n+1} x^n + \sum_{n=0}^{\infty} a_n x^n = \sum_{n=0}^{\infty} [(n+1)^2 a_{n+1} + a_n]x^n = 0.$$

Hence $a_{n+1} = -a_n/(n+1)^2$, $n = 0, 1, 2, \ldots$, which implies

$$a_1 = -a_0, \qquad a_2 = \frac{-a_1}{2^2} = \frac{a_0}{2^2},$$

$$a_3 = \frac{-a_2}{3^2} = \frac{-a_0}{3^2 \cdot 2^2}, \text{ etc.,}$$

and in general,

$$a_n = \frac{(-1)^n a_0}{n^2 \cdots 3^2 \cdot 2^2} = \frac{(-1)^n a_0}{(n!)^2}.$$

Since a_0 is arbitrary, it may be set equal to unity, and the first solution is

$$y_1(x) = x^{-1} \sum_{n=0}^{\infty} \frac{(-1)^n}{(n!)^2} x^n.$$

Usually writing down the first few terms of the series expansion, after combining those series which lead off with the same power of x, will give a clue as to how to shift the indices. Careful bookkeeping does the rest! □

The following exercises deal, for the most part, with the nonlogarithmic case. To obtain the second solution in the logarithmic case, a technique called *the method of reduction of order* is often used. That is the subject of the next section.

EXERCISES 6.3

1. The following equations have a regular singular point at $x = 0$. Find the indicial equation, determine its roots, and state what general form the series solutions will have.

a) $2x^2 y'' + (3x - 2x^2)y' - (x + 1)y = 0$

b) $x^2 y'' + xy' + (x^2 - \frac{1}{9})y = 0$

c) $4xy'' + 2y' - y = 0$

d) $xy'' + y' + x^2 y = 0$

e) $x^2 y'' + 5xy' + 4(\cos x)y = 0$

f) $x^2 y'' + (2x + x^2)y' - 2y = 0$

g) $16x^2 y'' + 24xy' - 3y = 0$

2. If a differential equation has a regular singular point at $x = a$, then it can be written in the form

$$y'' + \frac{p(x)}{x - a} y' + \frac{q(x)}{(x - a)^2} y = 0,$$

where $p(x)$ and $q(x)$ are smooth functions in a neighborhood of $x = a$. By assuming a solution of the form

$$y(x) = (x - a)^r \sum_{n=0}^{\infty} a_n (x - a)^n,$$

show that r must be a root of the indicial equation

$$r^2 + (p_0 - 1)r + q_0 = 0,$$

where $p_0 = p(a)$, $q_0 = q(a)$.

3. Find the regular singular points, the corresponding indicial equation, and its roots, and state what general form the series solution will have for the following:

a) $(x + 1)y'' + \frac{3}{2}y' + y = 0$

b) $(x - 1)y'' + xy' + y = 0$

c) $2(x - 2)^2 y'' - \frac{6(x - 2)}{x} y' + 3y = 0$

d) $(1 - x^2)y'' - 2xy' + 6y = 0$

4. The following equations have a regular singular point at $x = 0$ and are the case $r_1 - r_2 \neq 0$ or a positive integer. Find the recursion relation and the first four nonzero terms of the series expansion for each of the two linearly independent solutions. Find a general expression for each solution if possible.

a) $2xy'' + y' - y = 0$

b) $2x^2 y'' - xy' + (1 - x^2)y = 0$

c) $4xy'' + 2y' + y = 0$

d) $3x^2 y'' + 4xy' - 2y = 0$

e) $2x^2 y'' - 5(\sin x)y' + 3y = 0$; find only the first three nonzero terms of each solution.

5. The following equations have a regular singular point at $x = 0$ and $r_1 - r_2$ is a positive integer (the nonlogarithmic case). Both series solutions can often be found by substituting $y(x) = x^{r_2} \sum_{n=0}^{\infty} a_n x^n$, where r_2 is the *smallest* root of the indicial equation. Find the recursion relation and the first four nonzero terms of the series expansion for each of the two linearly independent solutions. Find a general expression for each solution if possible:

a) $xy'' + 2y' + x^2 y = 0$

b) $xy'' + (3 + x^3)y' + 3x^2 y = 0$

c) $xy'' + (4 + x)y' + 2y = 0$

d) $x(1 - x)y'' - (4 + x)y' + 4y = 0$

e) Same as (d) but solve for the regular singular point $x = 1$.

6.4 THE METHOD OF REDUCTION OF ORDER AND THE LOGARITHMIC CASE OF A REGULAR SINGULAR POINT

Suppose we are given the second order linear equation

$$y'' + p(x)y' + q(x)y = 0, \tag{6.4.1}$$

where $p(x)$ and $q(x)$ are continuous for all x in some neighborhood of $x = x_0$ but may be discontinuous at x_0 itself. For instance, this would be the case if x_0 is a regular singular point. Suppose further the happy circumstance that a nontrivial solution $y_1(x)$ of (6.4.1) is known. This solution could be the result of a fortuitous guess or of diligent labor, as in the case of a regular singular point, where $y_1(x)$ is the series solution corresponding to r_1, the largest root of the indicial equation.

The method of reduction of order deals with the following question.

Given a nontrivial solution $y_1(x)$ of (6.4.1), can one find a second linearly independent solution $y_2(x)$? The answer is YES, and the method is especially useful when $y_1(x)$ is given by an infinite series, and one wishes to obtain the first few terms of the series representation of $y_2(x)$. For the logarithmic case of the regular singular point it is an especially effective tool.

Suppose $y_1(x)$ is a solution of (6.4.1). The method of reduction of order, like the method of variation of parameters, assumes that the second solution can be expressed as $y_2(x) = y_1(x)\, u(x)$, where $u(x)$ is to be determined. Therefore

$$y_2 = y_1 u, \qquad y_2' = y_1 u' + y_1' u,$$
$$y_2'' = y_1 u'' + 2y_1' u' + y_1'' u,$$

and since $y_2(x)$ is assumed to be a solution, this implies that

$$y_2'' + p(x)y_2' + q(x)y_2$$
$$= [y_1'' + p(x)y_1' + q(x)y_1]u + y_1 u'' + (2y_1' + p(x)y_1)u' = 0.$$

Since $y_1(x)$ is a solution, the coefficient of u is zero, and so $u = u(x)$ must satisfy the relation

$$y_1 u'' + [2y_1' + p(x)y_1]u' = 0$$

or

$$u'' + \left[2\,\frac{y_1'}{y_1} + p(x) \right] u' = 0.$$

The last relation is a first order linear homogeneous equation in $w = u'$ and can be solved by using the methods of Chapter 1. The solution is

$$u'(x) = \exp\left[-2 \int^x \frac{y_1'(s)}{y_1(s)}\, ds - \int^x p(s)\, ds \right] = \frac{\exp\left[-\int^x p(s)\, ds \right]}{y_1(x)^2}.$$

(We have ignored the immaterial constant of integration.) Integrating once more gives $u(x)$ and therefore

$$y_2(x) = y_1(x) \int^x \frac{\exp\left[-\int^r p(s)\, ds \right]}{y_1(r)^2}\, dr, \qquad (6.4.2)$$

which is the *reduction of order* formula.

To show that $y_1(x)$ and $y_2(x)$ are a linearly independent pair of solutions, we compute their Wronskian. The formulas above for u' and y_2' imply that

$$y_2'(x) = y_1'(x) \int^x \frac{\exp\left[-\int^r p(s)\, ds \right]}{y_1(r)^2}\, dr + \frac{\exp\left[-\int^x p(s)\, ds \right]}{y_1(x)},$$

and consequently

$$W(y_1, y_2)(x) = y_1(x)y_2'(x) - y_1'(x)y_2(x) = \exp\left[-\int^x p(s)\ ds\right] \neq 0.$$

Therefore $y_1(x)$ and $y_2(x)$ are a linearly independent pair. Note that if $p(x) = 0$, so that the differential equation is

$$y'' + q(x)y = 0,$$

then the reduction of order formula is simply

$$y_2(x) = y_1(x) \int^x \frac{1}{y_1(r)^2}\ dr.$$

Example 1 Use the reduction of order formula to construct a second solution to each of the differential equations given below.

a) The constant coefficient equation

$$y'' + 2by' + b^2y = 0$$

has a solution $y_1(x) = e^{-bx}$ and is the case where the characteristic polynomial has a double root $\lambda = b$. Since $p(x) = 2b$, the reduction of order formula gives

$$y_2(x) = e^{-bx} \int^x \frac{\exp\left[-\int^r 2b\ ds\right]}{(e^{-br})^2}\ dr = e^{-bx} \int^x \frac{e^{-2br}}{e^{-2br}}\ dr = xe^{-bx}.$$

b) Given the Euler equation

$$x^2y'' - 7xy' + 16y = 0, \quad x \neq 0,$$

a trial solution of the type $y_1(x) = x^k$ yields $k = 4$. Rewriting the equation as

$$y'' - \frac{7}{x}y' + \frac{16}{x^2}y = 0, \quad x \neq 0,$$

we see that $p(x) = -7/x$ and

$$\exp\left[-\int^r \left(-\frac{7}{s}\right)ds\right] = \exp\left[7\ \ln r\right] = r^7.$$

Hence

$$y_2(x) = x^4 \int^x \frac{r^7}{(r^4)^2}\ dr = x^4\ \ln x.$$

c) A careful examination of the equation

$$y'' - \frac{2}{x^2 + 1} y = 0$$

reveals that $y_1(x) = x^2 + 1$ is a solution. Since $p(x) = 0$, the second linearly independent solution is

$$y_2(x) = (x^2 + 1) \int^x \frac{1}{(r^2 + 1)^2} \, dr = (x^2 + 1) \left[\frac{x}{2(x^2 + 1)} + \frac{1}{2} \tan^{-1} x \right]$$

$$= \frac{x}{2} + \frac{1}{2} (x^2 + 1) \tan^{-1} x. \quad \Box$$

It is often the case that the integral in the reduction of order formula cannot be evaluated. Nevertheless the expression for the second solution may still be useful, for instance to determine the asymptotic behavior of the general solution.

Returning to the topic of series solutions of linear differential equations, it was mentioned above that the first solution $y_1(x)$ could always be expressed as an infinite series. The use of the reduction of order formula to find the second linearly independent solution $y_2(x)$ will then involve the squaring and inverting of an infinite series, followed by an integration term by term. There is a systematic way to accomplish this and it depends on the following.

The Cauchy product formula for series. Given the two power series

$$\sum_{n=0}^{\infty} a_n(x - x_0)^n, \qquad \sum_{n=0}^{\infty} b_n(x - x_0)^n,$$

their product is given by the power series

$$\left[\sum_{n=0}^{\infty} a_n(x - x_0)^n \right] \left[\sum_{n=0}^{\infty} b_n(x - x_0)^n \right]$$

$$= \sum_{n=0}^{\infty} \left(\sum_{j=0}^{n} a_j b_{n-j} \right) (x - x_0)^n$$

$$= a_0 b_0 + (a_0 b_1 + a_1 b_0)(x - x_0) + (a_0 b_2 + a_1 b_1 + a_2 b_0)(x - x_0)^2$$

$$+ \cdots + (a_0 b_n + a_1 b_{n-1} + \cdots + a_n b_0)(x - x_0)^n + \cdots.$$

The formula is merely a systematic way of combining in the product all those terms that have the same power of $x - x_0$. If the two series are the same, this gives the formula for squaring a series:

$$\left[\sum_{n=0}^{\infty} a_n(x - x_0)^n\right]^2 = a_0^2 + 2a_0a_1(x - x_0) + (a_1^2 + 2a_0a_2)(x - x_0)^2$$

$$+ (2a_0a_3 + 2a_1a_2)(x - x_0)^3$$
$$+ (a_2^2 + 2a_0a_4 + 2a_1a_3)(x - x_0)^4 + \cdots$$
$$+ (a_n^2 + 2a_0a_{2n} + \cdots + 2a_{n-1}a_{n+1})(x - x_0)^{2n}$$
$$+ (2a_0a_{2n+1} + \cdots + 2a_na_{n+1})(x - x_0)^{2n+1} + \cdots.$$

In the reduction of order formula there appears an expression of the form $1/y_1(x)^2$, so if $y_1(x)$ is represented by a power series, it can be squared by using the above formula. But then it must be inverted or, equivalently, the series representation of an expression of the form

$$1\Big/ \sum_{n=0}^{\infty} b_n(x - x_0)^n$$

must be computed. To accomplish this, let the last expression equal the series $\sum_{n=0}^{\infty} q_n(x - x_0)^n$, where the q_n are to be determined. Therefore we have the relation

$$\left[\sum_{n=0}^{\infty} b_n(x - x_0)^n\right]\left[\sum_{n=0}^{\infty} q_n(x - x_0)^n\right] = 1,$$

and now use the Cauchy product formula on the left side and then equate the coefficients of like powers of $x - x_0$ to obtain

$$b_0q_0 = 1, \quad b_0q_1 + b_1q_0 = 0, \quad b_0q_2 + b_1q_1 + b_2q_0 = 0, \ldots,$$
$$b_0q_n + b_1q_{n-1} + \cdots + b_nq_0 = 0, \ldots.$$

Now solve for q_0 in the first equation, substitute its value in the second equation, and solve for q_1, use q_0 and q_1 in the third equation to find q_2, etc. The above formulas are a systematic recursive approach to what amounts to "long division."

It is assumed in the above division process that $b_0 \neq 0$. If $b_0, b_1, \ldots, b_{k-1}$ are all zero, but $b_k \neq 0$, then the above quotient may be written as

$$\frac{1}{\sum\limits_{n=0}^{\infty} b_n(x - x_0)^n} = \frac{1}{(x - x_0)^k} \frac{1}{\sum\limits_{n=0}^{\infty} b_{k+n}(x - x_0)^n},$$

and the division algorithm applied to the second term.

Example 2 Find the first few terms of $1/g(x)^2$, where

$$g(x) = 1 + \sum_{n=1}^{\infty} \frac{1}{n+2} x^n.$$

We have

$$a_0 = 1, \quad a_1 = \frac{1}{3}, \quad a_2 = \frac{1}{4}, \quad a_3 = \frac{1}{5}, \dots, \quad a_n = \frac{1}{n+2},$$

and

$$g(x)^2 = 1^2 + [2 \cdot 1 \cdot \tfrac{1}{3}]x + [(\tfrac{1}{3})^2 + 2 \cdot 1 \cdot \tfrac{1}{4}]x^2$$
$$+ [2 \cdot 1 \cdot \tfrac{1}{5} + 2 \cdot \tfrac{1}{3} \cdot \tfrac{1}{4}]x^3 + \cdots$$
$$= 1 + \tfrac{2}{3}x + \tfrac{11}{18}x^2 + \tfrac{17}{30}x^3 + \cdots.$$

To find the first few terms of $1/g(x)^2$, one must solve the equations

$$1 \cdot q_0 = 1, \quad 1 \cdot q_1 + \tfrac{2}{3}q_0 = 0, \quad 1 \cdot q_2 + \tfrac{2}{3}q_1 + \tfrac{11}{18}q_0 = 0,$$
$$1 \cdot q_3 + \tfrac{2}{3}q_2 + \tfrac{11}{18}q_1 + \tfrac{17}{30}q_0 = 0, \quad \dots$$

to obtain

$$q_0 = 1, \quad q_1 = -\tfrac{2}{3}, \quad q_2 = -\tfrac{1}{6}, \quad q_3 = -\tfrac{13}{270}, \quad \dots.$$

Therefore

$$\frac{1}{g(x)^2} = 1 - \frac{2}{3}x - \frac{1}{6}x^2 - \frac{13}{270}x^3 + \cdots. \quad \square$$

After the above detour into the multiplication and inversion of power series, it is time to return to our principal topic—series solutions of linear differential equations around a regular singular point. The method of reduction of order will be used to obtain the first few terms of the series representation of the second solution $y_2(x)$, when the roots r_1 and r_2 of the indicial equation satisfy $r_1 - r_2 = 0$, the logarithmic case. This will be accomplished via some examples.

Example 3 Find the first few terms of the logarithmic solution to

$$y'' + \frac{1}{x}y' + y = 0.$$

In Section 6.1 it was stated that a solution is

$$y_1(x) = J_0(x) = \sum_{n=0}^{\infty} \frac{(-1)^n}{2^{2n}(n!)^2} x^{2n} = 1 - \frac{1}{4}x^2 + \frac{1}{64}x^4 - \frac{1}{2304}x^6 + \cdots.$$

Since $p(x) = 1/x$, the reduction of order formula gives

$$y_2(x) = J_0(x) \int^x \frac{\exp\left[-\int^r \frac{1}{s}\,ds\right]}{J_0(r)^2}\,dr = J_0(x) \int^x \frac{1}{rJ_0(r)^2}\,dr.$$

By rewriting the equation in the form

$$y'' + \frac{1}{x}y' + \frac{x^2}{x^2}y = 0,$$

we see that $x = 0$ is a regular singular point, and since $p_0 = 1$, $q_0 = 0$, the indicial equation is $r^2 + (1 - 1)r + 0 = r^2 = 0$ with roots $r_1 = r_2 = 0$. This is the logarithmic case.

To find the first few terms of $y_2(x)$, first calculate $J_0(x)^2$. Since

$$a_0 = 1, \quad a_1 = 0, \quad a_2 = -\tfrac{1}{4}, \quad a_3 = 0, \quad a_4 = \tfrac{1}{64}, \quad \ldots,$$

we have

$$\begin{aligned}
J_0(x)^2 &= 1^2 + (2 \cdot 1 \cdot 0)x + [0^2 + 2 \cdot 1 \cdot (-\tfrac{1}{4})]x^2 \\
&\quad + [2 \cdot 1 \cdot 0 + 2 \cdot 0 \cdot (-\tfrac{1}{4})]x^3 \\
&\quad + [(-\tfrac{1}{4})^2 + 2 \cdot 1 \cdot \tfrac{1}{64} + 2 \cdot 0 \cdot 0]x^4 + \cdots \\
&= 1 - \tfrac{1}{2}x^2 + \tfrac{3}{32}x^4 + \cdots.
\end{aligned}$$

To find $1/J_0(x)^2$, the following relations have to be solved:

$$1 \cdot q_0 = 1, \quad 1 \cdot q_1 + 0 \cdot q_0 = 0, \quad 1 \cdot q_2 + 0 \cdot q_1 + (-\tfrac{1}{2})q_0 = 0,$$
$$1 \cdot q_3 + 0 \cdot q_2 + (-\tfrac{1}{2})q_1 + 0 \cdot q_0 = 0,$$
$$1 \cdot q_4 + 0 \cdot q_3 + (-\tfrac{1}{2})q_2 + 0 \cdot q_1 + \tfrac{3}{32}q_0 = 0, \ldots$$

to obtain

$$q_0 = 1, \quad q_1 = 0, \quad q_2 = \tfrac{1}{2}, \quad q_3 = 0, \quad q_4 = \tfrac{5}{32}, \ldots.$$

Therefore

$$\begin{aligned}
y_2(x) &= J_0(x) \int^x \frac{1}{r}\left(1 + \frac{1}{2}r^2 + \frac{5}{32}r^4 + \cdots\right) dr \\
&= J_0(x)\left[\ln x + \frac{1}{4}x^2 + \frac{5}{128}x^4 + \cdots\right]. \quad \square
\end{aligned}$$

Example 4 Find the first few terms of the logarithmic solution to

$$y'' + \frac{3}{x}y' + \frac{1+x}{x^2}y = 0.$$

In the final example of the previous section, a solution was found to be

$$y_1(x) = x^{-1} \left[1 - x + \frac{x^2}{4} - \frac{x^3}{36} + \cdots \right].$$

The point $x = 0$ is a regular singular point and the roots of the indicial equation are $r_1 = r_2 = -1$. Since $p(x) = 3/x$, then

$$\exp\left[-\int^r \frac{3}{s} ds \right] = \frac{1}{r^3},$$

and the second solution is

$$y_2(x) = y_1(x) \int^x \frac{1}{r^3 y_1(r)^2} dr = y_1(x) \int^x \frac{1}{r} \frac{1}{\left[1 - r + \frac{r^2}{4} - \frac{r^3}{36} + \cdots \right]^2} dr.$$

Letting $g(r)$ be the expression in the brackets, we can write

$$g(r)^2 = 1 - 2r + \tfrac{3}{2}r^2 - \tfrac{5}{9}r^3 + \cdots.$$

To find $1/g(r)^2$ we must solve the relations

$$1 \cdot q_0 = 1, \quad 1 \cdot q_1 + (-2)q_0 = 0,$$
$$1 \cdot q_2 + (-2)q_1 + (\tfrac{3}{2})q_0 = 0,$$
$$1 \cdot q_3 + (-2)q_2 + (\tfrac{3}{2})q_1 + (-\tfrac{5}{9})q_0 = 0,$$

to obtain

$$\frac{1}{g(r)^2} = 1 + 2r + \frac{5}{2} r^2 + \frac{23}{9} r^3 + \cdots.$$

Therefore

$$y_2(x) = y_1(x) \int^x \frac{1}{r} \left[1 + 2r + \frac{5}{2} r^2 + \frac{23}{9} r^3 + \cdots \right] dr$$

$$= y_1(x) \ln x + y_1(x) \left[2x + \frac{5}{4} x^2 + \frac{23}{27} x^3 + \cdots \right].$$

A further multiplication of series would give

$$y_2(x) = y_1(x) \ln x + \left[2 - \frac{3}{4} x + \frac{11}{108} x^2 + \cdots \right]. \quad \square$$

Example 5 Find a solution to $y'' + (2/x)y' + xy = 0$ that is singular at $x = 0$.

The indicial equation is $r^2 + r = 0$ with roots $r_1 = 0$, $r_2 = -1$. Since $r_1 - r_2 = 1$ (a positive integer), this may or may not be a logarithmic case. The solution $y_1(x)$ corresponding to $r_1 = 0$ has the series representation (see Exercise 5(a) of the previous section)

$$y_1(x) = \sum_{n=0}^{\infty} a_n x^n.$$

By the reduction of order formula

$$y_2(x) = y_1(x) \int^x \frac{1}{r^2 y_1(r)^2}\, dr = y_1(x) \int^x \frac{1}{r^2 \left(\displaystyle\sum_{n=0}^{\infty} a_n r^n\right)^2}\, dr$$

$$= y_1(x) \int^x \frac{1}{r^2} [b_0 + b_1 r + b_2 r^2 + \cdots]\, dr$$

$$= y_1(x)[-b_0 x^{-1} + b_1 \ln x + b_2 x + \cdots],$$

where

$$\frac{1}{\left(\displaystyle\sum_{n=0}^{\infty} a_n x^n\right)^2} = \sum_{n=0}^{\infty} b_n x^n.$$

There will be a logarithmic term if $b_1 \neq 0$; otherwise the solution will be of the form

$$y_2(x) = x^{-1} \sum_{n=0}^{\infty} c_n x^n$$

corresponding to the root $r_2 = -1$. □

EXERCISES 6.4

1. Given the following linear differential equations and one solution $y_1(x)$, use the reduction of order formula to find a second linearly independent solution $y_2(x)$.

   a) $y'' - 4y' + 13y = 0$, $y_1(x) = e^{2x} \cos 3x$

   b) $y'' - 9y = 0$, $y_1(x) = e^{3x}$

   c) $y'' - xy' + y = 0$, $y_1(x) = x$

   d) $x^2 y'' + 4xy' - 10y = 0$, $y_1(x) = x^2$

   e) $y'' - \dfrac{6}{x^2} y = 0$, $y_1(x) = \dfrac{1}{x^2}$

   f) $(1 - x^2)y'' - 2xy' + 2y = 0$, $y_1(x) = x$

   g) $y'' - 2(1 + \tan^2 x)y = 0$, $y_1(x) = \tan x$

2. Given the following series

$$P(x) = 1 + \sum_{n=1}^{\infty} \frac{1}{n^2} x^n,$$

$$Q(x) = \sum_{n=0}^{\infty} \frac{(-1)^n}{n!} x^n,$$

compute the first four nonzero terms of the following:

   a) $P(x)\, Q(x)$ b) $P(x)^2$

   c) $\dfrac{1}{P(x)^2}$ d) $\dfrac{Q(x)}{P(x)}$

   e) $P(x)\, Q^2(x)$ f) $\dfrac{1}{Q(x)^2}$

3. Given the following functions $p(x)$ and the series $y_1(x)$, compute the first three nonzero terms of the expression

$$\int^x \frac{\exp\left[-\int^r p(s)\,ds\right]}{y_1(r)^2}\,dr.$$

a) $p(x) = -\dfrac{2}{x}$, $\quad y_1(x) = 1 + \dfrac{1}{2}x + \dfrac{1}{4}x^2 + \dfrac{1}{16}x^3 + \cdots$

b) $p(x) = \dfrac{3}{x}$, $\quad y_1(x) = 1 - \dfrac{x^2}{2} + \dfrac{x^4}{24} - \dfrac{x^6}{720} + \cdots$

c) $p(x) = \dfrac{1}{x}$, $\quad y_1(x) = x - \dfrac{x^3}{3} + \dfrac{x^5}{9} - \cdots$.

4. The following equations have a regular singular point at $x = 0$ and are examples of the logarithmic case. Find the recursion relation and the first four nonzero terms of the series expansion of the solution corresponding to the root r_1. If possible, find a general expression for this solution. Then use the method of reduction of order to find the first few terms of the second linearly independent solution.

a) $xy'' + y' - xy = 0$

b) $x^2y'' + xy' + xy = 0$

c) $x^2y'' + xy' + (x - 1)y = 0$

d) $xy'' + y = 0$

e) $x^2y'' - 3xy' + (4x + 4)y = 0$

f) $x^2y'' + 5xy' + (3 - x^2)y = 0$

6.5 SOME SPECIAL FUNCTIONS AND TOOLS OF THE TRADE

The Gamma Function

This function arises so frequently in series representations of solutions that it is worth discussing briefly. The *Gamma function, $\Gamma(x)$*, is defined by the definite integral

$$\Gamma(x) = \int_0^\infty e^{-t}\,t^{x-1}\,dt$$

which converges for all *positive x*. One sees immediately that $\Gamma(1) = 1$, and a simple integration by parts shows that

$$\Gamma(x + 1) = \int_0^\infty e^{-t}\,t^x\,dt = \left. -e^{-t}\,t^x \right|_0^\infty + x \int_0^\infty e^{-t}\,t^{x-1}\,dt = 0 + x\,\Gamma(x).$$

The *fundamental identity* $\Gamma(x + 1) = x\,\Gamma(x)$ implies, in particular, that for $x = n$, a positive integer, the following holds:

$$\Gamma(n + 1) = n\,\Gamma(n) = n(n - 1)\Gamma(n - 1)$$
$$= \cdots = n(n - 1) \cdots 2 \cdot 1 = n!$$

since $\Gamma(1) = 1$. Therefore the Gamma function extends the definition of the factorial $x!$, previously defined only for $x = 1, 2, \ldots$, to a function $\Gamma(x)$ which is defined for all values of $x > 0$ and which agrees with $n!$ when $x = n + 1$. Its importance will become clearer in the next section when we discuss the Bessel function.

The fundamental identity also implies that one needs to know only the values of $\Gamma(x + 1)$ for $0 \le x \le 1$ to be able to compute $\Gamma(x)$ for any $x > 0$. For if $x > 2$, one can always find a positive integer n and real number α, $0 \le \alpha \le 1$, so that $x = n + \alpha + 1$. Then, by successively using the fundamental identity we obtain

$$\begin{aligned}
\Gamma(x) = \Gamma(n + \alpha + 1) &= (n + \alpha)\Gamma(n + \alpha) \\
&= (n + \alpha)(n + \alpha - 1)\Gamma(n + \alpha - 1) = \cdots \\
&= (n + \alpha)(n + \alpha - 1) \cdots (1 + \alpha)\Gamma(1 + \alpha),
\end{aligned} \qquad (6.5.1)$$

hence computing $\Gamma(x)$ becomes a matter of knowing $\Gamma(1 + \alpha)$ from a table, then a series of multiplications. Note that $\Gamma(x) = \Gamma(x + 1)/x$ and $\Gamma(1) = 1$ implies that $\Gamma(x) \to \infty$ as $x \to 0^+$.

Example 1 Find $\Gamma(3.77)$.

$$\begin{aligned}
\Gamma(3.77) = 2.77\,\Gamma(2.77) &= (2.77)(1.77)\,\Gamma(1.77) \\
&\approx (2.77)(1.77)(0.92376) = 4.52910
\end{aligned}$$

Extensive tables of $\Gamma(x)$ can be found, for instance, in [10]. □

Finally, observe that by using the fundamental identity in reverse, $\Gamma(x)$ can be defined for $x < 0$, where x is not a negative integer. For instance, to find $\Gamma(-\tfrac{5}{3})$, we write

$$\begin{aligned}
(-\tfrac{5}{3})\,\Gamma(-\tfrac{5}{3}) &= \Gamma(-\tfrac{5}{3} + 1) = \Gamma(-\tfrac{2}{3}), \\
(-\tfrac{2}{3})\,\Gamma(-\tfrac{2}{3}) &= \Gamma(-\tfrac{2}{3} + 1) = \Gamma(\tfrac{1}{3}),
\end{aligned}$$

and

$$(\tfrac{1}{3})\,\Gamma(\tfrac{1}{3}) = \Gamma(\tfrac{4}{3}).$$

All of this implies that

$$\Gamma(-\tfrac{5}{3}) = (-\tfrac{3}{5})(-3)(\tfrac{3}{2})\Gamma(\tfrac{4}{3}) = \tfrac{27}{10}\Gamma(\tfrac{4}{3}),$$

and the last value can be obtained from tables.

The Bessel Function

A differential equation that arises frequently in boundary value problems and problems in mathematical physics is the *Bessel equation*:

$$y'' + \frac{1}{x}y' + \frac{x^2 - \alpha^2}{x^2}y = 0, \qquad (6.5.2)$$

where α is a given parameter. For this discussion, α will be assumed to be a real number. The solutions of (6.5.2) occur in many types of potential problems involving cylindrical boundaries, as well as in such areas as elasticity theory, fluid mechanics, and electromagnetic field theory.

The Bessel equation is so important that literally hundreds of volumes of tables of its solutions, their derivatives, and their zeros for both integer and fractional values of α have been published. For instance, in the late 1940s the Harvard Computation Laboratory published 12 volumes, each of about 650 pages, giving values of solutions to (6.5.2) for $\alpha = 0, 1, 2, \ldots, 135$. This was done by means of computers, which by today's standards would be called prehistoric; today many of those tables have been replaced by stored computational packages and subroutines.

One sees immediately that $x = 0$ is a regular singular point of the Bessel equation and that $p(x) = 1$, $q(x) = -\alpha^2 + x^2$. Hence $p_0 = 1$, $q_0 = -\alpha^2$, and its indicial equation is

$$r^2 + (1 - 1)r - \alpha^2 = r^2 - \alpha^2 = 0.$$

Therefore its roots are $r_1 = \alpha$ and $r_2 = -\alpha$ and $r_1 - r_2 = 2\alpha$, which means there are two distinct cases:

1. $\alpha \neq 0$ and not a positive integer. Then 2α is not zero and not a positive integer, and there are two linearly independent solutions of the form

$$J_\alpha(x) = x^\alpha \sum_{n=0}^{\infty} a_n x^n, \qquad J_{-\alpha}(x) = x^{-\alpha} \sum_{n=0}^{\infty} b_n x^n$$

called *Bessel functions of fractional or nonintegral order α and $-\alpha$*.

2. $\alpha = 0$ or a positive integer, say $\alpha = m$. In this case one solution is of the form

$$J_m(x) = x^m \sum_{n=0}^{\infty} a_n x^n;$$

it is called the *Bessel function of order m*. The second linearly independent solution is of the logarithmic type and is denoted by $Y_m(x)$, the *Bessel function of the second kind of order m*.

In the earlier sections of this chapter we introduced $J_0(x)$. In what follows we will develop the series representation of $J_\alpha(x)$, $\mathrm{Re}(\alpha) \geq 0$, and list some of its important properties.

Write the Bessel equations as

$$x^2 y'' + xy' + (-\alpha^2 + x^2)y = 0,$$

let $y(x) = x^\alpha \sum_{n=0}^{\infty} a_n x^n = \sum_{n=0}^{\infty} a_n x^{n+\alpha}$, and substitute the series to obtain

$$x^2 \sum_{n=0}^{\infty} [(n + \alpha)(n + \alpha - 1)]a_n x^{n+\alpha-2}$$

$$+ x \sum_{n=0}^{\infty} (n + \alpha)a_n x^{n+\alpha-1} - \alpha^2 \sum_{n=0}^{\infty} a_n x^{n+\alpha} + x^2 \sum_{n=0}^{\infty} a_n x^{n+\alpha} = 0.$$

The first three series can be combined to get

$$\sum_{n=0}^{\infty} [(n + \alpha)^2 - \alpha^2] a_n x^{n+\alpha} + \sum_{n=0}^{\infty} a_n x^{n+\alpha+2} = 0.$$

Note that in the first series the term corresponding to $n = 0$ is zero. Therefore, write down the term for $n = 1$ and reindex to obtain

$$[(1 + \alpha)^2 - \alpha^2] a_1 x^{1+\alpha} + \sum_{n=0}^{\infty} \{[(n + 2 + \alpha)^2 - \alpha^2] a_{n+2} + a_n\} x^{n+\alpha+2} = 0.$$

Since $\alpha \geq 0$, this implies that $a_1 = 0$ and that

$$a_{n+2} = -\frac{a_n}{(n + 2 + \alpha)^2 - \alpha^2} = -\frac{a_n}{(n + 2)(n + 2 + 2\alpha)}, \quad n = 0, 1, \dots.$$

Hence $a_1 = a_3 = \cdots = a_{2n+1} = 0$ and

$$a_2 = -\frac{a_0}{2(2 + 2\alpha)} = -\frac{a_0}{2^2(1 + \alpha)},$$

$$a_4 = -\frac{a_2}{4(4 + 2\alpha)} = \frac{a_0}{2^4 \cdot 2(2 + \alpha)(1 + \alpha)},$$

$$a_6 = -\frac{a_4}{6(6 + 2\alpha)} = -\frac{a_0}{2^6 \cdot 3 \cdot 2(3 + \alpha)(2 + \alpha)(1 + \alpha)}, \dots,$$

$$a_{2n} = (-1)^n \frac{a_0}{2^{2n} n! (n + \alpha)(n - 1 + \alpha) \cdots (1 + \alpha)}, \quad n = 1, 2, \dots.$$

We conclude that

$$J_\alpha(x) = a_0 x^\alpha \sum_{n=0}^{\infty} \frac{(-1)^n}{2^{2n} n! (n + \alpha)(n - 1 + \alpha) \cdots (1 + \alpha)} x^{2n},$$

and convention requires that $a_0 = [2^\alpha \Gamma(\alpha + 1)]^{-1}$ (see relation 6.5.1) to obtain the compact form

$$J_\alpha(x) = \left(\frac{x}{2}\right)^\alpha \sum_{n=0}^{\infty} \frac{(-1)^n}{n! \Gamma(n + \alpha + 1)} \left(\frac{x}{2}\right)^{2n}.$$

In comparing this expression with the previous one we can see the notational convenience of using the Gamma function. We now examine the various possibilities for different values of α.

If $\alpha = m$, where $m = 0$ or m is a positive integer, then, since

$$\Gamma(n + m + 1) = (n + m)!,$$

we obtain

$$J_m(x) = \left(\frac{x}{2}\right)^m \sum_{n=0}^{\infty} \frac{(-1)^n}{n! (n + m)!} \left(\frac{x}{2}\right)^n,$$

the *Bessel function of order m.* From the series it follows

$$J_0(0) = 1, \quad J_m(0) = 0, \quad m = 1, 2, \ldots .$$

A more extensive analysis shows that $J_m(x)$ is an oscillatory function which approaches zero as $x \to \infty$, very similar to a damped cosine or sine function. For instance,

$$J_0(x_j) = 0 \quad \text{for} \quad x_j \approx 2.405, 5.520, 8.654, 11.97, \ldots$$

$$J_1(x_j) = 0 \quad \text{for} \quad x_j \approx 3.832, 7.016, 10.17, 13.32, \ldots$$

$$J_2(x_j) = 0 \quad \text{for} \quad x_j \approx 5.136, 8.417, 11.62, 14.80, \ldots$$

A graph of $J_m(x)$ for $m = 0, 1, 2$ is shown in Fig. 6.1; these three Bessel functions occur often in applications.

As mentioned above, when $\alpha = m$, $m = 0$ or a positive integer, the

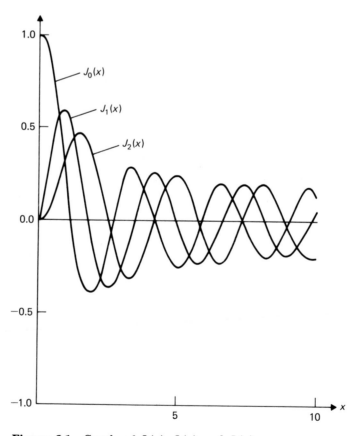

Figure 6.1 Graphs of $J_0(x)$, $J_1(x)$, and $J_2(x)$.

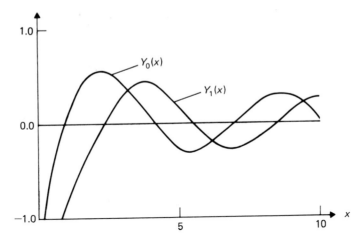

Figure 6.2 Graphs of $Y_0(x)$ and $Y_1(x)$.

second linearly independent solution of the Bessel equation is $Y_m(x)$, the *Bessel function of the second kind of order m,* and it is of logarithmic type. Its series representation is very complicated, but it is also an oscillatory function which, however, becomes unbounded as $x \to 0$. A graph of $Y_0(x)$ and $Y_1(x)$ is shown in Fig. 6.2.

There are many interesting properties and identities for the Bessel functions which we do not have time to discuss; the reader could browse through [5] or [8] to see the vast literature devoted to Bessel functions $J_m(x)$ and $Y_m(x)$. For this discussion we only need to state that when $\alpha = m$, $m = 0$ or a positive integer, a general solution of the Bessel equation (6.5.2) is

$$y(x) = aJ_m(x) + bY_m(x),$$

where a and b are arbitrary constants.

For the case $\alpha \neq 0$ or a positive integer, the theory of regular singular points tells us that

$$J_\alpha(x) = \left(\frac{x}{2}\right)^\alpha \sum_{n=0}^{\infty} \frac{(-1)^n}{n!\Gamma(n + \alpha + 1)} \left(\frac{x}{2}\right)^{2n}$$

and

$$J_{-\alpha}(x) = \left(\frac{x}{2}\right)^{-\alpha} \sum_{n=0}^{\infty} \frac{(-1)^n}{n!\Gamma(n - \alpha + 1)} \left(\frac{x}{2}\right)^{2n}$$

are a fundamental pair of solutions of the Bessel equation (6.5.2), and a general solution is

$$y(x) = aJ_\alpha(x) + bJ_{-\alpha}(x),$$

where a and b are arbitrary constants. The function $J_a(x)$ is oscillatory and approaches zero as $x \to \infty$, and for large x, $J_a(x)$ looks very much like a damped sine wave.

Finally, a frequently useful and time-saving fact is the following one: given an equation with a regular singular point at $x = 0$ in the form

$$x^2 y'' + (1 - 2s)xy' + [(s^2 - r^2\alpha^2) + a^2 r^2 x^{2r}]y = 0, \qquad (6.5.3)$$

where s, r, a, and α are given constants, every solution can be written in the form

$$y(x) = c_1 x^s J_\alpha(ax^r) + c_2 x^s J_{-\alpha}(ax^r),$$

where c_1 and c_2 are arbitrary constants. If $\alpha = 0$ or a positive integer, then we replace $J_{-\alpha}(ax^r)$ with $Y_\alpha(ax^r)$. This result is obtained by using a change of variable, and the verification is left to the reader in Exercise 8 (p. 333).

Example 2 **a)** In the previous section on regular singular points, the series expression for one solution of

$$x^2 y'' + 3xy' + (1 + x)y = 0$$

and the first few terms of the second solution (of logarithmic type) were developed. By comparison with the differential equation (6.5.3), we find:

$$1 - 2s = 3, \quad s^2 - r^2\alpha^2 = 1, \quad a^2 r^2 = 1, \quad 2r = 1,$$

which, when solved, gives $r = \frac{1}{2}$, $a = 2$, $s = -1$, $\alpha = 0$. Therefore a fundamental pair of solutions is

$$y_1(x) = x^{-1} J_0(2x^{1/2}), \quad y_2(x) = x^{-1} Y_0(2x^{1/2}).$$

b) Let $s = \frac{1}{2}$, $r = a = 1$ in (6.5.3) to obtain

$$x^2 y'' + [(\tfrac{1}{4} - \alpha^2) + x^2]y = 0$$

with solutions $x^{1/2} J_\alpha(x)$ and either $x^{1/2} J_{-\alpha}(x)$ or $x^{1/2} Y_\alpha(x)$. If $\alpha = \frac{1}{2}$, the above equation becomes $y'' + y = 0$, which means that $J_{1/2}(x)$ and $J_{-1/2}(x)$ can be expressed in terms of $x^{-1/2} \sin x$ and $x^{-1/2} \cos x$. In fact it can be shown that

$$J_{1/2}(x) = \sqrt{\frac{2}{\pi x}} \sin x, \qquad J_{-1/2}(x) = \sqrt{\frac{2}{\pi x}} \cos x,$$

so $J_{1/2}(x)$ and $J_{-1/2}(x)$ can be computed without the use of special tables. $\square$

The above examples show the advantage of the general form (6.5.3), namely that the solutions of a large class of equations with a regular singular point at $x = 0$ can be expressed in terms of well tabulated functions. Hence, the computation of complicated series solutions is avoided, which is a real benefit!

The Airy Functions

The *Airy functions* $\mathrm{Ai}(x)$, $\mathrm{Bi}(x)$ are the fundamental pair of solutions of *Airy's equation*

$$y'' - xy = 0, \qquad (6.5.4)$$

as was shown before. Similarly, $\mathrm{Ai}(-x)$, $\mathrm{Bi}(-x)$ denote the fundamental pair of solutions of

$$y'' + xy = 0, \qquad (6.5.5)$$

and both pairs of functions are tabulated, for example, in [10]. By writing (6.5.5) in the form

$$x^2 y'' + x^3 y = 0$$

we see that it is in the general form (6.5.3) with $1 - 2s = 0$, $s^2 - r^2\alpha^2 = 0$, $a^2 r^2 = 1$, and $2r = 3$. The solution of these equations is $r = \frac{3}{2}$, $s = \frac{1}{2}$, $a = \frac{2}{3}$, and $\alpha = \frac{1}{3}$; therefore $\mathrm{Ai}(-x)$ and $\mathrm{Bi}(-x)$ can be written as linear combinations of

$$\sqrt{x}\, J_{1/3}\left(\frac{2}{3} x^{3/2}\right) \qquad \text{and} \qquad \sqrt{x}\, J_{-1/3}\left(\frac{2}{3} x^{3/2}\right).$$

In fact, for $x > 0$,

$$\mathrm{Ai}(-x) = \frac{1}{3}\sqrt{x}\left[J_{1/3}\left(\frac{2}{3} x^{3/2}\right) + J_{-1/3}\left(\frac{2}{3} x^{3/2}\right) \right],$$

$$\mathrm{Bi}(-x) = -\sqrt{\frac{x}{3}}\left[J_{1/3}\left(\frac{2}{3} x^{3/2}\right) - J_{-1/3}\left(\frac{2}{3} x^{3/2}\right) \right],$$

and again we see how ubiquitous are the Bessel functions!

The Legendre Polynomials

The differential equation

$$(1 - x^2)y'' - 2xy' + \lambda y = 0, \qquad (6.5.6)$$

with λ a given parameter, is called *Legendre's equation;* it occurs frequently in the analysis of potential problems on spherical domains. One sees that $x = 0$ is an ordinary point since

$$p(x) = \frac{-2x}{1 - x^2}, \qquad q(x) = \frac{\lambda}{1 - x^2}$$

clearly are smooth functions for $|x| < 1$. From the equation in the form

$$y'' + \frac{1}{x-1}\frac{2x}{1+x}y' + \frac{1}{(x-1)^2}\frac{1-x}{1+x}\lambda y = 0 \qquad (6.5.7)$$

it follows that $x = 1$ is a regular singular point (and so is $x = -1$ by a similar rewriting).

For the regular singular point at $x = 1$ it is seen from (6.5.7) that

$$p(x) = \frac{2x}{1+x} \quad \text{and} \quad q(x) = \left(\frac{1-x}{1+x}\right)\lambda.$$

Therefore $p(1) = 1$, $q(1) = 0$, and the indicial equation is

$$r^2 + (1-1)r + 0 = r^2 = 0$$

with roots $r_1 = r_2 = 0$. This is the logarithmic case, and two linearly independent solutions valid in a deleted neighborhood of $x = 1$ are of the form

$$y_1(x) = \sum_{n=0}^{\infty} a_n(x-1)^n, \qquad y_2(x) = y_1(x)\beta \ln|x-1| + \sum_{n=0}^{\infty} b_n(x-1)^n.$$

A similar analysis can be made for $x = -1$, which is also a logarithmic case. The interesting case occurs when we examine $x = 0$.

Since $x = 0$ is an ordinary point, both linearly independent solutions have a series representation $y(x) = \sum_{n=0}^{\infty} a_n x^n$. Differentiating the series term by term and substituting it in (6.5.6) gives

$$\sum_{n=0}^{\infty} n(n-1)a_n x^{n-2} + \sum_{n=0}^{\infty} [-n(n-1) - 2n + \lambda]a_n x^n = 0.$$

Now shift the index by two in the first sum and combine terms to obtain

$$\sum_{n=0}^{\infty} \{(n+2)(n+1)a_{n+2} + [-n(n+1) + \lambda]a_n\}x^n = 0,$$

which leads to the recurrence relation

$$a_{n+2} = \frac{n(n+1) - \lambda}{(n+2)(n+1)} a_n, \quad n = 0, 1, 2, \ldots. \qquad (6.5.8)$$

It is seen that $a_2, a_4, \ldots, a_{2n} \ldots$ depend on a_0, while $a_3, a_5, \ldots, a_{2n+1}, \ldots$ depend on a_1, with a_0, a_1 arbitrary. This leads to two series, one in even powers of x and the other in odd powers of x, which are a linearly independent pair of solutions of Legendre's equation in a neighborhood of $x = 0$.

But the interesting case in many applications is when the parameter $\lambda = m(m+1)$ for some nonnegative integer m, in which case the relation (6.5.8) tells us that:

1. If m is even, then $a_2, a_4, \ldots, a_m$ will not be zero, but $a_{m+2}, a_{m+4}, \ldots$ will all be zero. Therefore one solution will be an even polynomial $P_m(x)$ of

degree m and the other solution will be a power series in odd powers of x.

2. If m is odd, then $a_3, a_5, \ldots, a_m$ will not be zero, but $a_{m+2}, a_{m+4}, \ldots$ will all be zero. Therefore one solution will be an odd polynomial $P_m(x)$ of degree m and the other solution will be a power series in even powers of x.

The polynomial solutions $P_m(x)$ described above are called *the Legendre polynomials* corresponding to $\lambda = m(m + 1)$. They are extremely important because they are the *only* solutions of Legendre's equation that are defined for $x = 0$ and are bounded at $x = \pm 1$ (the power series solutions diverge at $x = \pm 1$ like $\ln(1 \mp x)$).

The recursion relation (6.5.8) may be used to construct some Legendre polynomials, for instance,

a) If $\lambda = 20 = 4 \cdot 5$, then $m = 4$ and

$$a_{0+2} = a_2 = \frac{(0)(1) - 20}{(0 + 2)(0 + 1)} a_0 = -10a_0,$$

$$a_{2+2} = a_4 = \frac{(2)(3) - 20}{(2 + 2)(2 + 1)} a_2 = \frac{-14}{12}(-10a_0) = \frac{35}{3}a_0,$$

hence $P_4(x) = a_0(1 - 10x^2 + \tfrac{35}{3}x^4)$. It is conventional to choose a_0 so $P_4(1) = 1$; therefore

$$P_4(x) = \tfrac{3}{8}(1 - 10x^2 + \tfrac{35}{3}x^4).$$

b) If $\lambda = 30 = 5 \cdot 6$, then $m = 5$ and

$$a_{1+2} = a_3 = \frac{(1)(2) - 30}{(1 + 2)(1 + 1)} a_1 = -\frac{14}{3}a_1,$$

$$a_{3+2} = a_5 = \frac{(3)(4) - 30}{(3 + 2)(3 + 1)} a_3 = \frac{-18}{20}\frac{-14}{3}a_1 = \frac{21}{5}a_1,$$

and choosing a_1 so that $P_5(1) = 1$, we have

$$P_5(x) = \tfrac{15}{8}(x - \tfrac{14}{3}x^3 + \tfrac{21}{5}x^5).$$

The reader may check that

$$P_0(x) = 1, \quad P_1(x) = x, \quad P_2(x) = \tfrac{1}{2}(3x^2 - 1), \quad P_3(x) = \tfrac{1}{2}(\tfrac{5}{3}x^3 - x),$$

and for a further check, may wish to use the *Rodrigues' formula,*

$$P_m(x) = \frac{1}{2^m m!} \frac{d^m}{dx^m}(x^2 - 1)^m.$$

It generates all the Legendre polynomials by successive differentiation.

Finally, we remark that the Legendre polynomials are an example of a *family of orthogonal polynomials on* $-1 \le x \le 1$. By this we mean that

$$\int_{-1}^{1} P_n(x) P_m(x) \, dx = 0 \quad \text{if} \quad n \ne m.$$

This fact is extremely useful in boundary value problems and in approximation theory.

EXERCISES 6.5

1. Given

  a) $\Gamma(1.185) = 0.92229$; evaluate $\Gamma(5.185)$ and $\Gamma(-3.815)$.

  b) $\Gamma(1.910) = 0.96523$; evaluate $\Gamma(4.1910)$ and $\Gamma(-2.090)$.

2. Stirling's formula for an asymptotic approximation of $n! = \Gamma(n + 1)$ is

$$\Gamma(n + 1) \approx n^n e^{-n} \sqrt{2\pi n},$$

meaning that

$$\lim_{n \to \infty} \frac{\Gamma(n + 1)}{n^n e^{-n} \sqrt{2\pi n}} = 1.$$

This means the approximation is good as n gets large, in the sense of a small relative error *not* a small absolute error. Use Stirling's formula to find an approximation of $100! \approx 9.3326 \times 10^{157}$ and compute the relative and absolute errors. *Hint:* For computing, use the logarithmic form

$$\ln \Gamma(n + 1) \approx (n + \tfrac{1}{2}) \ln n - n + \tfrac{1}{2} \ln 2\pi.$$

3. The Maclaurin series for e^{-x} is

$$e^{-x} = \sum_{n=0}^{\infty} \frac{(-1)^n}{n!} x^n;$$

and since it is an alternating series, the error in stopping at N terms is less than the magnitude of the $(N + 1)$st term.

  a) Use Stirling's formula to approximate the error in estimating e^{-10} with the first 20, 25, and 30 terms of the series.

  b) How many terms would be needed to compute e^{-20} with an error of less than 10^{-10}?

One can infer from the above that for computational purposes, power series can be very inefficient!

4. By differentiating the series for the Bessel function term by term, show that

$$x J_p'(x) = p J_p(x) - x J_{p+1}(x),$$

$$x J_p'(x) = -p J_p(x) + x J_{p-1}(x).$$

5. The two relations in Exercise 4 imply the important recursion relation

$$J_{p+1}(x) = \frac{2p}{x} J_p(x) - J_{p-1}(x),$$

which allows one to compute higher order Bessel functions in terms of lower order ones. Use it and the expressions given on p. 328 for $J_{1/2}(x)$ and $J_{-1/2}(x)$ to obtain the values of

  a) $J_{3/2}(1.76)$,   b) $J_{-3/2}(0.587)$,

  c) $J_{5/2}(6.78)$.

6. Use the substitution $y = u/\sqrt{x}$ directly in Bessel's equation to obtain the differential equation

$$u'' + \left(1 - \frac{\alpha^2 - \frac{1}{4}}{x^2}\right) u = 0$$

with solutions $x^{1/2} J_\alpha(x)$.

7. The integral representation of $J_n(x)$,

$$J_n(x) = \frac{1}{\pi} \int_0^{\pi} \cos(n\theta - x \sin\theta) \, d\theta,$$

$$n = 0, 1, 2, \ldots,$$

was obtained by F. W. Bessel (1784–1846) in his study of astronomical orbits.

a) Use it to show that $J_{-n}(x) = (-1)^n J_n(x)$ and $J_0'(x) = -J_1(x)$.

b) Use it and a quadrature formula (e.g., Simpson's rule) to show that 2.405 is an approximate zero of $J_0(x)$.

c) Similarly, show that 3.833 is an approximate zero of $J_1(x)$.

8. Given the Bessel equation

$$z^2 w'' + zw' + (z^2 - \alpha^2)w = 0,$$

let $z = ax^r$, $y = x^s w$ and obtain the differential equation (6.5.3).

9. Express the solutions of the following differential equations in terms of Bessel functions by using (6.5.3):

a) $y'' + 4xy = 0$

b) $x^2 y'' + 5xy' + (3 + 4x^2)y = 0$

c) $2x^2 y'' - xy' + (1 + x^2)y = 0$

d) $x^2 y'' + 3xy' + (1 + x)y = 0$

e) $xy'' + y = 0$

10. Using Rodrigues' formula directly,

a) find $P_2(x)$, $P_3(x)$, and $P_4(x)$;

b) show that $\int_{-1}^{1} x^r P_m(x)\ dx = 0$, $r = 0, 1$, $\ldots$, $m - 1$. Hint: Use successive integration by parts.

11. Show that the Legendre equation can be written in the form

$$\frac{d}{dx}[(1 - x^2)y'] = -\lambda y$$

and use this to show the orthogonality relation

$$\int_{-1}^{1} P_n(x)P_m(x)\ dx = 0, \quad n \neq m.$$

Hint: Let $\lambda = m(m + 1)$ and $y(x) = P_m(x)$ in the differential equation, then multiply both sides by $P_n(x)$ and integrate on $-1 \leq x \leq 1$ by parts. Repeat the process with $\lambda = n(n + 1)$, etc.

12. The function $y(\theta) = P_n(\cos \theta)$, $0 < \theta < \pi$, where $P_n(x)$ is a Legendre polynomial, arises in potential theory for a spherical body. Show that $y(\theta)$ satisfies

$$y'' + (\cot \theta)y' + n(n + 1)y = 0.$$

MISCELLANEOUS EXERCISES

6.1 A solution $y(x)$ of an ordinary differential equation is said to be *oscillatory* if it vanishes infinitely often on a half line $x_0 \leq x < \infty$. The following comparison theorem is useful in determining whether solutions are oscillatory:

Given the two second order linear equations

$$y'' + p(x)y = 0 \qquad (1)$$
$$y'' + q(x)y = 0 \qquad (2)$$

with $q(x) \geq p(x)$ for $x \geq x_0$. If the solutions of (1) are oscillatory, then so are those of (2), and the zeros of the solutions of (2) are closer together than those of the solutions of (1).

Use the theorem to study the Airy equation

$$y'' + xy = 0, \quad x \geq 1,$$

and by comparison with equations of the form $y'' + k^2 y = 0$ show that its solutions are oscillatory. Furthermore show that:

a) The zeros of its solutions are less than π units apart.

b) The distance between successive zeros approaches zero as x approaches ∞.

c) There are either 1 or 2 zeros in the interval $1 \leq x \leq 4$. (A check with tables shows that $Ai(-x)$ vanishes once and $Bi(-x)$ vanishes twice in the interval.)

6.2 By transforming the Bessel equation to the form

$$y'' + \left[1 - \frac{\alpha^2 - 1/4}{x^2} \right] y = 0$$

(whose solutions are $x^{1/2} J_\alpha(x)$ and $x^{1/2} J_{-\alpha}(x)$ or $x^{1/2} Y_\alpha(x)$), use the comparison theorem above to show that:

a) The Bessel functions are oscillatory.

b) The zeros of $J_\alpha(x)$ are separated by more than π units if $\alpha^2 < \frac{1}{4}$.

6.3 Show that $x = 0$ is a regular singular point of the differential equation

$$x^2 y'' + (\sin x) y' - (\cos x) y = 0,$$

and that the roots of the indicial equation are $r_1 = 1$ and $r_2 = -1$. By expanding $\sin x$ and $\cos x$ in their Taylor series at $x = 0$ and by retaining enough terms, determine the first three nonzero terms of the series solution corresponding to r_1.

6.4 In the previous problem use the method of reduction of order to determine whether the second solution corresponding to $r_2 = -1$ is of logarithmic type or not.

6.5 Find the series solution of Laguerre's equation

$$xy'' + (1 - x) y' + \lambda y = 0$$

near the regular singular point $x_0 = 0$. Show that the solution reduces to a polynomial $L_m(x)$, called a Laguerre polynomial, if $\lambda = m$, a positive integer. Find $L_1(x)$, $L_2(x)$, and $L_3(x)$.

6.6 The Fourier–Legendre series expansion of a function $f(x)$, $-1 < x < 1$, is given by

$$f(x) = \sum_{n=0}^{\infty} a_n P_n(x), \qquad a_n = \frac{2n + 1}{2} \int_{-1}^{1} f(x) P_n(x) \, dx.$$

It will converge to $f(x)$, for instance, for any function $f(x)$ that is continuous and has a piecewise continuous first derivative.

a) Using the orthogonality relation (see Exercise 11 of Section 6.5) and the fact that $\int_{-1}^{1} P_n(x)^2 \, dx = 2/(2n + 1)$, obtain the above formula for a_n.

b) Show that if $f(x)$ is an even (odd) function, only even (odd) indexed terms will appear in the series.

   c) Use the formula above to compute the first three nonzero terms of the Fourier–Legendre series for $f(x) = |x|$, $-1 < x < 1$. Graph your result.

   d) Do the same as in (c) for $f(x) = e^x$, $-1 < x < 1$.

6.7 Find the solution of the boundary value problem

$$(1 - x)^2 y'' - 2xy' + 12y = 0, \qquad y(0) = 0, \qquad y(\tfrac{1}{2}) = 4.$$

6.8 Find the solution of the boundary value problem

$$y'' + \frac{1}{x} y' + \frac{x^2 - \tfrac{1}{4}}{x^2} y = 0, \qquad y(0) \text{ bounded}, \qquad y\left(\frac{\pi}{2}\right) = 1.$$

6.9 Using (6.5.3), determine for what values of α will the boundary value problem

$$x^2 y'' - 3xy' + \left(\frac{15}{4} + x^2\right) y = 0, \qquad y\left(\frac{\pi}{2}\right) = 0, \qquad y(\alpha\pi) = 0,$$

have nontrivial solutions or only the zero solution.

6.10 Derive the formula for the Laplace transform of t^α, where α is not necessarily an integer:

$$L\{t^\alpha\} = \frac{\Gamma(\alpha + 1)}{s^{\alpha+1}}; \quad \alpha > -1, \ \ \mathrm{Re}(s) > 0.$$

REFERENCES

A majority of books on ordinary differential equations discuss series solutions and special functions. For a more detailed discussion than is given here we could suggest

  1. W. E. Boyce and R. C. DiPrima, *Elementary Differential Equations*, 3d ed., Wiley, New York, 1977.

  2. E. A. Coddington, *An Introduction to Ordinary Differential Equations*, Prentice-Hall, Englewood Cliffs, New Jersey, 1961.

  3. E. D. Rainville, *Intermediate Differential Equations*, 2nd ed., Macmillan, New York, 1964.

  4. G. F. Simmons, *Differential Equations with Applications and Historical Notes*, McGraw-Hill, New York, 1972.

To get an idea of the depth of analysis devoted to special functions at the turn of this century, we recommend browsing through these two classics:

  5. G. N. Watson, *A Treatise on the Theory of Bessel Functions*, 2nd ed., Cambridge University Press, Cambridge, 1944.

  6. E. T. Whittaker and G. N. Watson, *A Course of Modern Analysis*, 4th ed., Cambridge University Press, Cambridge, 1965.

For a more accessible discussion of special functions we suggest

  7. H. Hochstadt, *Special Functions of Mathematical Physics*, Holt, New York, 1961.

8. H. Hochstadt, *The Functions of Mathematical Physics,* Wiley Interscience, New York, 1971.

9. F. W. J. Olver, *Introduction to Asymptotics and Special Functions,* Academic Press, New York, 1974.

For a book of tables and formulas of special functions the standby is

10. M. Abramowitz and I. A. Stegun, *Handbook of Mathematical Functions,* Applied Mathematics Series 55, National Bureau of Standards, Washington, D.C., 1964. (Also available from Dover Publications.)

Nonlinear Differential Equations

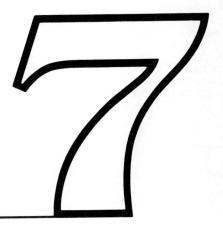

The study of nonlinear phenomena is an essential part of the education of an engineer, physical scientist, or mathematical biologist. Self-sustained oscillations, amplitude-dependent periodicities, and growth-limited populations are all nonlinear phenomena that can be modeled by nonlinear differential equations. This does not mean that the study of linear differential equations is unimportant. They successfully model low-current LCR circuits, small-amplitude mass–spring systems, and radioactive decay. They are also employed to approximate nonlinear differential equations. But much of nature is inherently nonlinear and must be modeled and studied accordingly. We shall start the study of nonlinear differential equations in this chapter. It is a fascinating and difficult subject but, fortunately, many of the essential ideas are accessible to undergraduates.

Most nonlinear differential equations cannot be solved explicitly, i.e., their solutions cannot be written in terms of a finite combination of elementary functions, such as polynomials or exponentials. Because of this, they are often studied by geometric methods as well as by numerical and analytical approximate methods. Geometric methods are employed to investigate qualitative properties, such as the behavior of solutions near equilibrium states (constant solutions), to characterize bounded and unbounded solution regimes, and to establish the existence of periodic solutions. Numerical methods are employed to construct approximate solutions to initial value problems and as an exploratory tool to help understand the qualitative properties of the solutions.

7.1 AN INTRODUCTORY EXAMPLE—THE PENDULUM

Our study begins with a description of the slightly damped or undamped motion of a simple pendulum. This motion is described by a nonlinear ordinary differential equation, and in this section we will describe (somewhat heuristically) two methods used to study nonlinear equations. In later sections the discussion of these methods will be greatly expanded.

A simple pendulum is a bob of mass m suspended from a fixed point 0 by a light rod of length a. If it is only allowed to swing in a vertical plane, its position is described by the angle x, the inclination of the rod to the downward vertical (Fig. 7.1). The angular momentum of the bob, $y = ma \, dx/dt$ and the angle x completely describe the *state* of the system at any time.

By Newton's second law, the time rate of change of the angular momentum is equal to the force acting perpendicular to the rod. If F represents the force due to friction and air resistance, then the total force is $-mg \sin x - F$. As in the case of the mass–spring systems discussed in Chapter 2, suppose we assume that the force F is proportional to the angular velocity and hence to the angular momentum. If the constant of proportionality is $2k$, $k > 0$, then the first order nonlinear system of differential equations

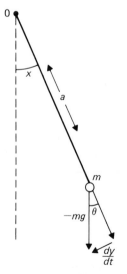

Figure 7.1   The simple pendulum.

governing the motion of the simple pendulum is

$$\frac{dx}{dt} = \frac{1}{ma} y, \qquad \frac{dy}{dt} = -mg \sin x - 2ky. \tag{7.1.1}$$

The system is nonlinear because of the presence of the term $\sin x$, a nonlinear function of the dependent variable x.

An equivalent second order nonlinear differential equation can be obtained from (7.1.1) by differentiating both sides of the first equation and substituting the second expression for dy/dt with y replaced by $ma \, dx/dt$. This gives

$$\frac{d^2x}{dt^2} = \frac{1}{ma} \frac{dy}{dt} = \frac{1}{ma} \left(-mg \sin x - 2kma \frac{dx}{dt} \right)$$

or

$$\frac{d^2x}{dt^2} + 2k \frac{dx}{dt} + \frac{g}{a} \sin x = 0, \tag{7.1.2}$$

which is called the *damped nonlinear simple pendulum equation.* We assume that the damping is light, which requires that the constant k satisfy the relation $0 < k^2 < g/a$.

An *equilibrium state* of the nonlinear system (7.1.1) is a constant solution, and immediate inspection shows that $x(t) = 0$, $y(t) = 0$ is an equilibrium state. It corresponds to the pendulum at rest with the bob directly below the suspension point. If it is assumed furthermore that the motion of the pendulum is restricted to small oscillations about the equilibrium state, then it seems reasonable to use the approximation $\sin x \approx x$ (valid for $|x| < 0.1$). This results in the approximate system

$$\frac{dx}{dt} = \frac{1}{ma} y, \qquad \frac{dy}{dt} = -mgx - 2ky, \tag{7.1.3}$$

that can be completely solved for given values of the constants m, g, a, and k by using the methods of Chapter 5.

The system (7.1.3) is the *linearized* system corresponding to the system (7.1.1) at the equilibrium point $(x, y) = (0, 0)$. The question is: do the solutions of the linearized system (7.1.3) approximate the solutions of the nonlinear system (7.1.1), and equally important, are their limiting qualitative behaviors the same? The answer is YES, and the particular conditions for which this is true will be discussed in the next section. Here we will give only some computer generated plots of $x(t)$ versus $y(t)$ for particular values of the parameters (see Fig. 7.2).

The variables x and y are called the *phase* of the motion and the xy plane is called the *phase plane*. The curves plotted above are called *phase paths*, or *orbits*, and the arrows indicate the evolution of the system, i.e., the direction of increasing time. The plots support the reasonable physical con-

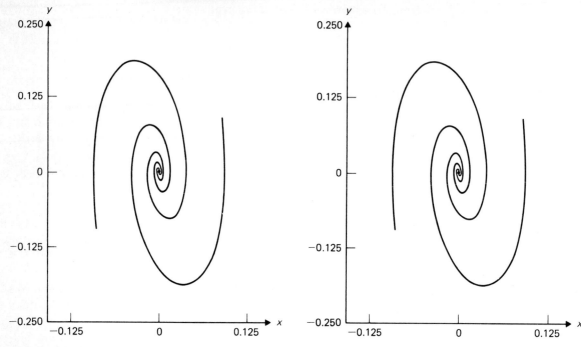

Figure 7.2(a) The nonlinear system ($k = 1$, $m = 2$, $a = 1$, $g = 32$).

Figure 7.2(b) The linearized system ($k = 1$, $m = 2$, $a = 1$, $g = 32$).

clusion that for small initial disturbances from rest, the pendulum will have a damped oscillatory motion which approaches the rest position. Thus we can say $(0, 0)$ is a *stable equilibrium state* (the precise notion of stability will be defined later).

It should be mentioned that there is a second equilibrium state $x(t) = \pi$, $y(t) = 0$. This can be thought of as the case where the pendulum is (very) carefully balanced on its end with the bob directly above the suspension point. Physical intuition tells us that $(\pi, 0)$ should be an *unstable equilibrium state* since a small disturbance results in a motion which does not return to the equilibrium state. Finally, because of the periodicity of sin x we should expect that the equilibrium states $(2n\pi, 0)$, $n = \pm 1, \pm 2, \ldots$, will be copies of $(0, 0)$.

Let us now return to the system (7.1.1) with $k = 0$ (i.e., the case of no damping or friction force) to demonstrate another quite distinct method for studying certain nonlinear systems. The system is therefore

$$\frac{dx}{dt} = \frac{1}{ma} y, \qquad \frac{dy}{dt} = -mg \sin x, \qquad (7.1.4)$$

which is equivalent to the simple undamped pendulum equation

$$\frac{d^2x}{dt^2} + \frac{g}{a}\sin x = 0. \tag{7.1.5}$$

Since the system is autonomous (does not depend explicitly on time), we can associate with it the first order differential equation

$$\frac{dy}{dx} = \frac{dy/dt}{dx/dt} = \frac{-mg\sin x}{y/ma}.$$

This equation is separable and has an implicit solution

$$E(x, y) = \frac{1}{2m}y^2 + mga(1 - \cos x) = \text{const.} \tag{7.1.6}$$

By graphing the curves described by (7.1.6) for various choices of the constant, a phase plane portrait of the system (7.1.4) can be obtained; it is shown in Fig. 7.3. A simple technique for obtaining the graphs shown will be given in a later section.

What is the physical significance of the curves shown in Fig. 7.3? By rewriting (7.1.6) as

$$E\left(x, \frac{dx}{dt}\right) = \frac{1}{2}m\left(a\frac{dx}{dt}\right)^2 + mg(1 - \cos x) = \text{const,}$$

we can see that the first term represents the *kinetic energy* of the pendulum, and the second term represents the *potential energy*. Hence E is simply the

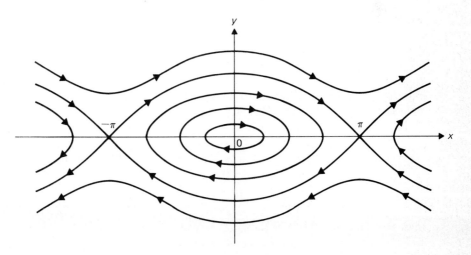

Figure 7.3 Phase plane portrait of the simple pendulum.

total energy of the pendulum, and the last expression, and consequently (7.1.6), simply expresses the fact that for the simple undamped pendulum, the total energy is always constant. It follows that the curves sketched in the phase plane portrait shown in Fig. 7.3 are simply curves of constant energy for various choices of the total energy value.

A system whose total energy is constant is called a *conservative system*, and the method described above to construct the phase plane portrait is called the *energy method*. It is typically used to study systems of the form

$$\dot{x} = y, \qquad \dot{y} = -g(x),$$

corresponding to second order undamped equations of the form

$$\frac{d^2x}{dt^2} + g(x) = 0.$$

Letting $g(x) = (g/a) \sin x$, we see that the simple undamped pendulum equation is of this type.

We have given a brief overview of two methods frequently used to study nonlinear systems of differential equations. The first consists of finding the equilibrium points of the system and studying the linearized system near each equilibrium point. The second is used for conservative systems and consists of finding the curves of constant energy. An amplified discussion of each method will be given in the following sections.

EXERCISES 7.1

1. For periodic motion, the average kinetic and potential energies are defined as

$$\overline{KE} = \frac{1}{T} \int_0^T \frac{1}{2} ma^2 \left(\frac{dx}{dt}\right)^2 dt,$$

$$\overline{PE} = \frac{1}{T} \int_0^T mga(1 - \cos x) \, dt,$$

where T is the period. Show that if (7.1.5) is replaced by

$$\frac{d^2x}{dt^2} + \omega^2 x = 0,$$

where

$$\omega^2 = \frac{g}{a},$$

then $T = 2\pi/\omega$ and $\overline{KE} = \overline{PE}$. (Note: $\cos x \approx 1 - x^2/2$ if x is small, where x is in radians.)

2. The lightly damped approximate pendulum

equation is given by

$$\frac{d^2x}{dt^2} + 2c\frac{dx}{dt} + \omega^2 x = 0, \quad c^2 < \omega^2, \quad 0 < c.$$

Its characteristic equation has roots

$$\lambda = -c + i\sigma, \lambda^* = -c - i\sigma,$$

where

$$\sigma = (\omega^2 - c^2)^{1/2}.$$

Let $T^* = 2\pi/\sigma$ denote the pseudoperiod. Replace T by T^* in the above formulas for $\overline{KE}$ and $\overline{PE}$ and compute $\Delta = \overline{KE} - \overline{PE}$. Take $\omega^2 = 1$ and plot Δ as a function of c.

3. Some of the orbits shown in Fig. 7.3 are closed curves which represent periodic motions of the simple undamped pendulum. If α is the value of x, where a closed orbit crosses the positive x axis (hence α is the maximum angular displacement), then it can be shown that the period T of the periodic motion is given by the expression

$$T = 4\left(\frac{a}{2g}\right)^{1/2} \int_0^\alpha (\cos x - \cos \alpha)^{-1/2} \, dx.$$

Show that an equivalent expression is

$$T = 2\left(\frac{a}{g}\right)^{1/2} \int_0^\alpha \left[\sin^2\left(\frac{\alpha}{2}\right) - \sin^2\left(\frac{x}{2}\right)\right]^{-1/2} \, dx,$$

and use this expression to show that as α approaches π, T approaches ∞.

4. In the second formula given in Exercise 3 let $\sin(x/2) = k \sin \phi$, $k = \sin(\alpha/2)$ and show that

$$T = 4\left(\frac{a}{g}\right)^{1/2} \int_0^{\pi/2} (1 - k^2 \sin^2 \phi)^{-1/2} \, d\phi.$$

The integral

$$K(k) = \int_0^{\pi/2} (1 - k^2 \sin^2 \phi)^{-1/2} \, d\phi$$

is called an *elliptic integral of the first kind*. It has been extensively studied and its values for selected k's can be found in most handbooks [14].

5. Using values from a handbook, plot $K(k) - \pi/2$ and $\alpha - \sin \alpha$ versus α for various values of α on the same graph. Comment on the relationship between the small-angle approximation $\sin \alpha \approx \alpha$ and $T \approx 2\pi(a/g)^{1/2}$.

6. The Schuler pendulum has period 24 hours and is of importance in the analysis of inertial guidance systems. Using $g = 9.80$ m/sec², plot the length (in meters) of a Schuler pendulum as a function of α, the maximum displacement.

7. Let $\eta = x/2$, $g/a = 1$ and transform the pendulum equation (7.1.5) into

$$\frac{d^2\eta}{dt^2} + \sin \eta \cos \eta = 0.$$

Select a value of the energy so that

$$\left(\frac{d\eta}{dt}\right)^2 = \cos^2 \eta$$

and derive formulas for η and $d\eta/dt$ as functions of t. Sketch the corresponding orbits in the phase plane.

8. If an external force is imposed on the pendulum, the differential equation governing the motion is

$$\frac{d^2x}{dt^2} + c\frac{dx}{dt} + \omega^2 \sin x = F(t).$$

Take $c = 0.1$, $\omega^2 = 4.0$, $F(t) = 2\cos t$, $x(0) = 0.1$, $\dot{x}(0) = 0$, and compute a numerical solution with RK4 on $0 \le t \le 10$ with $h = 0.1$. Plot $x(t)$ versus t and $\dot{x}(t)$ versus $x(t)$ on separate sheets of graph paper.

9. Show that the time rate of change of the total energy $E(x, \dot{x})$ for the damped nonlinear pendulum equation

$$\frac{d^2x}{dt^2} + c\frac{dx}{dt} + \omega^2 \sin x = 0$$

is given by

$$\frac{dE}{dt} = -c\left(\frac{dx}{dt}\right)^2 = -2c \text{ (Kinetic energy)},$$

where

$$E(x, \dot{x}) = \frac{1}{2}\dot{x}^2 + \omega^2(1 - \cos x).$$

If the pendulum is oscillating, the averaged total energy $\overline{E}$ is almost equal to one-half the averaged kinetic energy when c and x are small. Hence

$$\frac{d\overline{E}}{dt} \approx -c\overline{E} \quad \text{and} \quad \ln \overline{E} \approx -ct + \text{const.}$$

A plot of the logarithm of the *instantaneous energy* versus time should be a curve that oscillates around the plot of the logarithm of the *averaged energy* versus time. (The concept of a time-dependent averaged variable is developed in the method of averaging in [9].) Investigate this by integrating the damped nonlinear pendulum equation numerically with RK4 on $0 \le t \le 10$, $c = 0.1$, $\omega^2 = 4$, and $h = 0.1$. Take $x = x_0$, $\dot{x} = 0.5$ at $t = 0$, $x_0 = 0.0, 0.1, 0.5, 1.0$. Compute E and its logarithm at each step and plot $\ln E$ versus t for each choice of x_0. A sample plot is given in Fig. 7.4.

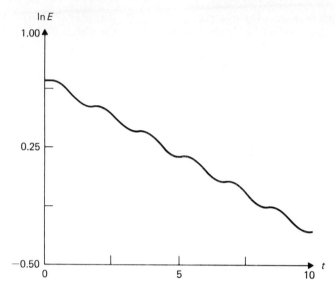

Figure 7.4   Plot of ln E versus t for Exercise 9 ($x_0 = 1.0$, $\dot{x}_0 = 0.5$, $c = 0.1$, $\omega^2 = 4.0$).

7.2 PHASE PATHS, EQUILIBRIUM STATES, AND ALMOST LINEAR SYSTEMS

In this section we begin the study of two-dimensional autonomous systems,

$$\frac{dx}{dt} = P(x, y), \qquad \frac{dy}{dt} = Q(x, y), \tag{7.2.1}$$

where $P(x, y)$, $Q(x, y)$ are continuously differentiable functions defined in a region of the xy plane. According to the existence theorem found in the Appendix 1, for any point (x_0, y_0) in the region of differentiability and for any t_0, there is a unique solution $x = x(t)$, $y = y(t)$ of (7.2.1) such that $x(t_0) = x_0$, $y(t_0) = y_0$. A curve in the *phase plane* (the xy plane) that is described parametrically by a solution is called a *trajectory:* on a trajectory, position is given as a function of time. More generally, the graph of a solution in the phase plane is a *phase path*, or *orbit*.

The following example should help distinguish between the concepts of a trajectory and a phase path. The equation

$$\frac{d^2x}{dt^2} + x = 0 \tag{7.2.2}$$

is equivalent to the system

$$\frac{dx}{dt} = y, \qquad \frac{dy}{dt} = -x, \qquad\qquad (7.2.3)$$

as is easily verified. A general solution of (7.2.2) is

$$x(t) = C \sin(t - \theta), \qquad\qquad (7.2.4)$$

where C and θ are arbitrary constants and t is the independent variable. Therefore, a general solution of the system (7.2.3) is

$$x(t) = C \sin(t - \theta), \qquad y(t) = C \cos(t - \theta), \quad -\infty < t < \infty. \quad (7.2.5)$$

Clearly, the graph of the solution (7.2.5) is the circle

$$x^2 + y^2 = C^2.$$

The circle is a phase path, while the graph of (7.2.5) is a trajectory. This may seem pedantic, but if you were tracking a satellite, its *trajectory* (position as a function of time) is far more important than its path. Other examples are easy to imagine.

For many problems, the constant solutions, if any, are significant. Constant solutions are called *critical points*, or *equilibrium points* of the system (7.2.1). Since the derivatives must vanish if the solution is constant, it follows that $(\bar{x}, \bar{y})$ are the coordinates of an equilibrium point if and only if

$$P(\bar{x}, \bar{y}) = 0 \qquad \text{and} \qquad Q(\bar{x}, \bar{y}) = 0. \qquad\qquad (7.2.6)$$

Example 1 Find the equilibrium points of the system

$$\frac{dx}{dt} = x(2 + x + y), \qquad \frac{dy}{dt} = y(2 - x + y).$$

The equilibrium points are the solutions to the nonlinear algebraic equations

$$x(2 + x + y) = 0, \qquad y(2 - x + y) = 0.$$

There are four possibilities:

a) $\bar{x} = 0, \quad \bar{y} = 0$

b) $\bar{x} = 0, \quad 2 - \bar{x} + \bar{y} = 0$

c) $2 + \bar{x} + \bar{y} = 0, \quad \bar{y} = 0$

d) $2 + \bar{x} + \bar{y} = 0, \quad 2 - \bar{x} + \bar{y} = 0$

Solving in each of the cases gives

a) $(0, 0)$, b) $(0, -2)$, c) $(-2, 0)$, d) $(0, -2)$;

hence there are three distinct equilibrium points. □

The solutions near an equilibrium point $(\bar{x}, \bar{y})$ can be studied by creating a linear differential equation that approximates the original differential equation near the equilibrium point. To do this, let $x = \bar{x} + u$, $y = \bar{y} + v$ and substitute them into the system (7.2.1) to obtain

$$\frac{dx}{dt} = 0 + \frac{du}{dt} = P(\bar{x} + u, \bar{y} + v),$$

$$\frac{dy}{dt} = 0 + \frac{dv}{dt} = Q(\bar{x} + u, \bar{y} + v).$$

It was assumed that $P(x, y)$ and $Q(x, y)$ are continuously differentiable and, consequently, the Taylor expansion gives

$$P(\bar{x} + u, \bar{y} + v) = P(\bar{x}, \bar{y}) + P_x(\bar{x}, \bar{y})u + P_y(\bar{x}, \bar{y})v + R(u, v),$$

$$Q(\bar{x} + u, \bar{y} + v) = Q(\bar{x}, \bar{y}) + Q_x(\bar{x}, \bar{y})u + Q_y(\bar{x}, \bar{y})v + S(u, v).$$

where the subscripts denote partial derivatives, and $R(u, v)$ and $S(u, v)$ are the remainder terms. Since $(\bar{x}, \bar{y})$ is an equilibrium point, then $P(\bar{x}, \bar{y}) = 0$, $Q(\bar{x}, \bar{y}) = 0$, and therefore u and v satisfy the differential equation

$$\frac{du}{dt} = au + bv + R(u, v),$$

$$\frac{dv}{dt} = cu + dv + S(u, v),$$

$$(7.2.7)$$

where the constants a, b, c, d are defined by

$$a = P_x(\bar{x}, \bar{y}), \qquad b = P_y(\bar{x}, \bar{y}),$$
$$c = Q_x(\bar{x}, \bar{y}), \qquad d = Q_y(\bar{x}, \bar{y}).$$

$$(7.2.8)$$

If $P(x, y)$ and $Q(x, y)$ have continuous second derivatives, then the remainders $R(u, v)$ and $S(u, v)$ are small near $u = 0$, $v = 0$. In fact, Taylor's theorem states that $R(u, v)$ and $S(u, v)$ vanish so rapidly that

$$\lim_{(u, v) \to (0, 0)} \frac{R(u, v)}{(u^2 + v^2)^{1/2}} = 0, \qquad \lim_{(u, v) \to (0, 0)} \frac{S(u, v)}{(u^2 + v^2)^{1/2}} = 0. \qquad (7.2.9)$$

If, in addition, the condition

$$ad - bc \neq 0 \qquad (7.2.10)$$

is satisfied, then it can be shown that the equilibrium point $(\bar{x}, \bar{y})$ is *isolated*. By this it is meant that there is some neighborhood of $(\bar{x}, \bar{y})$ containing no other equilibria.

> **Definition** A differential equation of the form (7.2.7), for which the conditions (7.2.9) and (7.2.10) are satisfied, is said to be *almost linear*.

The solutions to an almost linear differential equation are approximated in a neighborhood of $u = 0$, $v = 0$ by the solutions of the corresponding linear system

$$\frac{du}{dt} = au + bv, \qquad \frac{dv}{dt} = cu + dv. \qquad (7.2.11)$$

Therefore a description of the solutions of the linear system (7.2.11) near $(0, 0)$ will help analyze the behavior of solutions of the original nonlinear system (7.2.1) near the equilibrium point $(\bar{x}, \bar{y})$.

A precise formulation of the sense in which the solutions to an almost linear system are approximated by the solutions of the corresponding linear system can be given by using techniques developed in more advanced texts [3, 4, 5]. We shall illustrate it with a carefully chosen example of an almost linear system that can be solved explicitly—a rare occurrence!

Example 2 The system

$$\frac{dx}{dt} = -y(1 - x^2 - y^2), \qquad \frac{dy}{dt} = x(1 - x^2 - y^2)$$

has an isolated equilibrium point $\bar{x} = 0$, $\bar{y} = 0$, as well as a circle of equilibrium points, $\bar{x}^2 + \bar{y}^2 = 1$. The trivial change of variables, $x = 0 + u$, $y = 0 + v$ gives the almost linear system

$$\frac{du}{dt} = -v + v(u^2 + v^2), \qquad \frac{dv}{dt} = +u - u(u^2 + v^2),$$

and therefore $R(u, v) = v(u^2 + v^2)$ and $S(u, v) = -u(u^2 + v^2)$. The reader can easily check that both conditions (7.2.9) and (7.2.10) are satisfied. Now multiply the first equation by u, the second by v, and then add them to obtain

$$u\frac{du}{dt} + v\frac{dv}{dt} = -uv + vu + uv(u^2 + v^2) - vu(u^2 + v^2) = 0.$$

The last relation is equivalent to

$$\frac{1}{2}\frac{d}{dt}(u^2 + v^2) = 0,$$

which implies that $u^2 + v^2 = c^2$, a constant, i.e., the orbits of the almost linear system are circles. The orbits of the corresponding linear system

$$\frac{du}{dt} = -v, \qquad \frac{dv}{dt} = u$$

are also circles. It can be easily shown (see Exercise 17, p. 351) that

$$u_1(t) = \rho \cos(\omega t + \phi_0), \qquad v_1(t) = \rho \sin(\omega t + \phi_0),$$

where $\omega = 1 - \rho^2$, $0 \le \rho < 1$, is the solution to the almost linear system within the unit circle. Furthermore,

$$u_2(t) = \rho \cos(t + \phi_0), \qquad v_2(t) = \rho \sin(t + \phi_0)$$

is the corresponding solution to the linear system, and both solutions satisfy the same initial conditions

$$u(0) = u_0 = \rho \cos \phi_0, \qquad v(0) = v_0 = \rho \sin \phi_0.$$

Both solutions lie on the circle $u^2 + v^2 = \rho^2$, but the dependence of ω on ρ means that they will drift apart as t increases. However, since

$$|u_1(t) - u_2(t)| \le 2\rho, \qquad |v_1(t) - v_2(t)| \le 2\rho,$$

their difference is bounded by 2ρ, the diameter of the circle. Furthermore, if ρ is small, ω will be almost unity, so the solutions will drift apart very slowly. We conclude that the closer we are to the equilibrium point $(0, 0)$, the better the solutions of the linear system approximate the solutions of the almost linear system. $\square$

Example 3   Find the equilibrium points and construct approximate solutions by solving the corresponding linear differential equations for the system

$$\frac{dx}{dt} = x, \qquad \frac{dy}{dt} = y(1 - x + y).$$

Solving the equations

$$P(x, y) = x = 0, \qquad Q(x, y) = y(1 - x + y) = 0,$$

gives the two equilibrium points $(0, 0)$ and $(0, -1)$. Furthermore

$$P_x(\bar{x}, \bar{y}) = 1, \qquad P_y(\bar{x}, \bar{y}) = 0,$$
$$Q_x(\bar{x}, \bar{y}) = -\bar{y}, \qquad Q_y(\bar{x}, \bar{y}) = 1 - \bar{x} + 2\bar{y}.$$

We first study $(\bar{x}, \bar{y}) = (0, 0)$ and let $x = 0 + u$, $y = 0 + v$, which gives

$$a = P_x(0, 0) = 1, \qquad b = P_y(0, 0) = 0,$$
$$c = Q_x(0, 0) = 0, \qquad d = Q_y(0, 0) = 1.$$

Therefore

$$\frac{du}{dt} = u, \qquad \frac{dv}{dt} = v + (v^2 - uv)$$

and

$$\frac{du}{dt} = u, \qquad \frac{dv}{dt} = v,$$

are the almost linear and linear differential equations associated with the equilibrium point $(0, 0)$. To see that $(0, 0)$ is an isolated equilibrium point, the conditions (7.2.9) and (7.2.10) will be verified. Clearly, $ad - bc = 1 \neq 0$. To show that the remainder terms $R(u, v) = 0$, $S(u, v) = v^2 - uv$ satisfy (7.2.9), polar coordinates

$$u = \rho \cos \phi, \qquad v = \rho \sin \phi$$

are introduced. Then

$$\frac{S(u, v)}{(u^2 + v^2)^{1/2}} = \frac{S(\rho \cos \phi, \rho \sin \phi)}{\rho} = \frac{\rho^2 \sin^2 \phi - \rho^2 \cos \phi \sin \phi}{\rho}$$
$$= \rho(\sin^2 \phi - \cos \phi \sin \phi) \to 0 \quad \text{as} \quad \rho \to 0,$$

so both conditions are satisfied. The theory asserts that the solution to the linear system

$$u(t) = u_0 e^t, \qquad v(t) = v_0 e^t$$

is an approximate solution to the almost linear system for a limited time interval if $u_0^2 + v_0^2$ is small.

The second equilibrium point is $(\bar{x}, \bar{y}) = (0, -1)$, and since

$$a = P_x(0, -1) = 1, \qquad b = P_y(0, -1) = 0,$$
$$Q_x(0, -1) = 1, \qquad Q_y(0, -1) = -1,$$

substituting $x = 0 + u$, $y = -1 + v$ gives

$$\frac{du}{dt} = u, \qquad \frac{dv}{dt} = u - v + (v^2 - uv),$$

and

$$\frac{du}{dt} = u, \qquad \frac{dv}{dt} = u - v,$$

the almost linear and linear systems of differential equations associated with the equilibrium point $(0, 1)$. Observe that since $P(x, y)$ and $Q(x, y)$ in this problem are polynomials of degree not more than 2, then $R(u, v)$ and $S(u, v)$ do not depend on the coordinates of the equilibrium point being studied. The theory asserts that the solution to the linear system

$$u = u_0 e^t, \qquad v = \left(v_0 - \frac{u_0}{2} \right) e^{-t} + \frac{u_0}{2} e^t$$

is an approximate solution to the almost linear system for a limited time interval if $u_0^2 + v_0^2$ is small. $\square$

The *local nature* of the approximation of solutions near $(u, v) = (0, 0)$ of the almost linear system by the solutions of the linear system must be emphasized. Furthermore, it is not necessarily the case that the phase paths of the two systems are similar. For instance, the phase paths of the first could be a family of spirals focused at $(0, 0)$, while for the second system they could be a family of circles centered at $(0, 0)$. Hence local approximation would be good for short time intervals but not long ones (we will discuss this matter further in the next section). We shall also study the possible phase plane portraits of both linear and almost linear two-dimensional autonomous systems.

EXERCISES 7.2

Find the equilibrium points of the following systems:

1. $\dfrac{dx}{dt} = x + 2y + 1, \quad \dfrac{dy}{dt} = 3x + y - 2$

2. $\dfrac{dx}{dt} = 2x + 5y - 11, \quad \dfrac{dy}{dt} = x + y + 1$

3. $\dfrac{dx}{dt} = xy + x - 4y + 6, \quad \dfrac{dy}{dt} = x - y + 2$

4. $\dfrac{dx}{dt} = -x + y + 3, \quad \dfrac{dy}{dt} = 2xy + 8x - 12$

5. $\dfrac{dx}{dt} = 4x^2 + y^2 - 17, \quad \dfrac{dy}{dt} = x^2 + y^2 - 5$

6. $\dfrac{dx}{dt} = x^2 + y^2 - 2x - 2y, \quad \dfrac{dy}{dt} = x + y - 2$

Find the equilibrium points and construct approximate solutions near them by solving the corresponding system of linear differential equations.

7. $\dfrac{dx}{dt} = 2x - y - 1, \quad \dfrac{dy}{dt} = 3x + 6y - 24$

8. $\dfrac{dx}{dt} = 3x + 4y + 1, \quad \dfrac{dy}{dt} = 2x + y - 1$

9. $\dfrac{dx}{dt} = y, \quad \dfrac{dy}{dt} = 1 - e^x$

10. $\dfrac{dx}{dt} = 1 - y^3, \quad \dfrac{dy}{dt} = x$

11. $\dfrac{dx}{dt} = y + x^2, \quad \dfrac{dy}{dt} = x^2 - x - 2$

12. $\dfrac{dx}{dt} = 4 - y^2, \quad \dfrac{dy}{dt} = 2x$

Verify that the following systems are almost linear with respect to origin. *Hint:* Use polar coordinates.

13. $\dfrac{dx}{dt} = -4x + 5y + x \sin y, \quad \dfrac{dy}{dt} = x + 4y + y \sin x$

14. $\dfrac{dx}{dt} = 3x + 2y + x^2 + 2xy^4, \quad \dfrac{dy}{dt} = -2x + y + y^2$

Consider the initial value problem

$$\frac{dx}{dt} = P(x, y), \qquad \frac{dy}{dt} = Q(x, y),$$

$$x(0) = \bar{x} + u_0, \qquad y(0) = \bar{y} + v_0$$

where u_0 and v_0 are measurement errors and $(\bar{x}, \bar{y})$ is an equilibrium point of the system. Let

$$x(t) = \bar{x} + u(t), \qquad y(t) = \bar{y} + v(t),$$

then construct approximations for $u(t)$ and $v(t)$ by solving a suitable linear system of differential equations. Find conditions on the eigenvalues of the linear system so that the error is bounded for all positive t. Apply your analysis to the following initial value problems:

15. $\dfrac{dx}{dt} = -x - 3y + x \sin y, \quad x(0) = 0 + 10^{-4}$

$\dfrac{dy}{dt} = 3x - y + 1 - \cos y, \quad y(0) = 0 - 10^{-4}$

16. $\dfrac{dx}{dt} = x + 5y + xy, \quad x(0) = 0 + 10^{-10}$

$\dfrac{dy}{dt} = 5x + y + x^2 + y^2, \quad y(0) = 0$

17. Show that $x(t) = \rho \cos(\omega t + \phi), \; y(t) = \rho \sin(\omega t + \phi), \; \omega = 1 - \rho^2$, with ρ, ϕ arbitrary

constants, is a solution to the nonlinear system

$$\dfrac{dx}{dt} = -y(1 - x^2 - y^2), \quad \dfrac{dy}{dt} = x(1 - x^2 - y^2)$$

$$x(0) = x_0, \qquad y(0) = y_0, \qquad \rho^2 = x_0^2 + y_0^2.$$

18. Show that if $y_0 > x_0 \geq 0$, then the difference between the solutions of the two systems

$$\dfrac{dx}{dt} = x, \quad \dfrac{dy}{dt} = y$$

and

$$\dfrac{dx}{dt} = x, \quad \dfrac{dy}{dt} = y(1 - x + y),$$

where $x(0) = x_0, \; y(0) = y_0$, diverges as t increases. *Hint:* Let $y(t) = w(t)e^t, \; x(t) = x_0 e^t$ denote the solution to the nonlinear system. Derive the differential equation satisfied by w and show that $w(t) > y_0$ for t large.

7.3 THE PHASE PLANE AND STABILITY

The *phase plane* of solutions to a two-dimensional autonomous system

$$\dfrac{dx}{dt} = P(x, y), \quad \dfrac{dy}{dt} = Q(x, y) \tag{7.3.1}$$

can be considered to be generated by particles flowing in the plane with the velocity at the point (x, y) given by

$$\mathbf{V} = \dfrac{dx}{dt}\mathbf{i} + \dfrac{dy}{dt}\mathbf{j} = P(x, y)\,\mathbf{i} + Q(x, y)\,\mathbf{j}. \tag{7.3.2}$$

The curves generated by the particles are the graphs of a one-parameter family of solutions $x = x(t - s), \; y = y(t - s)$, where s is arbitrary. This *time shift invariance* of the solutions is a consequence of the system being autonomous, i.e., the independent variable t does not appear explicitly. To prove this invariance, observe that if

$$\dfrac{dx}{dt}(t) = P(x(t), y(t)), \quad \dfrac{dy}{dt}(t) = Q(x(t), y(t)),$$

then

$$\dfrac{d}{dt} x(t - s) = \dfrac{dx(t - s)}{d(t - s)} = P(x(t - s), y(t - s)).$$

$$\dfrac{d}{dt} y(t - s) = \dfrac{dx(t - s)}{d(t - s)} = Q(x(t - s), y(t - s)).$$

The phase plane of system (7.3.1) will also include equilibrium states, if any. Some will be stagnation points around which the flow swirls, while others will be sources from which all nearby paths exit or sinks into which they enter. There are also equilibrium states from which exactly two paths exit and into which two paths enter. Our understanding of the geometric structure of the phase paths is facilitated by constructing phase plane portraits consisting of sketches of selected phase paths and equilibrium states. It is convenient to extend the definition of a path, so that equilibrium states may be included. A path is said to *enter* an equilibrium point $(\bar{x}, \bar{y})$ if

$$\lim_{t \to +\infty} \frac{y(t) - \bar{y}}{x(t) - \bar{x}}$$

exists or diverges to $\pm\infty$. A path is said to *leave* an equilibrium state $(\bar{x}, \bar{y})$ if

$$\lim_{t \to -\infty} \frac{y(t) - \bar{y}}{x(t) - \bar{x}}$$

exists or diverges to $\pm\infty$. The reader should look ahead to Fig. 7.5(c) to see an example where two paths enter and two paths leave the (unstable) equilibrium state (0, 0).

Example 1 Consider the system

$$\frac{dx}{dt} = x, \qquad \frac{dy}{dt} = y.$$

The only equilibrium state is $\bar{x} = 0$, $\bar{y} = 0$, and there is a family of paths covering the entire xy plane (with exception of the origin) that leave the equilibrium state in every possible direction. These paths are the graphs of

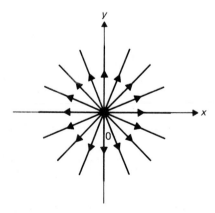

Figure 7.5

the family of solutions parametrized by α:

$$x(t) = (\cos \alpha)e^t, \qquad y(t) = (\sin \alpha)e^t, \quad 0 \le \alpha < 2\pi.$$

Clearly, $\lim\limits_{t \to -\infty} x(t) = \lim\limits_{t \to -\infty} y(t) = 0$ and

$$\lim_{t \to -\infty} \frac{(\sin \alpha)e^t - 0}{(\cos \alpha)e^t - 0} = \tan \alpha,$$

so the paths can leave in every possible direction. Figure 7.5 shows the phase plane of the system. □

Associated with the two-dimensional autonomous system (7.3.1) is the first order differential equation

$$\frac{dy}{dx} = \frac{Q(x, y)}{P(x, y)} \qquad \text{or} \qquad Q(x, y)\,dx - P(x, y)\,dy = 0. \qquad (7.3.3)$$

The graphs of solutions to (7.3.3) are called *integral curves* and an implicit solution $F(x,y) = $ const is called an *integral*. Unlike phase paths, integral curves are not directed since t, the independent variable in (7.3.1), has been eliminated in (7.3.3). The integral curves contain the phase paths, but the converse is not true. This is illustrated by the system in Example 1 whose associated first order equation

$$\frac{dy}{dx} = \frac{y}{x} \qquad \text{or} \qquad y\,dx - x\,dy = 0$$

has an implicit solution, $\arctan(y/x) = $ const or $y = mx$. Each choice of m determines a straight line through the origin that contains two phase paths from the family previously given. These are

$$x(t) = (\cos \alpha)e^t, \qquad y(t) = (\sin \alpha)e^t$$

and

$$x(t) = [\cos(\alpha + \pi)]e^t = -(\cos \alpha)e^t,$$
$$y(t) = [\sin(\alpha + \pi)]e^t = -(\sin \alpha)e^t,$$

where α, $0 \le \alpha < \pi$, is chosen to satisfy $m = (\sin \alpha)/(\cos \alpha)$ with the choice of $\alpha = \pi/2$ corresponding to the y axis.

A property of integral curves, and therefore of phase paths, is that they cannot intersect except at equilibrium states. Otherwise there would be two solutions of the first order equation (7.3.3), $y = y_1(x)$, $y = y_2(x)$, taking on the same initial value $y_0 = y_1(x_0)$, $y_0 = y_2(x_0)$, where (x_0, y_0) is the hypothesized point of intersection. This would violate the uniqueness of solutions of the initial value problem.

The easiest class of two-dimensional autonomous systems to study are linear constant coefficient systems that have the origin as an isolated equilibrium state

$$\frac{dx}{dt} = ax + by, \qquad \frac{dy}{dt} = cx + dy, \qquad ad - bc \neq 0. \qquad (7.3.4)$$

The eigenvalue–eigenvector method of Chapter 5 provides us with a systematic approach to the construction of phase plane portraits for these systems. Let $x = \alpha e^{\lambda t}$, $y = \beta e^{\lambda t}$. Then, after substituting and dividing by $e^{\lambda t}$, we have

$$\lambda \alpha = a\alpha + b\beta, \qquad \lambda \beta = c\alpha + d\beta.$$

The eigenvalues λ_1 and λ_2 are roots of the characteristic equation

$$0 = \begin{vmatrix} a - \lambda & b \\ c & d - \lambda \end{vmatrix} = \lambda^2 - (a + d)\lambda + (ad - bc).$$

Note that the assumption $ad - bc \neq 0$ implies that no eigenvalue can be zero.

The equilibrium state $\bar{x} = 0$, $\bar{y} = 0$ is classified by the type of eigenvalue. If the eigenvalues are real, there are five possibilities:

Real 1: $\lambda_1 < \lambda_2 < 0$,   Real 4: $\lambda_1 = \lambda_2 < 0$,

Real 2: $0 < \lambda_1 < \lambda_2$,   Real 5: $0 < \lambda_1 = \lambda_2$.

Real 3: $\lambda_2 < 0 < \lambda_1$,

If the eigenvalues are complex, $\lambda = p + iq$, $\lambda^* = p - iq$, there are three possibilities:

Complex 1: $p = \text{Re}(\lambda) < 0$,

Complex 2: $p = \text{Re}(\lambda) > 0$,

Complex 3: $p = \text{Re}(\lambda) = 0$.

Each case will be examined in some detail and a typical phase plane portrait will be sketched.

Real 1: $\lambda_1 < \lambda_2 < 0$. By using the theory developed in Chapter 5 it can be established that there are two linearly independent eigenvectors and that a general solution to (7.3.4) can be written in the form

$$x(t) = c_1 \alpha_1 e^{\lambda_1 t} + c_2 \alpha_2 e^{\lambda_2 t},$$

$$y(t) = c_1 \beta_1 e^{\lambda_1 t} + c_2 \beta_2 e^{\lambda_2 t}$$

or

$$\begin{bmatrix} x(t) \\ y(t) \end{bmatrix} = c_1 \begin{bmatrix} \alpha_1 \\ \beta_1 \end{bmatrix} e^{\lambda_1 t} + c_2 \begin{bmatrix} \alpha_2 \\ \beta_2 \end{bmatrix} e^{\lambda_2 t}.$$

The vectors in parentheses are the eigenvectors associated with λ_1 and λ_2, respectively. Points on the half-lines determined by the eigenvectors flow toward the origin as t increases since the exponentials $e^{\lambda_1 t}$ and $e^{\lambda_2 t}$ tend to zero. Points in the sectors between these half-lines flow toward the origin asymptotically to the half-line associated with λ_2 since $\lambda_1 < \lambda_2 < 0$ implies that $e^{\lambda_1 t}$ damps out more rapidly than $e^{\lambda_2 t}$. A more formal justification is given by the computation

$$\frac{y(t)}{x(t)} = \frac{e^{\lambda_2 t}(c_1\beta_1 e^{(\lambda_1-\lambda_2)t} + c_2\beta_2)}{e^{\lambda_2 t}(c_1\alpha_1 e^{(\lambda_1-\lambda_2)t} + c_2\alpha_2)} \to \frac{\beta_2}{\alpha_2}$$

as $t \to \infty$ since $\lambda_1 - \lambda_2 < 0$.

Example 2   Find the general solution and sketch a phase plane portrait for the system

$$\frac{dx}{dt} = -6x + 3y, \qquad \frac{dy}{dt} = -2x - y.$$

The characteristic equation

$$0 = \begin{vmatrix} -6 - \lambda & 3 \\ -2 & -1 - \lambda \end{vmatrix} = \lambda^2 + 7\lambda + 12 = (\lambda + 3)(\lambda + 4)$$

has roots $\lambda_1 = -4$, $\lambda_2 = -3$. The eigenvectors are found by solving the algebraic system

$$(-6 - \lambda)\alpha + 3\beta = 0, \qquad -2\alpha + (-1 - \lambda)\beta = 0$$

with $\lambda = \lambda_1 = -4$ and $\lambda = \lambda_2 = -3$. This leads to the equations

$$-2\alpha_1 + 3\beta_1 = 0 \qquad \text{and} \qquad -3\alpha_2 + 3\beta_2 = 0.$$

Nontrivial solutions are $\alpha_1 = 3$, $\beta_1 = 2$, and $\alpha_2 = 1$, $\beta_2 = 1$, hence a general solution is

$$\begin{bmatrix} x(t) \\ y(t) \end{bmatrix} = c_1 \begin{bmatrix} 3 \\ 2 \end{bmatrix} e^{-4t} + c_2 \begin{bmatrix} 1 \\ 1 \end{bmatrix} e^{-3t}.$$

If c_1 and c_2 are not both zero, the first term disappears more rapidly than the second, so as the phase path approaches the origin it becomes asymptotic to the line $y = x$ determined by the second eigenvector. A phase plane portrait is sketched in Fig. 7.6.

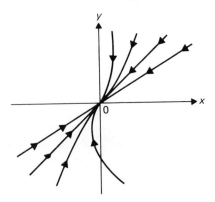

Figure 7.6

If all the phase paths enter the equilibrium state as in the case above, the equilibrium state is called a *stable node*. If all the phase paths leave the equilibrium state, the equilibrium state is called an *unstable node*. The latter is illustrated by the next type.

Real 2: $0 < \lambda_2 < \lambda_1$. Since the eigenvalues λ_1 and λ_2 are both positive, the phase paths enter the origin as $t \to -\infty$. The preceding phase plane portrait described for the case $\lambda_1 < \lambda_2 < 0$ is unchanged except that the arrows showing the direction of increasing time point outward instead of inward.

Example 3   By replacing t by $-t$ in Example 2 we obtain the system

$$\frac{dx}{dt} = 6x - 3y, \qquad \frac{dy}{dt} = 2x + y.$$

The eigenvalues are $\lambda_1 = 4$, $\lambda_2 = 3$, and a general solution is

$$\begin{bmatrix} x(t) \\ y(t) \end{bmatrix} = c_1 \begin{bmatrix} 3 \\ 2 \end{bmatrix} e^{4t} + c_2 \begin{bmatrix} 1 \\ 1 \end{bmatrix} e^{3t}.$$

A phase plane portrait is sketched in Fig. 7.7. $\square$

Real 3: $\lambda_2 < 0 < \lambda_1$. If the eigenvalues are real and of opposite sign, the equilibrium state is *unstable* and is called a *saddle point*. The points on the half-lines corresponding to the negative eigenvalue flow toward the origin, and those on the half-lines corresponding to the positive eigenvalue flow away from the origin as time increases. A general solution with $c_1, c_2 \neq 0$ is of the form

$$\begin{bmatrix} x(t) \\ y(t) \end{bmatrix} = c_1 \begin{bmatrix} \alpha_1 \\ \beta_1 \end{bmatrix} e^{\lambda_1 t} + c_2 \begin{bmatrix} \alpha_2 \\ \beta_2 \end{bmatrix} e^{\lambda_2 t},$$

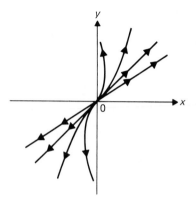

Figure 7.7

and therefore

$$\frac{y(t)}{x(t)} = \frac{e^{\lambda_1 t}(c_1\beta_1 + c_2\beta_2 e^{(\lambda_2 - \lambda_1)t})}{e^{\lambda_1 t}(c_1\alpha_1 + c_2\alpha_2 e^{(\lambda_2 - \lambda_1)t})} \rightarrow \frac{\beta_1}{\alpha_1}$$

as $t \rightarrow \infty$ since $\lambda_2 - \lambda_1 < 0$, and

$$\frac{y(t)}{x(t)} = \frac{e^{\lambda_2 t}(c_1\beta_1 e^{(\lambda_1 - \lambda_2)t} + c_2\beta_2)}{e^{\lambda_2 t}(c_1\alpha_1 e^{(\lambda_1 - \lambda_2)t} + c_2\alpha_2)} \rightarrow \frac{\beta_2}{\alpha_2}$$

as $t \rightarrow -\infty$ since $\lambda_1 - \lambda_2 > 0$. Therefore, points on the sectors between the half-lines flow away from the origin asymptotically toward the half-line associated with the positive eigenvalue as $t \rightarrow +\infty$ and asymptotically toward the half-line associated with the negative eigenvalue as $t \rightarrow -\infty$.

Example 4   Find a general solution and sketch the phase plane portrait for

$$\frac{dx}{dt} = x + 3y, \quad \frac{dy}{dt} = 3x + y.$$

The characteristic equation

$$0 = \begin{vmatrix} 1 - \lambda & 3 \\ 3 & 1 - \lambda \end{vmatrix} = \lambda^2 - 2\lambda - 8 = (\lambda + 2)(\lambda - 4)$$

has roots $\lambda_1 = 4$ and $\lambda_2 = -2$, and by computing the eigenvectors a general solution of the system is found to be

$$\begin{bmatrix} x(t) \\ y(t) \end{bmatrix} = c_1 \begin{bmatrix} 1 \\ 1 \end{bmatrix} e^{4t} + c_2 \begin{bmatrix} 1 \\ -1 \end{bmatrix} e^{-2t}. \quad \square$$

A phase plane portrait is sketched in Fig. 7.8. This sketch should be compared with the sketch of paths near the unstable equilibrium states $(\pi, 0)$ or $(-\pi, 0)$ of the pendulum equation in Fig. 7.3.

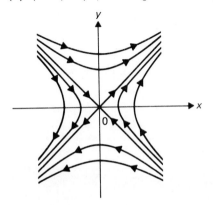

Figure 7.8

Real 4: $\lambda_1 = \lambda_2 < 0$. If the eigenvalues are real, equal, and negative, the equilibrium state is a *stable node*. If λ is the common value, then there are two subcases that require separate discussion:

a) There are two linearly independent eigenvectors;

b) There is one linearly independent eigenvector.

We consider first subcase (a). Let

$$\begin{bmatrix} \alpha_1 \\ \beta_1 \end{bmatrix} \quad \text{and} \quad \begin{bmatrix} \alpha_2 \\ \beta_2 \end{bmatrix}$$

be the two linearly independent eigenvectors. The algebraic system satisfied by the components of the eigenvectors,

$$\begin{cases} (a - \lambda)\alpha_1 + b\beta_1 = 0 \\ c\alpha_1 + (d - \lambda)\beta_1 = 0 \end{cases} \quad \text{and} \quad \begin{cases} (a - \lambda)\alpha_2 + b\beta_2 = 0 \\ c\alpha_2 + (d - \lambda)\beta_2 = 0 \end{cases}$$

can be regrouped (λ is the same throughout) to give

$$\begin{cases} (a - \lambda)\alpha_1 + b\beta_1 = 0 \\ (a - \lambda)\alpha_2 + b\beta_2 = 0 \end{cases} \quad \text{and} \quad \begin{cases} c\alpha_1 + (d - \lambda)\beta_1 = 0 \\ c\alpha_2 + (d - \lambda)\beta_2 = 0. \end{cases}$$

The linear independence of the eigenvectors implies that

$$\begin{vmatrix} \alpha_1 & \alpha_2 \\ \beta_1 & \beta_2 \end{vmatrix} = \alpha_1\beta_2 - \alpha_2\beta_1 \neq 0,$$

and therefore each of the last pairs of equations can only have trivial solutions, hence $a - \lambda = 0$, $b = 0$, $c = 0$, $d - \lambda = 0$. This means that when there is a double eigenvalue and two linearly independent eigenvectors, the system must be of the form

$$\frac{dx}{dt} = \lambda x, \qquad \frac{dy}{dt} = \lambda y.$$

A general solution is

$$\begin{bmatrix} x(t) \\ y(t) \end{bmatrix} = c_1 \begin{bmatrix} 1 \\ 0 \end{bmatrix} e^{\lambda t} + c_2 \begin{bmatrix} 0 \\ 1 \end{bmatrix} e^{\lambda t} = \begin{bmatrix} c_1 \\ c_2 \end{bmatrix} e^{\lambda t},$$

and since $\lambda < 0$, all points flow into the origin on straight lines determined by the choice of c_1 and c_2 (see Fig. 7.9).

If there is only one linearly independent eigenvector, a general solution can be written in the form

$$\begin{bmatrix} x(t) \\ y(t) \end{bmatrix} = c_1 \begin{bmatrix} \alpha \\ \beta \end{bmatrix} e^{\lambda t} + c_2 \left(\begin{bmatrix} \eta \\ \sigma \end{bmatrix} + t \begin{bmatrix} \alpha \\ \beta \end{bmatrix} \right) e^{\lambda t}.$$

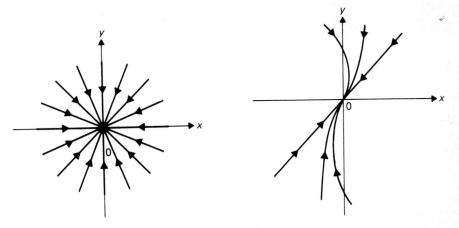

Figure 7.9   **Figure 7.10**

The points on the two half-lines determined by the one eigenvector flow toward the origin as t increases. All other points flow toward the origin asymptotically to these half-lines since the $te^{\lambda t}$ term in the general solution dominates:

$$\frac{y(t)}{x(t)} = \frac{e^{\lambda t}(c_1\beta + c_2\sigma + c_2 t\beta)}{e^{\lambda t}(c_1\alpha + c_2\eta + c_2 t\alpha)} = \frac{c_2\beta + (c_1\beta + c_2\sigma)/t}{c_2\alpha + (c_1\alpha + c_2\eta)/t} \to \frac{\beta}{\alpha}$$

as $t \to \infty$.

Example 5 Find a general solution and sketch the phase plane portrait for

$$\frac{dx}{dt} = -4x + y, \qquad \frac{dy}{dt} = -x - 2y.$$

The characteristic equation $\lambda^2 + 6\lambda + 9 = (\lambda + 3)^2 = 0$ has a double root $\lambda = -3$. Following the procedure discussed in Chapter 5, we set $\lambda = -3$ in the expressions

$$(-4 - \lambda)\alpha + \beta = 0, \qquad (-4 - \lambda)\eta + \sigma = \alpha,$$
$$-\alpha + (-2 - \lambda)\beta = 0, \qquad -\eta + (-2 - \lambda)\sigma = \beta,$$

and obtain $\alpha = 1, \beta = 1, \eta = 0, \sigma = 1$. A general solution is

$$\begin{bmatrix} x(t) \\ y(t) \end{bmatrix} = c_1 \begin{bmatrix} 1 \\ 1 \end{bmatrix} e^{-3t} + c_2 \left(\begin{bmatrix} 0 \\ 1 \end{bmatrix} + t \begin{bmatrix} 1 \\ 1 \end{bmatrix} \right) e^{-3t}.$$

A phase plane portrait is sketched in Fig. 7.10. □

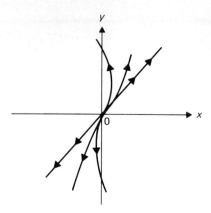

Figure 7.11

Real 5: $0 < \lambda_1 = \lambda_2$. If the common eigenvalue is positive, the phase paths enter the origin as $t \to -\infty$. Once again, the direction of the arrows of the preceding phase plane portraits is reversed (see Figs. 7.5 and 7.11).

Example 6 Solve

$$\frac{dx}{dt} = 4x - y, \qquad \frac{dy}{dt} = x + 2y.$$

The characteristic equation $\lambda^2 - 6\lambda + 9 = (\lambda - 3)^2 = 0$ has a double root $\lambda = 3$ and a general solution is

$$\begin{bmatrix} x(t) \\ y(t) \end{bmatrix} = c_1 \begin{bmatrix} 1 \\ 1 \end{bmatrix} e^{3t} + c_2 \left(\begin{bmatrix} 2 \\ 1 \end{bmatrix} + t \begin{bmatrix} 1 \\ 1 \end{bmatrix} \right) e^{3t}. \qquad \square$$

Let us now consider the remaining cases for which the eigenvalues are complex. To avoid being lost in a jungle of algebra, only systems of the form (called the *real canonical form*)

$$\frac{dx}{dt} = px - qy, \qquad \frac{dy}{dt} = qx + py$$

will be considered. It is known that any two-dimensional linear system with complex eigenvalues can be transformed into such a system by a change of coordinates. The characteristic equation of the above system is

$$\begin{vmatrix} p - \lambda & -q \\ q & p - \lambda \end{vmatrix} = (p - \lambda)^2 + q^2 = 0,$$

and the eigenvalues are $\lambda = p + iq$, $\lambda^* = p - iq$. Since there are no real eigenvectors, another approach to obtain the geometry of the phase plane is needed.

One approach that works is to write the system using polar coordinates. Let $x = r \cos \theta$, $y = r \sin \theta$; differentiating and substituting into the above system gives

$$\frac{dx}{dt} = \frac{dr}{dt} \cos \theta - r \sin \theta \frac{d\theta}{dt} = pr \cos \theta - qr \sin \theta,$$

$$\frac{dy}{dt} = \frac{dr}{dt} \sin \theta + r \cos \theta \frac{d\theta}{dt} = qr \cos \theta + pr \sin \theta.$$

To solve for dr/dt, multiply the first equation by $\cos \theta$, the second by $\sin \theta$ and add to get

$$\frac{dr}{dt} (\cos^2 \theta + \sin^2 \theta) = pr(\cos^2 \theta + \sin^2 \theta) \quad \text{or} \quad \frac{dr}{dt} = pr.$$

Proceeding in a similar manner to solve for $d\theta/dt$, one obtains

$$r \frac{d\theta}{dt} (\sin^2 \theta + \cos^2 \theta) = qr(\sin^2 \theta + \cos^2 \theta)$$

or, if $r \neq 0$,

$$\frac{d\theta}{dt} = q.$$

Therefore, a general solution is

$$r = r_0 e^{pt}, \qquad \theta = qt + \theta_0,$$

or

$$x(t) = r_0 e^{pt} \cos(qt + \theta_0), \qquad y(t) = r_0 e^{pt} \sin(qt + \theta_0),$$

where r_0 and θ_0 can be determined if initial conditions are given. The phase paths are spirals if $p \neq 0$; they are circles if $p = 0$. The direction of the flow is determined by q and is clockwise if q is negative, counterclockwise if q is positive. The complex eigenvalue cases can now be summarized.

Complex 1: p = Re(λ) < 0. The origin is a *stable spiral point*. The phase paths spiral in toward the origin with the orientation determined by the sign of q.

Complex 2: p = Re(λ) > 0. The origin is an *unstable spiral point*. The phase paths spiral out from the origin with the orientation determined by the sign of q.

Complex 3: p = Re(λ) = 0. The origin is a *center*. The phase paths are circles with their orientation determined by the sign of q. They are the graphs of periodic solutions to the differential equation.

Example 7 For each of the following differential equations sketch a phase plane portrait.

$$Complex\ 1: \quad \frac{dx}{dt} = -3x + y, \quad \frac{dy}{dt} = -y - 3x.$$

Since $p = -3$, the origin is a stable spiral point, and since $q = -1$, the flow is clockwise. A general solution is

$$x(t) = r_0 e^{-3t} \cos(-t + \theta_0), \quad y(t) = r_0 e^{-3t} \sin(-t + \theta_0),$$

and a phase plane portrait is sketched in Fig. 7.12(a).

$$Complex\ 2: \quad \frac{dx}{dt} = 3x + 2y, \quad \frac{dy}{dt} = -5x + y.$$

This system is not in the real canonical form that was discussed above, but once its eigenvalues are computed, the qualitative behavior of the phase paths is known. The characteristic equation

$$0 = \begin{vmatrix} 3 - \lambda & 2 \\ -5 & 1 - \lambda \end{vmatrix} = \lambda^2 - 4\lambda + 13$$

has roots $\lambda = 2 + 3i$, $\lambda^* = 2 - 3i$. Since $\text{Re}(\lambda) = 2 > 0$, the phase paths spiral outward. Because the system is not in canonical form, the direction of the flow cannot be determined by the sign of q since it is not known. However the direction is easily found by determining whether y is decreasing or increasing as a phase path crosses the x axis. In the second differential equation, if $x > 0$ and $y = 0$, then $dy/dt < 0$, hence y is decreasing. The flow must be clockwise. A phase plane portrait is sketched in Fig. 7.12(b).

$$Complex\ 3: \quad \frac{dx}{dt} = x - 2y, \quad \frac{dy}{dt} = 2x - y.$$

The characteristic equation

$$0 = \begin{vmatrix} 1 - \lambda & -2 \\ 2 & -1 - \lambda \end{vmatrix} = \lambda^2 + 3$$

has roots $\lambda = \sqrt{3}i$, $\lambda^* = -\sqrt{3}i$. Because $\text{Re}(\lambda) = 0$, the phase paths are not spirals. If the system were in real canonical form, we could conclude that the phase paths were circles. The transformation of coordinates, which takes the above system to the real canonical form and vice versa, is a linear one. This has the effect of stretching or shrinking the coordinate axes and rotating them. Consequently, the circular phase paths will be transformed into elliptical paths whose major and minor axes will not necessarily be the xy axes. Since $dy/dt > 0$ when $x > 0$, $y = 0$, the flow is counterclockwise.

In this case the equations of the family of ellipses can be determined by finding the equations of the integral curves. The implicit solution of the asso-

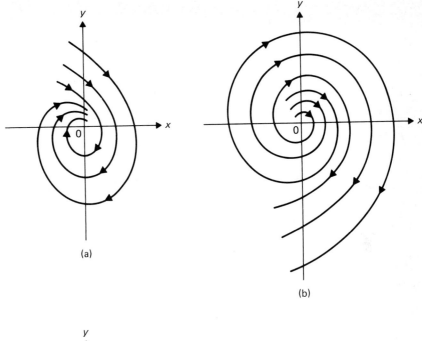

(a)

(b)

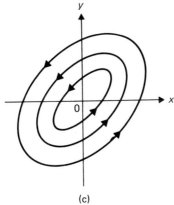

(c)

Figure 7.12

ciated differential equation

$$\frac{dy}{dx} = \frac{2x - y}{x - 2y} \qquad \text{or} \qquad 2x \, dx + 2y \, dy - (y \, dx + x \, dy) = 0$$

is $x^2 - xy + y^2 = $ const. By a rotation of axes it can be shown that this represents a family of ellipses centered at $(0, 0)$: these are the integral curves. A phase plane portrait is sketched in Fig. 7.12(c). $\quad\square$

Up to now the terms *stable* and *unstable* have been used in the classi-fication of equilibrium points without defining them precisely. We now pro-ceed to do so.

Definition An equilibrium point $(\bar{x}, \bar{y})$ of the system (7.3.1) is said to be *stable* if for each number $\epsilon > 0$ there is a number $\delta > 0$ such that if $x(t)$, $y(t)$ is any solution to (7.3.1) with

$$|x(t_0) - \bar{x}| + |y(t_0) - \bar{y}| < \delta,$$

then the solution exists for all $t \geq t_0$ and

$$|x(t) - \bar{x}| + |y(t) - \bar{y}| < \epsilon \quad \text{for all } t \geq t_0.$$

An equilibrium point $(\bar{x}, \bar{y})$ is said to be *unstable* if it is not stable.

Definition An equilibrium point $(\bar{x}, \bar{y})$ of the system (7.3.1) is said to be *asymptotically stable* if it is stable and if there is a number $\eta > 0$ such that if

$$|x(t_0) - \bar{x}| + |y(t_0) - \bar{y}| < \eta,$$

then

$$\lim_{t \to \infty} \{|x(t) - \bar{x}| + |y(t) - \bar{y}|\} = 0.$$

These concepts are illustrated in Fig. 7.13. If the above definitions are applied to the linear system we have just analyzed, the results for the equi-librium point (0, 0) can be summarized in Table 7.1.

The very important mathematical result for almost linear systems

$$\frac{dx}{dt} = ax + by + R(x, y), \qquad \frac{dy}{dt} = cx + dy + S(x, y),$$

is that the stability of the origin is determined (with one exception) by the associated linear system of differential equations. If the origin is an asymp-totically stable equilibrium point of the associated linear system, then it is an asymptotically stable equilibrium point of the almost linear system. If the origin is an unstable equilibrium point of the linear system, then it is an unstable equilibrium point of the almost linear system. The exceptional case is when the origin is a center of the linear system for it may or may not be a stable equilibrium point for the almost linear system. One can imagine the

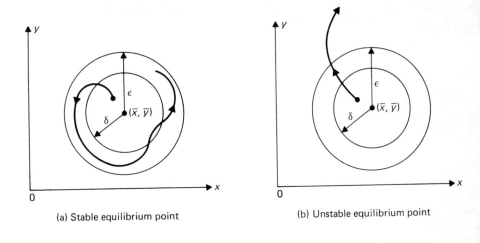

(a) Stable equilibrium point   (b) Unstable equilibrium point

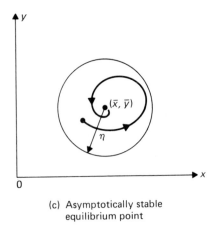

(c) Asymptotically stable
equilibrium point

Figure 7.13

family of ellipses centered at the origin being "smeared" in the almost linear system and possibly becoming a family of spirals which could be unstable or asymptotically stable. On the other hand, the center structure of the linear system could be preserved in the almost linear system, as Example 2 in Section 7.2 demonstrates.

Equally important is that the phase portrait near the origin of the almost linear system is locally approximated (in the sense described previously) by the phase portrait of the linear system in the unstable and asymptotically stable cases. These statements allow us to refer to the origin as a stable or unstable node, or spiral, or a saddle point of the almost linear system when it is of that type for the linear system. The justification for the above statements can be found in more advanced texts such as [4].

TABLE 7.1   **Classification of Equilibrium States**

Case		Type	Stability
Real(1)	$\lambda_1 < \lambda_2 < 0$	node	asymptotically stable
Real(2)	$0 < \lambda_2 < \lambda_1$	node	unstable
Real(3)	$\lambda_2 < 0 < \lambda_1$	saddle point	unstable
Real(4)	$\lambda_1 = \lambda_2 < 0$	node	asymptotically stable
Real(5)	$0 < \lambda_1 = \lambda_2$	node	unstable
Complex(1)	$\mathrm{Re}(\lambda) < 0$	spiral	asymptotically stable
Complex(2)	$\mathrm{Re}(\lambda) > 0$	spiral	unstable
Complex(3)	$\mathrm{Re}(\lambda) = 0$	center	stable

Example 8   The damped pendulum equation

$$\ddot{x} + c\dot{x} + \sin x = 0, \quad 0 < c < 2,$$

is equivalent to the system

$$\frac{dx}{dt} = p, \qquad \frac{dp}{dt} = -\sin x - cp.$$

Its equilibrium states are either asymptotically stable or unstable. To see this, first linearize with respect to the equilibrium states $x = 0$, $p = 0$, and then $x = \pi$, $p = 0$, and verify in each case that it is an almost linear system. Letting $x = 0 + u$, $p = 0 + v$, we get

$$\frac{du}{dt} = v, \qquad \frac{dv}{dt} = -u - cv + (u - \sin u),$$

and therefore $R(u, v) = 0$ and $S(u, v) = u - \sin u$. Since

$$\frac{|u - \sin u|}{\sqrt{u^2 + v^2}} \le \frac{|u - \sin u|}{|u|} = \left| 1 - \frac{\sin u}{u} \right| \to 0$$

as $u \to 0$, the system is almost linear. The same system would have been directly obtained by using the Taylor expansion for $\sin u$ near $u = 0$ as follows:

$$\frac{dv}{dt} = -cv - \sin u = -cv - \left(u - \frac{u^3}{3!} + \frac{u^5}{5!} \cdots \right)$$

$$= -u - cv + S(u, v).$$

The characteristic equation of the associated linear system is

$$\begin{vmatrix} 0 - \lambda & 1 \\ -1 & -c - \lambda \end{vmatrix} = \lambda^2 + c\lambda + 1 = 0.$$

If $c^2 < 4$, the eigenvalues are

$$\lambda = -\frac{c}{2} + i\frac{(4 - c^2)^{1/2}}{2}, \qquad \lambda^* = -\frac{c}{2} - i\frac{(4 - c^2)^{1/2}}{2}.$$

Since $c > 0$, the origin is an asymptotically stable spiral point of the linear system. Therefore, the origin is an asymptotically stable spiral point of the damped pendulum. To analyze the equilibrium point $(\pi, 0)$, set $x = \pi + u$, $y = 0 + v$ and obtain

$$\frac{du}{dt} = v, \qquad \frac{dv}{dt} = -cv - \sin(u + \pi) = u - cv + (\sin u - u).$$

The characteristic equation of the associated linear system is

$$\begin{vmatrix} 0 - \lambda & 1 \\ 1 & -c - \lambda \end{vmatrix} = \lambda^2 + c\lambda - 1 = 0.$$

The eigenvalues are

$$\lambda_2 = -\frac{c}{2} - \frac{(c^2 + 4)^{1/2}}{2} < 0 < -\frac{c}{2} + \frac{(c^2 + 4)^{1/2}}{2} = \lambda_1,$$

and $(u, v) = (\pi, 0)$ is a saddle point of the linear system. By the theory, the point $(\pi, 0)$ is therefore an unstable equilibrium point (a saddle) of the damped pendulum. $\square$

Given a two-dimensional autonomous nonlinear system that cannot be solved explicitly, how does one analyze it? We suggest the following systematic approach:

1. Find and classify the equilibrium states.
2. Solve the linear systems associated with the equilibrium states to obtain approximate solutions to the nonlinear system and to determine stability.
3. Sketch a local phase plane portrait, i.e., sketch phase paths near the equilibrium states.
4. Seek implicit solutions to the corresponding first order differential equation to obtain integral curves (usually difficult).
5. (Optional) Use a computer graphics package to sketch a phase portrait. A rough sketch can often be made by determining regions where dx/dt and dy/dt are univalent.

Example 9 Analyze

$$\frac{dx}{dt} = y - x^3, \qquad \frac{dy}{dt} = 1 - xy.$$

To find the equilibrium points solve the equations

$$P(x, y) = y - x^3 = 0, \qquad Q(x, y) = 1 - xy.$$

Eliminating y from both equations, one obtains $0 = 1 - x^4$ with solutions $\bar{x} = 1$ and $\bar{x} = -1$. The corresponding values of y are $\bar{y} = 1$ and $\bar{y} = -1$, so the two equilibrium points are $(1, 1)$ and $(-1, -1)$. Furthermore,

$$P_x(\bar{x}, \bar{y}) = -3\bar{x}^2, \qquad P_y(\bar{x}, \bar{y}) = 1,$$

$$Q_x(\bar{x}, \bar{y}) = -\bar{y}, \qquad Q_y(\bar{x}, \bar{y}) = -\bar{x}.$$

To analyze the point $(1, 1)$, let $(\bar{x}, \bar{y}) = (1, 1)$ and obtain the associated linear system

$$\frac{du}{dt} = P_x(1, 1)u + P_y(1, 1)v = -3u + v,$$

$$\frac{dv}{dt} = Q_x(1, 1)u + Q_y(1, 1)v = -u - v.$$

The characteristic equation

$$0 = \begin{vmatrix} -3 - \lambda & 1 \\ -1 & -1 - \lambda \end{vmatrix} = \lambda^2 + 4\lambda + 4 = (\lambda + 2)^2$$

has a double root, $\lambda = -2$, and therefore $(0, 0)$ is an asymptotically stable node of the linear system. Consequently, the equilibrium state $(1, 1)$ is an asymptotically stable node. Since $x = 1 + u$, $y = 1 + v$, an approximate solution is of the form

$$\begin{bmatrix} x(t) \\ y(t) \end{bmatrix} = \begin{bmatrix} 1 \\ 1 \end{bmatrix} + c_1 \begin{bmatrix} \alpha \\ \beta \end{bmatrix} e^{-2t} + c_2 \left(\begin{bmatrix} \eta \\ \sigma \end{bmatrix} + t \begin{bmatrix} \alpha \\ \beta \end{bmatrix} \right) e^{-2t},$$

where α, β, η, and σ satisfy the system of equations

$$[-3 - (-2)]\alpha + \beta = 0, \qquad [-3 - (-2)]\eta + \sigma = \alpha,$$
$$-\alpha + [-1 - (-2)]\beta = 0, \qquad -\eta + [-1 - (-2)]\sigma = \beta.$$

A solution is $\alpha = \beta = \sigma = 1$ and $\eta = 0$, and therefore an approximate solution to the nonlinear system near $(1, 1)$ is

$$\begin{bmatrix} x(t) \\ y(t) \end{bmatrix} = \begin{bmatrix} 1 \\ 1 \end{bmatrix} + c_1 \begin{bmatrix} 1 \\ 1 \end{bmatrix} e^{-2t} + c_2 \left(\begin{bmatrix} 0 \\ 1 \end{bmatrix} + t \begin{bmatrix} 1 \\ 1 \end{bmatrix} \right) e^{-2t}.$$

To analyze the point $(-1, -1)$, let $\bar{x} = -1$, $\bar{y} = -1$, and proceed as above to obtain the associated linear system

$$\frac{du}{dt} = -3u + v, \qquad \frac{dv}{dt} = u + v.$$

The characteristic equation

$$0 = \begin{vmatrix} -3 - \lambda & 1 \\ 1 & 1 - \lambda \end{vmatrix} = \lambda^2 + 2\lambda - 4$$

has roots $\lambda_1 = -1 + \sqrt{5}$, $\lambda_2 = -1 - \sqrt{5}$, and therefore the origin is a saddle point of the linear system. Consequently, the equilibrium state $(-1, -1)$ is a saddle point and, therefore, is an unstable state. It is left for the reader to show that an approximate solution to the nonlinear system near $(-1, -1)$ is

$$\begin{bmatrix} x(t) \\ y(t) \end{bmatrix} = \begin{bmatrix} -1 \\ -1 \end{bmatrix} + c_1 \begin{bmatrix} 1 \\ 2 + \sqrt{5} \end{bmatrix} e^{(-1+\sqrt{5})t} + c_2 \begin{bmatrix} 1 \\ 2 - \sqrt{5} \end{bmatrix} e^{(-1-\sqrt{5})t}.$$

A local phase plane portrait is sketched in Fig. 7.14. The curves $y = x^3$ and $xy = 1$ on which $dx/dt = 0$ and $dy/dt = 0$, respectively, are included. In this example it is possible to follow the paths leaving the saddle $(-1, -1)$ that flow up and to the right. They must cross the $dy/dt = 0$ curve, turn

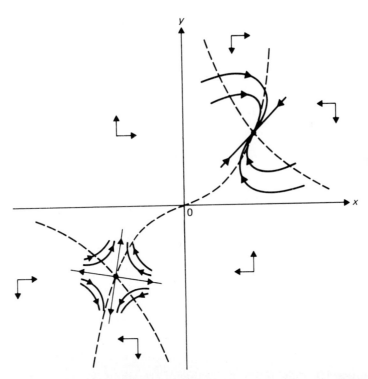

Figure 7.14 Phase plane portrait of $\dot{x} = y - x^3$, $\dot{y} = 1 - xy$. The arrow pairs indicate whether x and y are increasing or decreasing in each region determined by the curves $y = x^3$ and $xy = 1$.

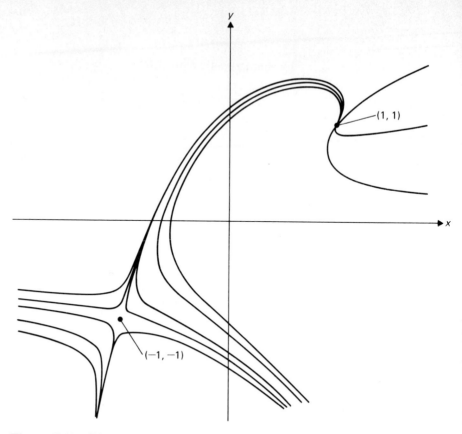

Figure 7.15 Plot generated using RKF45 of $\dot{x} = y - x^3$, $\dot{y} = 1 - xy$.

down, cross the $dx/dt = 0$ curve, and flow into the attracting node $(1, 1)$. This is confirmed in the computer generated phase plane portrait shown in Fig. 7.15.

The reader may have noticed that in the discussion we dealt only with the associated linear system and did not verify the conditions

$$\lim_{(u,\,v)\to(0,\,0)} \frac{R(u, v)}{\sqrt{u^2 + v^2}} = \lim_{(u,\,v)\to(0,\,0)} \frac{S(u, v)}{\sqrt{u^2 + v^2}} = 0$$

on the remainder terms for the almost linear system. These conditions are automatically satisfied whenever the right-hand side of the original differential equation is a continuously twice differentiable function in a neighborhood of the equilibrium point. Certainly this is the case if it consists of a linear part plus polynomial terms in x and y of degree 2 or higher. □

Example 10 The following system of equations occurs in the analysis of a forced nonlinear oscillator:

$$\frac{dx}{dt} = x(4 - x^2 - y^2) - h, \qquad \frac{dy}{dt} = y(4 - x^2 - y^2),$$

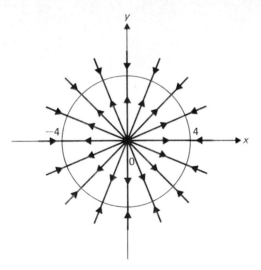

Figure 7.16   Phase portrait of $\dot{r} = r(4 - r^2)$, $\dot{\theta} = 0$.

where h is the amplitude of a periodic forcing function. Describe its equilibrium points and sketch a phase plane portrait.

If $h = 0$, the origin is an unstable node since the corresponding linear system is

$$\frac{du}{dt} = 4u, \qquad \frac{dv}{dt} = 4v.$$

When $h = 0$, the circle $x^2 + y^2 = 4$ is a set of stable equilibrium points; this is easily shown if polar coordinates are employed. Set $x = r \cos \theta$, $y = r \sin \theta$, then differentiate and substitute to obtain

$$\frac{dx}{dt} = \frac{dr}{dt} \cos \theta - r \frac{d\theta}{dt} \sin \theta = (r \cos \theta)(4 - r^2),$$

$$\frac{dy}{dt} = \frac{dr}{dt} \sin \theta + r \frac{d\theta}{dt} \cos \theta = (r \sin \theta)(4 - r^2).$$

With appropriate multiplication by $\sin \theta$ and $\cos \theta$, this system can be solved for dr/dt and $d\theta/dt$:

$$\frac{dr}{dt} = r(4 - r^2), \qquad \frac{d\theta}{dt} = 0.$$

That the circle of equilibria $r^2 = 4$ is stable can be seen by noticing that $dr/dt > 0$ for $0 < r < 4$ and $dr/dt < 0$ for $r > 4$. A global phase plane portrait for the case $h = 0$ is sketched in Fig. 7.16.

If $h \neq 0$, the coordinates of the equilibrium points satisfy the algebraic equations

$$0 = x(4 - x^2 - y^2) - h, \qquad 0 = y(4 - x^2 - y^2).$$

Therefore, $\bar{y} = 0$ and $\bar{x}$ is a root of $x(4 - x^2) - h = 0$. If $|h| < 16/(3\sqrt{3})$, then, as illustrated in Fig. 7.17 where we have graphed $z = x(4 - x^2)$, there

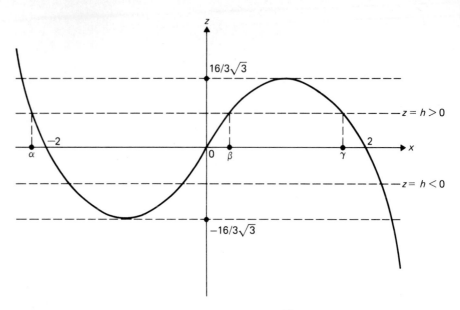

Figure 7.17 Graph of $z = x(4 - x^2)$. (The roots $\bar{x} = \alpha, \beta$, and γ are marked for the case $h > 0$.)

are three real roots: α, β, γ. If $h > 0$, the roots satisfy

$$\alpha < -2, \quad 0 < \beta < \gamma < 2,$$

whereas if $h < 0$, then

$$-2 < \alpha < \beta < 0, \quad 2 < \gamma.$$

Since $P(x, y) = 4x - x^3 - xy^2 - h$ and $Q(x, y) = 4y - x^2y - y^3$, then

$$P_x(x, 0) = 4 - 3x^2, \qquad P_y(x, 0) = 0,$$
$$Q_x(x, 0) = 0, \qquad Q_y(x, 0) = 4 - x^2.$$

The linear system associated with the equilibrium point $(\bar{x}, 0)$ is

$$\frac{d}{dt}\begin{bmatrix} u \\ v \end{bmatrix} = \begin{bmatrix} 4 - 3\bar{x}^2 & 0 \\ 0 & 4 - \bar{x}^2 \end{bmatrix}\begin{bmatrix} u \\ v \end{bmatrix},$$

with eigenvalues and eigenvectors

$$\lambda_1 = 4 - 3\bar{x}^2, \quad \begin{bmatrix} 1 \\ 0 \end{bmatrix}; \qquad \lambda_2 = 4 - \bar{x}^2, \quad \begin{bmatrix} 0 \\ 1 \end{bmatrix}.$$

Assume that $0 < h < 16/(3\sqrt{3})$ and let $\bar{x}$ take on successively the values α, β, and γ. Since the slope of $z = \bar{x}(4 - \bar{x}^2)$ is $z' = 4 - 3\bar{x}^2$, the value $\bar{x} = \alpha < -2$ corresponds to a point with negative slope, and therefore

$$\lambda_1 = 4 - 3\alpha^2 < 0, \qquad \lambda_2 = 4 - \alpha^2 < 0.$$

We conclude that point $(\alpha, 0)$ is a stable node. The second root, $\bar{x} = \beta$, where

$0 < \beta < 2$, corresponds to a point with positive slope, hence,

$$\lambda_1 = 4 - 3\beta^2 > 0, \qquad \lambda_2 = 4 - \beta^2 > 0$$

and $(\beta, 0)$ is an unstable node. For the third root, $\bar{x} = \gamma$, where $0 < \gamma < 2$, the slope is negative, so

$$\lambda_1 = 4 - 3\gamma^2 < 0, \qquad \lambda_2 = 4 - \gamma^2 > 0$$

and $(\gamma, 0)$ is a saddle point.

If $-16/(3\sqrt{3}) < h < 0$, a similar analysis shows that $(\alpha, 0)$ is a saddle point, $(\beta, 0)$ is an unstable node, and $(\gamma, 0)$ is a stable node. Thus $(\beta, 0)$ stays the same regardless of the sign of h, whereas the points $(\alpha, 0)$ and $(\gamma, 0)$ exchange types. This interchange of the stable node and the saddle point is due to the change in sign of λ_2 as h goes from positive to negative.

A sketch of the local phase plane portrait for $0 < h < 16/(3\sqrt{3})$ is shown in Fig. 7.18. A computer generated plot of the phase plane portrait, when $h = 1$, is shown in Fig. 7.19. ☐

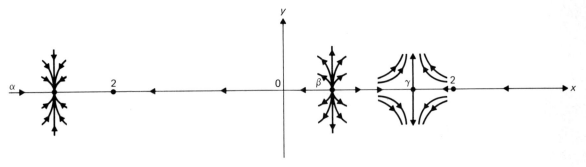

Figure 7.18 Local phase plane portrait of $\dot{x} = x(4 - x^2 - y^2) - h$, $\dot{y} = y(4 - x^2 - y^2)$ for $0 < h < 16/(3\sqrt{3})$.

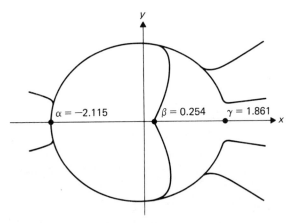

Figure 7.19 Plot generated using RKF45 of $\dot{x} = x(4 - x^2 - y^2) - 1$, $\dot{y} = y(4 - x^2 - y^2)$.

EXERCISES 7.3

Use Table 7.1 to describe the equilibrium state $(0, 0)$. Sketch a phase plane portrait.

1. $\dfrac{dx}{dt} = x + 12y$, $\quad \dfrac{dy}{dt} = 3x + y$

2. $\dfrac{dx}{dt} = 5x - 2y$, $\quad \dfrac{dy}{dt} = 4x - y$

3. $\dfrac{dx}{dt} = -4x + 3y$, $\quad \dfrac{dy}{dt} = -2x + y$

4. $\dfrac{dx}{dt} = 2x + 5y$, $\quad \dfrac{dy}{dt} = x - 2y$

5. $\dfrac{dx}{dt} = x - 4y$, $\quad \dfrac{dy}{dt} = x + y$

6. $\dfrac{dx}{dt} = 3x + 2y$, $\quad \dfrac{dy}{dt} = -5x + y$

7. $\dfrac{dx}{dt} = -2x + y$, $\quad \dfrac{dy}{dt} = -4x + 3y$

8. $\dfrac{dx}{dt} = 3x - 4y$, $\quad \dfrac{dy}{dt} = 2x - 3y$

9. $\dfrac{dx}{dt} = -3x + 4y$, $\quad \dfrac{dy}{dt} = -x + y$

10. $\dfrac{dx}{dt} = -y$, $\quad \dfrac{dy}{dt} = x - 2y$

11. $\dfrac{dx}{dt} = x - y$, $\quad \dfrac{dy}{dt} = x + 3y$

12. $\dfrac{dx}{dt} = 2x - 5y$, $\quad \dfrac{dy}{dt} = x - 2y$

13. $\dfrac{dx}{dt} = -8y$, $\quad \dfrac{dy}{dt} = 2x$

14. $\dfrac{dx}{dt} = -2x + y$, $\quad \dfrac{dy}{dt} = -3x - 6y$

Verify that the given point is an equilibrium of each of the following systems, derive and solve the corresponding linear equations, classify the equilibrium point, and sketch a phase plane portrait near the equilibrium point.

15. $\dfrac{dx}{dt} = 2x + 3y - xy - 6$, $\quad \dfrac{dy}{dt} = y + x^2 - 3$,
given $(\bar{x}, \bar{y}) = (1, 2)$

16. $\dfrac{dx}{dt} = -x + 6y + 2xy - 5$,
$\dfrac{dy}{dt} = 6x + y + x^2 + 4$, given $(\bar{x}, \bar{y}) = (-1, 1)$

17. $\dfrac{dx}{dt} = 6x - 2y + \cos y - 2\pi$,
$\dfrac{dy}{dt} = 2x + y - \dfrac{3\pi}{2} \sin x$, given $(\bar{x}, \bar{y}) = \left(\dfrac{\pi}{2}, \dfrac{\pi}{2}\right)$

18. $\dfrac{dx}{dt} = 3x + 2y + \cos y + 1 - \pi$,
$\dfrac{dy}{dt} = 5x + 5y + \sin x$, given that $(\bar{x}, \bar{y})$
$= (\pi, -\pi)$

19. $\dfrac{dx}{dt} = 2x + 2y + 5e^{-x} - 5$, $\quad \dfrac{dy}{dt} = -x - y$,
given $(\bar{x}, \bar{y}) = (0, 0)$

20. $\dfrac{dx}{dt} = 3x + 5y + e^x - 1$, $\quad \dfrac{dy}{dt} = -2x - 2y$,
given $(\bar{x}, \bar{y}) = (0, 0)$

Find all equilibrium points, derive and solve the corresponding associated linear equations, classify the equilibrium points, and sketch a local phase plane portrait.

21. $\dfrac{dx}{dt} = x - 4y + 2$, $\quad \dfrac{dy}{dt} = 4x - 7y - 1$

22. $\dfrac{dx}{dt} = y + 2$, $\quad \dfrac{dy}{dt} = -x + 2y + 7$

23. $\dfrac{dx}{dt} = x^2 + y$, $\quad \dfrac{dy}{dt} = -x + y^2$

24. $\dfrac{dx}{dt} = x^2 - 5y$, $\quad \dfrac{dy}{dt} = x - \sqrt{5}y^2$

25. $\dfrac{dx}{dt} = x - y$, $\quad \dfrac{dy}{dt} = xy - 1$

26. $\dfrac{dx}{dy} = x + y$, $\quad \dfrac{dy}{dt} = xy + y$

27. $\dfrac{dx}{dt} = x - y$, $\quad \dfrac{dy}{dt} = xy + 4y + 3$

28. $\dfrac{dx}{dt} = xy$, $\quad \dfrac{dy}{dt} = -x + 2y + 1$

29. Sketch a local phase plane portrait for Example 10 in the case $|h| \geq 16/(3\sqrt{3})$.

30. A two-dimensional linear system is said to be degenerate if it has a zero eigenvalue. Show that if it is nontrivial, i.e., the coefficients are not all zero, then it has a line of equilibrium points.

31. Find the general solution and sketch a phase plane portrait of the degenerate system

$$\frac{dx}{dt} = x + y, \qquad \frac{dy}{dt} = x + y.$$

32. Find an integral of the nonlinear system

$$\frac{dx}{dt} = x + y,$$

$$\frac{dy}{dt} = x + y - (x + y)^2.$$

33. In the preceding exercise, make the change of dependent variables $z = x + y$, $w = e^x$; derive and solve the resulting nonlinear system.

7.4 ENERGY METHODS FOR SYSTEMS WITH ONE DEGREE OF FREEDOM

A one-degree-of-freedom mechanical system with forces dependent on position can be modeled by a second order differential equation

$$\frac{d^2x}{dt^2} + g(x) = 0, \tag{7.4.1}$$

where x denotes a position coordinate and $-g(x)$ denotes the net force per unit mass. The system is said to be *conservative* since there are no dissipative forces acting. The simple pendulum and the frictionless mass–spring system can be modeled in this way.

The second order differential equation (7.4.1) can be written as a two-dimensional system

$$\frac{dx}{dt} = y, \qquad \frac{dy}{dt} = -g(x). \tag{7.4.2}$$

The equilibrium states are points $(\bar{x}, 0)$ where $g(\bar{x}) = 0$, and they could be analyzed by the methods discussed in the preceding section. But in this section we will analyze the equation in its original form (7.4.1). The significant advantage of retaining the second order equation is that energy methods can be employed to reveal the global geometric structure of its solutions.

Let $E(x, y)$ denote the total energy per unit mass of the system:

$$E(x, y) = \frac{1}{2} y^2 + G(x), \tag{7.4.3}$$

where $G(x) = \int^x g(s)\, ds$ denotes the potential energy per unit mass and

$$\frac{1}{2} y^2 = \frac{1}{2} \left(\frac{dx}{dt} \right)^2$$

is the kinetic energy per unit mass. The potential energy function, $G(x)$, is determined up to an additive constant; the constant will be chosen so that the total energy at an equilibrium point is zero. Note that the statement "the total energy is constant for conservative systems" means that the energy function is constant along trajectories. This follows directly since, if $x = x(t)$ is a solution to (7.4.1) and $y = dx(t)/dt$, then

$$
\begin{aligned}
\frac{d}{dt} E\left(x(t), \frac{dx}{dt}(t)\right) &= \frac{d}{dt}\left[\frac{1}{2}\left(\frac{dx}{dt}(t)\right)^2 + G(x(t))\right] \\
&= \frac{dx}{dt}(t) \frac{d^2x}{dt^2}(t) + \frac{d}{dx} G(x(t)) \frac{dx}{dt}(t) \\
&= \frac{dx}{dt}(t)\left[\frac{d^2x}{dt^2}(t) + g(x(t))\right] = 0, \qquad (7.4.4)
\end{aligned}
$$

since $g(x) = G'(x)$. But this implies that

$$
E\left(x(t), \frac{dx}{dt}(t)\right) = \text{const.}
$$

The *level curves* of the energy surface

$$
z = E(x, y) \qquad (7.4.5)
$$

are the curves $E(x, y) = K = \text{const}$. The relation (7.4.4) shows that they are *integral curves* of the differential equation (7.4.1). They can also be considered as the graphs of the implicit solution

$$
\frac{y^2}{2} + G(x) = K, \quad K = \text{const}
$$

to the associated first order differential equation

$$
\frac{dy}{dx} = \frac{-g(x)}{y} \qquad \text{or} \qquad g(x)\, dx + y\, dy = 0.
$$

The equation $E(x, y) = K$ is called the *energy integral*. The direction of flow on the phase paths corresponding to the integral curves is governed by the two-dimensional system (7.4.2).

Example 1 The point $(0, 0)$ is an equilibrium point of the system (7.4.2) corresponding to each of the following:

a) $\dfrac{d^2x}{dt^2} + x = 0$. Here $g(x) = x$ and $G(x) = x^2/2$ satisfies $G'(x) = x$ and $G(0) = 0$, so the total energy is

$$
E(x, y) = \frac{y^2}{2} + \frac{x^2}{2}.
$$

b) $\dfrac{d^2x}{dt^2} + \sin x = 0$. Here $g(x) = \sin x$ and $G(x) = 1 - \cos x$ satisfies $G'(x) = \sin x$ and $G(0) = 0$. Hence,

$$E(x, y) = \frac{y^2}{2} + 1 - \cos x.$$

c) $\dfrac{d^2x}{dt^2} + x - \dfrac{x^4}{8} = 0$. The point $(2, 0)$ is also an equilibrium point, so we could use

$$G(x) = \frac{x^2}{2} - \frac{x^5}{40} - \frac{6}{5},$$

which satisfies $G'(x) = x - (x^4/8)$ and $G(2) = 0$. In this case,

$$E(x, y) = \frac{y^2}{2} + \frac{x^2}{2} - \frac{x^5}{40} - \frac{6}{5}. \quad \Box$$

The level curves of the energy surface are constructed by taking the intersection of planes $z = K$ (in the xyz space) and the surface and then projecting the intersection set onto the xy plane. This is shown in Fig. 7.20 for Example 1(a), where the energy surface is a paraboloid $z = \frac{1}{2}(x^2 + y^2)$ and the level curves are the circles $x^2 + y^2 = 2K$, corresponding to periodic solutions.

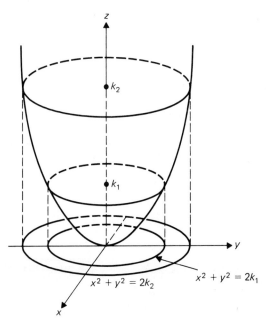

Figure 7.20 Level curves of $z = \frac{1}{2}(x^2 + y^2)$.

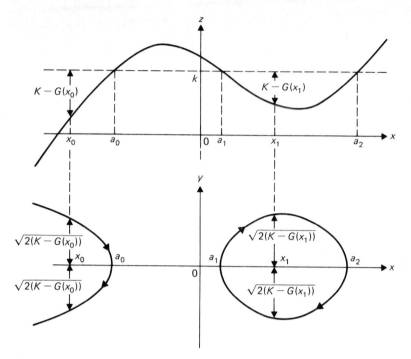

Figure 7.21   The potential plane and the phase plane (in the diagram, $y = \pm\sqrt{2[K - G(x)]}$ is not defined for $a_0 < x < a_1$ and $a_2 < x$).

Familiar examples of level curves are the isobars of weather maps and the contour lines of topographic maps.

We shall exploit the special structure of the energy surfaces by sketching the level curves in the *phase plane* (the xy plane) with the aid of the graph of the potential energy function $G(x)$ in the *potential plane* (the xz plane). These two planes can be represented in one diagram by sketching the z axis of the xz plane and the y axis of the xy plane on the same vertical line (Fig. 7.21). In the potential plane, draw the potential energy curve, $z = G(x)$, and a horizontal line at height K. Since a level curve satisfies

$$E(x, y) = \frac{y^2}{2} + G(x) = K,$$

then

$$y = \pm\sqrt{2[K - G(x)]}. \tag{7.4.6}$$

Clearly, the velocity is real only if $K - G(x)$ is nonnegative.

If, for fixed K, the velocity is real only in certain intervals, then the real solution can only exist on these intervals. Specifically, if $G(x) - K > 0$ only on the interval $a < x < b$, then any solution satisfying $x(t_0) = x_0$ for some x_0 in the interval will satisfy $a < x(t) < b$ for all t. The *integral curves* in the phase plane corresponding to a fixed K are symmetric with respect to the x axis. They cross the x axis at any point x_0 where $G(x_0) = K$ and, since this implies that $dy/dx = -g(x)/y = \pm\infty$ at such a point, the curves cross the x axis at right angles.

Example 2   Consider $d^2x/dt^2 + 4x = 0$ and $E(x, y) = y^2/2 + 2x^2$. For a given K_0, the integral curves are given by solving $E(x, y) = K_0$ to obtain $y = \pm\sqrt{2(K_0 - 2x^2)}$. They are defined only for $-\sqrt{K_0/2} \le x \le \sqrt{K_0/2}$. By squaring both sides, one obtains the ellipse

$$\frac{x^2}{K_0/2} + \frac{y^2}{2K_0} = 1.$$

It is the locus of all trajectories

$$(x(t), y(t)) = \left(x(t), \frac{dx(t)}{dt}\right), \quad -\infty < t < \infty,$$

for which the energy integral (or total energy) equals K_0. The solutions of the original equation are $x = A\sin(2t - \alpha)$, hence

$$y = \dot{x} = 2A\cos(2t - \alpha),$$

and it is easy to check that the ellipse is the locus of all solutions for which $2A^2 = K_0$. $\square$

Let us now consider the integral curves resulting from a simple *minimum* of the potential energy function $G(x)$. Therefore at some point $x = \bar{x}$,

$$G'(\bar{x}) = g(\bar{x}) = 0 \quad \text{and} \quad G''(\bar{x}) = g'(\bar{x}) > 0$$

(see Fig. 7.22). In the phase plane, $(\bar{x}, 0)$ is an equilibrium point of the system (7.4.2). But instead of finding the linearized system one can work directly with the second order differential equation. Since $g(\bar{x}) = 0$, then if we let $x = \bar{x} + u$, Taylor's theorem implies

$$g(x) = g(\bar{x} + u) = g'(\bar{x})u + R(u),$$

where $R(u)$ is the remainder term. Hence

$$\frac{d^2x}{dt^2} + g(x) = \frac{d^2u}{dt^2} + g'(\bar{x})u + R(u) = 0,$$

and neglecting $R(u)$ and letting $g'(\bar{x}) = \omega^2 > 0$, we get

$$\frac{d^2u}{dt^2} + \omega^2 u = 0.$$

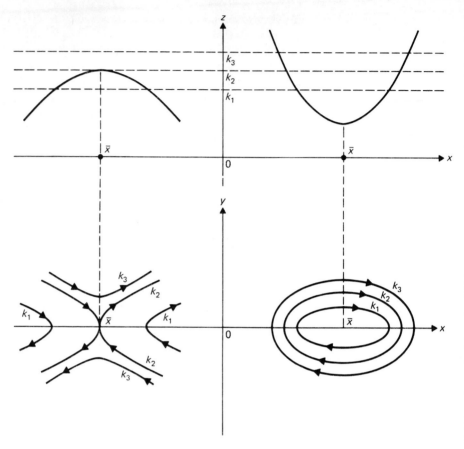

Figure 7.22 The phase plane diagram when the potential energy function has a local maximum or minimum.

The roots of the characteristic equation are $\lambda_1 = i\omega$, $\lambda_2 = -i\omega$. In the phase plane of (7.4.2), the point $(\bar{x}, 0)$ is a center and the approximate level curves are ellipses

$$\frac{y^2}{2} + \frac{\omega^2}{2}(x - \bar{x})^2 = K.$$

This implies the following very important fact: in the neighborhood of an equilibrium point $(\bar{x}, 0)$, where $G(\bar{x})$ is a minimum, there is a family of closed level curves. These correspond to periodic solutions of the original differential equation. If, instead, the potential energy function has a simple maximum at some $x = \bar{x}$, then $G'(\bar{x}) = g(\bar{x}) = 0$ and $G''(\bar{x}) = g'(\bar{x}) < 0$. By the same line of reasoning as above, the associated second order linear differential equation is

$$\frac{d^2u}{dt^2} - \omega^2 u = 0, \qquad -\omega^2 = g'(\bar{x}) < 0.$$

The roots of the characteristic equation are $\lambda_2 = -\omega < 0 < \lambda_1 = \omega$; in the phase plane of (7.4.2) the point $(\bar{x}, 0)$ is a saddle point. The approximate level curves are hyperbolas

$$\frac{y^2}{2} - \frac{\omega^2}{2}(x - \bar{x})^2 = K,$$

and straight lines $y = \pm\omega(x - \bar{x})$. This implies that in the phase plane of the original system the equilibrium point $(\bar{x}, 0)$ will be a saddle point.

If the potential energy curve $z = G(x)$ has a horizontal tangent at a simple point of inflection (Fig. 7.23), then at $x = \bar{x}$, we get $G'(\bar{x}) = g(\bar{x}) =$

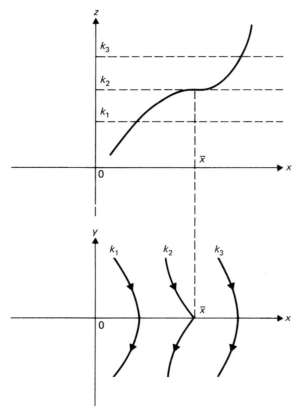

Figure 7.23   The phase plane diagram when the potential energy function has a point of inflection.

0, $G''(\bar{x}) = g'(\bar{x}) = 0$, $G'''(\bar{x}) \neq 0$. The degenerate equilibrium state $(\bar{x}, 0)$ is unstable. Its associated linear differential equation, $d^2u/dt^2 = 0$, is of little value, but the approximate potential energy function

$$G(x) = G'''(\bar{x}) \frac{(x - \bar{x})^3}{6} = g''(\bar{x}) \frac{(x - \bar{x})^3}{6}$$

and approximate energy integral

$$\frac{y^2}{2} + \frac{g''(\bar{x})}{6} (x - \bar{x})^3 = K$$

have a monotonic family of level curves. The curve through $(\bar{x}, 0)$ is distinguished by its *cusp* (see Fig. 7.23).

Example 3   Find and classify the equilibrium points of

$$\frac{d^2x}{dt^2} + x - \frac{x^3}{4} = 0.$$

Derive and solve the corresponding linear equations and sketch a local phase plane portrait.

First solve

$$0 = x - \frac{x^3}{4}$$

and obtain $\bar{x} = 0$, $\bar{x} = 2$, $\bar{x} = -2$. Hence, there are three equilibrium points: $(0, 0)$, $(2, 0)$, $(-2, 0)$. A potential energy function $G(x)$ that satisfies $G'(x) = g(x)$ will be of the form

$$G(x) = \tfrac{1}{2}x^2 - \tfrac{1}{16}x^4 + C,$$

and therefore $G'(\bar{x}) = 0$ for $\bar{x} = 0$, ± 2. Furthermore,

$$G''(\bar{x}) = g'(\bar{x}) = 1 - \tfrac{3}{4}\bar{x}^2$$

and, consequently, $G''(0) = 1 > 0$ and $G''(\pm 2) = -2 < 0$; hence $x = 0$ is a local minimum and $x = \pm 2$ are local maxima for the potential energy function. We conclude that the points $(0, 0)$ and $(\pm 2, 0)$ in the phase plane are a center and saddle points respectively.

   The linearized second order equations are given by

$$\frac{d^2u}{dt^2} + g'(\bar{x})u = 0,$$

where $\bar{x} = 0$, ± 2. Consequently at $(0, 0)$ we have

$$\frac{d^2u}{dt^2} + u = 0, \qquad u(t) = A \cos t + B \sin t,$$

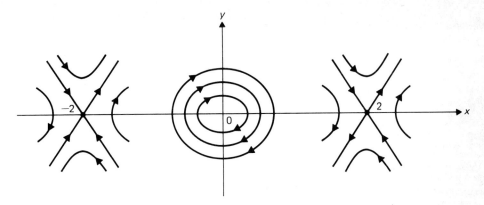

Figure 7.24 Local phase plane portrait of $\ddot{x} + x - x^3/4 = 0$.

and at $(\pm 2, 0)$,

$$\frac{d^2u}{dt^2} - 2u = 0, \qquad u(t) = A \cosh(\sqrt{2}t) + B \sinh(\sqrt{2}t).$$

The local phase plane portrait is sketched in Fig. 7.24. □

A physically attractive approach to one-degree-of-freedom problems is to associate a bead sliding on the potential curve $z = G(x)$ with a particle moving in the phase plane. The velocity of the bead is given by $v = \pm\sqrt{2[K - G(x)]}$. Suppose that it is released at (x_0, K), which is the point of intersection of the line $z = K$ and the curve $z = G(x)$. The corresponding point in the phase plane is $(x_0, 0)$. If the bead moves downhill to the right and approaches the boundary of its interval of motion, it either slows (in infinite time) to a stop ($K = K_2$ and $x = x_3$ in Fig. 7.25), since it has just enough energy to reach the top of the hill, or, if it pauses on the slope ($K = K_1$ and $x = x_2$), it will reverse its direction and slide down. The bead with energy level K_1 will move to the right, slide up the hill, pause and repeat the pattern. It is said to be in a *potential well*. The corresponding paths traced out in the phase plane are also sketched in Fig. 7.25—they are closed curves.

We conclude that a particle in a potential well corresponds to a periodic solution to the differential equation (7.4.1). A formula for its period can be derived from the energy integral, $E(x, \dot{x}) = K$. On one excursion from $x_{\min}$ (the minimum value of x) to $x_{\max}$ (the maximum value of x) t is a monotonic function of x. After replacing y by dx/dt in the energy integral, we can obtain a formula for dt/dx, namely,

$$\frac{dt}{dx} = \{2[K - G(x)]\}^{-1/2}.$$

Hence,

$$\frac{T}{2} = \int_{x_{min}}^{x_{max}} \{2[K - G(x)]\}^{-1/2} \, dx \qquad (7.4.7)$$

is the time needed to transverse from x_{min} to x_{max}. By symmetry, the return time is the same; hence, T is the *period*. The *frequency* of the motion is $2\pi/T$ and, in general, the period and, consequently, the frequency are functions of the amplitude.

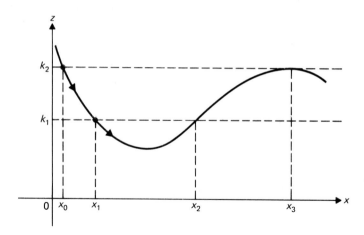

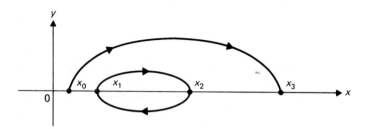

Figure 7.25 The particle starting at $(x_0, 0)$ with energy level k_2 will reach $(x_3, 0)$ in infinite time. The particle starting at $(x_1, 0)$ with energy level k_2 will reach $(x_2, 0)$, then return to $(x_1, 0)$ in finite time.

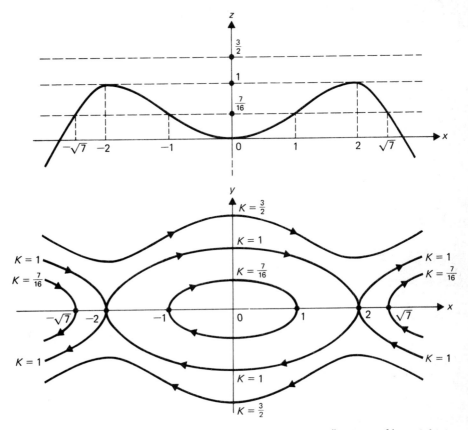

Figure 7.26 The potential and phase plane portraits of $\ddot{x} + x - x^3/4 = 0$ for energy levels $K = 7/16$, 1, and 3/2.

Example 4   Consider the example discussed previously:

$$\frac{d^2x}{dt^2} + x - \frac{x^3}{4} = 0, \qquad G(x) = \frac{x^2}{2} - \frac{x^4}{16};$$

the motion will be analyzed for three values of the energy level. Potential and phase plane portraits are sketched in Fig. 7.26 (the reader should glance at them as the discussion proceeds).

a) $K = \dfrac{7}{16}$: $y = \dfrac{dx}{dt} = \pm\sqrt{2\left(\dfrac{7}{16} - \dfrac{x^2}{2} + \dfrac{x^4}{16}\right)}$

$$= \pm\sqrt{\frac{1}{8}(x^2 - 1)(x^2 - 7)}.$$

The bounded motion corresponds to the portion of the level curve that is trapped in the potential well. If the initial conditions are $x(0) = 1$, $\dot{x}(0) = 0$, then in the potential plane this would correspond to the bead being released at $(x, z) = (1, G(1)) = (1, 17/16)$ on the path $z = G(x)$. It will slide down and to the left with negative velocity

$$\frac{dx}{dt} = -\sqrt{\frac{1}{8}(x^2 - 1)(x^2 - 7)}$$

until it reaches $(x, z) = (-1, 17/16)$, at which point its velocity will again be zero. But its acceleration will be

$$\frac{d^2x}{dt^2} = \frac{(-1)^3}{4} - (-1) = \frac{3}{4},$$

so it will move down and to the right with positive velocity

$$\frac{dx}{dt} = \sqrt{\frac{1}{8}(x^2 - 1)(x^2 - 7)}$$

until it reaches $(1, 17/16)$ again. In the phase plane the corresponding particle will start at $(1, 0)$ and move to the left and down until it reaches the point $(0, -\sqrt{7/8})$. Then it will move to the left and upward until it reaches the point $(-1, 0)$, then travels back to $(1, 0)$ in the upper half plane on the mirror image of its path in the lower half plane. Its period could be estimated by approximating the integral

$$T = 2 \int_{-1}^{1} \frac{1}{\sqrt{\frac{1}{8}(x^2 - 1)(x^2 - 7)}} \, dx.$$

The unbounded motion on the right in the potential plane would correspond to the bead sliding up the wire from the right ($x(0) > \sqrt{7}$, $dx/dt < 0$) until it reaches the point $(x, z) = (\sqrt{7}, 7/16)$. But $d^2x/dt^2 > 0$, so it will slide back down with positive velocity. In the phase plane this motion would be represented by a symmetric curve open to the right with vertex at $(\sqrt{7}, 0)$. A similar analysis can be done for the unbounded motion to the left.

b) $K = \dfrac{3}{2}:$ $\quad y = \dfrac{dx}{dt} = \pm\sqrt{2\left(\dfrac{3}{2} - \dfrac{x^2}{2} + \dfrac{x^4}{16}\right)}$

$$= \pm\sqrt{\frac{1}{8}(24 - 8x^2 + x^4)}.$$

The discriminant of $q^2 - 8q + 24$ is $8^2 - 4(24) = -32 < 0$, which implies that $24 - 8x^2 + x^4$ will never vanish and will always have the same positive sign. The two level curves correspond to beads moving either to the left or right with enough energy so as to never come to rest. The phase plane path will be a similar motion.

c) $K = 1$: $y = \dfrac{dx}{dt} = \pm\sqrt{2\left(1 - \dfrac{x^2}{2} + \dfrac{x^4}{16}\right)}$

$$= \pm\frac{1}{\sqrt{8}}(x - 2)(x + 2).$$

There will be six distinct phase paths. Two of them connect the equilibrium states (saddle points) $(-2, 0)$ and $(2, 0)$ and are called *separatrices*. The other four are unbounded and flow either into or out of the equilibrium states. ☐

The last example is a special case of the general *nonlinear spring equation*

$$\frac{d^2x}{dt^2} + \alpha x + \beta x^3 = 0, \quad \alpha > 0, \tag{7.4.8}$$

which, for appropriate values of the parameters, can be used to model a number of distinct physical phenomena. For instance, since $\sin x \approx x - x^3/6$, the choice $\alpha = 1$, $\beta = -1/6$ gives an approximate nonlinear pendulum equation.

The nonlinear spring equation is a good example of an equation whose solutions change character with a parameter. If $\alpha > 0$ is fixed and $\beta < 0$, there are three equilibrium points: a center at $(0, 0)$ and saddle points at $(\pm[\alpha/(-\beta)]^{1/2}, 0)$. As $\beta \to 0$, the potential well deepens and the saddle points recede to infinity. If $\beta \geq 0$, there is only one equilibrium point, a stable center, and all solutions are periodic (see Fig. 7.27).

In the remainder of this section we shall deal with nonconservative sys-

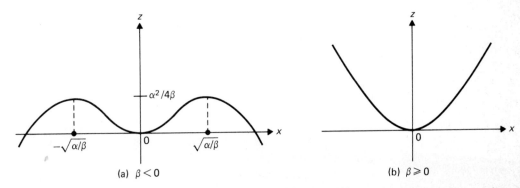

Figure 7.27 The potential energy function $z = \alpha x^2/2 - \beta x^4/4$, $\alpha > 0$, corresponding to the nonlinear spring equation $\ddot{x} + \alpha x + \beta x^3 = 0$.

tems, i.e., problems in which damping forces play an essential role. The phase paths of the differential equation

$$\frac{d^2x}{dt^2} + h\left(x, \frac{dx}{dt}\right)\frac{dx}{dt} + g(x) = 0 \tag{7.4.9}$$

are, in general, no longer level curves of an energy function. However, the potential plane sketch is still useful if the horizontal constant-energy lines are replaced by curves with slope given by

$$\frac{d}{dt} E\left(x(t), \frac{dx}{dt}(t)\right) = \frac{d}{dt}\left[\frac{1}{2}\left(\frac{dx}{dt}(t)\right)^2 + G(x(t))\right]$$

$$= \frac{dx}{dt}(t)\left[\frac{d^2x}{dt^2}(t) + g(x(t))\right] \tag{7.4.10}$$

$$= -h\left(x(t), \frac{dx}{dt}(t)\right)\left[\frac{dx}{dt}(t)\right]^2.$$

Letting $y = dy/dt$, we can write this more simply as

$$\frac{d}{dt} E(x, y) = -h(x, y)y^2.$$

For sketching purposes, precise values of dE/dt are not needed and rough estimates are sufficient. This is illustrated with the damped pendulum equation

$$\frac{d^2x}{dt^2} + c\frac{dx}{dt} + \sin x = 0, \quad 0 < c < 2. \tag{7.4.11}$$

For the undamped pendulum ($c = 0$), the total energy is

$$E(x, y) = \frac{y^2}{2} + 1 - \cos x,$$

and (7.4.10) gives for the damped pendulum

$$\frac{d}{dt} E(x, y) = -cy^2,$$

so the total energy is always a decreasing function of time.

By considering the two-dimensional system corresponding to (7.4.11),

$$\dot{x} = y, \quad \dot{y} = -\sin x - cy,$$

one sees that the points $(0, 0)$ and $(n\pi, 0)$, $n = \pm 1, \pm 2, \ldots$, in the phase plane are its equilibrium points. An analysis of the linearized system at each point shows that the points $(0, 0)$, $(2k\pi, 0)$, $k = \pm 1, \pm 2, \ldots$, are stable spiral points, and the points $((2k + 1)\pi, 0)$, $k = 0, \pm 1, \pm 2, \ldots$, are saddle points. A local phase plane portrait is sketched in Fig. 7.28.

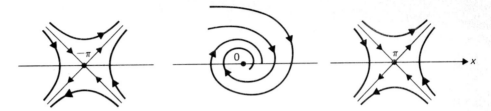

Figure 7.28

In a typical nonconservative system modeled by equation (7.4.9), $h(x, \dot{x})$ is positive and, consequently, dE/dt is negative. Therefore the system is dissipative and any nontrivial motion eventually ceases. However, there is a fascinating phenomenon called *self-sustained oscillations* in which the energy oscillates as it seemingly is pumped into and out of the system even though the modeling differential equation is autonomous. A well known example is that of a nonlinear electric oscillator that is modeled by the *van der Pol equation*

$$\frac{d^2x}{dt^2} + \epsilon(x^2 - 1)\frac{dx}{dt} + x = 0. \tag{7.4.12}$$

The dependent variable x is proportional to a grid voltage and ϵ is a positive parameter definable in terms of circuit elements. As before, we set $E(x, \dot{x}) = \dot{x}^2/2 + x^2/2$, and (7.4.10) implies that

$$\frac{dE}{dt} = -\epsilon(x^2 - 1)\left(\frac{dx}{dt}\right)^2.$$

Let us assume that $\dot{x} = 0$ only for isolated values of t. If x^2 is greater than one, the energy is decreasing, and if x^2 is less than one, the energy is increasing. A balance between the energy loss and gain would result in self-sustained oscillations. A mathematical proof of this is not elementary, but it has been shown (see [6], pp. 252–256) that for any value of $\epsilon > 0$ the van der Pol equation has an isolated periodic solution, called a *limit cycle*, toward which all other nontrivial solutions tend with increasing time. In the phase plane this would be represented by an isolated closed curve; this is illustrated in Fig. 7.29. If the parameter ϵ is large, a two-phase phenomenon called a *relaxation oscillation* is observed. During the first phase energy is slowly stored until a critical threshold is reached when the energy is discharged almost instantaneously (see Fig. 7.30). The corresponding periodic solution would be a very jerky oscillation with abrupt changes in the sign of the amplitude.

The van der Pol equation was first developed in the study of electrical circuits. Self-sustained oscillations also occur in biological organisms, such as

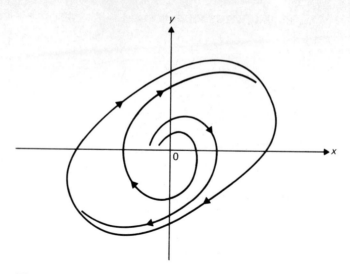

Figure 7.29 A limit cycle of the van der Pol equation when ϵ is small.

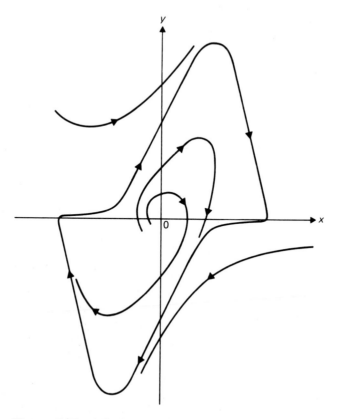

Figure 7.30 A limit cycle of the van der Pol equation when ϵ is large.

Figure 7.31 Each container has an opening on the top and side. When water flowing from the faucet fills the uppermost container the system inverts its position. Then water will flow out of the first container while the second is being filled.

the timing of mitosis in slime molds (see [10], p. 184) and in autocatalytic chemical reactions (see [11], p. 146). Relaxation oscillations would typically model physical systems where there are thresholds and abrupt self-correction. An example of a simple system showing this type of behavior is described in Fig. 7.31.

EXERCISES 7.4

For the following differential equations, find and classify the equilibrium points, derive and solve the corresponding linear equations, and sketch *local* phase plane portraits.

1. $\dfrac{d^2x}{dt^2} + x^3 + x^2 - 2x = 0$

2. $\dfrac{d^2x}{dt^2} + x^3 - 2x^2 - 3x = 0$

3. $\dfrac{d^2x}{dt^2} + x - x^3 = 0$

4. $\dfrac{d^2x}{dt^2} + 3x^2 + 5x + 2 = 0$

For the following differential equations, sketch potential plane and global phase plane portraits.

5. $\dfrac{d^2x}{dt^2} + x^3 - \dfrac{x}{4} = 0$

6. $\dfrac{d^2x}{dt^2} + 2x^2 - 5x - 3 = 0$

7. $\dfrac{d^2x}{dt^2} + \sin x - \dfrac{\sqrt{8}}{\pi} x = 0$

8. $\dfrac{d^2x}{dt^2} + \dfrac{\sinh x}{\sinh 1} - x = 0$

9. $\dfrac{d^2x}{dt^2} + \dfrac{x}{x - 1} = 0$

10. $\dfrac{d^2x}{dt^2} + \dfrac{x}{x^2 - 1} = 0$

In the following problems find and classify the equilibrium points, derive and solve the corresponding linear equations, and sketch potential plane and global phase plane portraits. The reader is urged to consult the cited references for more details.

11. According to the general theory of relativity, the path of a particle moving along a timelike geodesic in the gravitational field of a spherically symmetric heavy body is governed by

$$\frac{d^2u}{d\theta^2} + u = 1 + \lambda u^2,$$

where u is inversely proportional to the distance from the heavy body, θ is an angle in the plane of motion, and λ is a parameter proportional to the square of the speed of light, $0 < \lambda \ll 1$.

Study the effect of increasing λ on the potential plane and phase plane portraits and observe the disappearance of the potential well at $\lambda = \frac{1}{4}$ (see [12]).

12. The motion of an elastically restrained current-carrying bus bar in the magnetic field of a long fixed parallel wire is governed by

$$m\frac{d^2x}{dt^2} + k\left(x - \frac{\lambda}{L-x}\right) = 0.$$

The parameter λ is proportional to the product of the currents in the two conductors, k is the elastic constant, L is the distance from the fixed wire to a supporting wall, and m is the mass of the bus bar. Study the effect of varying λ on the phase plane portrait (see [2], pp. 50–54).

13. By using carefully selected units, we can model a linear spring–magnet system by

$$\frac{d^2x}{dt^2} + x - \frac{36}{(x-7)^2} = 0.$$

The magnets are assumed to be long enough that the attractive force is inversely proportional to the distance between them (see [8], pp. 268–269). Sketch the phase plane portrait.

14. The Sine–Gordon partial differential equation

$$\frac{\partial^2 \phi}{\partial x^2} - \frac{\partial^2 \phi}{\partial t^2} = \sin\phi$$

was constructed to illustrate nonlinear wave phenomena. To study the existence of traveling waves, one studies an associated ordinary differential equation obtained as follows: let $\phi(x, t) = f(x - ct)$, where c is the wave velocity, and substitute it into the Sine–Gordon equation. For instance, if we let $z = x - ct$, then

$$\frac{\partial\phi}{\partial t}(x, t) = \frac{\partial}{\partial t}f(x - ct) = \frac{df}{dz}(z)\frac{dz}{dt}$$

$$= -c\frac{df}{dz}, \text{ etc.}$$

Analyze the ordinary differential equation you obtain according to the above instructions. Show that the upper separatrix in the phase plane is the graph of a solution to

$$\frac{df}{dz} = \frac{2}{\sqrt{1-c^2}}\sin\frac{f}{2}.$$

Solve this equation with initial condition $f(0) = \pi/2$. Its solution is called a "kink" soliton (see [13]). Why is $c = 1$ an exceptional value?

Linear damping changes centers into spiral points but saddle points remain saddle points. Sketch potential plane and global phase plane portraits for the following equations. Assume that $0 < c \ll 1$.

15. $\dfrac{d^2x}{dt^2} + c\dfrac{dx}{dt} + x^3 - \dfrac{x}{4} = 0$

16. $\dfrac{d^2x}{dt^2} + c\dfrac{dx}{dt} + 2x^2 - 5x - 3 = 0$

17. $\dfrac{d^2x}{dt^2} + c\dfrac{dx}{dt} + \sin x - \dfrac{\sqrt{8}}{\pi}x = 0$

18. $\dfrac{d^2x}{dt^2} + c\dfrac{dx}{dt} + 3x^2 + 5x + 2 = 0$

19. $\dfrac{d^2x}{dt^2} + c\dfrac{dx}{dt} + x = 1 + \dfrac{1}{8}x^2$

20. $m\dfrac{d^2x}{dt^2} + c\dfrac{dx}{dt} + k\left(x - \dfrac{\lambda}{L-x}\right) = 0$

21. Show that $\ddot{x} + \dot{x}^2 + g(x) = 0$ has an integral,

$$E(x, \dot{x}) = (e^{2x})\frac{\dot{x}^2}{2} + e^{2x}F(x) = \text{const.}$$

with $F'(x) + 2F(x) = g(x)$.

a) Set $g(x) = x$ and show that $(0, 0)$ is a center by showing that the nearby level curves of $E(x, \dot{x})$ are closed.

b) Set $g(x) = \sin x$. Is there a level curve of $E(x, \dot{x})$ joining two unstable equilibrium points? Justify your answer.

c) Introduce a new independent variable s by setting $ds/dt = e^{-x}$ and derive the differential equation

$$x'' + e^{2x}g(x) = 0, \text{ where } (\ )' = d(\ )/ds.$$

Sketch a global phase plane portrait for the integral curves of

$$x'' + xe^{2x} = 0.$$

Are the integral curves the same as those for

$$\ddot{x} + \dot{x}^2 + x = 0?$$

Hint: Are the coordinate systems of the two phase planes the same?

22. Compute the period of the van der Pol equation

$$\ddot{x} + \epsilon(x^2 - 1)\dot{x} + x = 0$$

with $\epsilon = 0.1$ by integrating it using RK4 on $0 \leq t \leq 30$, $h = 0.1$, $x(0) = 2$, $\dot{x}(0) = 0$. Estimate the period by computing the time between suc-

cessive crossings of the positive x axis *after* the integral curve is near the limit cycle. To avoid excessive output, print t, x, $\dot{x}$ only for x near 2 and $\dot{x}$ near 0. Interpolate to find the crossing time. Repeat the computation with $\epsilon = 0.4$, 0.8, 1.6, 2.0, and plot the period versus ϵ.

7.5 MATHEMATICAL MODELS OF TWO POPULATIONS

In Chapter 1 we discussed the mathematical modeling of the growth of one population. The model developed there was based on certain principles that could be satisfied by the per capita rate of growth. This analysis will now be extended to the study of two populations. Much of what we have to say is applicable to models of more than two populations, but the advantage of having the phase plane to describe the qualitative behavior of the models given favors the study of two populations.

Let X and Y represent two populations whose size at time t is given by the functions $x(t)$ and $y(t)$. Both of these will be assumed to be nonnegative (since negative population sizes make no sense), and any time one or both of the components of the vector $(x(t), y(t))$ becomes zero, the study is over, since one or both of the populations has expired. The per capita rates of growth of each population at time t are then $\dot{x}(t)/x(t)$ and $\dot{y}(t)/y(t)$ for populations X and Y, respectively.

The next task is to give plausible conditions which these rates must satisfy. We must first assume that there is some form of interaction between the two populations X and Y, and therefore the per capita rates of growth must depend on both the present size $x(t)$ of X and the present size $y(t)$ of Y. This immediately suggests that there exist functions f and g of two variables such that

$$\frac{\dot{x}(t)}{x(t)} = f(x(t), y(t)), \qquad \frac{\dot{y}(t)}{y(t)} = g(x(t), y(t)),$$

or, equivalently, that the model is described by the autonomous system of two differential equations

$$\dot{x} = x\,f(x, y), \qquad \dot{y} = y\,g(x, y). \tag{7.5.1}$$

There is no inherent biological reason why f or g should not also depend on t, but the obvious advantage of using the phase plane to analyze the qualitative behavior favors an autonomous model.

What should be the nature of f and g? First of all, in the absence of one population, the per capita rate of growth of the remaining population should satisfy whatever principles are decided to be appropriate for the growth of that population without interaction. Since the absence of population Y

means $y = 0$, we could have, for instance,

$$f(x, 0) = a > 0, \quad \text{or} \quad f(x, 0) = -a < 0, \quad \text{or} \quad f(x) = r\left(1 - \frac{x}{K}\right),$$

$$r, K > 0,$$

which are exponential growth, exponential decay, and logistic growth of X, respectively. Similarly, $g(0, y)$ should satisfy whatever principles the per capita rate of growth of population Y should follow in the absence of population X.

What should be the nature of f and g when both x and y are not zero? To answer this question one has to specify the type of interaction between the two populations X and Y. There are two general classifications:

Competition models. Both populations are in competition for the same shared resources (e.g., food, water, space). Consequently, an increase in Y will decrease the per capita growth rate of X. Similarly, an increase in X will decrease the per capita growth rate of Y.

Predator–prey models. Population Y is a predator population which has X as its food source, or prey. Consequently, an increase in X will result in an increase in the per capita growth rate of Y, whereas an increase in Y results in a decrease in the per capita growth rate of X.

In the predator–prey case it is sometimes assumed that the only food source for Y is X and that Y will perish in its absence. This would mean that $g(0, y)$ is strictly negative, for example, $g(0, y) = -ay$, $a > 0$, corresponding to exponential decay. While this makes for lovely phase plane portraits, few ecological systems are so simple. Each class of models will now be discussed in more detail.

Competition Models

Since the per capita growth rate of X is $f(x, y)$, the competition assumption suggests that f should decrease as y increases for fixed x and $x > 0$ and $y > 0$. This leads to the requirement that $\partial f/\partial y$ be negative for $x > 0$ and $y > 0$. The same reasoning applied to Y implies that $\partial g/\partial x$ is also negative for $x > 0$ and $y > 0$. Consequently, our model is

$$\dot{x} = x f(x, y), \qquad \dot{y} = y g(x, y) \tag{7.5.2}$$

$$\frac{\partial f}{\partial y}(x, y) < 0, \quad \frac{\partial g}{\partial x}(x, y) < 0, \quad x > 0, \quad y > 0,$$

where $f(x, 0)$ and $g(0, y)$ are suitable population growth laws for X and Y, respectively.

To analyze a model of the type (7.5.2), one must first find all the equilibrium points in the upper right-hand quadrant $x \geq 0$, $y \geq 0$, and then analyze their stability. These equilibrium points will be of three types and each will be discussed briefly and some possible phase plane configurations given before analyzing a specific model.

Types of Equilibrium Points for Competition Models

1. Point $(0, 0)$. This is always an equilibrium point, and if the system is linearized, one obtains

$$\dot{u} = f(0, 0)\, u, \qquad \dot{v} = g(0, 0)\, v.$$

The point $(0, 0)$ will be asymptotically stable if both $f(0, 0)$ and $g(0, 0)$ are negative. This means that for small initial sizes, the population X will diminish in the absence of Y and the population Y will diminish in the absence of X. Since it is usually assumed that X and Y are growing populations in the absence of competition, $f(0, 0)$ and $g(0, 0)$ are assumed positive and, as a consequence, $(0, 0)$ will be an unstable equilibrium point.

2. Points $(K, 0)$ and/or $(0, J)$, if any, where $f(K, 0) = 0$ and/or $g(0, J) = 0$. The values K and J would represent the natural carrying capacities of the X and Y population, respectively, in the absence of competition. If $(K, 0)$ were asymptotically stable, it would mean that if $x(t)$ were near K and $y(t)$ was small, then competition would favor X and it would survive (with limiting value K) whereas Y would perish. If $(K, 0)$ were unstable, then for X near its carrying capacity and the size of Y small, competition favors Y and it increases in size. This obviously suggests a saddle-point type behavior.

3. Points (x_∞, y_∞), $x_\infty > 0$, $y_\infty > 0$, if any, which are solutions of the two equations

$$f(x, y) = 0, \qquad g(x, y) = 0.$$

If such a point is asymptotically stable, it means that populations X and Y have positive limiting values x_∞ and y_∞, respectively. This is the case of *stable coexistence*. If such a point were unstable or did not exist, it will usually mean that a point of the type $(K, 0)$ or $(J, 0)$ will be asymptotically stable and only one population will survive competition. This is the case of *competitive exclusion*. Another interesting case is where (x_∞, y_∞) is unstable but is enclosed by a closed curve which is a stable limit cycle. Competition will then create a limiting cyclic configuration.

Some possible phase plane portraits for the competition model are shown in Fig. 7.32.

A Logistic Competition Model

A simple but frequently used model for competition is where one assumes the following:

1. The growth of populations X and Y in the absence of competition is logistic, and

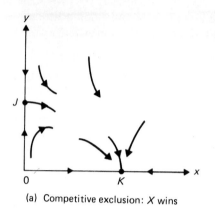

(a) Competitive exclusion: X wins

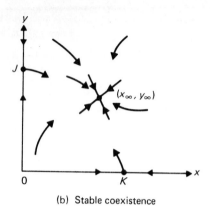

(b) Stable coexistence

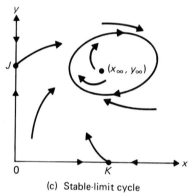

(c) Stable-limit cycle

Figure 7.32

2. The effect of competition between X and Y is to reduce the per capita growth rate of each in proportion to the number of competitors present.

This leads to the model

$$\dot{x} = xr\left(1 - \frac{x}{K}\right) - \alpha xy, \qquad \dot{y} = ys\left(1 - \frac{y}{J}\right) - \beta xy, \qquad (7.5.3)$$

where r, K, α, s, J, and β are all positive numbers.

The intrinsic growth rate of X is r, its carrying capacity is K, and its competition coefficient is α. A similar statement applies to the population Y and the constants s, J, and β. The functions f and g are the linear functions

$$f(x, y) = r\left(1 - \frac{x}{K}\right) - \alpha y, \qquad g(x, y) = s\left(1 - \frac{y}{J}\right) - \beta x,$$

and since $f_y = -\alpha < 0$, $g_x = -\beta < 0$, the hypotheses for a competition model are satisfied.

It is a straightforward calculation, which we leave to the reader as an exercise, to verify the following properties of the system (7.5.3):

1. Point $(0, 0)$ is an unstable node;

2. Points $(K, 0)$ and $(0, J)$ are equilibrium points and
   a) Point $(K, 0)$ is a stable node if $K > s/\beta$, whereas it is a saddle point if $K < s/\beta$.
   b) Point $(0, J)$ is a stable node if $J > r/\alpha$, whereas it is a saddle point if $J < r/\alpha$.

One now can graph the lines $f(x, y) = 0$ and $g(x, y) = 0$ and determine that

3. They will have a point of intersection (x_∞, y_∞), where $x_\infty > 0$, $y_\infty > 0$, if
   a) $J > r/\alpha$ and $K > s/\beta$, in which case $(K, 0)$ and $(0, J)$ are stable nodes, or
   b) $J < r/\alpha$ and $K < s/\beta$, in which case $(K, 0)$ and $(0, J)$ are saddle points;

4. They will have no point of intersection in the upper right-hand quadrant if
   a) $J > r/\alpha$ and $K < s/\beta$, which implies that $(0, J)$ is a stable node and $(K, 0)$ is a saddle point. Competition favors the population Y and it will tend towards its natural carrying capacity J, whereas population X will become extinct.
   b) $J < r/\alpha$ and $K > s/\beta$, which gives the same qualitative behavior as in (a), with the roles of X and Y reversed.

A graphical representation of the two cases 4(a) and (b), which represent competitive exclusion, is shown in Fig. 7.33.

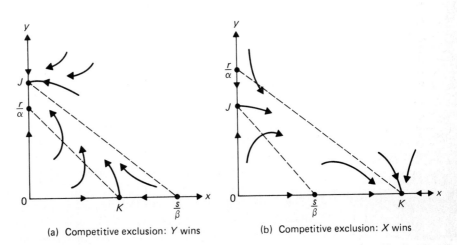

(a) Competitive exclusion: Y wins   (b) Competitive exclusion: X wins

Figure 7.33

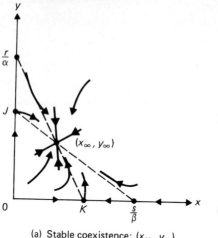
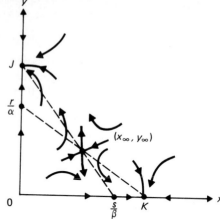

(a) Stable coexistence: (x_∞, y_∞)
is a stable node

(b) Unstable coexistence: (x_∞, y_∞) is
a saddle point and initial conditions
will determine which population wins

Figure 7.34

Finally, we must analyze the case 3 where there is an equilibrium point
(x_∞, y_∞) with $x_\infty > 0$, $y_\infty > 0$. If the system (7.5.3) is linearized about the
equilibrium point, then, since (x_∞, y_∞) satisfy the equations

$$r - \frac{r}{K}\, x_\infty - \alpha y_\infty = 0, \qquad s - \frac{s}{J}\, y_\infty - \beta x_\infty = 0, \qquad (7.5.4)$$

the coefficient matrix of the linearized system will be

$$\begin{bmatrix} \dfrac{-r}{K}\, x_\infty & -\alpha x_\infty \\[2mm] -\beta y_\infty & \dfrac{-s}{J}\, y_\infty \end{bmatrix}.$$

Its characteristic equation is

$$\lambda^2 + \left(\frac{r}{K}\, x_\infty + \frac{s}{J}\, y_\infty \right)\lambda + x_\infty y_\infty \left(\frac{rs}{KJ} - \alpha\beta \right) = 0,$$

and since $x_\infty y_\infty > 0$, it will have roots with negative real parts if and only if

$$\frac{r}{K}\, x_\infty + \frac{s}{J}\, y_\infty > 0 \qquad \text{and} \qquad \Delta = \frac{rs}{KJ} - \alpha\beta > 0. \qquad (7.5.5)$$

The first condition is always satisfied since all the quantities are positive; a
biological interpretation of the condition $\Delta > 0$ will be given below.

Suppose $(K, 0)$ and $(0, J)$ are unstable; then case 3(b) above implies that

$\alpha < r/J$ and $\beta < s/K$, hence

$$\alpha\beta < \frac{r}{J} \cdot \frac{s}{K},$$

and therefore $\Delta > 0$. Hence if $(K, 0)$ and $(0, J)$ are unstable, then (x_∞, y_∞) is asymptotically stable. A similar analysis shows that if $(K, 0)$ and $(0, J)$ are stable, then (x_∞, y_∞) is unstable. The phase plane plots for the two possible cases are shown in Fig. 7.34.

Example 1 Consider the competitive system

$$\dot{x} = x\left[(0.5) - \frac{1}{400}x - \frac{1}{10^3}y\right], \qquad \dot{y} = y\left[(0.8) - \frac{1}{500}y - \frac{1}{10^3}x\right],$$

which can be written as

$$\dot{x} = x(0.5)\left[1 - \frac{1}{200}x\right] - \frac{1}{10^3}xy, \qquad \dot{y} = y(0.8)\left[1 - \frac{1}{400}y\right] - \frac{1}{10^3}xy,$$

and therefore

$$r = 0.5, \qquad K = 200, \qquad \alpha = \frac{1}{10^3},$$

$$s = 0.8, \qquad J = 400, \qquad \beta = \frac{1}{10^3}.$$

The point $(0, 0)$ is an unstable node, and since

$$J = 400 < \frac{r}{\alpha} = (0.5)10^3 = 500,$$

$$K = 200 < \frac{s}{\beta} = (0.8)10^3 = 800,$$

the equilibrium points $(200, 0)$ and $(0, 400)$ are saddle points, and there will be a stable equilibrium point (x_∞, y_∞). By solving the pair of equations

$$\frac{1}{400}x + \frac{1}{10^3}y = \frac{1}{2}, \qquad \frac{1}{10^3}x + \frac{1}{500}y = \frac{8}{10}$$

we find that $(x_\infty, y_\infty) = (50, 375)$. The coefficient matrix of the linearized system at (x_∞, y_∞) is

$$\begin{bmatrix} -\dfrac{0.5}{200}(50) & -\dfrac{1}{1000}(50) \\ -\dfrac{1}{1000}(375) & -\dfrac{0.8}{400}(375) \end{bmatrix} = \begin{bmatrix} -\dfrac{1}{8} & -\dfrac{1}{20} \\ -\dfrac{3}{8} & -\dfrac{3}{4} \end{bmatrix}.$$

Its characteristic polynomial is $\lambda^2 + \frac{6}{5}\lambda + \frac{9}{40}$ which has distinct negative roots, so the point $(50, 375)$ is a stable node. $\square$

If the original system of differential equations is written in the form

$$\dot{x} = rx - \frac{r}{K}x^2 - \alpha xy, \qquad \dot{y} = sy - \frac{s}{J}y^2 - \beta xy,$$

the coefficient r/K can be interpreted as a measure of how much population X controls its own growth. The coefficient α can be interpreted as a measure of how much population Y controls X's growth, and similar interpretations can be given for s/J and β. Therefore the condition $\Delta > 0$ is equivalent to

$$\left(\frac{r}{K}\right)\left(\frac{s}{J}\right) > \alpha\beta,$$

which says that stability can occur when species control their own growth stronger than when they try to control their competitor's. Put another way, $\Delta < 0$ means that stability will not occur when species compete too strongly.

Predator–Prey Models

In the general classification of types of population models, predator–prey models were characterized by the assumptions that for $x > 0$ and $y > 0$,

1. The per capita rate of growth $f(x, y)$ of population X (the prey) must decrease as the size y of population Y (the predator) increases, and

2. The per capita rate of growth $g(x, y)$ of Y must increase as the size x of population X increases.

This leads to the model

$$\dot{x} = x\,f(x, y), \qquad \dot{y} = y\,g(x, y), \qquad (7.5.6)$$

$$\frac{\partial f}{\partial y}(x, y) < 0, \qquad \frac{\partial g}{\partial x}(x, y) > 0, \qquad x > 0, \qquad y > 0.$$

The types of equilibrium points for the predator–prey model (7.5.6) are the same as for the competitive model, namely, point $(0, 0)$, possibly points of the type $(K, 0)$ or $(0, J)$, and of the type (x_∞, y_∞), where $x_\infty > 0$, $y_\infty > 0$. By linearizing the system about each of them and studying their asymptotic behavior one tries to put together a phase plane portrait of the system.

But the real interest in predator–prey models arises from the fact that they often exhibit periodic behavior. In the phase plane this would be represented by a closed curve or families of closed curves with an equilibrium point in their interior. If we take such a curve and at points on it put an arrow pointing up or down if Y is increasing or decreasing, and another arrow

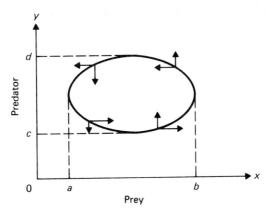

Figure 7.35 A periodic predator–prey model. The prey population size will oscillate between the values a and b; the predator between c and d.

pointing to the right or left if X is increasing or decreasing, we will obtain the picture given in Fig. 7.35.

There is a certain poetic simplicity in such pictures (rabbits and lynxes living together in harmonious balance but out of phase!), but often such models are based upon simplified assumptions which makes their biological significance minimal.

A classical predator–prey model is given by the Lotka–Volterra system of equations[1]

$$\dot{x} = x(r - \alpha y), \qquad \dot{y} = y(-s + \beta x), \tag{7.5.7}$$

where r, s, α, and β are positive constants. It is based on the assumptions that

1. The prey X grows exponentially in the absence of predator Y; that is, $f(x, 0) = r > 0$;
2. The predator Y will become extinct exponentially in the absence of prey; that is, $g(0, y) = -s < 0$.

Since $\partial f/\partial y = -\alpha < 0$, $\partial g/\partial x = \beta > 0$, we see that the effect of predation is to reduce or increase the per capita rate of growth in proportion to the numbers of predators or prey present.

---

[1] Named after the American biophysicist Alfred J. Lotka (1880–1949) and the Italian mathematician Vito Volterra (1860–1940) who independently developed and analyzed the model.

The Lotka–Volterra model (7.5.7) has two equilibrium points: $(0, 0)$, which is easily seen to be a saddle point since $r > 0$ and $-s < 0$, and $(x_\infty, y_\infty) = (s/\beta, r/\alpha)$. If the system is linearized about the latter, the coefficient matrix of the linearized system is

$$\begin{bmatrix} 0 & -\alpha s/\beta \\ \beta r/\alpha & 0 \end{bmatrix}.$$

Its characteristic equation is $\lambda^2 + rs = 0$ and since $rs > 0$, the roots are pure imaginary, which says that $(s/\beta, r/\alpha)$ is a center for the linearized system.

As was discussed in a previous section, the case where an equilibrium point of a nonlinear system is a center for the linearized system is an ambiguous case. The nonlinear system may have a family of closed curves (periodic solutions) around the equilibrium point, as in the linear case, or the equilibrium point may be a stable or unstable spiral point.

It is a fact that for the Lotka–Volterra model, given any point (x_0, y_0), $x_0 > 0$, $y_0 > 0$, there is a closed curve, centered at the equilibrium point $(s/\beta, r/\alpha)$ that represents a periodic solution of the system and passes through (x_0, y_0). More simply said, the center structure of the linearization is preserved in the nonlinear model. The proof of this statement depends on the fact that one can easily solve the equivalent first order equation

$$\frac{dy}{dx} = \frac{y(-s + \beta x)}{x(r - \alpha y)}, \qquad y(x_0) = y_0 \tag{7.5.8}$$

since it is separable. The implicitly defined solution will describe the integral curve passing through (x_0, y_0). The proof that the curve is closed is left for the reader as a problem.

The integral curves of the Lotka–Volterra equations are elliptically shaped near the equilibrium point and become more and more distorted as one moves away from it. Graphs of a few of them are shown in Fig. 7.36.

A disturbing consequence of the model is that no matter how small the initial prey population is and how large the initial predator population is, both populations will survive in some sort of cyclic growth curve. Imagine a scenario for the case of 2 rabbits and 4000 lynxes and you will see the consequences of the oversimplified assumptions on which it is based. What if both rabbits were males?

Another problem with the Lotka–Volterra model is its lack of *structural stability* which can be loosely defined as robustness of the phase portrait under small perturbations—a desired property of any good model. To show the lack of structural stability suppose instead that the prey population grows logistically and has a huge carrying capacity $K = r/\epsilon$, where $\epsilon > 0$ is very small. Then the system of equations becomes

$$\dot{x} = xr\left(1 - \frac{x}{K}\right) - \alpha xy = x(r - \epsilon x - \alpha y),$$

$$\dot{y} = y(-s + \beta x), \tag{7.5.9}$$

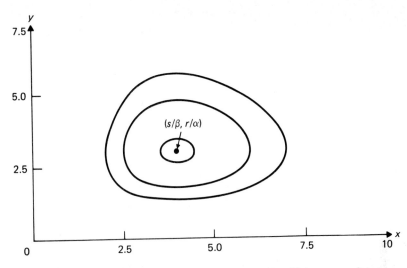

Figure 7.36 A phase plane portrait of the Lotka–Volterra predator–prey system generated with RKF45.

which are certainly small perturbations of the original Lotka–Volterra equations (7.5.7).

The point $(0, 0)$ is an equilibrium point of (7.5.9) and is still a saddle point. The second equilibrium point is

$$(x_\infty, y_\infty) = \left(\frac{s}{\beta}, \frac{r}{\alpha} - \frac{\epsilon s}{\beta \alpha} \right)$$

and can be made as near to $(s/\beta, r/\alpha)$ as desired by appropriate choice of ϵ. If the perturbed system (7.5.9) is linearized about (x_∞, y_∞), the coefficient matrix of the linearized system is

$$\begin{bmatrix} -\dfrac{\epsilon s}{\beta} & -\dfrac{\alpha s}{\beta} \\[2ex] \dfrac{\beta r}{\alpha} - \dfrac{\epsilon s}{\alpha} & 0 \end{bmatrix}$$

The characteristic equation of the matrix is

$$\lambda^2 + \epsilon \frac{s}{\beta} \lambda + \left(rs - \frac{\epsilon s^2}{\beta} \right) = 0$$

with roots

$$\lambda_1, \lambda_2 = \frac{-\epsilon s}{2\beta} \pm \sqrt{\left(\frac{\epsilon s}{2\beta} \right)^2 - \left(rs - \frac{\epsilon s^2}{\beta} \right)}.$$

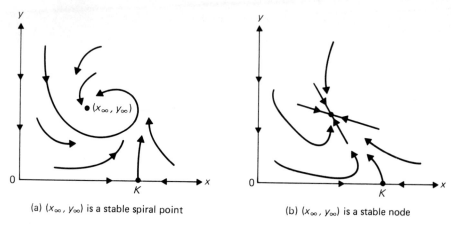

(a) (x_∞, y_∞) is a stable spiral point

(b) (x_∞, y_∞) is a stable node

Figure 7.37

One sees that for the linear system and locally for the nonlinear system (7.5.9) the equilibrium point will be

1. A saddle point if $rs - \epsilon s^2/\beta < 0$ (the reader may wish to check that this implies $y_\infty < 0$, which makes no sense);
2. A stable spiral point if $rs - \epsilon s^2/\beta > (\epsilon s/2\beta)^2$;
3. A stable node if $0 < rs - \epsilon s^2/\beta \le (\epsilon s/2\beta)^2$.

A sketch of the phase plane for some of these possible configurations when $\epsilon > 0$ is in marked contrast to the phase plane portrait of the Lotka–Volterra equations corresponding to $\epsilon = 0$ (see Fig. 7.37). Note that $(K, 0)$ is also an equilibrium point and it will be a saddle point if $K = r/\epsilon > s/\beta$, which will always be the case if ϵ is small.

Example 2   **a)** Consider the Lotka–Volterra system

$$\dot{x} = x(1 - \tfrac{1}{5}y) = x - \tfrac{1}{5}xy,$$
$$\dot{y} = y(-1 + \tfrac{1}{10}x) = -y + \tfrac{1}{10}xy.$$

The origin $(0, 0)$ is an equilibrium point and the linearized equations at $(0, 0)$ are

$$\dot{x} = x, \qquad \dot{y} = -y,$$

so it is a saddle point. The other equilibrium point, $(10, 5)$, is obtained by solving

$$1 - \tfrac{1}{5}y = 0, \qquad -1 + \tfrac{1}{10}x = 0.$$

The linearized system at $(10, 5)$ is

$$\dot{x} = -2y, \qquad \dot{y} = \tfrac{1}{2}x,$$

and the coefficient matrix has the characteristic polynomial $\lambda^2 + 1 = 0$ with roots $\pm i$. Since (10, 5) is a center of the nonlinear system as well, the integral curves very near (10, 5) will look like ellipses given by the solutions

$$x(t) = A \cos(t - \theta), \qquad y(t) = \tfrac{1}{2}A \sin(t - \theta)$$

of the linear system. They will become more distorted as one moves away from (10, 5).

b) Consider instead the following perturbed variation of (a):

$$\dot{x} = x - \epsilon x^2 - \tfrac{1}{5}xy, \qquad \dot{y} = -y + \tfrac{1}{10}xy,$$

where $\epsilon > 0$ is a small parameter. The point (0, 0) is still a saddle point, and the second equilibrium point is $(10, 5 - 50\epsilon)$ obtained by solving

$$1 - \epsilon x - \tfrac{1}{5}y = 0, \qquad -1 + \tfrac{1}{10}x = 0.$$

If the system is linearized at the point $(10, 5 - 50\epsilon)$, one obtains the system

$$\dot{x} = -10\epsilon x - 2y, \qquad \dot{y} = (\tfrac{1}{2} - 5\epsilon)x.$$

The characteristic equation of the coefficient matrix is

$$\lambda^2 + 10\epsilon\lambda + (1 - 10\epsilon) = 0$$

and its roots are

$$\lambda_1, \lambda_2 = -5\epsilon \pm \sqrt{25\epsilon^2 - (1 - 10\epsilon)}.$$

If $1 - 10\epsilon < 0$ or $\epsilon > \tfrac{1}{10}$, then the equilibrium point $(10, 5 - 50\epsilon)$ will be a saddle point, but its y coordinate will be negative. If $25\epsilon^2 < 1 - 10\epsilon$, the equilibrium point will be a stable spiral, and if $0 < 1 - 10\epsilon \leq 25\epsilon^2$, the point will be a stable node. Since $25\epsilon^2 = 1 - 10\epsilon$ for $\epsilon_0 = -\tfrac{1}{5}(1 - \sqrt{2}) \simeq 0.08284$, one can graph the relationship between the value of $\epsilon > 0$ and the nature of the equilibrium point as shown in Fig. 7.38. $\square$

In general, for both predator–prey and competitive-population models the presence of families of periodic solutions should not be expected. If a periodic solution, i.e., a closed curve in the phase plane does exist, it is more likely to be an isolated one, that is, a limit cycle. It will correspond to a particular set of initial conditions or values of certain parameters, and if it is stable, nearby solutions will tend to it as t approaches infinity. The "balance of nature" should in some sense be thought of as a limiting situation rather than a realizable one.

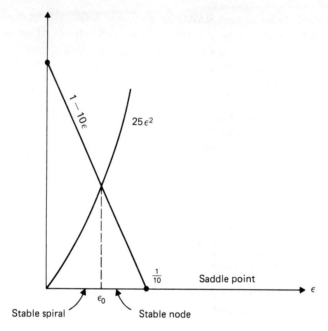

Figure 7.38 A stability diagram for the perturbed Lotka–Volterra system.

EXERCISES 7.5

1. A simple model of competing populations is

$$\dot{x} = x(r - \alpha y), \qquad \dot{y} = y(s - \beta x),$$

where r, s, α, and β are positive constants. Find the equilibrium points, determine their stability, and sketch a phase plane diagram.

2. In the model given in Exercise 1, suppose that the population Y is being harvested at a constant rate $H > 0$. Such a harvesting could correspond to the effects of disease, hunting, or fishing, or emigration. This leads to the model

$$\dot{x} = x(r - \alpha y), \qquad \dot{y} = y(s - \beta x) - H.$$

For the case of $r = s = 1$, $\alpha = \frac{1}{10}$, $\beta = \frac{1}{5}$, sketch the phase plane for $H = 4$ and $H = 8$. What happens as H approaches 10?

3. For the general logistic competition model (7.5.3) show that

   a) $(0, 0)$ is an unstable node

   b) $(K, 0)$ is a saddle point if $K < s/\beta$

   c) $(0, J)$ is a stable node if $J > r/\alpha$

4. For the following logistic competition models determine whether there is stable coexistence or competitive exclusion. If the former, determine the nature of the equilibrium point (x_∞, y_∞). If the latter, determine if one population wins the competition. In all cases sketch the phase plane diagram.

   a) $\dot{x} = x(2 - \frac{1}{3}x - \frac{1}{6}y),$
      $\dot{y} = y(1 - \frac{1}{10}y - \frac{1}{8}x)$

   b) $\dot{x} = 2x(1 - \frac{1}{20}x) - \frac{1}{25}xy,$
      $\dot{y} = 4y(1 - \frac{1}{40}y) - \frac{1}{10}xy$

   c) $\dot{x} = x(1 - \frac{1}{20}x - \frac{1}{8}y),$
      $\dot{y} = y(1 - \frac{1}{12}y - \frac{1}{16}x)$

   d) $\dot{x} = 2x(1 - \frac{1}{100}x) - \frac{1}{40}xy,$
      $\dot{y} = 10y(1 - \frac{1}{50}y) - \frac{1}{8}xy$

5. A simpler model for competition is where one population grows logistically and the other exponentially in the absence of competition. This leads to the model

$$\dot{x} = xr\left(1 - \frac{x}{K}\right) - \alpha xy, \qquad \dot{y} = sy - \beta xy,$$

where r, K, α, s, and β are all positive numbers. Show that in this case there is no possibility of stable coexistence.

6. If one population of a competing pair of populations is harvested, the equilibrium structure can be radically changed. For instance, stable coexistence could become competitive exclusion or vice versa. In the following competition models, H is a constant harvest rate. Find the equilibrium points of the system for each given value of H, analyze their stability, and sketch the phase plane.

a) $\dot{x} = x(1 - \frac{1}{10}x - \frac{1}{10}y) - H$,
$\dot{y} = y(1 - \frac{1}{6}y - \frac{1}{8}x), \quad H = 0, \frac{3}{4}$,

b) $\dot{x} = x(1 - \frac{1}{6}x - \frac{1}{10}y) - H$,
$\dot{y} = y(1 - \frac{1}{6}y - \frac{1}{12}x), \quad H = 0, \frac{1}{12}, \frac{1}{2}$

7. For the following Lotka–Volterra equations find the equilibrium points, then use the code RK4 with step size $h = 0.1$ to approximate the periodic solutions of the given initial value problems. If possible, graph your solutions in the phase plane.

a) $\dot{x} = x(2 - \frac{1}{2}y), \quad \dot{y} = y(-2 + x)$
  (i) $x(0) = 1.75, \quad y(0) = 4.00, \quad 0 \le t \le 3.5$
  (ii) $x(0) = 2.00, \quad y(0) = 2.00, \quad 0 \le t \le 3.5$

b) $\dot{x} = x(1 - \frac{1}{2}y), \quad \dot{y} = y(-1 + \frac{1}{3}x)$
  (i) $x(0) = 5.00, \quad y(0) = 2.25, \quad 0 \le t \le 7.0$
  (ii) $x(0) = 3.00, \quad y(0) = 2.00, \quad 0 \le t \le 7.0$

8. To prove that the integral curves of the Lotka–Volterra equations are closed curves, consider the equivalent first order equation:

$$\frac{dy}{dx} = \frac{y(-s + \beta x)}{x(r - \alpha y)}, \qquad y(x_0) = y_0,$$

where r, s, α, and β are positive constants.

a) Using separation of variables, show that the implicit solution of the equation is given by the expression

$$y^r e^{-\alpha y} = C_0 x^{-s} e^{\beta x},$$

where $C_0 > 0$ if x_0, $y_0 > 0$.

b) Consider the function

$$F(y) = y^r e^{-\alpha y}, \quad y \ge 0.$$

Show that $F(0) = 0$, $\lim_{y \to \infty} F(y) = 0$, and that $F(y)$ has a positive maximum for $y = r/\alpha$.

c) Show that the result of (b) implies that the equation

$$F(y) = K, K > 0, K \ne F\left(\frac{r}{\alpha}\right)$$

can only have two positive solutions or no positive solution.

The results of (c) apply as well to the equation

$$G(x) = x^s e^{-\beta x} = K, \quad K > 0, \quad K \ne G\left(\frac{s}{\beta}\right).$$

Recalling that $(s/\beta, r/\alpha)$ is the equilibrium point of the Lotka–Volterra system, show why (a), (b), and (c) imply that the integral curves in the positive xy quadrant are closed curves.

9. For any competition or predator–prey model

$$\dot{x} = x\, f(x, y), \qquad \dot{y} = y\, g(x, y)$$

with an equilibrium point (x_∞, y_∞) with $x_\infty\, y_\infty \ne 0$, show that the matrix of the linearized system at (x_∞, y_∞) is

$$\begin{bmatrix} x_\infty f_x(x_\infty, y_\infty) & x_\infty f_y(x_\infty, y_\infty) \\ y_\infty g_x(x_\infty, y_\infty) & y_\infty g_y(x_\infty, y_\infty) \end{bmatrix}.$$

10. Find all equilibrium points of the following predator–prey systems and analyze their stability. Sketch the phase plane diagram.

a) $\dot{x} = x\left[2\left(1 - \frac{x}{20}\right) - \frac{1}{4} y \right]$,

$\dot{y} = y[-1 + \beta x], \quad \beta = \frac{1}{5}$ and $\beta = \frac{1}{40}$

b) $\dot{x} = x\left[1 - \dfrac{x}{K} - \dfrac{1}{8}y\right],$

$\dot{y} = y\left[-1 + \dfrac{1}{16}x\right], \quad K = 12, 18, \text{ and } 20$

c) $\dot{x} = x\left[\dfrac{1}{2}\left(1 - \dfrac{x}{20}\right) - \dfrac{1}{16}y\right],$

$\dot{y} = y\left[\left(1 - \dfrac{y}{J}\right) + \dfrac{1}{8}x\right],$

$J = 4 \text{ and } J = 10$

11. Examples are given below of predator–prey systems that have been suggested by mathematical biologists; the authors have assigned values to certain parameters. For each of the models find all equilibrium points and analyze their stability. *Hint:* The result of Exercise 9 is useful.

a) $\dot{x} = x\left[\left(1 - \dfrac{x}{K}\right) - \dfrac{y}{x + 10}\right],$

$\dot{y} = \dfrac{y}{2}\left[\dfrac{x - 10}{x + 10}\right]. \text{ Let } K = 20 \text{ and } K = 40.$

b) $\dot{x} = x\left[\dfrac{R}{x} - \dfrac{y}{x + 10}\right], \quad \dot{y} = \dfrac{y}{2}\left[\dfrac{x - 10}{x + 10}\right].$

Let $R = 20$ and $R = 60$.

c) $\dot{x} = x\left[2\left(1 - \dfrac{x}{K}\right) - \dfrac{y}{x}(1 - e^{-x/10})\right],$

$\dot{y} = y[e^{-2} - e^{-x/10}]. \text{ Let } K = 25, K = 40,$

and $K = 60$.

12. In a predator–prey system containing a parameter K, which represents the natural carrying capacity of the prey, enlarging K would correspond to enrichment (e.g., increasing the prey's food supply). This enrichment could, for instance, destabilize an equilibrium point and create oscillations represented by limit cycles.

On the other hand, harvesting the predator will certainly destabilize the system if the harvest rate is too large and could result in extinction of the predator. The following numerical experiment shows the effect on a predator–prey system of enrichment and harvesting.

Consider the system

$\dot{x} = x\left[2\left(1 - \dfrac{x}{K}\right) - \dfrac{y}{x}(1 - e^{x/10})\right],$

$\dot{y} = y[e^{-2} - e^{-x/10}] - H,$

where K is the carrying capacity of the prey and H is the constant rate of harvesting. Use RK4 with $h = 0.2$, $0 \le t \le 35$, to compute the solution for the following parameter values and initial conditions. Have your program print out at unit steps of t, that is, $t = 1, 2, \ldots$, and also include a statement in your program to STOP if $y(t) \le 0$. If possible, graph your results. In all cases use $x(0) = 50$, $y(0) = 10$.

a) $K = 25; H = 0, H = 0.280$

b) $K = 40; H = 0, H = 1.10$

c) $K = 60; H = 0, H = 1.10, H = 1.4$

(For reference, see F. Brauer, A. C. Soudack, H. S. Jarosch, *International Journal of Control*, 23 (1976), pp. 553–573.)

MISCELLANEOUS EXERCISES

7.1 In Example 4.4 of Section 7.4, the following expression is given for the period T of the periodic solution corresponding to the energy level $K = 7/16$:

$$T = 2 \int_{-1}^{1} \dfrac{1}{\sqrt{\tfrac{1}{6}(x^2 - 1)(x^2 - 7)}} \, dx.$$

To approximate it, first let $x = \sin\theta$ and show that

$$T = 4\sqrt{\dfrac{8}{7}} \int_{0}^{\pi/2} \left(1 - \dfrac{1}{7}\sin^2\theta\right)^{-1/2} d\theta \approx 2\pi\sqrt{\dfrac{8}{7}}\left(1 + \dfrac{1}{28} + \dfrac{9}{3136}\right).$$

Hint: Use the binomial expansion

$$(1 - u)^{-1/2} = 1 + \tfrac{1}{2}u + \tfrac{3}{8}u^2 + \cdots$$

to approximate the integrand and obtain an estimate for T.

7.2 The idea given in the previous problem can be used to approximate for small k the values of the elliptic integral of the first kind

$$K(k) = \int_0^{\pi/2} (1 - k^2 \sin^2 \theta)^{-1/2} \, d\theta.$$

Do this for $k = 0.05, 0.10$, and 0.50, and compare your answers with the values of $K(k)$ from a table.

The following computer problems can be done by modifying the sample program using RK4 given in Chapter 3. The more advanced algorithm RKF45 given in Chapter 8 may also be employed. Study each problem carefully and design your print statements so that you have adequate but not excessive output.

7.3 Employing a generalization of the van der Pol equation, W. S. Krogdahl (*Astrophysical Journal*, 122 (1955), pp. 43–51) has had considerable success in modeling the velocity curves observed in the pulsation of variable stars of the Cepheid type. Denoting the variable radius by $r(\tau)$ and the variable velocity by $v(\tau)$, with τ a scaled time variable, he modeled r and v by

$$r(\tau) = [1 + q(\tau)]^{1/3},$$

$$v(\tau) = r'(\tau) = \frac{q'(\tau)/3}{[1 + q(\tau)]^{2/3}},$$

where $q(\tau) = \lambda Q(\tau)$ and

$$Q'' = -Q + \tfrac{2}{3}\lambda Q^2 - \tfrac{14}{27}\lambda^2 Q^3 + \mu(1 - Q^2)Q' + \tfrac{2}{3}\lambda(1 - \lambda Q)Q'^2 \, ;$$

λ and μ are empirical constants. If $\lambda = 0$, the differential equation for $Q(\tau)$ reduces to the van der Pol equation. Use RK4 with step $h = 0.05$ to approximate $r(\tau)$ and $v(\tau)$ for $0 \le \tau \le 10$ when

a) $\mu = 0.01, \lambda = 0.1, Q(0) = 1.05, Q'(0) = -0.05$

b) $\mu = 0.1, \lambda = 0.15, Q(0) = 0, Q'(0) = 1.9$

In each case plot $v(\tau)$ versus $r(\tau)$ in the phase plane and also $v(\tau)$ versus τ. The velocity curves should have the general shape shown in Fig. 7.39.

7.4 The differential equation $y'' + t^2 e^y = 0$, $t > 0$, occurred in a study on non-isothermal flow of a Newtonian fluid between parallel plates. It may be possible to solve it exactly (see *SIAM Review*, Vol. 23, No. 4, Oct. 1981, p. 524). Even without an exact solution, we can observe that $y'' < -t^2$ if y is positive, and therefore y should become and stay negative as t increases. But then $t^2 e^y$ should approach zero and y' should become constant. Test this conjecture by rewriting the differential equation as

$$y_1' = y_2, \qquad y_2' = -t^2 e^{y_1},$$

and constructing several approximate solutions with different initial values in the square $|y_1| \le 3$, $|y_2| \le 3$ by using RK4 with $h = 0.1$, $0 \le t \le 10$ (or less).

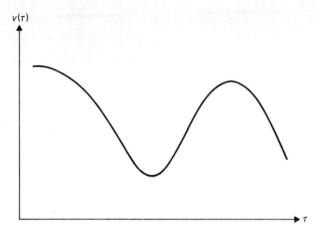

Figure 7.39

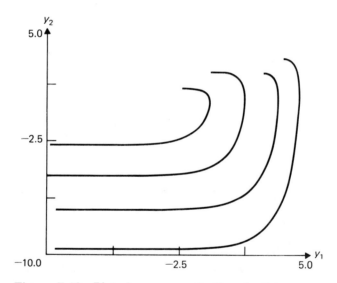

Figure 7.40 Plot of y_1 versus y_2 for Exercise 7.4.

Stop your competition if $y_1 < -10$. Plot your results in the y_1y_2 plane. A computer generated plot of several solution curves is given in Fig. 7.40.

7.5 The mass–spring system of Example 3 of Section 5.1 of Chapter 5 has a mirror image normal-mode solution, $y_1(t) = -y_2(t)$ (see Fig. 7.41) even if the middle linear spring is replaced by a nonlinear spring with spring constant

$$k' = k + \epsilon(y_2 - y_1)^2.$$

If ϵ is positive, the nonlinear spring is said to be *stiff* since k' increases with displacement.

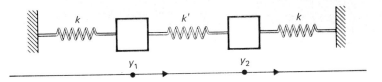

Figure 7.41

The differential equations governing the motion are

$$\frac{dy_1}{dt} = y_3, \qquad\qquad \frac{dy_2}{dt} = y_4,$$

$$\frac{dy_3}{dt} = \frac{-k}{m} y_1 + \frac{k'}{m}(y_2 - y_1), \qquad \frac{dy_4}{dt} = \frac{-k}{m} y_2 - \frac{k'}{m}(y_2 - y_1).$$

To investigate the effect of the nonlinearity, study the mirror image normal mode of vibration with $m = 1$, $k = 1$, $\epsilon = 0.02$, and initial conditions

$$y_1(0) = -y_2(0) = a, \qquad y_3(0) = y_4(0) = 0$$

for several values of a, the initial displacement. The resulting motion is periodic with an amplitude-dependent period. For each choice of a, estimate the period by integrating the differential equations with RK4, $h = 0.025$, and determine the time between successive attainments of the initial configuration. A typical output is given in Table 7.2 and a graph of the period as a function of a is given in Fig. 7.42.

Table 7.2

a	Time	y_1	y_1'	$t^* = $ period	$y(t^*) \approx a$
0.5	0.05	0.4981E+00	−0.7590E−01	−0.1304E−03	0.5019E+00
	3.55	0.4973E+00	0.9043E−01	0.3610E+01	0.5027E+00
	3.57	0.4991E+00	0.5256E−01	0.3610E+01	0.5009E+00
	3.60	0.4999E+00	0.1460E−01	0.3610E+01	0.5001E+00
	3.62	0.4998E+00	−0.2340E−01	0.3610E+01	0.5002E+00
	3.65	0.4988E+00	−0.6135E−01	0.3610E+01	0.5012E+00
	3.67	0.4968E+00	−0.9918E−01	0.3609E+01	0.5033E+00
1.0	3.55	0.9999E+00	0.2295E−01	0.3557E+01	0.1000E+01
	3.57	0.9995E+00	−0.5604E−01	0.3557E+01	0.1000E+01
1.5	3.47	0.1500E+01	0.4252E−01	0.3475E+01	0.1500E+01
	3.48	0.1500E+01	−0.4148E−01	0.3475E+01	0.1500E+01
2.0	3.36	0.2000E+01	0.5080E−01	0.3369E+01	0.2000E+01
	3.37	0.2000E+01	−0.4020E−01	0.3369E+01	0.2000E+01
2.5	3.24	0.2500E+01	0.7333E−01	0.3247E+01	0.2500E+01
	3.25	0.2500E+01	−0.2666E−01	0.3247E+01	0.2500E+01
3.0	3.11	0.3000E+01	0.9147E−01	0.3115E+01	0.3000E+01
	3.12	0.3000E+01	−0.1953E−01	0.3115E+01	0.3000E+01

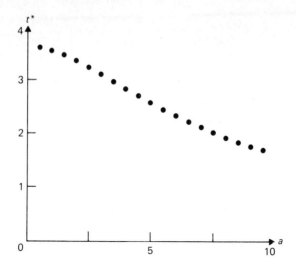

Figure 7.42

7.6 A biological oscillator subject to a constant positive input b can be modified by a system of nonlinear differential equations (see [7], p. 67):

$$\dot{x} = H(x)(x - ay + b), \qquad \dot{y} = x\,H(x) - cy,$$

where $a > c > 0$, $b > 0$, and $H(x)$ is the Heavyside step function: $H(x) = 1$, $x > 0$; $H(x) = 0$, $x \le 0$. If the equilibrium point in the first quadrant is an unstable spiral point, then there is a limit cycle surrounding it that contains a portion of the positive y axis.

a) Show that if $1 - c > 0$, $4a > (1 + c)^2$, the equilibrium point is an unstable spiral point.

b) Show that $(0, b/a)$ is on the arc of the limit cycle described by the linear system

$$\dot{x} = x - ay + b, \qquad \dot{y} = x - cy.$$

c) Show that the period of the limit cycle is independent of b.

d) Set $c = \frac{1}{2}$, $a = \frac{25}{16}$, $b = 1$, and use RK4 with $h = 0.1$ to compute the limit cycle. Since the approximate curve will penetrate the $x < 0$ half-plane, insert "IF(Y(1).LT.0.0) Y(1) = 0.0" just before the return statement in RK4.

e) The local error is not $O(h^5)$ at the points where the limit cycle enters and leaves the y axis. Why? This difficulty can be overcome by integrating with different independent variables on different segments of the limit cycle. Write a program to
  (i) Integrate with $h = 0.1$

$$\frac{dx}{dt} = x - a + b, \qquad \frac{dy}{dt} = x - cy$$

from $t = 0$, $x = 0$, $y = b/a$ to $t = t_1$, $x = x_1$, $y = y_1$, where $dx/dt \approx -0.5$ at this point;

(ii) Integrate with $h = -0.1$

$$\frac{dy}{dx} = \frac{x - cy}{x - ay + b}, \qquad \frac{dt}{dx} = \frac{1}{x - ay + b}$$

from $x = x_1$, $t = t_1$, $y = y_1$ to $x = 0$, $t = t_2$, $y = y_2$;

(iii) Integrate with $h = -0.1$

$$\frac{dt}{dy} = -\frac{1}{cy}, \qquad \frac{dx}{dy} = 0$$

from $y = y_2$, $t = t_2$, $x = 0$ to $y = b/a$, $t =$ the period, $x = 0$.

In your program use three different subroutines for computing the right sides of the differential equations, F1(TI,X,XP), F2(XI,Y,YP), F3(YI,T,TP) with

$$\begin{array}{llll}
\text{TI} = t, & \text{X}(1) = x, & \text{X}(2) = y, & \text{NEQN} = 2, \\
\text{XI} = x, & \text{Y}(1) = y, & \text{Y}(2) = t, & \text{NEQN} = 2, \\
\text{YI} = y, & \text{T}(1) = t, & & \text{NEQN} = 1.
\end{array}$$

REFERENCES

Classics containing superb examples:

1. N. Minorsky, *Introduction to Nonlinear Mechanics*, J. W. Edwards, Ann Arbor, 1947.

2. J. J. Stoker, *Nonlinear Vibrations in Mechanical and Electrical Systems*, Interscience, New York, 1950.

3. L. Cesari, *Asymptotic Behavior and Stability Problems in Ordinary Differential Equations*, Academic Press, New York, 1963.

Advanced textbooks on the theory of ordinary differential equations:

4. E. A. Coddington and N. Levinson, *Theory of Ordinary Differential Equations*, McGraw-Hill, New York, 1955.

5. W. Hurewicz, *Lectures on Ordinary Differential Equations*, M.I.T. Press, Cambridge, Mass., 1958.

Intermediate textbooks on ordinary differential equations:

6. F. Brauer and J. Nobel, *Qualitative Theory of Ordinary Differential Equations*, W. A. Benjamin, New York, 1969.

7. D. W. Jordan and P. Smith, *Nonlinear Ordinary Differential Equations*, Clarendon Press, Oxford, 1977.

8. H. K. Wilson, *Ordinary Differential Equations*, Addison-Wesley, Reading, Mass., 1971.

Applications:

9. N. Bogolyubov and Yu. A. Mitropolsky, *Asymptotic Methods in the Theory of Nonlinear Oscillations*, Gordon and Breach, New York, 1961.

10. J. McIntosh and R. McIntosh, *Mathematical Modeling and Computers in Endocrinology,* Springer-Verlag, Berlin, 1980.

11. J. D. Murray, *Nonlinear Differential Equation Models in Biology,* Oxford University Press, New York, 1977.

12. W. M. Smart, *Celestial Mechanics,* Longmans, Green, and Co., New York, 1953.

13. G. B. Whitham, *Linear and Nonlinear Waves,* John Wiley, New York, 1974.

Tables:

14. H. B. Dwight, *Tables of Integrals and other Mathematical Data,* Macmillan, New York, 1947.

More on
Numerical
Methods

8.1 INTRODUCTION

In Chapter 3 some elementary numerical methods for solving initial value problems for ordinary differential equations were described. There it was mentioned that errors were present in numerical solutions but no attempt was made to identify them or to compensate for their effect. In this chapter these errors will be examined in more detail, and a method will be presented for estimating and controlling them to some extent.

The computer program RKF45, which solves quite general initial value problems for first order systems of differential equations, will also be introduced. This program, based on the Runge–Kutta formulas discussed in Chapter 3, automatically estimates local error and chooses an appropriate step size h to insure an accuracy specified by the user. Several numerical examples that illustrate the use of RKF45 are given.

In what follows, the theory and algorithms are developed for the scalar case. The results can be extended to higher order equations and to systems of first order equations, just as they were in Chapter 3.

8.2 ERRORS, LOCAL AND GLOBAL

Errors enter into the numerical solution of initial value problems from two sources. The first is *discretization error;* it is a property of the method being used. The second is *roundoff error;* it depends on the arithmetic being used. If all calculations were done in exact arithmetic (no roundoff errors), discretization error would be the only error present. In general, one tries to carry enough digits in the calculations (at least one or two more than the accuracy desired) so that roundoff error does not become a problem.

Two types of discretization errors, *local* and *global errors,* should be distinguished. If we are solving the initial value problem

$$y' = f(t, y), \quad a \le t \le b, \tag{8.2.1}$$

$$y(a) = A, \tag{8.2.2}$$

using the one step method

$$y_{k+1} = y_k + h\theta(t_k, y_k), \tag{8.2.3}$$

$$t_{k+1} = t_k + h, \tag{8.2.4}$$

with $h = (b - a)/n$, where n is a positive integer, then the global error at $t = t_{k+1}$ is the quantity

$$\text{g.e.} = y(t_{k+1}) - y_{k+1}.$$

This is the difference between the actual solution $y(t_{k+1})$ at $t = t_{k+1}$ and our numerical approximation to it, y_{k+1}, obtained by using the one-step method (8.2.3, 8.2.4). The local error, on the other hand, is the error that is made in trying to follow the solution to the differential equation over just one step of length h, that is, from $t = t_k$ to $t = t_{k+1}$. To make this more precise, let $u(t)$ satisfy the initial value problem

$$u' = f(t, u), \tag{8.2.5}$$

$$u(t_k) = y_k. \tag{8.2.6}$$

Then the local error at $t = t_{k+1}$ is just the quantity

$$\text{l.e.} = u(t_{k+1}) - y_{k+1}. \tag{8.2.7}$$

The local and global errors at $t = t_{k+1}$ are illustrated graphically in Fig. 8.1.

Example 1 Suppose the initial value problem

$$y' = 5(y + t), \tag{8.2.8}$$

$$y(0) = -\tfrac{1}{6} \tag{8.2.9}$$

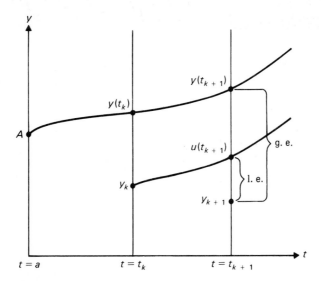

Figure 8.1 Local and global errors at $t = t_{k+1}$.

has been solved by using Euler's method with $h = 0.1$, and the results are given in Table 8.1.

To develop expressions for the local and global errors, first solve the differential equation (8.2.8) subject to the arbitrary initial data $y(a) = A$ to get

$$y(t) = (A + a + \tfrac{1}{5})e^{5(t-a)} - t - \tfrac{1}{5}.$$

The solution to (8.2.8, 8.2.9) is then

$$y(t) = -t - \tfrac{1}{5},$$

TABLE 8.1	k	t_k	y_k
	0	0.00	−0.2000
	1	0.10	−0.3000
	2	0.20	−0.4000
	⋮	⋮	⋮
	98	9.80	−5.1085
	99	9.90	−2.7632
	100	10.00	+0.8053

whereas the solution with initial data at any point (t_k, y_k) is

$$u(t) = (y_k + t_k + \tfrac{1}{5})e^{5(t-t_k)} - t - \tfrac{1}{5}.$$

Hence, the local error at $t = t_{k+1}$ is given by

$$\text{l.e.} = [(y_k + t_k + \tfrac{1}{5})e^{5(t_{k+1}-t_k)} - t_{k+1} - \tfrac{1}{5}] - y_{k+1},$$

and the global error is given by

$$\text{g.e.} = [-t_{k+1} - \tfrac{1}{5}] - y_{k+1}.$$

Referring to the above data, we see that the local and global errors at $t = 10.00$ are

$$\text{l.e.} = [(-2.7632 + 9.90 + 0.2)e^{5(0.1)} - 10.0 - 0.2] - 0.8053 = 1.0910$$

and

$$\text{g.e.} = [-10.0 - 0.2] - 0.8053 = -11.0053. \quad \square$$

It is interesting to note that the global error at a given point is not necessarily equal to the sum of all the local errors up to that point. In fact, in this last example the global error at $t = 10.0$ is -11.0053, while the sum of all local errors from $t = 0.10$ to $t = 10.0$ is -3.2734. The difference between these two quantities is a measure of the robustness of the differential equation itself. For a more detailed discussion of this point, see [1] and [4].

Although we will not develop the details here, something can be said about the effect of roundoff error on a solution of (8.2.1, 8.2.2) by using a one-step method (8.2.3, 8.2.4) of order p. If the error in evaluating $\theta(t_k, y_k)$ is ϵ_k with $|\epsilon_k| \leq \epsilon$ for all k and the error in forming $y_k + h\theta(t_k, y_k)$ is ρ_k with $|\rho_k| \leq \rho$ for all k, then it can be shown that

$$\max_{1 \leq i \leq n} |y(t_i) - y_i| \leq |y(t_0) - y_0|e^{L(b-a)} + \frac{e^{L(b-a)}}{L}\left(\frac{ch^p}{2} + \epsilon + \frac{\rho}{h}\right), \quad (8.2.10)$$

where c and L are constants depending on $y(t)$ and its derivatives. An examination of the estimate shows that the global error at a fixed point in $[a, b]$ has the same qualitative behavior as shown in Fig. 3.3 of Chapter 3. Values of h too small cause roundoff error problems, while discretization errors become pronounced for h values too large. It seems logical, not only for the sake of efficiency but also to avoid roundoff error difficulties, that the largest value of h consistent with some specified accuracy goal should be used. In the next section we will discuss a method for estimating local errors in one step methods and devise a procedure for choosing a proper step size h to achieve a given accuracy requirement.

Example 2   Solve the initial value problem

$$y' = \frac{y^2 + 2ty}{t^2}, \qquad y(1) = 1, \quad t > 0$$

TABLE 8.2	h	$|\text{error}|_{t=1.5}$
	5×10^{-1}	1.22222
	5×10^{-2}	0.05101
	5×10^{-3}	0.00034
	5×10^{-4}	0.00043
	5×10^{-5}	0.00314
	5×10^{-6}	0.06421
	5×10^{-7}	1.93206

using the improved Euler method with h values of $5 \times 10^{-1}, 5 \times 10^{-2}, \ldots,$ 5×10^{-7}. Tabulate the absolute error in the solution at $t = 1.5$.

First of all, the initial value problem has the solution

$$y(t) = \frac{t^2}{2 - t},$$

which, when evaluated at $t = 1.5$, gives 4.5. Using a program written in single precision arithmetic on an IBM 370 computer we obtain Table 8.2. From these data it appears that discretization error is dominant for h values larger than about 5×10^{-3}, while roundoff error tends to become dominant for smaller h values. $\square$

EXERCISES 8.2

1. Using Euler's method with a step size $h = \frac{1}{4}$, obtain an approximate solution to the initial value problem

   $$y' - 10y = t^2, \qquad y(0) = 0$$

   at the points $t = \frac{1}{4}, \frac{1}{2}, \frac{3}{4}, 1$. Calculate the local error at each of these points. What is the global error at $t = 1$ and how does it compare with the sum of the four local errors up to that point?

2. Write a computer program to use the improved Euler algorithm to approximate the solution to

   $$y' + y = e^{-t}, \qquad y(0) = 0$$

   at $t = 1$. Starting with a step size of $h = \frac{1}{2}$, decrease h successively by a factor of $\frac{1}{2}$ until the roundoff error becomes more pronounced than the discretization error.

8.3 ESTIMATING LOCAL ERRORS

Local error is the natural quantity to try to estimate and control. Of course, it is hoped that by controlling local error in a one step method, global errors will also be controlled. Fortunately this is the case if the function $f(t, y)$ is smooth, although, as we noted in the last section, the global error at a given point can grow faster than the sum of the local errors up to that point.

A general procedure for estimating local errors is to compare the results of formulas of different orders. To make this idea concrete, suppose we are solving the initial value problem

$$y' = f(t, y), \qquad y(t_0) = y_0,$$

and wish to approximate $y(t_1)$, where $t_1 = t_0 + h$. Assuming that $f(t, y)$ is sufficiently differentiable, we can expand the solution $y(t_1)$ in a Taylor series:

$$y(t_1) = y_0 + hy'(t_0) + \frac{h^2}{2} y''(t_0) + \frac{h^3}{6} y'''(t_0) + \cdots$$

$$= y_0 + hf(t_0, y_0) + \frac{h^2}{2} f^{(1)}(t_0, y_0) + \frac{h^3}{6} f^{(2)}(t_0, y_0) + \cdots. \quad (8.3.1)$$

Suppose y_1 is computed by using the first two terms of the series (Taylor's method of order $p = 1$, the Euler algorithm):

$$y_1 = y_0 + hf(t_0, y_0) \qquad (8.3.2)$$

and the first three terms (Taylor's method of order $p = 2$)

$$\bar{y}_1 = y_0 + hf(t_0, y_0) + \frac{h^2}{2} f^{(1)}(t_0, y_0). \qquad (8.3.3)$$

The difference

$$\bar{y}_1 - y_1 = \frac{h^2}{2} f^{(1)}(t_0, y_0)$$

is an approximation of the local error made in using the Euler algorithm. In fact, if $y(t)$ has at least three continuous derivatives on some interval containing t_0 and t_1, then at $t = t_1$ we get:

$$\text{l.e.} = \bar{y}_1 - y_1 + O(h^3). \qquad (8.3.4)$$

If h is small, the term $O(h^3)$ will be negligible and the quantity $\bar{y}_1 - y_1$ will be a good approximation to the local error.

This method of assessing local error obviously extends to higher order Taylor algorithms and, in fact, to one step methods in general. The next example illustrates how the improved Euler algorithm can be used to estimate the local error in the Euler algorithm.

Example 1   Solve the initial value problem

$$y' = t + y, \, y(0) = 1.0,$$

by using Euler's method with $h = 0.1$. Use the improved Euler algorithm to estimate the local error in the Euler approximation and compare the results with the actual local error.

TABLE 8.3	t_k	y_k	Estimated local error (8.3.5)	True local error (8.3.6)
	0.0	1.0		
	0.1	1.10000	0.10×10^{-1}	0.10×10^{-1}
	0.2	1.22000	0.11×10^{-1}	0.11×10^{-1}
	0.3	1.36200	0.12×10^{-1}	0.13×10^{-1}
	0.4	1.52820	0.13×10^{-1}	0.14×10^{-1}
	0.5	1.72102	0.15×10^{-1}	0.15×10^{-1}
	0.6	1.94312	0.16×10^{-1}	0.17×10^{-1}
	0.7	2.19743	0.18×10^{-1}	0.18×10^{-1}
	0.8	2.48718	0.19×10^{-1}	0.20×10^{-1}
	0.9	2.81590	0.21×10^{-1}	0.22×10^{-1}
	1.0	3.18748	0.24×10^{-1}	0.24×10^{-1}

The Euler and improved Euler methods in this case are

$$y_{k+1} = y_k + h(t_k + y_k)$$

and

$$\bar{y}_{k+1} = y_k + \frac{h}{2}[(t_k + y_k) + (t_k + h + y_k + h(t_k + y_k))].$$

So the estimate of the local error to within $O(h^3)$ is given by

$$r = \bar{y}_{k+1} - y_{k+1} = \frac{h}{2}[h + h(t_k + y_k)]. \tag{8.3.5}$$

The solution to the initial value problem

$$u' = t + u, \qquad u(t_k) = y_k$$

is

$$u(t) = (y_k + t_k + 1)e^{t-t_k} - t - 1,$$

hence the *true* local error at $t = t_{k+1}$ is

$$\text{l.e.} = u(t_{k+1}) - y_{k+1}$$
$$= [(y_k + t_k + 1)e^{(t_{k+1}-t_k)} - t_{k+1} - 1] - y_{k+1}. \tag{8.3.6}$$

Some results are summarized in Table 8.3. □

8.4 A STEP SIZE STRATEGY

We have discussed a procedure for estimating the local error in a typical one step method. The next order of business is to use this estimate to select a step size h, so that the error that is actually committed in taking one step is less than some preassigned value. How this selection is made will be illustrated by using the Euler and the improved Euler algorithms.

Suppose we have just computed y_1 from y_0 by using Euler's method and have also computed an improved Euler approximation $\bar{y}_1$. From the discussion in the previous section it follows that the local error in the Euler approximation can be expressed in the form

$$\text{l.e.} = h^2\tau_0 + O(h^3)$$

for some constant τ_0, and the term $h^2\tau_0$ can be represented within $O(h^3)$ by

$$r = \bar{y}_1 - y_1 = \frac{h}{2}[f(t_0 + h, y_0 + hf(t_0, y_0)) - f(t_0, y_0)].$$

(The reader might wish to check these formulas.)

Let

$$\text{est.} = \frac{r}{h}$$

be the estimate of local error relative to h. If we had started at $t = t_0$ with a step size of length γh, the local error would have been

$$\text{l.e.} = (\gamma h)^2\tau_0 + O(h^3),$$

and so

$$|\text{l.e.}| \approx |\gamma^2(h^2\tau_0)| = \gamma^2|r| = \gamma^2(h|\text{est.}|).$$

If $\epsilon > 0$ has been selected as an allowable local error per unit change in t, then the allowable error in a step of length γh is $\gamma h\epsilon$. What are the criteria for choosing γ? Clearly, since y_1 was computed with step size h, then

1. If $|\text{est.}| > \epsilon$, reject y_1 and compute a new y_1 with a smaller step size γh, $0 < \gamma < 1$;

2. If $|\text{est.}| \leq \epsilon$, accept y_1 and now proceed to compute y_2 by using a step size γh, $\gamma \geq 1$.

But since we want $|\text{l.e.}| \simeq \gamma^2 h|\text{est.}| \leq \gamma h\epsilon$, make the last inequality an equality, which implies

$$\gamma = \frac{\epsilon}{|\text{est.}|}.$$

This is the criterion used to choose γ.

Observe that the above criterion depends on two quantities, a preassigned relative error ϵ, *which the user selects,* and an approximate local relative error $|\text{est.}| = |r/h|$, *which the numerical scheme produces* at each step by applying two algorithms of different order. In practice, a fraction of γ, usually 0.9γ, is used to avoid frequent overshoot, which is clearly undesirable.

A simple Fortran program is given in Fig. 8.2 which implements the above step size strategy for Euler's method by using a local error estimator

```
C
C   PROGRAM TO SOLVE D/DT Y = Y USING EULER'S METHOD
C   WITH A LOCAL ERROR ESTIMATOR BASED ON THE IMPROVED
C   EULER ALGORITHM.
C
C
C   SET INITIAL VALUES AND TOLERANCES.
C
        F(T,Y)=Y
        T0=0.0
        Y0=1.0
        EPS=1.0E-3
        T=T0
        Y=Y0
        TFINAL=1.0
        H=0.1
C
C   BASIC LOOP
C
      1 IF(T+H.GT.TFINAL) H=TFINAL-T
        YE=Y+H*F(T,Y)
        R=0.5*H*(F(T+H,YE)-F(T,Y))
        EST=R/H
        GAMMA=EPS/ABS(EST)
        HE=0.9*GAMMA*H
        IF(ABS(EST).GT.EPS)GO TO 2
        Y=YE
        T=T+H
        H=HE
        GO TO 3
      2 H=HE
        Y=Y+H*F(T,Y)
        T=T+H
      3 IF(T.LT.TFINAL)GO TO 1
C
        WRITE(6,10)T,Y
     10 FORMAT(/1X,' T = ',F10.4,3X,' Y = ',F10.5)
        STOP
        END

        OUTPUT
T =    1.0000   Y =     2.71611
```

Figure 8.2 Sample program for using Euler's method with a local error estimator based on the improved Euler method to solve $y' = y$, $y(0) = 1$.

based on the improved Euler method. The program solves the initial value problem

$$y' = y, \qquad y(0) = 1,$$

and prints out the solution at $t = 1.0$. The value $\epsilon = 0.01$ has been used as the desired value of the local error per unit change in t.

It should be pointed out that the above program is not "robust," but merely a straightforward implementation of a fairly simple algorithm based on Euler's method. For example, in a sophisticated program to control step size, one should introduce safeguards for the case where the step size h becomes too small for machine precision or when it suddenly becomes large, indicating a possible breakdown in the step size estimate. Besides, lower order schemes, such as Euler's method, are not practical for solving most problems.

8.5 THE SUBROUTINE RKF45

As we pointed out earlier in Chapter 3, Runge–Kutta formulas of order $p = 4$ are probably the most common of the Runge–Kutta type formulas and require four function evaluations per step. If we are going to use the above ideas to approximate the local error in the $p = 4$ method, then a second approximation of order $p \geq 5$ is needed. A $p = 5$ formula would in general require six additional function evaluations per step. These evaluations are not necessarily independent of those required in the $p = 4$ formulas, e.g., in going from $t = t_k$ to $t = t_{k+1}$ both would require the evaluation of $f(t_k, y_k)$. Can a set of $p = 4$ and $p = 5$ formulas that use more points in common be found to minimize the number of function evaluations at each step? E. Fehlberg did about the best that one could hope for and produced a set of Runge–Kutta formulas involving six function evaluations per step, which gives both a fourth and fifth order approximation [7]. The formulas are quite lengthy and will not be given here; for details see [7].

The Fehlberg formulas have been implemented into the highly efficient Fortran subroutine RKF45 by H. A. Watts and L. F. Shampine [8]. A complete listing of the subroutine is given in Appendix 4. Since RKF45 is extensively annotated, no effort will be made here to duplicate the material in those comments. A few remarks about the use of the routine, however, are in order.

The subroutine RKF45 is designed to integrate a system of first order ordinary differential equations of the form

$$\frac{d\mathbf{y}}{dt} = \mathbf{f}(t, \mathbf{y}),$$

where $\mathbf{y}$ and $\mathbf{f}$ are vectors of the same length. The *call list* of the routine is

F,NEQN,Y,T,TOUT,RELERR,ABSERR,IFLAG,WORK,IWORK

where the parameters represent:

> F — subroutine F(T,Y,YP) to evaluate the derivatives $YP(I) = dY(I)/dt$, $I = 1,2, \ldots ,NEQN$, at T;
>
> NEQN — number of equations to be integrated;
>
> Y — solution vector at T;
>
> T — independent variable;
>
> TOUT — value of independent variable at which the solution is desired;
>
> RELERR,ABSERR — relative and absolute error tolerances for local error test;
>
> IFLAG — indicator for status of integration;
>
> WORK — work vector of dimension at least $3 + 6(NEQN)$;
>
> IWORK — work vector of dimension at least 5.

The parameter IFLAG is an important control variable and must be set to 1 before the first call to the subroutine. In the event that the integration reached TOUT with no problems, IFLAG is set to 2 by RKF45 and should be left at that value for continued integration. If a value of IFLAG = 3 is returned by RKF45, the relative error requested (RELERR) was unreasonably small; for instance, the number of working digits might be insufficient for the error requested. The value of RELERR is appropriately increased by the subroutine and integration can be continued. Values of 4 or 7 for IFLAG are warnings that the subroutine has done a significant amount of work or has taken a lot of steps to obtain the accuracy requested. IFLAG values of 5 or 6 indicate that error tolerances must be changed before continuing. IFLAG = 8 indicates improper input parameters. For specifics of each IFLAG value, see the comments at the beginning of the subroutine.

The parameters RELERR and ABSERR, the relative and absolute error tolerances for local error, must be chosen with care. These parameters measure local error by

$$|R| \leq (RELERR)(|Y|) + ABSERR$$

for each component of estimated local error R and each component of the solution vector Y. Note that this error test is a pure *relative error test* if ABSERR = 0 and a pure *absolute error test* if RELERR = 0; it is called a *mixed error test* if both ABSERR and RELERR are not zero. To avoid limiting precision difficulties, the subroutine requires RELERR to be larger than an internally computed relative error parameter, which is machine dependent. A pure absolute error test is not permitted and a pure relative error test should be avoided if the solution is expected to vanish for any value of T in the interval of integration. Reasonable values of RELERR and ABSERR, of course, depend on the accuracy desired in the solution, on the particular computer being used, and in a practical problem on the noise level of the data. For the IBM370 series in single precision, the values

$$\text{RELERR} = 1.0E-5 \text{ and } \text{ABSERR} = 1.0E-5$$

are appropriate for most of the examples in the book.

Perhaps the best way to introduce the subroutine RKF45 is through an example.

Example 1   Solve van der Pol's equation

$$y'' + \epsilon(y^2 - 1)y' + y = 0$$

with the initial conditions

$$y(0) = 1, \qquad y'(0) = 1$$

and the parameter $\epsilon = 1$. This equation arises quite often in nonlinear mechanics.

The first thing that must be done is to change the problem to an equivalent first order system. Let $y_1 = y$, $y_2 = y'$ to get

$$y_1' = y_2, \qquad y_2' = -y_1 - (y_1^2 - 1)y_2, \qquad y_1(0) = 1, \qquad y_2(0) = 1. \quad (8.5.1)$$

The sample program in Fig. 8.3 solves the problem (8.5.1) and tabulates the solution and its derivative at $t = 0.0, 0.5, \ldots, 10.0$. A phase plane plot (a plot in the yy' plane) of the solution is shown in Fig. 8.4.   □

```
C
C   SAMPLE PROGRAM FOR RKF45 — SOLVES VAN DER POL'S EQUATION
C
      SUBROUTINE & DP(T,Y,YP)
      REAL Y(2),YP(2)
      EPS = 1.0
      YP(1) = Y(2)
      YP(2) = -Y(1)-EPS*(Y(1)**2-1.0)*Y(2)
      RETURN
      END
C
      EXTERNAL VDP
      REAL WORK(15),Y(2)
      INTEGER IWORK(5)
      T = 0.0
      NEQN = 2
      Y(1) = 1.0
      Y(2) = 1.0
```

```
      RELERR=1.0E−5
      ABSERR=1.0E−5
      TF=10.0
      DT=0.5
      IFLAG=1
      TOUT=T
C
    1 CALL RKF45(VDP,NEQN,Y,T,TOUT,RELERR,ABSERR,IFLAG,WORK, IWORK)
      WRITE(6,5)T,Y(1),Y(2)
    5 FORMAT(1X,F6.2,2F15.4)
      IF (IFLAG.NE.2) GO TO 6
      TOUT=T+DT
      IF(T.LT.TF) GO TO 1
      STOP
C
    6 WRITE (6,10) IFLAG
   10 FORMAT (1X,'IFLAG =',12,1X,'; CHECK RKF45 COMMENTS FOR DETAILS)
      STOP
      END
```

T	Y(1)	Y(2)
0.	1.0000	1.0000
0.50	1.3313	0.2800
1.00	1.2985	−0.3670
1.50	0.9928	−0.8519
2.00	0.4212	−1.4890
2.50	−0.5499	−2.3666
3.00	−1.6348	−1.4855
3.50	−1.9288	0.0799
4.00	−1.7440	0.5689
4.50	−1.3920	0.8414
5.00	−0.8787	1.2581
5.50	−0.0607	2.1025
6.00	1.1871	2.5218
6.50	1.9625	0.5030
7.00	1.9330	−0.4068
7.50	1.6528	−0.6865
8.00	1.2456	−0.9632
8.50	0.6485	−1.4914
9.00	−0.3296	−2.4672
9.50	−1.5765	−1.9331
10.00	−2.0083	−0.0342

Figure 8.3 Sample program and output obtained by using RKF45 to solve van der Pol's equation.

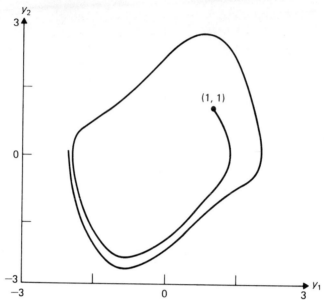

Figure 8.4 Phase plane plot of the solution of van der Pol's equation of Example 1.

In our discussion of numerical methods for solving initial value problems, we have up to this point considered only one step methods. There is another wide class of procedures known as multistep methods. For an excellent discussion of such procedures, as well as for a highly efficient computer program see [4]. The subroutine RKF45 is adequate for most problems the reader will encounter in practice. Generally speaking, Runge–Kutta type codes are very effective for solving initial value problems where the derivatives are reasonably inexpensive to evaluate and moderate accuracy is required (say, less than 10^{-6}).

8.6 SOME EXAMPLES

In this section we present some reasonably sophisticated examples leading to differential equations for which analytical solutions cannot be found conveniently, if at all. A subroutine like RKF45, which adjusts its step size automatically to control error, is essential for problems of this type, so that the user can have some confidence in the numerical results obtained.

Example 1 The orbit of the planet Mercury around the Sun can be represented as a solution to the differential equation

$$\frac{d^2u}{d\theta^2} + u = \frac{\mu}{h^2}(1 + \epsilon u^2), \qquad (8.6.1)$$

where $u = 1/r$ and r denotes the distance from the Sun to Mercury. Here θ is an angle in the plane of the orbit, μ is a gravitational constant, h is the angular momentum, and ϵ is a parameter determined by the effects of other planets on Mercury as well as the Sun's oblateness, and a correction required by the general theory of relativity. For simplicity, the second order differential equation is replaced by the system

$$\frac{dv}{d\theta} = w, \qquad \frac{dw}{d\theta} = 1 - v + \gamma v^2,$$

where $v = h^2 u/\mu$, $\gamma = \epsilon\mu^2/h^4$.

To illustrate the phenomenon of *precession*, we choose $\gamma = 0.1$, $v(0) = 2$, and $w(0) = 0$ (these values do not necessarily correspond to observed values), and integrate the system using RKF45 over several revolutions. As illustrated in Fig. 8.5, where $x = (1/v)\cos\theta$, $y = (1/v)\sin\theta$, Mercury moves on an ellipse that is slowly rotating in the orbital plane. The points of closest approach to the Sun are called perihelia. The *precession* of these points is due to the *perturbing* nonlinearity in the differential equation. The observed *precession of the perihelion of Mercury* could not be explained by Newtonian mechanics and remained a puzzle for many years. The close agreement between observations and the orbit modeled by the differential equation (8.6.1) with ϵ containing a relativistic correction is one of the major experimental confirmations of Einstein's theory of general relativity. □

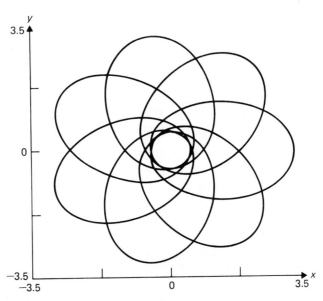

Figure 8.5   Rotating elliptical orbits of Mercury.

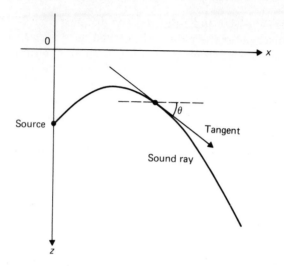

Figure 8.6 The path of a sound ray in water.

Example 2[1]   According to Fermat's principle applied to the ray theory of sound, sound propagates from one point to another along the curve for which the transit time is the smallest [7, pp. 19, 37]. In water, the speed of sound varies with depth. Thus, its path can be described by

$$\frac{dz}{dx} = \tan \theta$$

together with

$$\frac{\cos \theta}{C(z)} = \text{const.}$$

(by Fermat's principle), where z denotes the depth, x is the horizontal distance, θ is the angle from the horizontal, and $C(z)$ is the speed of sound as a function of depth (see Fig. 8.6). To derive a relationship between θ and x, we differentiate the second equation and replace dz/dx by $\tan \theta$ to get

$$
\begin{aligned}
0 &= \frac{d}{dx}\frac{\cos \theta}{C(z)} = \frac{-\sin \theta}{C(z)}\frac{d\theta}{dx} - \frac{C'(z)}{C(z)^2}\frac{dz}{dx}\cos \theta \\
&= \frac{-\sin \theta}{C(z)}\frac{d\theta}{dx} - \frac{C'(z)}{C^2(z)}\tan \theta \cos \theta \\
&= \frac{-\sin \theta}{C(z)}\left[\frac{d\theta}{dx} + \frac{C'(z)}{C(z)}\right].
\end{aligned}
$$

---

[1] This example is based on a problem given in [2, p. 173] and on classroom notes written by Stanly Steinberg in collaboration with W. T. Kyner and Cleve Moler (used with permission).

It follows that either $\sin \theta = 0$ and the sound propagates horizontally, or the path is governed by the pair of equations

$$\frac{dz}{dx} = \tan \theta, \qquad \frac{d\theta}{dx} = \frac{-C'(z)}{C(z)}.$$

In the ocean, the speed of sound depends on the temperature, salinity, pressure, and other depth-dependent parameters. Because of the difficulty in deriving an analytic expression of $C(z)$, it is common to approximate the speed at various depths with a suitable function. Typical data are given in Table 8.4 [2]. The construction of an empirical function for $C(z)$ must be done carefully since $C'(z)$ is also needed. The following representation due to S. Steinberg works well (see Fig. 8.7):

$$C(z) = 4779 + 0.01668z + \frac{160295}{z + 600}.$$

An inspection of the data reveals that the speed of sound can decrease with depth, attain a minimum value, and then increase. This permits the phenomenon of channeling of sound so that the direction of propagation oscillates about the horizontal and thereby energy is not lost by contact with either the surface or the bottom of the ocean. It is claimed that whales exploit this channeling in order to communicate over large distances. The optimal depth for such channeling is that at which the speed of sound is a minimum. From our data, this depth is nearly 2500 feet.

	Depth (ft)	Speed of sound (ft/sec)
TABLE 8.4	0	5042
	500	4995
	1000	4948
	1500	4887
	2000	4868
	2500	4863
	3000	4865
	3500	4869
	4000	4875
	5000	4875
	6000	4887
	7000	4905
	8000	4918
	9000	4933
	10000	4949
	11000	4973
	12000	4991

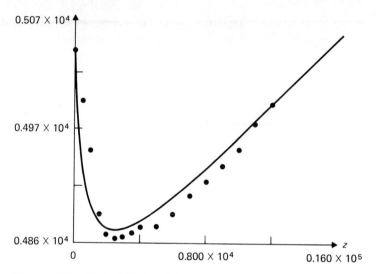

Figure 8.7 Plot of $C(z)$ and data.

Numerical solutions to the differential equations were investigated with RKF45 by assuming the initial conditions: $x = 0$, $z = 2500.016$ (the value where $C(z)$ is a minimum) and $\theta = \pm\alpha$, where the value $\alpha = 0.264$ radians was picked so that the ray grazed the surface of the water. An upper bound was thereby obtained on the cone of rays that can be propagated without energy loss caused by the contact with the surface of the ocean. A graph of the path of one of these bounding rays is given in Fig. 8.8. □

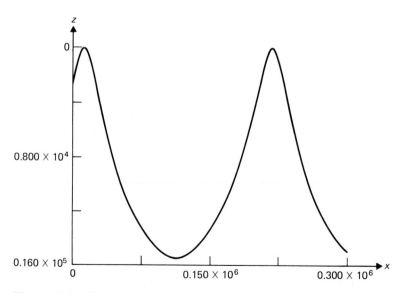

Figure 8.8 The graph of a bounding ray.

Example 3   In this example, a numerical technique for solving the two-point boundary value problem

$$y'' + p(t)\, y' + q(t)\, y = r(t), \qquad (8.6.2)$$

$$y(a) = A, \qquad y(b) = B, \qquad (8.6.3)$$

is considered. Geometrically, a function $y(t)$ is sought which satisfies the differential equation (8.6.2) and whose graph passes through the points (a, A) and (b, B). The approach to solving the problem is to try and find the value $y'(a)$, for then we would have an initial value problem and RKF45 could be used directly to solve it. Therefore, let $y'(a) = s$, and the task is to find a value $s = s^*$ so that the resulting solution, denoted by $y(t; s^*)$ satisfies $y(b; s^*) = B$. The method that is employed is called *shooting* and is described in the following algorithm:

The shooting algorithm.   The problem is to find a zero of

$$u(s) = y(b; s) - B.$$

1. Choose $s = s_1$ and solve the differential equation (8.6.2) with the initial conditions

$$y(a) = A, \qquad y'(a) = s_1;$$

denote the resulting solution by $y(x; s_1)$. If

$$u(s_1) = y(b; s_1) - B = 0,$$

set $s^* = s_1$ and stop. The solution is $y(x; s_1)$.

2. Choose $s = s_2 \neq s_1$ and solve (8.6.2) with

$$y(a) = A, \qquad y'(a) = s_2;$$

denote the solution by $y(x; s_2)$. If

$$u(s_2) = y(b; s_2) - B = 0,$$

set $s^* = s_2$ and stop. The solution is $y(x; s_2)$.

3. Calculate the value of s for which $u(s) = 0$:

$$s^* = s_1 - \frac{s_1 - s_2}{u(s_1) - u(s_2)} u(s_1).$$

This can be done because the function $u(s) = y(b; s) - B$ is a linear function of s and its zero is easy to calculate. The linearity of $u(s)$ follows from the fact that (8.6.2) is a linear equation (see Exercise 2).

4. Solve (8.6.2) with the initial conditions

$$y(a) = A, \qquad y'(a) = s^*$$

to get the desired solution.

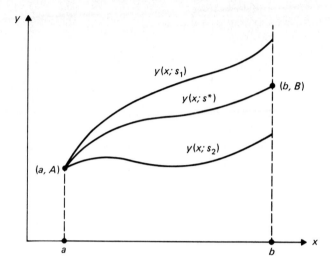

Figure 8.9 Graphic illustration of the shooting algorithm.

The process is illustrated graphically in Fig. 8.9.

To illustrate the algorithm, consider the problem

$$y'' + \frac{2}{t} y' + y = 0 \qquad (8.6.4)$$

$$y(1) = 1, \qquad y(2) = 5. \qquad (8.6.5)$$

The choice $s_1 = 1$ leads to the initial value problem (8.6.4) with initial conditions

$$y(1) = 1, \qquad y'(1) = 1.$$

This is solved at $t = 2$ using RKF45:

$$y(2; 1) = 1.111622,$$

$$u(1) = 1.111622 - 5 = -3.888378.$$

The choice $s_2 = 2$ leads to the initial conditions

$$y(1) = 1, \qquad y'(1) = 2,$$

and using RKF45 again gives

$$y(2; 2) = 1.532358,$$

$$u(2) = 1.532358 - 5 = -3.467642.$$

From the formula in step 3 it follows that

$$s^* = 1 - \frac{1 - 2}{(-3.888378) - (-3.467642)}(-3.888378) = 10.241848.$$

TABLE 8.5	t	$y(t)$	$y'(t)$
	1.00	1.000000	10.241848
	1.10	1.924833	8.328199
	1.20	2.677897	6.784327
	1.30	3.290407	5.502938
	1.40	3.784888	4.414373
	1.50	4.178141	3.472055
	1.60	4.483100	2.644097
	1.70	4.710023	1.908235
	1.80	4.867286	1.248683
	1.90	4.961922	0.654103
	2.00	4.999996	0.116265

The solution of (8.6.4) with initial data

$$y(1) = 1, \qquad y'(1) = 10.241848$$

is shown in Table 8.5. □

EXERCISES 8.6

1. Problem on propagation of sound in water.

a) Set $B = \dfrac{C(z)}{\cos \theta}$ and show that

$$B^2 = C^2(z) + \left[C(z) \frac{dz}{dx} \right]^2$$

is an integral for the system of differential equations

$$\frac{dz}{dx} = \tan \theta, \qquad \frac{d\theta}{dx} = -\frac{C'(z)}{C(z)}.$$

b) Define a new variable, $y = C(z) \tan \theta$ and show that

$$\frac{dy}{dx} = -C'(z), \qquad \frac{dz}{dx} = \frac{y}{C(z)}.$$

Derive and solve the linear differential equations corresponding to the equilibrium point, $y = 0$, $z = 2500.016$. Use the integral, $B^2 = C^2(z) + y^2$ and sketch a global phase plane portrait of the integral curves in the $z > 0$ half-plane of the nonlinear differential equations.

c) Integrate the y, z differential equations with initial conditions $x = 0$, $z = 2500.016$, $y =$

β, and by trial and error find β so that the ray corresponding to the solution grazes the surface of the water.

d) If the depth of the water is less than 15,486 feet, parameter β can be selected so that the ray is not reflected by the surface of the water and just grazes the bottom. Assume that the depth is 12,000 feet and find β by trial and error.

e) If two submarines are at a depth of 2500 feet and are 200,000 feet apart, they can communicate by acoustic signals. Find the rays bounding the signals, i.e., two solutions to the two-point boundary value problem with boundary conditions

$$z(0) = z(200,000) = 2500.$$

2. Show that $u(s)$ defined by

$$u(s) = y(x; s) - B$$

in the shooting algorithm of Example 3 is a linear function of s.

3. Write a computer program to implement the algorithm of Example 3 for solving two-point boundary value problems. Your program should

use RKF45 or any other comparable routine for solving initial value problems. Approximate the solution to each of the following problems at the points indicated.

a) $y'' + y' + xy = 0$,
   $y(0) = 1$, $y(1) = 0$;
   $t = 0.1, 0.2, \ldots, 1.0$

b) $y'' + \dfrac{2}{t}y' - \dfrac{2}{t^2}y = \dfrac{\sin t}{t^2}$,
   $y(1) = 1$, $y(2) = 2$;
   $t = 1.1, 1.2, \ldots, 2.0$

c) $y'' + 4y = \cos t$,
   $y(0) = 0$, $y\left(\dfrac{\pi}{4}\right) = 0$;
   $t = \dfrac{\pi}{32}, \dfrac{\pi}{16}, \dfrac{3\pi}{32}, \ldots, \dfrac{\pi}{4}$

d) $y'' + \sqrt{t}\,y' + y = e^t$,
   $y(0) = 0$, $y(1) = 0$;
   $t = 0.1, 0.2, \ldots, 1.0$

e) $y'' + 4ty' + (4t^2 + 2)y = 0$,
   $y(0) = 0$, $y(1) = 1/e$;
   $t = 0.1, 0.2, \ldots, 1.0$

f) $y'' - \dfrac{2t}{1 - t^2}y' + \dfrac{12}{1 - t^2}y = 0$,
   $y(0) = 0$, $y(0.5) = 4$;
   $t = 0.1, 0.2, \ldots, 0.5$

4. Modify the algorithm in Example 3 to solve the problem

$$(t^2 + 1)y'' - 2y = 0, \quad y(0) = 1, \quad y'(1) = 2.$$

Write a computer program to implement your procedure and find $y'(0)$ and $y(1)$.

5. a) Explain why your computer program in Exercise 3 cannot be used to solve the boundary value problem

$$y'' + e^{-y} = 0, \quad y(0) = 0, \quad y(1) = 0.$$

b) How would you modify the algorithm of Example 3 so as to obtain an approximate solution to this problem?

6. The differential equation

$$\ddot{x} + c\dot{x} + k \sin x = L$$

describes the rotation of a synchronous motor.

The dependent variable x is measured from an axis rotating with the electric field, $\ddot{x}$ represents the inertia torque due to the mass of the motor and its connected load, $c\dot{x}$ is due to "slippage," $k \sin x$ is due to the angle between the field of the rotor and armature, and the term L represents the torque of the external load on the motor. An extensive discussion of the mathematical problems arising from the operation of alternating-current motors is given in [11].

One problem of interest is the determination of the *critical damping factor*, the value of c such that there is an integral curve joining two unstable equilibrium states [12]. Remarkably, the same differential equation and the critical damping factor problem occur in celestial mechanics in the study of the spin-orbit coupling of the planet Mercury [13].

Take $k = 1$, $L = 0.1$, $\sin x_j = L$ with

$$x_0 \approx 0.1002, \qquad x_1 \approx 3.041, \qquad x_{-1} \approx -3.242.$$

a) Show that $(0, x_0)$ is a stable equilibrium point and $(0, x_{-1})$, $(0, x_1)$ are the neighboring unstable equilibrium points.

b) Sketch a phase plane portrait including the three equilibrium points with (i) $c = 0$, (ii) $0 < c \ll 0.1$. Note that there is no integral curve joining the two unstable equilibrium points x_{-1} and x_1.

c) Find λ_{-1} and λ_1, $\lambda_1 < 0 < \lambda_{-1}$, so that

$$x(t) = x_{-1} + \delta \exp(\lambda_{-1}t),$$

$$x(t) = x_1 - \delta \exp(\lambda_1 t)$$

are solutions to the linear differential equations corresponding to the two unstable equilibrium points $(x_{-1}, 0)$ and $(x_1, 0)$ with initial data $(x_{-1} + \delta, \delta\lambda_{-1})$ and $(x_1 - \delta, \delta\lambda_1)$. The parameter δ is arbitrary.

d) Let $y = \dot{x}$; then the second order differential equation governing the rotation can be replaced by an equivalent system of differential equations

$$\frac{dx}{dt} = y, \qquad \frac{dy}{dt} = -cy - k \sin x + L.$$

The integral curves of this system are also

the graphs of the solutions to a corresponding first order differential equation

$$\frac{dy}{dx} = \frac{-cy - k \sin x + L}{y}.$$

Set $c = 0$, $\delta = 0.05$, and use RKF45 to integrate this equation from $x = x_{-1} + \delta$ to $x = x_0$ and then from $x = x_1 - \delta$ to $x = x_0$. Why must δ be positive? Interpolate to get the two values of y at x_0 from the computed values. They will be different since the two curves corresponding to the solutions are not part of the same integral curve. Increase c until the two values of y are the same to three significant figures. (You might use the bisection method of numerical analysis to help you pick reasonable values of c.) The resulting value of c is less than 0.1 and is a good approximation to the critical damping factor. Sketch the corresponding integral curve in the phase plane. (*Remark:* Your programs should be written so that the two integrations are carried out sequentially. Only the two interpolated values of y and their differences should be printed at x_0.)

7. A hypothetical reaction exhibiting some of the characteristics of real biochemical oscillators was proposed by I. Prigogine and R. Lefever (a good treatment of this topic is given in [10], Chap. 4). The system of reactions is

$$A \xrightarrow{k_1} X, \qquad B + X \xrightarrow{k_2} Y + D,$$
$$2X + Y \xrightarrow{k_3} 3X, \qquad X \xrightarrow{k_4} E.$$

The reactants A, B, D, and E are kept constant, so $X(t)$ and $Y(t)$ are to be found.

From the law of mass action, we get

$$\frac{dX}{dt} = k_1 A - k_2 BX + k_3 YX^2 - k_4 X,$$

$$\frac{dY}{dt} = k_2 BX - k_3 YX^2.$$

Let

$$s = k_4 t, \qquad x = \frac{k_4 X}{k_1 A}, \qquad y = \frac{k_4 Y}{k_1 A},$$

$$a = \frac{k_3 (k_1 A)^2}{k_4^3}, \qquad b = \frac{k_2 B}{k_4};$$

then

$$\frac{dx}{ds} = 1 - (b + 1)x + ax^2 y, \qquad \frac{dy}{ds} = bx - ax^2 y.$$

There is only one equilibrium state, $x = 1$, $y = b/a$, with both x and y positive.

a) Let $x = 1 + u$, $y = v + b/a$; show that

$$\frac{du}{ds} = (b - 1)u + av + g(u, v),$$

$$\frac{dv}{ds} = -bu - av - g(u, v), \qquad a > 0, \quad b > 0,$$

where $g(u, v) = u^2(b + av) + 2auv$. Verify that the eigenvalues of the corresponding linear system are

$$\lambda = \frac{b - a - 1 \pm [(b - a - 1)^2 - 4a]^{1/2}}{2}.$$

b) If $b - a < 1$, the equilibrium state is stable, while if $b - a > 1$, it is unstable. The system of differential equations is of considerable interest since, as $b - a$ varies from the stable to the unstable range, a periodic solution (a limit cycle) appears. This is an example of a *Hopf bifurcation*. Set $a = 1$, $b = 3$ and use RKF45 to compute the limit cycle and estimate its period.

c) If a sustained external rhythm is imposed on the internal rhythm of the limit cycle, the internal rhythm is changed until it matches the imposed rhythm. This is called *entrainment*. In the above system, this can be modeled by the differential equations

$$\frac{dx}{ds} = 1 - (b + 1)x + ax^2 y,$$

$$\frac{dy}{ds} = bx - ax^2 y + c \sin qs,$$

with $a = 1$, $b = 3$, $c = 1$, $q = 1$. Start at $x = 1$, $y = 2$ and integrate the system using RKF45 on $0 \le s \le 30$ and estimate the period of the solution. Compare it to that of the forcing function and the period of the limit cycle.

REFERENCES

General

1. L. F. Shampine and R. C. Allen, *Numerical Computing: An Introduction*, W. B. Saunders Co., Philadelphia, 1973.
2. G. Forsythe, G. Malcolm, and C. B. Moler, *Computer Methods for Mathematical Computations*, Prentice-Hall, Englewood Cliffs, 1977.
3. C. W. Gear, *Numerical Initial Value Problems in Ordinary Differential Equations*, Prentice-Hall, Englewood Cliffs, 1971.

A good discussion of Adams multistep methods is

4. L. F. Shampine and M. K. Gordon, *Computer Solution of Ordinary Differential Equations: The Initial Value Problem*, W. H. Freeman and Co., San Francisco, 1975.

More advanced discussions of numerical solutions of differential equations and error propagation are

5. J. W. Daniel and R. E. Moore, *Computation and Theory in Ordinary Differential Equations*, W. H. Freeman and Co., San Francisco, 1970.
6. P. Henrici, *Discrete Variable Methods in Ordinary Differential Equations*, John Wiley and Sons, Inc., New York, 1962.

The Runge–Kutta Fehlberg formulas can be found in

7. E. Fehlberg, "Klassische Runge–Kutta Formeln vierter und neidrigerer Ordnung mit Schrittweitten-Kontrolle und ihre Anwendung Auf Wärmeleitungsprobleme," *Computing*, 6 (1970), pp. 61–71.

The original report in which RKF45 was developed is

8. L. F. Shampine and H. A. Watts, *Practical Solution of Ordinary Differential Equations by Runge–Kutta Methods*, Sandia Laboratories Report, SAND76-0585.

Applications

9. I. M. Gelfand and S. V. Fomin, *Calculus of Variations*, Prentice-Hall, Englewood Cliffs, 1963.
10. J. D. Murray, *Nonlinear Differential Equation Models in Biology*, Clarendon Press, Oxford Press, 1977.
11. J. J. Stoker, *Nonlinear Vibrations in Mechanical and Electrical Systems*, Interscience, New York, 1950.
12. W. V. Lyon and H. E. Edgerton, "Transient Torque-Angle Characteristics of Synchronous Machines," *Trans. Amer. Inst. Electr. Eng.*, vol. 49, 1930.
13. T. J. Burns, "On the Rotation of Mercury," *Celestial Mechanics*, vol. 19, 1979.

The Fundamental
Local Existence
and Uniqueness
Theory

Appendix

We have had many occasions in the preceding chapters to invoke existence and uniqueness theorems for solutions to differential equations. Such theorems answer questions such as: If I cannot exhibit a solution to a differential equation, how do I know there is one? We shall now prove an existence and uniqueness theorem for first order scalar differential equations and then indicate how the result can be extended to systems and higher order equations.

Mathematics is built on examples, and confusion can be avoided by examining several of them before plunging into a topic that may seem overly complicated and theoretical. The equation

$$\frac{dy}{dt} = y$$

is familiar and nonthreatening. Its solutions, $y(t) = Ae^t$, exist for all t and if initial data are given, there is only one choice of A to be made. Therefore, there is one and only one solution to an initial value problem. On the other hand, the innocuous equation

$$\frac{dy}{dt} = y^2$$

has only one solution that exists for all t, the trivial one, $y(t) = 0$. All others are of the form

$$y(t) = \frac{y_0}{[1 - y_0(t - t_0)]},$$

and for any initial condition $y(t_0) = y_0 \neq 0$ there is a unique solution, but it becomes unbounded for finite t. Finally, the initial value problem

$$\frac{dy}{dt} = y^{1/2}, \qquad y(0) = y_0$$

has no (real) solution if y_0 is negative and has many solutions if y_0 is zero. They are

$$y(t) = \begin{cases} 0 & \text{for } 0 \leq t \leq c, \\ \tfrac{1}{4}(t - c)^2 & \text{for } c \leq t, \end{cases}$$

where c is any positive constant.

It is easy to state reasonable conditions that insure the existence of a solution and prohibit multiple solutions. It will be shown that the initial value problem

$$\frac{dy}{dt} = f(t, y), \qquad y(t_0) = y_0 \tag{1}$$

has one and only one solution if the function $f(t, y)$ is continuously differentiable on a rectangle $R: |t - t_0| \leq a, |y - y_0| \leq b$, where a and b are positive numbers. But the problem of the existence of a solution defined for all t or even just for $t_0 \leq t < \infty$ is much more difficult. There are many simple, but still interesting, nonlinear differential equations whose solutions can only be defined on finite time intervals. Therefore, we shall be forced to ask for solutions defined near the initial data and accept the possibility of them becoming unbounded later.

The method that will be used to study the initial value problem (1) is attributed to the French mathematician Emile Picard (1856–1941) and is called *successive approximations*. As its name implies, it is an iterative process that starts with an approximate solution and attempts to improve it again and again. In particular, one starts with an initial approximation $y_0(t) = y_0$, then solves

$$\frac{dy_1}{dt} = f(t, y_0), \qquad y_1(t_0) = y_0$$

by integrating to get

$$y_1(t) = y_0 + \int_{t_0}^{t} f(s, y_0)\, ds$$

as the first iterate. Then one solves

$$\frac{dy_2}{dt} = f(t, y_1(t)), \qquad y_2(t_0) = y_0$$

again by integrating to get

$$y_2(t) = y_0 + \int_{t_0}^{t} f(s, y_1(s)) \, ds,$$

for the second iterate.

Given $y_n(t)$, the typical step is to solve

$$\frac{dy_{n+1}}{dt} = f(t, y_n(t)), \qquad y_{n+1}(t_0) = y_0$$

for

$$y_{n+1}(t) = y_0 + \int_{t_0}^{t} f(s, y_n(s)) \, ds, \quad n = 0, 1, 2, \ldots. \tag{2}$$

For this method to succeed, the constructed sequence $y_1(t), y_2(t), y_3(t), \ldots$ must be bounded and the graph of each iterate must be contained in the rectangle R. This may not be true. For example, if we define

$$y_{n+1}(t) = 1 + \int_{0}^{t} [y_n(s)]^2 \, ds, \quad |t| \leq 1, \quad |y| \leq 1,$$

with $y_0(t) \equiv y_0 = 1$, then

$$y_1(t) = 1 + t,$$

$$y_2(t) = 1 + \int_{0}^{t} (1 + s)^2 \, ds = 1 + t + t^2 + \frac{t^3}{3},$$

and all subsequent iterates violate the bound $|y| \leq 1$. This example, which was easy to construct, corresponds to the initial value problem

$$\frac{dy}{dt} = y^2, \qquad y(0) = 1.$$

As noted earlier, its solution $y(t) = (1 - t)^{-1}$ becomes unbounded, so in fact even if R were the rectangle $|t| \leq 1$, $|y| \leq 1000$, some $y_n(t)$ would eventually escape from it.

In order to show that the sequence $y_n(t)$ defined in (2) is well behaved, a bound is needed on $f(t, y)$ and, perhaps, a restriction on the interval $|t - t_0| \leq a$. Let us assume that

$$|f(t, y)| \leq M \quad \text{for} \quad |t - t_0| \leq a, \quad |y - y_0| \leq b.$$

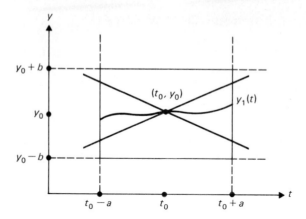

Figure A1.1

(Such a bound exists since $f(t, y)$ is continuous on the closed finite rectangle R.) Then the first iterate $y_1(t)$ satisfies

$$\left|\frac{dy_1}{dt}\right| = |f(t, y_0)| \le M,$$

and

$$|y_1(t) - y_0| \le \left|\int_{t_0}^{t} f(s, y_0)\, ds\right| \le \left|\int_{t_0}^{t} |f(s, y_0)|\, ds\right| \le M|t - t_0| \le Ma.$$

Therefore, the graph of $y_1(t)$ must lie between the lines $y = y_0 \pm M(t - t_0)$. If $Ma \le b$, these lines intersect the vertical boundaries of the rectangle R (see Fig. A1.1). But if $Ma > b$, these lines intersect the horizontal boundaries of R and $y_1(t)$ may leave the rectangle before $t = t_0 + a$ or $t = t_0 - a$ (see Fig. A1.2).

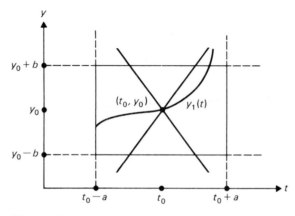

Figure A1.2

If this were to happen, the right side of

$$\frac{dy_2}{dt} = f(t, y_1(t)), \qquad y_2(t_0) = y_0$$

might not be defined after $y_1(t)$ leaves the rectangle, and the iteration process would stop.

Example 1   Show that the method of successive approximations fails for the problem

$$\frac{dy}{dt} = \sqrt{y}, \qquad y(0) = 1,$$

where R is the rectangle $|t| \leq 2$, $|y - 1| \leq 1$.

Starting with $y_0 = 1$, the first iteration gives

$$y_1(t) = 1 + \int_0^t \sqrt{1} \, dt = 1 + t, \quad -2 \leq t \leq 2.$$

But $y_1(-\tfrac{3}{2}) = -\tfrac{1}{2}$ is outside the domain of definition of the differential equation, whereas $y_1(\tfrac{3}{2}) = \tfrac{5}{2}$ is outside the rectangle. □

To avoid the difficulty of a poorly defined sequence, the t-interval must be restricted. Let h be the smaller of the numbers a and b/M and require that $|t - t_0| \leq h$. Let R_0 denote the rectangle $|t - t_0| \leq h$, $|y - y_0| \leq b$. Since R_0 is contained in R, $|f(t, y)| \leq M$ for all (t, y) in R_0. Then

$$|y_1(t) - y_0| \leq M|t - t_0| \leq Mh \leq b$$

and $y_2(t)$ is well defined and is subject to the bounds

$$\left| \frac{dy_2(t)}{dt} \right| = |f(t, y_1(t))| \leq M,$$

and for $|t - t_0| \leq h$ we obtain

$$|y_2(t) - y_0| \leq \left| \int_{t_0}^t f(s, y_1(s)) \, ds \right| \leq M|t - t_0| \leq Mh \leq b.$$

Therefore the graph of $y_2(t)$ is in R_0 and in fact the functions of the sequence $y_1(t), y_2(t), y_3(t), \ldots$ are subject to the same bounds. This is a consequence of the following lemma.

Lemma 1 Let $f(t, y)$ be a continuous function defined on the rectangle R: $|t - t_0| \leq a$, $|y - y_0| \leq b$, where a, b are positive numbers, and let $|f(t, y)| \leq M$ for all (t, y) in R. If h is the smaller of a and b/M, then the functions $y_n(t)$ given by (2) are defined on the interval $|t - t_0| \leq h$, and on this interval

$$|y_n(t) - y_0| \leq M|t - t_0| \leq Mh \leq b, \quad n = 0, 1, 2, \ldots. \quad (3)$$

Hence, the graph of each $y_n(t)$ is contained in the rectangle R_0: $|t - t_0| \leq h$, $|y - y_0| \leq b$. Furthermore, each $y_n(t)$ is continuously differentiable and

$$\left| \frac{dy_n}{dt}(t) \right| \leq M \quad \text{for} \quad |t - t_0| \leq h. \quad (4)$$

Proof The proof is by induction. First consider $n = 0$. The constant function $y_0(t) \equiv y_0$, defined on the interval $|t - t_0| \leq h$, is continuously differentiable and trivially satisfies (3) and (4). Now assume that for $j \leq n$, each $y_j(t)$ is continuously differentiable and satisfies (3) and (4). Our task is to show that $y_{n+1}(t)$ has the same properties. By (2), the function $y_{n+1}(t)$ is defined on $|t - t_0| \leq h$ and its derivative $dy_{n+1}/dt = f(t, y_n(t))$ is continuous. Hence, $y_{n+1}(t)$ is continuously differentiable and

$$\left| \frac{dy_{n+1}}{dt}(t) \right| = |f(t, y_n(t))| \leq M \quad \text{for} \quad |t - t_0| \leq h.$$

(The bounds (3) insure that $(t, y_n(t))$ is in R_0 and, therefore, is in R, the rectangle on which $f(t, y)$ is defined and bounded.) It follows from the definition of $y_{n+1}(t)$ that

$$|y_{n+1}(t) - y_0| = \left| \int_{t_0}^{t} f(s, y_n(s)) \, ds \right| \leq M|t - t_0| \leq Mh \leq b,$$

and therefore $(t, y_{n+1}(t))$ is in R_0. This completes the proof of the lemma. ∎

So far so good: the sequence of successive approximations $\{y_n(t)\}$ is well defined, properly bounded, and all the graphs are contained in R_0. Picard was a truly accomplished mathematician, so even the most sceptical reader should be willing to grant that the sequence *might* converge. If it does, it is likely that the limit function $y(t)$ is continuously differentiable and that

$$y(t) = \lim_{n \to \infty} y_{n+1}(t) = y_0 + \int_{t_0}^{t} \lim_{n \to \infty} f(s, y_n(s)) \, ds = y_0 + \int_{t_0}^{t} f(s, y(s)) \, ds.$$

In other words, the hypothetical limit function might be a solution to the *integral* equation

$$y(t) = y_0 + \int_{t_0}^{t} f(s, y(s)) \, ds. \tag{5}$$

This has not yet been proved. It has only been asserted that the method of successive approximations leads naturally to the conjecture that there is a solution to (5). But we were seeking a solution to the initial value problem (1) not to an integral equation. Fortunately, the two are equivalent.

Lemma 2 If $y(t)$ is a continuously differentiable solution to the initial value problem (1) for $|t - t_0| \leq h$, then $y(t)$ satisfies (5). Conversely, if $y(t)$ satisfies (5) it is a solution of (1).

Proof If $y(t)$ is a continuously differentiable solution of (1), then

$$\frac{dy}{dt}(t) = f(t, y(t)), \quad |t - t_0| \leq h.$$

Integrating from t_0 to t, one obtains

$$y(t) - y(t_0) = \int_{t_0}^{t} f(s, y(s)) \, ds.$$

Since $y(t_0) = y_0$, this gives (5).

Conversely, if $y(t)$ is a continuously differentiable solution to the integral equation (5), then, from the fundamental theorem of calculus, the derivative of $y(t)$ must equal $f(t, y(t))$, that is,

$$\frac{dy}{dt}(t) = \frac{d}{dt} \left\{ y_0 + \int_{t_0}^{t} f(s, y(s)) \, ds \right\} = f(t, y(t)).$$

Finally, set $t = t_0$ in (5) to obtain $y(t_0) = t_0$. Hence $y(t)$ is a solution to the initial value problem (1) as required. This completes the proof of Lemma 2. ∎

The next, and hardest, task is to prove that the sequence of successive approximations converges. To do this, the difference between two successive iterates must be estimated. That difference is given by the expression

$$y_{n+1}(t) - y_n(t) = \int_{t_0}^{t} [f(s, y_n(s)) - f(s, y_{n-1}(s))] \, ds,$$

and therefore we must estimate

$$|f(s, y_n(s)) - f(s, y_{n-1}(s))|.$$

By the mean value theorem of calculus, if (t, u) and (t, v) are two points in R, then

$$f(t, u) - f(t, v) = \frac{\partial f}{\partial y}(t, w)(u - v),$$

where w is some number between u and v. Since $\partial f(t, y)/\partial y$ is assumed to be continuous in R, it is bounded there. Hence

$$\left|\frac{\partial f}{\partial y}(t, w)\right| \le L \text{ for all } (t, w) \text{ in } R,$$

and therefore for any two points (t, u) and (t, v) in R

$$|f(t, u) - f(t, v)| \le L|u - v|. \tag{6}$$

The inequality (6) is called a *Lipschitz condition*, after the German mathematician Rudolf Lipschitz (1832–1903). With the aid of (5) and (6), we have

$$|y_{n+1}(t) - y_n(t)| \le \left|\int_{t_0}^{t} |f(s, y_n(s)) - f(s, y_{n-1}(s))| \, ds\right|$$

$$\le L\left|\int_{t_0}^{t} |y_n(s) - y_{n-1}(s)| \, ds\right|, \quad n = 1, 2, 3, \ldots. \tag{7}$$

The estimate is needed in the proof of the following fundamental local existence theorem.

Theorem Suppose that $f(t, y)$ and $\partial f(t, y)/\partial y$ are continuous on the rectangle $R: |t - t_0| \le a, |y - y_0| \le b$, where a and b are positive numbers, and that

$$|f(t, y)| \le M, \quad \left|\frac{\partial f}{\partial y}(t, y)\right| \le L \quad \text{for all } (t, y) \text{ in } R.$$

Then if h is the smaller of a and b/M, the successive approximations y_n, given by (2), converge to a continuously differentiable solution to the initial value problem (1) for $|t - t_0| \le h$.

Proof   The convergence of y_n will be demonstrated by setting

$$y_n(t) = y_0 + [y_1(t) - y_0] + [y_2(t) - y_1(t)]$$
$$+ \cdots + [y_n(t) - y_{n-1}(t)] \tag{8}$$

and showing that the corresponding infinite series is dominated by a convergent series of positive constants. To do this, estimates of the difference $|y_{n+1}(t) - y_n(t)|$ are needed. We will prove by induction that

$$|y_{n+1}(t) - y_n(t)| \le ML^n \frac{|t - t_0|^{n+1}}{(n+1)!} \le \frac{ML^n h^{n+1}}{(n+1)!}, \tag{9}$$

where $n = 0, 1, 2, \ldots, |t - t_0| \le h$.

If $n = 0$, then

$$|y_1(t) - y_0(t)| \le \left| \int_{t_0}^{t} f(s, y_0) \, ds \right|$$

$$\le \left| \int_{t_0}^{t} |f(s, y_0)| \, ds \right| \le M|t - t_0| \le Mh.$$

Now assume that (9) is valid for $j \le n$. This implies that

$$|y_n(s) - y_{n-1}(s)| \le ML^{n-1} \frac{|s - t_0|^n}{n!},$$

and then by (7)

$$|y_{n+1}(t) - y_n(t)| \le L \left| \int_{t_0}^{t} ML^{n-1} \frac{|s - t_0|^n}{n!} \, ds \right|$$

$$= ML^n \frac{|t - t_0|^{n+1}}{(n+1)!} \le \frac{ML^n h^{n+1}}{(n+1)!}, \quad |t - t_0| \le h,$$

as required.

It follows that the series (8) is dominated by the series

$$|y_0| + Mh + \frac{MLh^2}{2!} + \frac{ML^2 h^3}{3!} + \cdots + ML^n \frac{h^{n+1}}{(n+1)!} + \cdots$$

$$= |y_0| + \frac{M}{L} \left\{ (Lh) + \frac{(Lh)^2}{2!} + \cdots + \frac{(Lh)^{n+1}}{(n+1)!} + \cdots \right\}$$

$$= |y_0| + \frac{M}{L} \{e^{Lh} - 1\}.$$

Therefore by the Weierstrass M-test (which the reader can find in any advanced calculus text) the series converges uniformly and, consequently, converges to a continuous function $y(t)$. It follows that

$$\lim_{n \to \infty} y_n(t) = y_0 + \lim_{n \to \infty} \sum_{k=1}^{n} [y_k(t) - y_{k-1}(t)] = y(t),$$

and consequently, the successive approximations converge to a continuous function. (The concept of uniform convergence and its consequences are discussed in most advanced calculus texts.)

It will now be shown by deriving an estimate on $|y(t) - y_k(t)|$ that the graph of $y(t)$ is in R_0. By definition,

$$y(t) = y_0 + \sum_{n=0}^{\infty} [y_{n+1}(t) - y_n(t)].$$

Using the identity

$$y_k(t) = y_0 + \sum_{m=0}^{k-1} [y_{m+1}(t) - y_m(t)],$$

we have

$$y(t) - y_k(t) = \sum_{n=k}^{\infty} [y_{n+1}(t) - y_n(t)],$$

and hence by (9)

$$|y(t) - y_k(t)| \leq \sum_{n=k}^{\infty} \frac{ML^n h^{n+1}}{(n+1)!} \leq \frac{M}{L} \frac{(Lh)^{k+1}}{(k+1)!} \sum_{n=0}^{\infty} \frac{(Lh)^n}{n!}$$

or

$$|y(t) - y_k(t)| \leq \frac{M}{L} \frac{(Lh)^{k+1}}{(k+1)!} e^{Lh}, \quad |t - t_0| \leq h. \tag{10}$$

Note that $(Lh)^{k+1}/(k+1)! \to 0$ as $k \to \infty$. It is now easy to show that the graph of $y(t)$ is in R_0. Write

$$|y(t) - y_0| = |y(t) - y_k(t) + y_k(t) - y_0|$$

and use the triangle inequality[1] and the estimates (3), (10) to get

$$|y(t) - y_0| \leq |y(t) - y_k(t)| + |y_k(t) - y_0|$$

$$\leq \frac{M}{L} \frac{(Lh)^{k+1}}{(k+1)!} e^{Lh} + b|t - t_0| \leq h.$$

Letting $k \to \infty$, we obtain

$$|y(t) - y_0| \leq b, \quad |t - t_0| \leq h,$$

and therefore the graph of $y(t)$ is in R_0.

The next step is to show with the aid of the estimates that the continuous function $y(t)$ is a solution to the integral equation (5). By using the

---

[1] The triangle inequality is used repeatedly in mathematical analysis: $|a + b| \leq |a| + |b|$ for any numbers a, b.

triangle inequality and estimates (7) and (10), we obtain

$$0 \le \left| y(t) - y_0 - \int_{t_0}^t f(s, y(s)) \, ds \right|$$

$$= \left| y(t) - y_0 - \int_{t_0}^t f(s, y_{k-1}(s)) \, ds + \int_{t_0}^t f(s, y_{k-1}(s)) \, ds \right.$$

$$\left. - \int_{t_0}^t f(s, y(s)) \, ds \right|$$

$$\le |y(t) - y_k(t)| + \left| \int_{t_0}^t |f(s, y_{k-1}(s)) - f(s, y(s))| \, ds \right|$$

$$\le |y(t) - y_k(t)| + L \left| \int_{t_0}^t |y_{k-1}(s) - y(s)| \, ds \right|$$

$$\le |y(t) - y_k(t)| + L \left| \int_{t_0}^t \frac{M}{L} \frac{(Lh)^k e^{Lh}}{k!} \, ds \right|$$

$$\le \frac{M}{L} \frac{(Lh)^{k+1}}{(k+1)!} e^{Lh} + M|t - t_0| \frac{(Lh)^k}{k!} e^{Lh}$$

$$\le \frac{M}{L} \frac{(Lh)^{k+1}}{k!} e^{Lh} \left(\frac{1}{k+1} + 1 \right).$$

This bound tends to zero as k tends to infinity, and hence

$$0 \le \left| y(t) - y_0 - \int_{t_0}^t f(s, y(s)) \, ds \right| \le 0,$$

so $y(t)$ is a continuous solution of the integral equation as required.

Finally, we note that $f(t, y(t))$ is a continuous function of t, and therefore the right-hand side of the expression

$$y(t) = y_0 + \int_{t_0}^t f(s, y(s)) \, ds$$

is a continuously differentiable function. It follows that $y(t)$ is a solution to the initial value problem (1). This completes the proof of the theorem. ∎

Corollary (Uniqueness of solutions of the initial value problem) There is only one continuous differentiable solution to the initial value problem (1) on the interval $|t - t_0| \le h$.

Proof Suppose that $y(t)$ and $z(t)$ are two solutions to (1). Then, by Lemma 2, for $|t - t_0| \le h$, we have

$$y(t) = y_0 + \int_{t_0}^t f(s, y(s)) \, ds$$

and

$$z(t) = y_0 + \int_{t_0}^{t} f(s, z(s)) \, ds,$$

from which it follows that

$$|y(t) - z(t)| \le \left| \int_{t_0}^{t} |f(s, y(s)) - f(s, z(s))| \, ds \right|.$$

Using the Lipschitz condition (6), we have

$$|y(t) - z(t)| \le L \left| \int_{t_0}^{t} |y(s) - z(s)| \, ds \right|. \tag{11}$$

This inequality is valid for $|t - t_0| \le h$ or $-h \le t - t_0 \le h$. It is convenient to consider separately the intervals $-h \le t - t_0 \le 0$ and $0 \le t - t_0 \le h$. Assume that $-h \le t - t_0 \le 0$ or, equivalently, $t_0 - h \le t \le t_0$, and set

$$S(t) = \int_{t}^{t_0} |y(s) - z(s)| \, ds = -\int_{t_0}^{t} |y(s) - z(s)| \, ds$$

$$= \left| \int_{t_0}^{t} |y(s) - z(s)| \, ds \right|.$$

Then by the fundamental theorem of calculus,

$$|y(t) - z(t)| = -\frac{dS}{dt}(t),$$

and hence for $-h \le t - t_0 \le 0$, it follows from (11) that

$$-\frac{dS}{dt}(t) \le LS(t) \qquad \text{or} \qquad 0 \le \frac{dS}{dt}(t) + LS(t). \tag{12}$$

Guided by our experience with first order differential equations, we multiply the differential inequality (12) by e^{Lt} to obtain

$$0 \le e^{Lt} \left[\frac{dS}{dt}(t) + LS(t) \right] = \frac{d}{dt} [e^{Lt} S(t)].$$

Since $t \le t_0$, one can integrate from t to t_0 without destroying the sense of this inequality. Thus,

$$0 \le \int_{t}^{t_0} \frac{d}{ds} [e^{Ls} S(s)] \, ds = e^{Lt_0} S(t_0) - e^{Lt} S(t)$$

or

$$e^{Lt} S(t) \le e^{Lt_0} S(t_0).$$

But since by definition $0 \le S(t)$ and $S(t_0) = 0$, it follows from the last inequality that $S(t) \equiv 0$ for $-h \le t - t_0 \le 0$. From the definition of $S(t)$ and (11) it follows that $y(t) = z(t)$ for $-h \le t - t_0 \le 0$. The proof for $0 \le t - t_0 \le h$ is similar and is left to the reader. ∎

The scalar local existence and uniqueness theorem extends easily to systems of ordinary differential equations. We shall sketch the proof for two-dimensional systems

$$\frac{dy_1}{dt} = f_1(t, y_1, y_2), \qquad \frac{dy_2}{dt} = f_2(t, y_1, y_2)$$

with initial conditions $y_1(t_0) = y_1^0$, $y_2(t_0) = y_2^0$. In vector notation,

$$\frac{d\mathbf{y}}{dt} = \mathbf{f}(t, \mathbf{y}), \qquad \mathbf{y}(t_0) = \mathbf{y}^0, \tag{13}$$

where

$$\mathbf{y} = \begin{bmatrix} y_1 \\ y_2 \end{bmatrix}, \qquad \mathbf{f}(t, \mathbf{y}) = \begin{bmatrix} f_1(t, y_1, y_2) \\ f_2(t, y_1, y_2) \end{bmatrix}.$$

The corresponding integral equations are

$$y_1(t) = y_1^0 + \int_{t_0}^{t} f_1(s, y_1(s), y_2(s))\, ds,$$

$$y_2(t) = y_2^0 + \int_{t_0}^{t} f_2(s, y_1(s), y_2(s))\, ds,$$

or

$$\mathbf{y}(t) = \mathbf{y}^0 + \int_{t_0}^{t} \mathbf{f}(s, \mathbf{y}(s))\, ds.$$

It is more convenient to define the distance between two vectors y and x by

$$\|\mathbf{y} - \mathbf{x}\| = |y_1 - x_1| + |y_2 - x_2|$$

than by the Euclidean distance

$$|\mathbf{y} - \mathbf{x}| = [(y_1 - x_1)^2 + (y_2 - x_2)^2]^{1/2}.$$

The Lipschitz condition is then

$$\|\mathbf{f}(t, \mathbf{y}) - \mathbf{f}(t, \mathbf{x})\| \le L\|\mathbf{y} - \mathbf{x}\|. \tag{14}$$

It can be shown that if $\mathbf{f}(t, \mathbf{y})$ is continuously differentiable in a box R: $|t - t_0| \le a$, $\|\mathbf{y} - \mathbf{y}_0\| \le b$, and if the partial derivatives $\partial f_i(t, y_1, y_2)/\partial y_j$ satisfy

$$\left| \frac{\partial f_i}{\partial y_j} (t, y_1, y_2) \right| \le L,$$

$$i, j = 1, 2, \quad |t - t_0| \le a, \quad |y_1 - y_1^0| + |y_2 - y_2^0| \le b,$$

then (14) is satisfied.

The proof of the existence of a continuously differentiable solution of (13) mimics the earlier proof, with $\|\mathbf{y}\|$ and $\|\mathbf{f}\|$ replacing $|y|$ and $|f|$. The sequence $\{\mathbf{y}^n(t)\}$ is defined as before by letting $\mathbf{y}_0(t) \equiv \mathbf{y}^0$ and

$$\mathbf{y}^n(t) = \mathbf{y}^0 + \int_{t_0}^t \mathbf{f}(s, \mathbf{y}^{n-1}(s)) \, ds, \quad n = 1, 2, \ldots. \tag{15}$$

Then the estimates

$$\left\| \frac{d\mathbf{y}^n}{dt}(t) \right\| \le M, \qquad \|\mathbf{y}^{n+1}(t) - \mathbf{y}^0\| \le M|t - t_0|,$$

and

$$\|\mathbf{y}^{n+1}(t) - \mathbf{y}^n(t)\| \le \frac{ML^n h^{n+1}}{(n+1)!}$$

are obtained and enable one to prove convergence. The uniqueness of the solution follows from the inequality

$$\|\mathbf{y}(t) - \mathbf{z}(t)\| \le L \left| \int_{t_0}^t \|\mathbf{y}(s) - \mathbf{z}(s)\| \, ds \right|.$$

The theorem for N-dimensional systems is the following:

Theorem Suppose that $\mathbf{f}(t, \mathbf{y})$ and $\partial f_i(t, \mathbf{y})/\partial y_j$ ($i, j = 1, 2, \ldots, N$) are continuous in the box R: $|t - t_0| \le a$, $\|\mathbf{y} - \mathbf{y}^0\| \le b$, where a and b are positive numbers, and that

$$\|\mathbf{f}(t, \mathbf{y})\| \le M, \qquad \left| \frac{\partial f_i}{\partial y_j}(t, \mathbf{y}) \right| \le L \quad (i, j = 1, 2, \ldots, N)$$

for all $(t, \mathbf{y})$ in R. Then if h is the smaller of a and b/M, the successive approximations $\mathbf{y}^n(t)$ given by (15) converge to a continuously differentiable solution to the initial value problem (13) for $|t - t_0| \le h$.

Corollary There is only one continuously differentiable solution to the initial value problem (13) on the interval $|t - t_0| \le h$.

The interval of existence estimated by this theorem, $t_0 - h \le t \le t_0 + h$, is conservative; it may be much larger. Unfortunately, it is not easy to give a more accurate estimate, except in the important case of systems of linear differential equations, where the interval of existence of solutions is the interval of continuity of the coefficient matrix. We shall now prove this result.

The Linear Existence and Uniqueness Theorem Let $A(t)$ be an $n \times n$ matrix and $\mathbf{f}(t)$ be an n-dimensional vector that are continuous on the interval $a \leq t \leq b$. Then, if $\mathbf{y}^0$ is an arbitrary n-vector, there exists a unique solution $\mathbf{y}(t)$ to the initial value problem

$$\frac{d\mathbf{y}}{dt} = A(t)\,\mathbf{y} + \mathbf{f}(t), \qquad \mathbf{y}(t_0) = \mathbf{y}^0, \qquad a \leq t_0 \leq b$$

that is defined on the interval $a \leq t \leq b$. Furthermore,

$$\mathbf{y}(t) = \Phi(t)\left\{ \mathbf{y}^0 + \int_{t_0}^{t} \Phi^{-1}(s)\,\mathbf{f}(s)\,ds \right\},$$

where $\Phi(t)$, a fundamental solution matrix, is a solution to

$$\frac{d\Phi}{dt} = A(t)\,\Phi, \qquad \Phi(t_0) = I, \quad a \leq t \leq b,$$

and I is the identity matrix (see Chapter 5 for more details).

Proof If $\dfrac{d\Phi}{dt} = A(t)\,\Phi,\ \Phi(t_0) = I,\ a \leq t \leq b$, and

$$\mathbf{y}(t) = \Phi(t)\left\{ \mathbf{y}^0 + \int_{t_0}^{t} \Phi^{-1}(s)\,\mathbf{f}(s)\,ds \right\},$$

then

$$\frac{d\mathbf{y}}{dt}(t) = A(t)\,\Phi(t)\left\{ \mathbf{y}^0 + \int_{t_0}^{t} \Phi^{-1}(s)\,\mathbf{f}(s)\,ds \right\} + \Phi(t)\frac{d}{dt}\int_{t_0}^{t} \Phi^{-1}(s)\,\mathbf{f}(s)\,ds$$

$$= A(t)\,\mathbf{y}(t) + \mathbf{f}(t).$$

Therefore, it is sufficient to prove the existence of $\Phi(t)$.

Define the sequence of matrices $\{\Phi_k(t)\}$ by

$$\Phi_0(t) = I, \qquad \Phi_{k+1}(t) = I + \int_{t_0}^{t} A(s)\,\Phi_k(s)\,ds,$$

$$a \leq t \leq b, \quad k = 1, 2, 3, \ldots.$$

Set

$$\Phi_N(t) - I = [\Phi_1(t) - I] + [\Phi_2(t) - \Phi_1(t)] + \cdots + [\Phi_N(t) - \Phi_{N-1}(t)].$$

Our goal is to prove that

$$\lim_{N \to \infty} \Phi_N(t) = \Phi(t), \quad a \leq t \leq b.$$

It is helpful to employ a *matrix norm* defined by

$$\|C\| = \sum_{i,j=1}^{n} |C_{ij}|,$$

where C_{ij} is a typical entry of the matrix C. The following properties of the norm are needed:

$$\|C + D\| \leq \|C\| + \|D\|, \qquad \|CD\| \leq \|C\| \cdot \|D\|.$$

(A verification is left to the reader.)

It follows that

$$\|\Phi_N - I\| \leq \|\Phi_1(t) - I\| + \|\Phi_2(t) - \Phi_1(t)\| + \cdots + \|\Phi_N(t) - \Phi_{N-1}(t)\|,$$

and

$$\|\Phi_1(t) - I\| = \left\| \int_{t_0}^{t} A(s) I \, ds \right\|$$

$$= \left\| \int_{t_0}^{t} A(s) \, ds \right\| \leq M|t - t_0|, \quad a \leq t \leq b,$$

where

$$\|A(t)\| \leq M, \quad a \leq t \leq b.$$

The bound on $\|A(t)\|$ exists since each entry of $A(t)$ is a continuous function on the closed interval $a \leq t \leq b$. Hence,

$$\|\Phi_{k+1}(t) - \Phi_k(t)\| = \left\| \int_{t_0}^{t} A(s)[\Phi_k(s) - \Phi_{k-1}(s)] \, ds \right\|$$

$$\leq M \left| \int_{t_0}^{t} \|\Phi_k(s) - \Phi_{k-1}(s)\| \, ds \right|$$

$$\leq M \left| \int_{t_0}^{t} M^k \frac{(s - t_0)^k}{k!} \, ds \right| = M^{k+1} \frac{|t - t_0|^{k+1}}{(k + 1)!},$$

$$a \leq t \leq b, \quad k = 1, 2, 3, \ldots, N,$$

and

$$\|\Phi_N(t) - I\| \leq M|t - t_0| + M^2 \frac{|t - t_0|^2}{2!} + \cdots + M^N \frac{|t - t_0|^N}{N!}$$

$$\leq M(b - a) + M^2 \frac{(b - a)^2}{2!} + \cdots + M^N \frac{(b - a)^N}{N!}.$$

We conclude that the series $\sum_{k=1}^{\infty} [\Phi_k(t) - \Phi_{k-1}(t)]$ is dominated by a constant convergent series and therefore converges to a continuous function $\Phi(t)$, $a \le t \le b$, that satisfies the integral equation

$$\Phi(t) = I + \int_{t_0}^{t} A(s)\,\Phi(s)\,ds,$$

and the equivalent initial value problem

$$\frac{d\Phi}{dt} = A(t)\,\Phi, \qquad \Phi(t_0) = I.$$

Uniqueness follows from the argument in the corollary to the general uniqueness theorem. ∎

The following result is needed in Chapter 2.

Corollary   The initial value problem

$$\alpha(t)\,y'' + \beta(t)\,y' + \gamma(t)\,y = g(t), \qquad a \le t \le b,$$

$$y(t_0) = r, \qquad y'(t_0) = s,$$

where $\alpha(t), \beta(t), \gamma(t), g(t)$ are continuous on $a \le t \le b$, $\alpha(t) \ne 0$, has a unique continuous solution defined on $a \le t \le b$.

Proof   Let $y = y_1$, $y' = y_2$,

$$A(t) = \begin{bmatrix} 0 & 1 \\ -\dfrac{\gamma(t)}{\alpha(t)} & -\dfrac{\beta(t)}{\alpha(t)} \end{bmatrix}$$

$$\mathbf{f}(t) = \begin{bmatrix} 0 \\ \dfrac{g(t)}{\alpha(t)} \end{bmatrix}, \qquad \mathbf{y}^0 = \begin{bmatrix} r \\ s \end{bmatrix}.$$

Then the initial value problem

$$\frac{d\mathbf{y}}{dt} = A(t)\,\mathbf{y} + \mathbf{f}(t), \qquad \mathbf{y}(t_0) = \mathbf{y}^0$$

is equivalent to the second order initial value problem. The result follows from the preceding theorem. ∎

Note that the first component of

$$\mathbf{y}(t) = \Phi(t) \left\{ \mathbf{y}^0 + \int_{t_0}^{t} \Phi^{-1}(u)\, \mathbf{f}(u)\, du \right\}$$

can be written as

$$y(t) = rp(t) + sq(t) + z(t),$$

where $z(t)$ is a solution to

$$\alpha(t)\, y'' + \beta(t)\, y' + \gamma(t)\, y = g(t)$$

and $p(t)$, $q(t)$ are solutions to the corresponding homogeneous differential equation.

EXERCISES A.1

Calculate the first three Picard iterates for the following initial value problems.

1. $\dfrac{dy}{dt} = t + y, \quad y(2) = 1$

2. $\dfrac{dy}{dt} = t + y^2, \quad y(-1) = 0$

3. $\dfrac{dx}{dt} = x + y, \quad x(0) = 0, \quad \dfrac{dy}{dt} = y, \quad y(0) = 2$

4. $\dfrac{dx}{dt} = x + 2y, \quad x(0) = 2,$

$\quad \dfrac{dy}{dt} = 3x + 2y, \quad y(0) = 3$

For the following functions, select a rectangle on which a Lipschitz condition is satisfied and find a Lipschitz constant L. Then select a rectangle in which no Lipschitz condition is satisfied.

5. $f(t, y) = t^2 y^{1/2}$

6. $f(t, y) = \sin\left(\dfrac{1}{t}\right) \sin\left(\dfrac{1}{y}\right)$

7. $f(t, y) = \dfrac{y}{t^2 + y^2}$

8. $f(t, y) = \arctan\left(\dfrac{y}{t}\right)$

9. Show that the integral equation

$$y(t) = \epsilon \int_{0}^{t} e^{-a(t-s)} f(s, y(s))\, ds$$

is equivalent to the initial value problem

$$\dfrac{dy}{dt} + ay = \epsilon f(t, y), \qquad y(0) = 0.$$

10. Show that the integral equation

$$x(t) = \epsilon \int_{0}^{t} \sin(t - s)\, f(s, x(s))\, ds$$

is equivalent to the initial value problem

$$\dfrac{d^2 x}{dt^2} + x = \epsilon f(t, x), \qquad x(0) = 0, \qquad \dfrac{dx}{dt}(0) = 0.$$

11. Estimate $|y_{n+1}(t) - y_n(t)|$, where $y_0(t) \equiv 0$,

$$y_{n+1}(t) = \epsilon \int_{0}^{t} e^{-2(t-s)} [s^2 + y_n^2(s)]\, ds,$$

$$n = 0, 1, 2, \ldots,$$

$$-1 \le t \le 1, \quad -1 \le y \le 1, \quad 0 < \epsilon \le \epsilon_0.$$

It may be necessary to restrict ϵ_0.

12. Estimate $|x_{n+1}(t) - x_n(t)|$, where $x_0(t) \equiv 0$,

$$x_{n+1}(t) = \epsilon \int_{0}^{t} \sin(t - s)\, [s^2 + x_n^2(s)]\, ds,$$

$$n = 0, 1, 2, \ldots,$$

$$-1 \le t \le 1, \quad -1 \le x \le 1, \quad 0 < \epsilon \le \epsilon_0.$$

It may be necessary to restrict ϵ_0.

PASCAL
Programs

Appendix
2

PROGRAM THAT USES EULER'S METHOD TO SOLVE $y' = ty^{1/3}$, $y(1) = 1$

```
PROGRAM EULER1(OUTPUT);
(*****************************************************************
**    THIS PROGRAM USES THE EULER METHOD TO                    **
**    SOLVE THE INITIAL VALUE PROBLEM:                         **
**                                                             **
**        DY/DT = F(T,Y)      Y(T0) = Y0                       **
*****************************************************************)

VAR I,N : INTEGER;
    T0,Y0,TF,T,Y,H : REAL;

FUNCTION F(T,Y:REAL) : REAL;
    BEGIN
    F = T*EXP(LN(Y)/3.0)
    END; (* FUNCTION F *)
```

```
BEGIN (* MAIN *)

(* SET INITIAL CONDITIONS *)
T0 := 1.0; Y0 := 1.0;
(* SET FINAL TIME AND NUMBER OF STEPS TO BE TAKEN *)
TF := 2.0; N := 10;
(* EULER LOOP *)
T := T0; Y := Y0;
H := (TF − T0)/N;
WRITELN(' ',T:6:2,Y:10:5);
FOR I := 1 TO N DO
    BEGIN
    Y := Y + H*F(T,Y);
    T := T + H;
    WRITELN(' ',T:6:2,Y:10:5)
    END

END. (* MAIN *)
```

PROGRAM THAT USES THE IMPROVED EULER METHOD TO SOLVE $y' = ty^{1/3}$, $y(1) = 1$

```
PROGRAM EULER2(OUTPUT);
(****************************************************************
**    THIS PROGRAM USES THE IMPROVED EULER METHOD           **
**    TO SOLVE THE INITIAL VALUE PROBLEM:                   **
**                                                          **
**                DY/DT = F(T,Y)     Y(T0) = Y0             **
****************************************************************)

VAR I,N : INTEGER;
    T0,Y0,TF,T,Y,H,S1,S2 : REAL;

FUNCTION F(T,Y:REAL) : REAL;
    BEGIN
    F = T*EXP(LN(Y)/3.0)
    END; (* FUNCTION F *)

BEGIN (* MAIN *)
```

```
(* SET INITIAL CONDITIONS *)
T0 := 1.0; Y0 := 1.0;
(* SET FINAL TIME AND NUMBER OF STEPS TO BE TAKEN *)
TF := 2.0; N := 10;
(* IMPROVED EULER LOOP *)
T := T0; Y := Y0;
H = (TF − T0)/N;
WRITELN(' ',T:6:2,Y:10:5);
FOR I := 1 TO N DO
    BEGIN
    S1 := F(T,Y);
    S2 := F(T+H,Y+H*S1);
    Y := Y + H*(S1 + S2)/2.0;
    T := T + H;
    WRITELN(' ',T:6:2,Y:10:5)
    END

END. (* MAIN *)
```

PROGRAM THAT USES THE IMPROVED EULER METHOD TO SOLVE THE SECOND ORDER LINEAR EQUATION
$a(t)y''(t) + b(t)y'(t) + c(t)y(t) = q(t)$

```
PROGRAM EULER2S(OUTPUT);

(*****************************************************************
** PROGRAM USES THE IMPROVED EULER METHOD TO             **
** SOLVE THE INITIAL VALUE PROBLEM:                      **
**                                                        **
**      A(T)*Y''(T) + B(T)*Y'(T) + C(T)*Y(T) = Q(T)      **
**      Y(T0) = Y0,   Y'(T0) = Z0                         **
*****************************************************************)

VAR T0,Y0,Z0,TF,T,Y,Z,H,YBAR,ZBAR,GYZ : REAL;
    N,I : INTEGER;

(*DEFINE FUNCTIONS A(T),B(T),C(T),Q(T),G(Y,Z,T)*)
FUNCTION A(T:REAL) : REAL;
    BEGIN
    A := 1.0
    END;
```

```
      FUNCTION B(T:REAL) : REAL;
          BEGIN
          B := 0.0
          END;

      FUNCTION C(T:REAL) : REAL;
          BEGIN
          C := −T
          END;

      FUNCTION Q(T:REAL) : REAL;
          BEGIN
          Q := 0.0
          END;

      FUNCTION G(Y,YP,T:REAL) : REAL;
          BEGIN
          G := (−C(T)*Y−B(T)*YP + Q(T))/A(T)
          END;

  BEGIN (* MAIN *)

  (* SET INITIAL CONDITIONS, FINAL TIME AND NUMBER OF
  STEPS TO BE TAKEN *)

  T0 := 0.0; Y0 := 0.35503;
  Z0 := 1.0;
  TF := 1.0; N:= 10;
  (* BASIC LOOP *)
  T := T0; Y := Y0;
  Z := Z0; H := (TF − T0)/N;
  WRITELN(' ',T:6:2,Y:10:5,Z:10:5);
  FOR I := 1 TO N DO
      BEGIN
      GYZ := G(Y,Z,T);
      YBAR := Y + H*Z;
      ZBAR := Z + H*GYZ;
      Y := Y + H*(Z + ZBAR)/2.0;
      Z := Z + H*(GYZ + G(YBAR,ZBAR,T+H))/2.0;
      T := T + H;
      WRITELN(' ',T:6:2,Y:10:5,Z:10:5)
      END

  END. (*MAIN*)
```

PROGRAM THAT USES EULER'S METHOD WITH AUTOMATIC STEP SIZE ADJUSTMENT TO SOLVE $y' = y$, $y_1(0) = 0$ AT $t = 1$

```
PROGRAM EULER(OUTPUT);

(****************************************************************
**    PROGRAM TO SOLVE D/DT(Y) = Y USING EULER'S              **
**    METHOD                                                   **
**    WITH A LOCAL ERROR ESTIMATOR BASED                       **
**    ON THE IMPROVED EULER ALGORITHM.                         **
****************************************************************)

VAR T0,Y0,EPS,T,Y,TFINAL,H : REAL;
    R,EST,GAMMA,YE,HE : REAL;

FUNCTION F(T,Y:REAL) : REAL;
    BEGIN
    F := Y
    END;

BEGIN (*MAIN*)

T0 := 0.0; Y0 := 1.0;
EPS := 0.001;
T := T0; Y := Y0;
TFINAL := 1.0;
H := 0.1;
REPEAT
    IF (T+H > TFINAL) THEN H := TFINAL−T;
    YE := Y + H*F(T,Y);
    R := 0.5*H*(F(T+H,YE) − F(T,Y));
    EST := R/H;
    GAMMA := EPS/ABS(EST);
    HE := 0.9*GAMMA*H;
    IF (ABS(EST) <= EPS) THEN
      BEGIN
      Y := YE;
      T := T+H;
      H := HE
      END
    ELSE
      BEGIN
      H := HE;
      Y := Y + H*F(T,Y);
```

```
            T := T + H
            END
        UNTIL (T > = TFINAL);
        WRITELN(' T = ',T:10:4,'Y = ',Y:10:5)

        END. (*MAIN*)
```

SUBROUTINE RK4 AND SAMPLE DRIVER SET-UP
TO SOLVE $y_1' = y_2$, $y_1(0) = 1$; $y_2' = y_1$, $y_2(0) = 1$

```
        PROGRAM DRIVER (OUTPUT);

        (***********************************************************
        **   SAMPLE PROGRAM TO ILLUSTRATE PROCEDURE RK4      **
        ***********************************************************)

        TYPE SVECTOR = ARRAY[1..2] OF REAL;

        VAR NP,NPRINT,I,NEQN : INTEGER;
            T,H : REAL;
            Y : SVECTOR;

        PROCEDURE F(T : REAL; N : INTEGER; Y : SVECTOR;
                VAR YP : SVECTOR);
            BEGIN
            YP[1] := Y[2];
            YP[2] := Y[1]
            END; (* F *)

        PROCEDURE RK4(H : REAL; N : INTEGER;
                VAR T : REAL;
                VAR Y : SVECTOR);
        (***********************************************************
        **   PROCEDURE RK4 IMPLEMENTS THE CLASSICAL 4TH      **
        **   ORDER RUNGE–KUTTA FORMULAS OVER                 **
        **   ONE STEP OF LENGTH H.                           **
        **                                                   **
        **   INPUT:                                          **
        **                                                   **
        **      F — NAME OF A PROCEDURE DEFINING             **
        **      THE DIFFERENTIAL EQUATIONS.                  **
        **      F MUST HAVE THE FORM                         **
```

```
**                                                          **
**          PROCEDURE F(T,N,Y,YP);                          **
**          BEGIN                                            **
**            YP(1)  = ...                                   **
**                        .                                  **
**                        .                                  **
**                        .                                  **
**            YP(N) = ...                                    **
**          END                                              **
**                                                          **
**    H — STEP SIZE TO BE USED (H#0.0)                       **
**    N — NUMBER OF COMPONENTS IN SOLUTION VECTOR  **
**    T — VALUE OF THE INDEPENDENT VARIABLE                  **
**    Y — SOLUTION VECTOR OF LENGTH N EVALUATED             **
**    AT T                                                   **
**                                                          **
**  OUTPUT:                                                  **
**    T — NEW VALUE OF INDEPENDENT VARIABLE (INPUT  **
**    VALUE PLUS H)                                          **
**    Y — SOLUTION VECTOR EVALUATED AT T                     **
***************************************************************)


VAR K1,K2,K3,K4,YTEMP : SVECTOR;
    TEMP : REAL;
    I : INTEGER;

    BEGIN
    F(T,N,Y,K1);
    FOR I := 1 TO N DO
      YTEMP[I] := Y[I] + 0.5*H*K1[I];
    TEMP := T + 0.5*H;
    F(TEMP,N,YTEMP,K2);
    FOR I := 1 TO N DO
      YTEMP[I] := Y[I] + 0.5*H*K2[I];
    F(TEMP,N,YTEMP,K3);
    FOR I := 1 TO N DO
      YTEMP[I] := Y[I] + H*K3[I];
    TEMP := T + H;
    F(TEMP,N,YTEMP,K4);
    FOR I := 1 TO N DO
      Y[I] := Y[I] + H*(K1[I] + 2.0*(K2[I] + K3[I]) +K4[I]) / 6.0;
    T := TEMP;
    END; (* RK4 *)
```

```
BEGIN (* MAIN *)

Y[1] := 1.0; Y[2] := 1.0;
T := 0.0; H := 0.1;
WRITELN(' ',T:10:2,Y[1]:10:2,Y[2]:10:2);
NP := 0; NPRINT := 5;
NEQN := 2;
FOR I := 1 TO 10 DO
    BEGIN
    NP := NP + 1;
    RK4(H,NEQN,T,Y);
    IF (NP >= NPRINT) THEN
      BEGIN
      WRITELN(' ',T:10:2,Y[1]:10:2,Y[2]:10:2);
      NP := 0
      END
    END

END. (*MAIN*)
```

Listing of Subroutine RKF45

Appendix 3

```
      SUBROUTINE RKF45(F,NEQN,Y,T,TOUT,RELERR,ABSERR,IFLAG,WORK,IWORK)
C
C     FEHLBERG FOURTH-FIFTH ORDER RUNGE-KUTTA METHOD
C
C     WRITTEN BY H.A.WATTS AND L.F.SHAMPINE
C
C
C
C     RKF45 IS PRIMARILY DESIGNED TO SOLVE NON-STIFF AND MILDLY STIFF
C     DIFFERENTIAL EQUATIONS WHEN DERIVATIVE EVALUATIONS ARE INEXPENSIVE.
C     RKF45 SHOULD GENERALLY NOT BE USED WHEN THE USER IS DEMANDING
C     HIGH ACCURACY.
C
C ABSTRACT
C
C     SUBROUTINE  RKF45  INTEGRATES A SYSTEM OF NEQN FIRST ORDER
C     ORDINARY DIFFERENTIAL EQUATIONS OF THE FORM
C             DY(I)/DT = F(T,Y(1),Y(2),...,Y(NEQN))
C             WHERE THE Y(I) ARE GIVEN AT T .
C     TYPICALLY THE SUBROUTINE IS USED TO INTEGRATE FROM T TO TOUT BUT IT
C     CAN BE USED AS A ONE-STEP INTEGRATOR TO ADVANCE THE SOLUTION A
C     SINGLE STEP IN THE DIRECTION OF TOUT.  ON RETURN THE PARAMETERS IN
C     THE CALL LIST ARE SET FOR CONTINUING THE INTEGRATION. THE USER HAS
C     ONLY TO CALL RKF45 AGAIN (AND PERHAPS DEFINE A NEW VALUE FOR TOUT).
C     ACTUALLY, RKF45 IS AN INTERFACING ROUTINE WHICH CALLS SUBROUTINE
C     RKFS FOR THE SOLUTION.  RKFS IN TURN CALLS SUBROUTINE  FEHL WHICH
C     COMPUTES AN APPROXIMATE SOLUTION OVER ONE STEP.
C
```

```
C       RKF45   USES THE RUNGE-KUTTA-FEHLBERG (4,5)  METHOD DESCRIBED
C       IN THE REFERENCE
C       E.FEHLBERG , LOW-ORDER CLASSICAL RUNGE-KUTTA FORMULAS WITH STEPSIZE
C                   CONTROL , NASA TR R-315 (ALSO IN COMPUTING,6(1970),
C                   PP. 61-71)
C
C       THE PERFORMANCE OF RKF45 IS ILLUSTRATED IN THE REFERENCE
C       L.F.SHAMPINE,H.A.WATTS,S.DAVENPORT, SOLVING NON-STIFF ORDINARY
C                   DIFFERENTIAL EQUATIONS-THE STATE OF THE ART ,
C                   SIAM REVIEW,18(1976),PP. 376-411
C
C       THE PARAMETERS REPRESENT-
C         F -- SUBROUTINE F(T,Y,YP) TO EVALUATE DERIVATIVES YP(I)=DY(I)/DT
C         NEQN -- NUMBER OF EQUATIONS TO BE INTEGRATED
C         Y(*) -- SOLUTION VECTOR AT T
C         T -- INDEPENDENT VARIABLE
C         TOUT -- OUTPUT POINT AT WHICH SOLUTION IS DESIRED
C         RELERR,ABSERR -- RELATIVE AND ABSOLUTE ERROR TOLERANCES FOR LOCAL
C              ERROR TEST. AT EACH STEP THE CODE REQUIRES THAT
C                      ABS(LOCAL ERROR) .LE. RELERR*ABS(Y) + ABSERR
C              FOR EACH COMPONENT OF THE LOCAL ERROR AND SOLUTION VECTORS
C         IFLAG -- INDICATOR FOR STATUS OF INTEGRATION
C         WORK(*) -- ARRAY TO HOLD INFORMATION INTERNAL TO RKF45 WHICH IS
C              NECESSARY FOR SUBSEQUENT CALLS. MUST BE DIMENSIONED
C              AT LEAST   3+6*NEQN
C         IWORK(*) -- INTEGER ARRAY USED TO HOLD INFORMATION INTERNAL TO
C              RKF45 WHICH IS NECESSARY FOR SUBSEQUENT CALLS. MUST BE
C              DIMENSIONED AT LEAST   5
C
C
C   FIRST CALL TO RKF45
C
C       THE USER MUST PROVIDE STORAGE IN HIS CALLING PROGRAM FOR THE ARRAYS
C       IN THE CALL LIST  -       Y(NEQN) , WORK(3+6*NEQN) , IWORK(5)  ,
C       DECLARE F IN AN EXTERNAL STATEMENT, SUPPLY SUBROUTINE F(T,Y,YP) AND
C       INITIALIZE THE FOLLOWING PARAMETERS-
C
C         NEQN -- NUMBER OF EQUATIONS TO BE INTEGRATED.  (NEQN .GE. 1)
C         Y(*) -- VECTOR OF INITIAL CONDITIONS
C         T -- STARTING POINT OF INTEGRATION , MUST BE A VARIABLE
C         TOUT -- OUTPUT POINT AT WHICH SOLUTION IS DESIRED.
C              T=TOUT IS ALLOWED ON THE FIRST CALL ONLY, IN WHICH CASE
C              RKF45 RETURNS WITH IFLAG=2 IF CONTINUATION IS POSSIBLE.
C         RELERR,ABSERR -- RELATIVE AND ABSOLUTE LOCAL ERROR TOLERANCES
C              WHICH MUST BE NON-NEGATIVE. RELERR MUST BE A VARIABLE WHILE
C              ABSERR MAY BE A CONSTANT. THE CODE SHOULD NORMALLY NOT BE
C              USED WITH RELATIVE ERROR CONTROL SMALLER THAN ABOUT 1.E-8 .
C              TO AVOID LIMITING PRECISION DIFFICULTIES THE CODE REQUIRES
C              RELERR TO BE LARGER THAN AN INTERNALLY COMPUTED RELATIVE
C              ERROR PARAMETER WHICH IS MACHINE DEPENDENT. IN PARTICULAR,
C              PURE ABSOLUTE ERROR IS NOT PERMITTED. IF A SMALLER THAN
C              ALLOWABLE VALUE OF RELERR IS ATTEMPTED, RKF45 INCREASES
C              RELERR APPROPRIATELY AND RETURNS CONTROL TO THE USER BEFORE
C              CONTINUING THE INTEGRATION.
C         IFLAG -- +1,-1  INDICATOR TO INITIALIZE THE CODE FOR EACH NEW
C              PROBLEM. NORMAL INPUT IS +1. THE USER SHOULD SET IFLAG=-1
C              ONLY WHEN ONE-STEP INTEGRATOR CONTROL IS ESSENTIAL. IN THIS
C              CASE, RKF45 ATTEMPTS TO ADVANCE THE SOLUTION A SINGLE STEP
C              IN THE DIRECTION OF TOUT EACH TIME IT IS CALLED. SINCE THIS
C              MODE OF OPERATION RESULTS IN EXTRA COMPUTING OVERHEAD, IT
C              SHOULD BE AVOIDED UNLESS NEEDED.
C
C
C   OUTPUT FROM RKF45
C
C       Y(*) -- SOLUTION AT T
```

```
C          T -- LAST POINT REACHED IN INTEGRATION.
C          IFLAG = 2 -- INTEGRATION REACHED TOUT. INDICATES SUCCESSFUL RETURN
C                       AND IS THE NORMAL MODE FOR CONTINUING INTEGRATION.
C                =-2 -- A SINGLE SUCCESSFUL STEP IN THE DIRECTION OF TOUT
C                       HAS BEEN TAKEN. NORMAL MODE FOR CONTINUING
C                       INTEGRATION ONE STEP AT A TIME.
C                = 3 -- INTEGRATION WAS NOT COMPLETED BECAUSE RELATIVE ERROR
C                       TOLERANCE WAS TOO SMALL. RELERR HAS BEEN INCREASED
C                       APPROPRIATELY FOR CONTINUING.
C                = 4 -- INTEGRATION WAS NOT COMPLETED BECAUSE MORE THAN
C                       3000 DERIVATIVE EVALUATIONS WERE NEEDED. THIS
C                       IS APPROXIMATELY 500 STEPS.
C                = 5 -- INTEGRATION WAS NOT COMPLETED BECAUSE SOLUTION
C                       VANISHED MAKING A PURE RELATIVE ERROR TEST
C                       IMPOSSIBLE. MUST USE NON-ZERO ABSERR TO CONTINUE.
C                       USING THE ONE-STEP INTEGRATION MODE FOR ONE STEP
C                       IS A GOOD WAY TO PROCEED.
C                = 6 -- INTEGRATION WAS NOT COMPLETED BECAUSE REQUESTED
C                       ACCURACY COULD NOT BE ACHIEVED USING SMALLEST
C                       ALLOWABLE STEPSIZE. USER MUST INCREASE THE ERROR
C                       TOLERANCE BEFORE CONTINUED INTEGRATION CAN BE
C                       ATTEMPTED.
C                = 7 -- IT IS LIKELY THAT RKF45 IS INEFFICIENT FOR SOLVING
C                       THIS PROBLEM. TOO MUCH OUTPUT IS RESTRICTING THE
C                       NATURAL STEPSIZE CHOICE. USE THE ONE-STEP INTEGRATOR
C                       MODE.
C                = 8 -- INVALID INPUT PARAMETERS
C                       THIS INDICATOR OCCURS IF ANY OF THE FOLLOWING IS
C                       SATISFIED -   NEQN .LE. O
C                                     T=TOUT  AND  IFLAG .NE. +1 OR -1
C                                     RELERR OR ABSERR .LT. O.
C                                     IFLAG .EQ. O OR .LT. -2 OR .GT. 8
C          WORK(*),IWORK(*) -- INFORMATION WHICH IS USUALLY OF NO INTEREST
C                       TO THE USER BUT NECESSARY FOR SUBSEQUENT CALLS.
C                       WORK(1),...,WORK(NEQN) CONTAIN THE FIRST DERIVATIVES
C                       OF THE SOLUTION VECTOR Y AT T. WORK(NEQN+1) CONTAINS
C                       THE STEPSIZE H TO BE ATTEMPTED ON THE NEXT STEP.
C                       IWORK(1) CONTAINS THE DERIVATIVE EVALUATION COUNTER.
C
C
C  SUBSEQUENT CALLS TO RKF45
C
C   SUBROUTINE RKF45 RETURNS WITH ALL INFORMATION NEEDED TO CONTINUE
C  THE INTEGRATION. IF THE INTEGRATION REACHED TOUT, THE USER NEED ONLY
C  DEFINE A NEW TOUT AND CALL RKF45 AGAIN. IN THE ONE-STEP INTEGRATOR
C  MODE (IFLAG=-2) THE USER MUST KEEP IN MIND THAT EACH STEP TAKEN IS
C  IN THE DIRECTION OF THE CURRENT TOUT. UPON REACHING TOUT (INDICATED
C  BY CHANGING IFLAG TO 2),THE USER MUST THEN DEFINE A NEW TOUT AND
C  RESET IFLAG TO -2 TO CONTINUE IN THE ONE-STEP INTEGRATOR MODE.
C
C   IF THE INTEGRATION WAS NOT COMPLETED BUT THE USER STILL WANTS TO
C  CONTINUE (IFLAG=3,4 CASES), HE JUST CALLS RKF45 AGAIN. WITH IFLAG=3
C  THE RELERR PARAMETER HAS BEEN ADJUSTED APPROPRIATELY FOR CONTINUING
C  THE INTEGRATION. IN THE CASE OF IFLAG=4 THE FUNCTION COUNTER WILL
C  BE RESET TO O AND ANOTHER 3000 FUNCTION EVALUATIONS ARE ALLOWED.
C
C   HOWEVER,IN THE CASE IFLAG=5, THE USER MUST FIRST ALTER THE ERROR
C  CRITERION TO USE A POSITIVE VALUE OF ABSERR BEFORE INTEGRATION CAN
C  PROCEED. IF HE DOES NOT,EXECUTION IS TERMINATED.
C
C   ALSO,IN THE CASE IFLAG=6, IT IS NECESSARY FOR THE USER TO RESET
C  IFLAG TO 2 (OR -2 WHEN THE ONE-STEP INTEGRATION MODE IS BEING USED)
C  AS WELL AS INCREASING EITHER ABSERR,RELERR OR BOTH BEFORE THE
C  INTEGRATION CAN BE CONTINUED. IF THIS IS NOT DONE, EXECUTION WILL
C  BE TERMINATED. THE OCCURRENCE OF IFLAG=6 INDICATES A TROUBLE SPOT
C  (SOLUTION IS CHANGING RAPIDLY,SINGULARITY MAY BE PRESENT) AND IT
C  OFTEN IS INADVISABLE TO CONTINUE.
```

```
C
C      IF IFLAG=7 IS ENCOUNTERED, THE USER SHOULD USE THE ONE-STEP
C      INTEGRATION MODE WITH THE STEPSIZE DETERMINED BY THE CODE OR
C      CONSIDER SWITCHING TO THE ADAMS CODES DE/STEP,INTRP. IF THE USER
C      INSISTS UPON CONTINUING THE INTEGRATION WITH RKF45, HE MUST RESET
C      IFLAG TO 2 BEFORE CALLING RKF45 AGAIN. OTHERWISE,EXECUTION WILL BE
C      TERMINATED.
C
C      IF IFLAG=8 IS OBTAINED, INTEGRATION CAN NOT BE CONTINUED UNLESS
C      THE INVALID INPUT PARAMETERS ARE CORRECTED.
C
C      IT SHOULD BE NOTED THAT THE ARRAYS WORK,IWORK CONTAIN INFORMATION
C      REQUIRED FOR SUBSEQUENT INTEGRATION. ACCORDINGLY, WORK AND IWORK
C      SHOULD NOT BE ALTERED.
C
C
       INTEGER NEQN,IFLAG,IWORK(5)
       REAL Y(NEQN),T,TOUT,RELERR,ABSERR,WORK(1)
C      IF COMPILER CHECKS SUBSCRIPTS, CHANGE WORK(1) TO WORK(3+6*NEQN)
C
       EXTERNAL F
C
       INTEGER K1,K2,K3,K4,K5,K6,K1M
C
C
C      COMPUTE INDICES FOR THE SPLITTING OF THE WORK ARRAY
C
       K1M=NEQN+1
       K1=K1M+1
       K2=K1+NEQN
       K3=K2+NEQN
       K4=K3+NEQN
       K5=K4+NEQN
       K6=K5+NEQN
C
C      THIS INTERFACING ROUTINE MERELY RELIEVES THE USER OF A LONG
C      CALLING LIST VIA THE SPLITTING APART OF TWO WORKING STORAGE
C      ARRAYS. IF THIS IS NOT COMPATIBLE WITH THE USERS COMPILER,
C      HE MUST USE RKFS DIRECTLY.
C
       CALL RKFS(F,NEQN,Y,T,TOUT,RELERR,ABSERR,IFLAG,WORK(1),WORK(K1M),
      1          WORK(K1),WORK(K2),WORK(K3),WORK(K4),WORK(K5),WORK(K6),
      2          WORK(K6+1),IWORK(1),IWORK(2),IWORK(3),IWORK(4),IWORK(5))
C
       RETURN
       END
       SUBROUTINE RKFS(F,NEQN,Y,T,TOUT,RELERR,ABSERR,IFLAG,YP,H,F1,F2,F3,
      1                F4,F5,SAVRE,SAVAE,NFE,KOP,INIT,JFLAG,KFLAG)
C
C      FEHLBERG FOURTH-FIFTH ORDER RUNGE-KUTTA METHOD
C
C
C      RKFS INTEGRATES A SYSTEM OF FIRST ORDER ORDINARY DIFFERENTIAL
C      EQUATIONS AS DESCRIBED IN THE COMMENTS FOR RKF45 .
C      THE ARRAYS YP,F1,F2,F3,F4,AND F5 (OF DIMENSION AT LEAST NEQN) AND
C      THE VARIABLES H,SAVRE,SAVAE,NFE,KOP,INIT,JFLAG,AND KFLAG ARE USED
C      INTERNALLY BY THE CODE AND APPEAR IN THE CALL LIST TO ELIMINATE
C      LOCAL RETENTION OF VARIABLES BETWEEN CALLS. ACCORDINGLY, THEY
C      SHOULD NOT BE ALTERED. ITEMS OF POSSIBLE INTEREST ARE
C          YP - DERIVATIVE OF SOLUTION VECTOR AT T
C          H  - AN APPROPRIATE STEPSIZE TO BE USED FOR THE NEXT STEP
C          NFE- COUNTER ON THE NUMBER OF DERIVATIVE FUNCTION EVALUATIONS
C
C
       LOGICAL HFAILD,OUTPUT
C
       INTEGER  NEQN,IFLAG,NFE,KOP,INIT,JFLAG,KFLAG
```

```
      REAL   Y(NEQN),T,TOUT,RELERR,ABSERR,H,YP(NEQN),
     1    F1(NEQN),F2(NEQN),F3(NEQN),F4(NEQN),F5(NEQN),SAVRE,
     2    SAVAE
C
      EXTERNAL F
C
      REAL   A,AE,DT,EE,EEOET,ESTTOL,ET,HMIN,REMIN,RER,S,
     1    SCALE,TOL,TOLN,U26,EPSP1,EPS,YPK
C
      INTEGER   K,MAXNFE,MFLAG
C
      REAL   ABS,AMAX1,AMIN1,SIGN
C
C     REMIN IS THE MINIMUM ACCEPTABLE VALUE OF RELERR.   ATTEMPTS
C     TO OBTAIN HIGHER ACCURACY WITH THIS SUBROUTINE ARE USUALLY
C     VERY EXPENSIVE AND OFTEN UNSUCCESSFUL.
C
      DATA REMIN/1.D-12/
C
C
C     THE EXPENSE IS CONTROLLED BY RESTRICTING THE NUMBER
C     OF FUNCTION EVALUATIONS TO BE APPROXIMATELY MAXNFE.
C     AS SET, THIS CORRESPONDS TO ABOUT 500 STEPS.
C
      DATA MAXNFE/3000/
C
C
C     CHECK INPUT PARAMETERS
C
C
      IF (NEQN .LT. 1) GO TO 10
      IF ((RELERR .LT. 0.0E0)   .OR.   (ABSERR .LT. 0.0E0)) GO TO 10
      MFLAG=IABS(IFLAG)
      IF (MFLAG .EQ. 0) GO TO 10
      IF (MFLAG .GT. 8) GO TO 10
      IF (MFLAG .NE. 1) GO TO 20
C
C     FIRST CALL, COMPUTE MACHINE EPSILON
      EPS = 1.0E0
    5 EPS = EPS/2.0E0
      EPSP1 = EPS + 1.0E0
      IF (EPSP1 .GT. 1.0E0) GO TO 5
      U26 = 26.0E0*EPS
      GO TO 50
C
C     INVALID INPUT
   10 IFLAG=8
      RETURN
C
C     CHECK CONTINUATION POSSIBILITIES
C
   20 IF ((T .EQ. TOUT) .AND. (KFLAG .NE. 3)) GO TO 10
      IF (MFLAG .NE. 2) GO TO 25
C
C     IFLAG = +2 OR -2
      IF (KFLAG .EQ. 3) GO TO 45
      IF (INIT .EQ. 0) GO TO 45
      IF (KFLAG .EQ. 4) GO TO 40
      IF ((KFLAG .EQ. 5)  .AND.  (ABSERR .EQ. 0.0E0)) GO TO 30
      IF ((KFLAG .EQ. 6)  .AND.  (RELERR .LE. SAVRE)  .AND.
     1    (ABSERR .LE. SAVAE)) GO TO 30
      GO TO 50
C
C     IFLAG = 3,4,5,6,7 OR 8
   25 IF (IFLAG .EQ. 3) GO TO 45
      IF (IFLAG .EQ. 4) GO TO 40
      IF ((IFLAG .EQ. 5) .AND. (ABSERR .GT. 0.0E0)) GO TO 45
```

```
C
C       INTEGRATION CANNOT BE CONTINUED SINCE USER DID NOT RESPOND TO
C       THE INSTRUCTIONS PERTAINING TO IFLAG=5,6,7 OR 8
   30 STOP
C
C       RESET FUNCTION EVALUATION COUNTER
   40 NFE=0
      IF (MFLAG .EQ. 2) GO TO 50
C
C       RESET FLAG VALUE FROM PREVIOUS CALL
   45 IFLAG=JFLAG
      IF (KFLAG .EQ. 3) MFLAG=IABS(IFLAG)
C
C       SAVE INPUT IFLAG AND SET CONTINUATION FLAG VALUE FOR SUBSEQUENT
C       INPUT CHECKING
   50 JFLAG=IFLAG
      KFLAG=0
C
C       SAVE RELERR AND ABSERR FOR CHECKING INPUT ON SUBSEQUENT CALLS
      SAVRE=RELERR
      SAVAE=ABSERR
C
C       RESTRICT RELATIVE ERROR TOLERANCE TO BE AT LEAST AS LARGE AS
C       2*EPS+REMIN TO AVOID LIMITING PRECISION DIFFICULTIES ARISING
C       FROM IMPOSSIBLE ACCURACY REQUESTS
C
      RER=2.0E0*EPS+REMIN
      IF (RELERR .GE. RER) GO TO 55
C
C       RELATIVE ERROR TOLERANCE TOO SMALL
      RELERR=RER
      IFLAG=3
      KFLAG=3
      RETURN
C
   55 DT=TOUT-T
C
      IF (MFLAG .EQ. 1) GO TO 60
      IF (INIT .EQ. 0) GO TO 65
      GO TO 80
C
C       INITIALIZATION --
C                         SET INITIALIZATION COMPLETION INDICATOR,INIT
C                         SET INDICATOR FOR TOO MANY OUTPUT POINTS,KOP
C                         EVALUATE INITIAL DERIVATIVES
C                         SET COUNTER FOR FUNCTION EVALUATIONS,NFE
C                         ESTIMATE STARTING STEPSIZE
C
   60 INIT=0
      KOP=0
C
      A=T
      CALL F(A,Y,YP)
      NFE=1
      IF (T .NE. TOUT) GO TO 65
      IFLAG=2
      RETURN
C
C
   65 INIT=1
      H=ABS(DT)
      TOLN=0.
      DO 70 K=1,NEQN
        TOL=RELERR*ABS(Y(K))+ABSERR
        IF (TOL .LE. 0.) GO TO 70
        TOLN=TOL
        YPK=ABS(YP(K))
        IF (YPK*H**5 .GT. TOL) H=(TOL/YPK)**0.2E0
```

```
      70 CONTINUE
         IF (TOLN .LE. 0.0E0) H=0.0E0
         H=AMAX1(H,U26*AMAX1(ABS(T),ABS(DT)))
         JFLAG=ISIGN(2,IFLAG)
C
C
C        SET STEPSIZE FOR INTEGRATION IN THE DIRECTION FROM T TO TOUT
C
      80 H=SIGN(H,DT)
C
C        TEST TO SEE IF RKF45 IS BEING SEVERELY IMPACTED BY TOO MANY
C        OUTPUT POINTS
C
         IF (ABS(H) .GE. 2.0E0*ABS(DT)) KOP=KOP+1
         IF (KOP .NE. 100) GO TO 85
C
C        UNNECESSARY FREQUENCY OF OUTPUT
         KOP=0
         IFLAG=7
         RETURN
C
      85 IF (ABS(DT) .GT. U26*ABS(T)) GO TO 95
C
C        IF TOO CLOSE TO OUTPUT POINT,EXTRAPOLATE AND RETURN
C
         DO 90 K=1,NEQN
      90   Y(K)=Y(K)+DT*YP(K)
         A=TOUT
         CALL F(A,Y,YP)
         NFE=NFE+1
         GO TO 300
C
C
C        INITIALIZE OUTPUT POINT INDICATOR
C
      95 OUTPUT= .FALSE.
C
C        TO AVOID PREMATURE UNDERFLOW IN THE ERROR TOLERANCE FUNCTION,
C        SCALE THE ERROR TOLERANCES
C
         SCALE=2.0E0/RELERR
         AE=SCALE*ABSERR
C
C
C        STEP BY STEP INTEGRATION
C
     100 HFAILD= .FALSE.
C
C        SET SMALLEST ALLOWABLE STEPSIZE
C
         HMIN=U26*ABS(T)
C
C        ADJUST STEPSIZE IF NECESSARY TO HIT THE OUTPUT POINT.
C        LOOK AHEAD TWO STEPS TO AVOID DRASTIC CHANGES IN THE STEPSIZE AND
C        THUS LESSEN THE IMPACT OF OUTPUT POINTS ON THE CODE.
C
         DT=TOUT-T
         IF (ABS(DT) .GE. 2.0E0*ABS(H)) GO TO 200
         IF (ABS(DT) .GT. ABS(H)) GO TO 150
C
C        THE NEXT SUCCESSFUL STEP WILL COMPLETE THE INTEGRATION TO THE
C        OUTPUT POINT
C
         OUTPUT= .TRUE.
         H=DT
         GO TO 200
C
     150 H=0.5E0*DT
```

```
C
C
C
C           CORE INTEGRATOR FOR TAKING A SINGLE STEP
C
C           THE TOLERANCES HAVE BEEN SCALED TO AVOID PREMATURE UNDERFLOW IN
C           COMPUTING THE ERROR TOLERANCE FUNCTION ET.
C           TO AVOID PROBLEMS WITH ZERO CROSSINGS,RELATIVE ERROR IS MEASURED
C           USING THE AVERAGE OF THE MAGNITUDES OF THE SOLUTION AT THE
C           BEGINNING AND END OF A STEP.
C           THE ERROR ESTIMATE FORMULA HAS BEEN GROUPED TO CONTROL LOSS OF
C           SIGNIFICANCE.
C           TO DISTINGUISH THE VARIOUS ARGUMENTS, H IS NOT PERMITTED
C           TO BECOME SMALLER THAN 26 UNITS OF ROUNDOFF IN T.
C           PRACTICAL LIMITS ON THE CHANGE IN THE STEPSIZE ARE ENFORCED TO
C           SMOOTH THE STEPSIZE SELECTION PROCESS AND TO AVOID EXCESSIVE
C           CHATTERING ON PROBLEMS HAVING DISCONTINUITIES.
C           TO PREVENT UNNECESSARY FAILURES, THE CODE USES 9/10 THE STEPSIZE
C           IT ESTIMATES WILL SUCCEED.
C           AFTER A STEP FAILURE, THE STEPSIZE IS NOT ALLOWED TO INCREASE FOR
C           THE NEXT ATTEMPTED STEP. THIS MAKES THE CODE MORE EFFICIENT ON
C           PROBLEMS HAVING DISCONTINUITIES AND MORE EFFECTIVE IN GENERAL
C           SINCE LOCAL EXTRAPOLATION IS BEING USED AND EXTRA CAUTION SEEMS
C           WARRANTED.
C
C
C           TEST NUMBER OF DERIVATIVE FUNCTION EVALUATIONS.
C           IF OKAY,TRY TO ADVANCE THE INTEGRATION FROM T TO T+H
C
  200 IF (NFE .LE. MAXNFE) GO TO 220
C
C           TOO MUCH WORK
      IFLAG=4
      KFLAG=4
      RETURN
C
C           ADVANCE AN APPROXIMATE SOLUTION OVER ONE STEP OF LENGTH H
C
  220 CALL FEHL(F,NEQN,Y,T,H,YP,F1,F2,F3,F4,F5,F1)
      NFE=NFE+5
C
C           COMPUTE AND TEST ALLOWABLE TOLERANCES VERSUS LOCAL ERROR ESTIMATES
C           AND REMOVE SCALING OF TOLERANCES. NOTE THAT RELATIVE ERROR IS
C           MEASURED WITH RESPECT TO THE AVERAGE OF THE MAGNITUDES OF THE
C           SOLUTION AT THE BEGINNING AND END OF THE STEP.
C
      EEOET=0.0E0
      DO 250 K=1,NEQN
        ET=ABS(Y(K))+ABS(F1(K))+AE
        IF (ET .GT. 0.0E0) GO TO 240
C
C           INAPPROPRIATE ERROR TOLERANCE
        IFLAG=5
        RETURN
C
  240   EE=ABS((-2090.0E0*YP(K)+(21970.0E0*F3(K)-15048.0E0*F4(K)))+
     1                          (22528.0E0*F2(K)-27360.0E0*F5(K)))
  250   EEOET=AMAX1(EEOET,EE/ET)
C
      ESTTOL=ABS(H)*EEOET*SCALE/752400.0E0
C
      IF (ESTTOL .LE. 1.0E0) GO TO 260
C
C
C           UNSUCCESSFUL STEP
C                         REDUCE THE STEPSIZE , TRY AGAIN
C                         THE DECREASE IS LIMITED TO A FACTOR OF 1/10
```

```
C
        HFAILD= .TRUE.
        OUTPUT= .FALSE.
        S=0.1E0
        IF (ESTTOL .LT. 59049.0E0) S=0.9E0/ESTTOL**0.2E0
        H=S*H
        IF (ABS(H) .GT. HMIN) GO TO 200
C
C     REQUESTED ERROR UNATTAINABLE AT SMALLEST ALLOWABLE STEPSIZE
        IFLAG=6
        KFLAG=6
        RETURN
C
C
C     SUCCESSFUL STEP
C                          STORE SOLUTION AT T+H
C                          AND EVALUATE DERIVATIVES THERE
C
  260 T=T+H
        DO 270 K=1,NEQN
  270   Y(K)=F1(K)
        A=T
        CALL F(A,Y,YP)
        NFE=NFE+1
C
C
C                     CHOOSE NEXT STEPSIZE
C                     THE INCREASE IS LIMITED TO A FACTOR OF 5
C                     IF STEP FAILURE HAS JUST OCCURRED, NEXT
C                          STEPSIZE IS NOT ALLOWED TO INCREASE
C
        S=5.0E0
        IF (ESTTOL .GT. 1.889568D-4) S=0.9E0/ESTTOL**0.2E0
        IF (HFAILD) S=AMIN1(S,1.0E0)
        H=SIGN(AMAX1(S*ABS(H),HMIN),H)
C
C     END OF CORE INTEGRATOR
C
C
C     SHOULD WE TAKE ANOTHER STEP
C
        IF (OUTPUT) GO TO 300
        IF (IFLAG .GT. 0) GO TO 100
C
C
C     INTEGRATION SUCCESSFULLY COMPLETED
C
C     ONE-STEP MODE
        IFLAG=-2
        RETURN
C
C     INTERVAL MODE
  300 T=TOUT
        IFLAG=2
        RETURN
C
        END
        SUBROUTINE FEHL(F,NEQN,Y,T,H,YP,F1,F2,F3,F4,F5,S)
C
C     FEHLBERG FOURTH-FIFTH ORDER RUNGE-KUTTA METHOD
C
C     FEHL INTEGRATES A SYSTEM OF NEQN FIRST ORDER
C     ORDINARY DIFFERENTIAL EQUATIONS OF THE FORM
C            DY(I)/DT=F(T,Y(1),---,Y(NEQN))
C     WHERE THE INITIAL VALUES Y(I) AND THE INITIAL DERIVATIVES
C     YP(I) ARE SPECIFIED AT THE STARTING POINT T. FEHL ADVANCES
C     THE SOLUTION OVER THE FIXED STEP H AND RETURNS
```

```
C     THE FIFTH ORDER (SIXTH ORDER ACCURATE LOCALLY) SOLUTION
C     APPROXIMATION AT T+H IN ARRAY S(I).
C     F1,---,F5 ARE ARRAYS OF DIMENSION NEQN WHICH ARE NEEDED
C     FOR INTERNAL STORAGE.
C     THE FORMULAS HAVE BEEN GROUPED TO CONTROL LOSS OF SIGNIFICANCE.
C     FEHL SHOULD BE CALLED WITH AN H NOT SMALLER THAN 13 UNITS OF
C     ROUNDOFF IN T SO THAT THE VARIOUS INDEPENDENT ARGUMENTS CAN BE
C     DISTINGUISHED.
C
C
      INTEGER   NEQN
      REAL    Y(NEQN),T,H,YP(NEQN),F1(NEQN),F2(NEQN),
     1  F3(NEQN),F4(NEQN),F5(NEQN),S(NEQN)
C
      REAL   CH
      INTEGER   K
C
      CH=H/4.0E0
      DO 221 K=1,NEQN
  221   F5(K)=Y(K)+CH*YP(K)
      CALL F(T+CH,F5,F1)
C
      CH=3.0E0*H/32.0E0
      DO 222 K=1,NEQN
  222   F5(K)=Y(K)+CH*(YP(K)+3.0E0*F1(K))
      CALL F(T+3.0E0*H/8.0E0,F5,F2)
C
      CH=H/2197.0E0
      DO 223 K=1,NEQN
  223   F5(K)=Y(K)+CH*(1932.0E0*YP(K)+(7296.0E0*F2(K)-7200.0E0*F1(K)))
      CALL F(T+12.0E0*H/13.0E0,F5,F3)
C
      CH=H/4104.0E0
      DO 224 K=1,NEQN
  224   F5(K)=Y(K)+CH*((8341.0E0*YP(K)-845.0E0*F3(K))+
     1                  (29440.0E0*F2(K)-32832.0E0*F1(K)))
      CALL F(T+H,F5,F4)
C
      CH=H/20520.0E0
      DO 225 K=1,NEQN
  225   F1(K)=Y(K)+CH*((-6080.0E0*YP(K)+(9295.0E0*F3(K)-
     1        5643.0E0*F4(K)))+(41040.0E0*F1(K)-28352.0E0*F2(K)))
      CALL F(T+H/2.0E0,F1,F5)
C
C     COMPUTE APPROXIMATE SOLUTION AT T+H
C
      CH=H/7618050.0E0
      DO 230 K=1,NEQN
  230   S(K)=Y(K)+CH*((902880.0E0*YP(K)+(3855735.0E0*F3(K)-
     1        1371249.0E0*F4(K)))+(3953664.0E0*F2(K)+
     2        277020.0E0*F5(K)))
C
      RETURN
      END
```

Power Series, Complex Numbers, and Euler's Formula

Appendix

This appendix contains a brief review of material that is sometimes not stressed in calculus courses but is needed in a study of differential equations, namely, power series, complex numbers, and Euler's formula that represents the complex exponential in terms of real trigonometric functions. A more extensive treatment is given in [1].

Power series are used to represent functions as infinite series and are defined as a series of the form

$$a_0 + a_1 t + a_2 t^2 + \cdots = \sum_{j=0}^{\infty} a_j t^j.$$

They arise when it is desired to approximate a given function $f(t)$ by a sequence of polynomials

$$f_n(t) = a_0 + a_1 t + a_2 t^2 + \cdots + a_n t^n.$$

Such an approximation will, in general, be of value only if t is small. The coefficients $a_0, a_1, a_2, \ldots$ are chosen so that $f_n(0) = f(0)$, and the first n derivatives of $f_n(t)$ are equal to the derivatives of the given function $f(t)$ at $t = 0$. Since $f_n^{(j)}(0) = j! \, a_j$, we have

$$a_0 = f(0), \, a_1 = f'(0), \, a_2 = \frac{f''(0)}{2!}, \, a_3 = \frac{f^{(3)}(0)}{3!}, \ldots, a_n = \frac{f^{(n)}(0)}{n!}.$$

If the coefficients a_j are chosen this way, then $f_n(t)$ is called the *Taylor polynomial* of degree n of $f(t)$.

If the function $f(t)$ and its first $n + 1$ derivatives are continuous on an interval containing 0 and t, then it can be shown that

$$f(t) = f_n(t) + R_n(0, t),$$

where the remainder term is given by

$$R_n(0, t) = f^{(n+1)}(\tau) \frac{t^{n+1}}{(n + 1)!}, \, \tau \text{ between } 0 \text{ and } t.$$

The above formula is called *Taylor's formula with remainder*. (A proof of this result can be found in [1], Chap. 16.)

A slight generalization of this result is valuable. Replace t by $s = t + k$ and consider the function

$$f(s) = f(t + k) = q(k),$$

where t is held fixed and k is the independent variable. It follows that

$$q^{(j)}(k) = f^{(j)}(s) \quad \text{and} \quad \text{at } k = 0 \quad q^{(j)}(0) = f^{(j)}(t).$$

If we now approximate $f(t + k) = q(k)$ by Taylor polynomials, we obtain

$$f(s) = f(t + k)$$

$$= f(t) + f'(t) \, k + f''(t) \frac{k^2}{2!} + f^{(3)}(t) \frac{k^3}{3!} + \cdots + f^{(n)}(t) \frac{k^n}{n!} + R_n(t, k),$$

where the remainder term is

$$R_n(t, k) = f^{(n+1)}(\tau) \frac{t^{n+1}}{(n + 1)!}, \quad \tau \text{ between } t \text{ and } t + k.$$

This is most frequently used with $n = 1$ to construct the linear approximation

$$f(t + k) \simeq f(t) + f'(t)k.$$

In Chapters 3 and 7, extension of these formulas to functions of two variables is needed, for example,

$$P(x + u, y + v) \simeq P(x, y) + \frac{\partial P}{\partial x}(x, y) \, u + \frac{\partial P}{\partial y}(x, y) \, v.$$

This can be obtained from the single-variable approximation by setting

$$q(s) = P(x + su, y + sv)$$

and employing the linear approximation at $s = 0$. Since

$$q(s) \simeq q(0) + q'(0) s$$

and since by the chain rule

$$q'(s) = \frac{\partial P}{\partial x} (x + su, y + sv) u + \frac{\partial P}{\partial y} (x + su, y + sv) v,$$

then

$$q'(0) = \frac{\partial P}{\partial x} (x, y) u + \frac{\partial P}{\partial y} (x, y) v$$

and the desired formula is obtained letting $s = 1$.

The representation of a function by an infinite series results from taking the limit

$$f(t) = \lim_{n \to \infty} f_n(t) = \sum_{j=0}^{\infty} f^{(j)}(0) \frac{t^j}{j!},$$

where the convention $f^{(0)}(t) = f(t)$ is followed. If, instead of approximating $f(t)$ near $t = 0$, we approximate it near $t = a$, the Taylor polynomials and series will contain powers of $(t - a)$:

$$f_n(t) = a_0 + a_1(t - a) + a_2(t - a)^2 + \cdots + a_n(t - a)^n$$

with $a_j = f^{(j)}(a)/j!$, and

$$f(t) = f(a) + f'(a) (t - a) + f''(a) \frac{(t - a)^2}{2} + \cdots = \sum_{j=0}^{\infty} f^{(j)}(a) \frac{(t - a)^j}{j!}.$$

The simplest example of a function represented by an infinite series is offered by the exponential function, $f(t) = e^t$. Since $f^{(j)}(0) = 1$, we obtain $a_j = 1/j!$, and therefore

$$e^t = 1 + t + \frac{t^2}{2!} + \frac{t^3}{3!} + \cdots = \sum_{j=0}^{\infty} \frac{t^j}{j!}.$$

While this expression is valid for all values of t, the approximating polynomials

$$f_n(t) = 1 + t + \frac{t^2}{2!} + \cdots + \frac{t^n}{n!}$$

are only useful for small values of n if t is small (see Fig. A4.1).

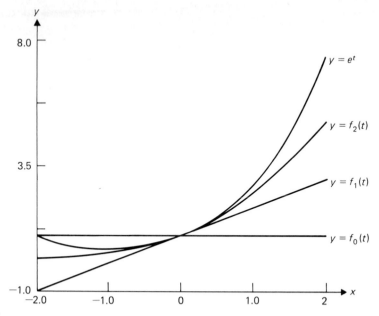

Figure A4.1 Plot of $y = f(t) = e^t$, $y = f_0(t) = 1$, $y = f_1(t) = 1 + t$, $y = f_2(t) = 1 + t + t^2/2$.

Other examples of functions represented by power series are

$$\sin t = t - \frac{t^3}{3!} + \frac{t^5}{5!} - \frac{t^7}{7!} + \cdots = \sum_{j=0}^{\infty} (-1)^j \frac{t^{2j+1}}{(2j+1)!};$$

$$\cos t = 1 - \frac{t^2}{2!} + \frac{t^4}{4!} - \frac{t^6}{6!} + \cdots = \sum_{j=0}^{\infty} (-1)^j \frac{t^{2j}}{(2j)!};$$

$$\sinh t = \frac{e^t - e^{-t}}{2} = \frac{1}{2} \sum_{j=0}^{\infty} \frac{t^j}{j!} - \frac{1}{2} \sum_{j=0}^{\infty} \frac{(-t)^j}{j!} = \sum_{j=0}^{\infty} \frac{t^{2j+1}}{(2j+1)!};$$

$$\cosh t = \frac{e^t + e^{-t}}{2} = \frac{1}{2} \sum_{j=0}^{\infty} \left[\frac{t^j}{j!} + \frac{(-t)^j}{j!} \right] = \sum_{j=0}^{\infty} \frac{t^{2j}}{(2j)!}.$$

The last two series illustrate that, within their common domain of convergence, power series can be added, subtracted, or multiplied by constants, and the result is a convergent power series. Moreover, it is also true that a power series can be differentiated or integrated term by term within its domain of convergence to produce a convergent power series. This permits the construction of power series solutions to differential equations, a topic considered at length in Chapter 6. (A good exposition of operations on power series is in [2], Chap. 9.)

Our next topic is *complex numbers, complex variables* and *functions, and power series of complex variables.* Complex numbers are extensions of real numbers that permit the solution of quadratic equations, such as $x^2 + 1 = 0$, and the uniform treatment of linear differential equations as well as the development of powerful mathematical theories, such as two-dimensional fluid flow and potential theory. The complex-number system consists of the complex numbers written as $x + iy$, where x and y are real, together with the rules by which they are added, multiplied, and divided. Rather than giving a list of such rules, we shall take an informal approach and say that the operations are the same as those for real numbers except that $i^2 = -1$. Hence

$$(x + iy) + (u + iv) = (x + u) + i(y + v);$$
$$(x + iy) \cdot (u + iv) = xu + iyu + ixv + i^2yv$$
$$= (xu - yv) + i(xv + yu).$$

When convenient, we write $x + iy$ as $x + yi$; then x is called the *real part* and y the *imaginary part* of $x + iy$ (or $x + yi$). The notation

$$x = \text{Re}(x + iy), \quad y = \text{Im}(x + iy)$$

is frequently employed.

The *complex conjugate* of $x + iy$ is defined as $x - iy$. It is customary to write $z^* = x - iy$ to denote the complex conjugate of $z = x + iy$. (The notation $\bar{z} = x - iy$ is also common.) The complex conjugate satisfies the relation

$$zz^* = (x + iy)(x - iy) = x^2 + y^2,$$

which is a real quantity, and if $z = x + iy$, $w = u + iv$, then

$$\frac{z}{w} = \frac{zw^*}{ww^*} = \frac{(x + iy)(u - iv)}{(u + iv)(u - iv)} = \frac{(xu + yv) + i(yu - xv)}{u^2 + v^2}.$$

This is the formula for division of two complex numbers.

The complex number $z = x + iy$ can be represented both as a point in the z plane and as a vector from the origin to the point with coordinates (x, y). Addition of complex numbers then corresponds to the parallelogram rule for addition of vectors. For instance, the addition of $(2 + 3i)$ and $(1 - i)$ can be represented as the result of adding the corresponding vectors (see Fig. A4.2).

The length of a vector corresponding to a complex number is the *absolute value* of the complex number:

$$|z| = |x + iy| = \sqrt{x^2 + y^2} = \sqrt{zz^*}.$$

In terms of polar coordinates

$$x = r \cos \theta, \quad y = r \sin \theta,$$

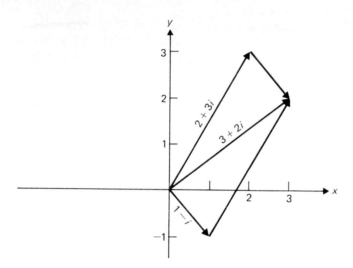

Figure A4.2   Addition of $1 - i$ to $2 + 3i$ to give $3 + 2i$.

where $\theta = \arctan (y/x)$, we have

$$z = x + iy = r(\cos \theta + i \sin \theta)$$

with $r = |z|$. The angle θ is called the *argument* of z. Since

$$
\begin{aligned}
z_1 z_2 &= r_1(\cos \theta_1 + i \sin \theta_1) \cdot r_2(\cos \theta_2 + i \sin \theta_2) \\
&= r_1 r_2[(\cos \theta_1 \cos \theta_2 - \sin \theta_1 \sin \theta_2) + i(\cos \theta_1 \sin \theta_2 + \sin \theta_1 \cos \theta_2)] \\
&= r_1 r_2[\cos (\theta_1 + \theta_2) + i \sin(\theta_1 + \theta_2)],
\end{aligned}
$$

this gives a geometrical interpretation of multiplication of two complex numbers. The product vector is rotated, so that its argument is the sum of the arguments of the two factors and its length is the product of the two lengths.

Example   Let $z_1 = \sqrt{3} + i$, $z_2 = 1 - i$. Compute the lengths and arguments of z_1, z_2, z_1^*, z_2^*, $z_1 z_2$, z_1/z_2, $z_1 + z_2$.

For $z_1 = \sqrt{3} + i$, we have $x_1 = \sqrt{3}$ and $y_1 = 1$; hence

$$|z_1| = \sqrt{(\sqrt{3})^2 + 1^2} = 2 \quad \text{and} \quad \theta_1 = \arg z_1 = \arctan \frac{1}{\sqrt{3}} = \frac{\pi}{6}.$$

For $z_2 = 1 - i$, we have $x_2 = 1$, $y_2 = -1$; hence

$$|z_2| = \sqrt{1^2 + (-1)^2} = \sqrt{2} \quad \text{and} \quad \theta_2 = \arg z_2 = \arctan \frac{-1}{1} = -\frac{\pi}{4}.$$

Therefore,

$$z_1 = 2\left(\cos\frac{\pi}{6} + i\sin\frac{\pi}{6}\right), \qquad z_2 = \sqrt{2}\left(\cos\frac{\pi}{4} - i\sin\frac{\pi}{4}\right).$$

A similar calculation for $z_1^* = \sqrt{3} - i$ and $z_2^* = 1 + i$ gives $|z_1^*| = 2$, $\arg z_1^* = -\pi/6 = -\arg z_1$, $|z_2^*| = \sqrt{2}$, $\arg z_2^* = \pi/4 = -\arg z_2$; hence

$$z_1^* = 2\left(\cos\frac{\pi}{6} - i\sin\frac{\pi}{6}\right), \qquad z_2^* = \sqrt{2}\left(\cos\frac{\pi}{4} + i\sin\frac{\pi}{4}\right).$$

Furthermore,

$$|z_1 z_2| = |z_1||z_2| = 2\sqrt{2}$$

and

$$\arg(z_1 z_2) = \arg z_1 + \arg z_2 = \frac{\pi}{6} + \left(-\frac{\pi}{4}\right) = -\frac{\pi}{12},$$

hence

$$z_1 z_2 = 2\sqrt{2}\left(\cos\frac{\pi}{12} - i\sin\frac{\pi}{12}\right)$$

is the polar representation of $z_1 z_2 = (1 + \sqrt{3}) + (1 - \sqrt{3})i$. For the quotient z_1/z_2 we have

$$\frac{z_1}{z_2} = \frac{z_1 z_2^*}{z_2 z_2^*} = \frac{z_1 z_2^*}{|z_2|^2} = \frac{|z_1|}{|z_2|}[\cos(\theta_1 - \theta_2) + i\sin(\theta_1 - \theta_2)],$$

and therefore

$$\frac{z_1}{z_2} = \frac{2}{\sqrt{2}}\left(\cos\frac{5\pi}{12} + i\sin\frac{5\pi}{12}\right)$$

is the polar representation of

$$\frac{z_1}{z_2} = \frac{1}{2}[(\sqrt{3} - 1) + (1 + \sqrt{3})i].$$

Finally, $z_1 + z_2 = 1 + \sqrt{3}$, hence $|z_1 + z_2| = 1 + \sqrt{3}$ and $\arg(z_1 + z_2) = 0$. $\square$

Exponential functions e^z of a complex variable z are employed in the study of linear differential equations. They can be defined in terms of power series which are limits of Taylor polynomials of a complex variable. Therefore define

$$e^z = \lim_{n\to\infty}\sum_{j=0}^{n}\frac{z^j}{j!} = \sum_{j=0}^{\infty}\frac{z^j}{j!},$$

and compute (the details are messy and are omitted)

$$e^{x+iy} = \sum_{j=0}^{\infty} \frac{(x + iy)^j}{j!} = \left(\sum_{j=0}^{\infty} \frac{x^j}{j!}\right)\left(\sum_{j=0}^{\infty} \frac{(iy)^j}{j!}\right) = e^x e^{iy}$$

and

$$e^{iy} = \sum_{j=0}^{\infty} \frac{(iy)^j}{j!} = \sum_{j=0}^{\infty} (-1)^j \frac{y^{2j}}{(2j)!} + i \sum_{j=0}^{\infty} (-1)^j \frac{y^{2j+1}}{(2j + 1)!} = \cos y + i \sin y.$$

In the last computation, the formulas $i^{2j} = (-1)^j$, $i^{2j+1} = i(-1)^j$ were used. The relation

$$e^{iy} = \cos y + i \sin y$$

is called *Euler's formula*. It represents the complex exponential of a purely imaginary variable in terms of real trigonometric functions.

Our final topic is the observation that a *complex solution to a real linear homogeneous equation is equivalent to two real solutions.* For example, $z(t) = e^{it} = \cos t + i \sin t$ is a solution of the linear homogeneous differential equation

$$y'' + y = 0$$

since

$$(e^{it})'' + e^{it} = (i^2 + 1)e^{it} = (-1 + 1)e^{it} = 0.$$

Furthermore

$$(\cos t + i \sin t)'' + (\cos t + i \sin t)$$
$$= [(\cos t)'' + \cos t] + i[(\sin t)'' + \sin t]$$
$$= [-\cos t + \cos t] + i[-\sin t + \sin t] = 0,$$

and hence $\cos t = \mathrm{Re}(e^{it})$ and $\sin t = \mathrm{Re}(e^{it})$ are also both solutions.

REFERENCES

1. G. B. Thomas and R. L. Finney, *Calculus and Analytic Geometry*, Addison-Wesley, Reading, 1979.

2. D. V. Widder, *Advanced Calculus*, Prentice-Hall, Englewood Cliffs, 1961.

Answers to Selected Exercises

Chapter 1

Section 1.1

1. a) Ordinary   **c)** Partial   **e)** Ordinary

2. a) 1   **c)** 2   **e)** 2

Section 1.2

1. a) All t **c)** All t **e)** $-\infty < t < 2$,

5. a) Any t_0, $y = 1$ **c)** Any t_0, $1 \le y \le 2$

Section 1.3

1. $y(t) = \dfrac{t^5}{5} + 2t + c$ **3.** $y(t) = \tfrac{2}{3} \ln|t + 2| + \tfrac{2}{3} \ln|t - 4| + c$

5. $y(t) = \tfrac{1}{2}(t^{4/3} - 1)^{3/2} + c$ **7.** $y(t) = \ln \left| \dfrac{\pi}{4} + \dfrac{t}{2} \right| + c$

9. $y(t) = \dfrac{t}{2} - \cos t + 3$

11. $y(t) = \tfrac{5}{6} \ln(t + 5) + \tfrac{1}{6} \ln(1 - t) - \tfrac{5}{6} \ln 5 + 1,\ -5 < t < 1$

13. $y(t) = \dfrac{t^2}{2}\left[\ln 2t - \tfrac{1}{2}\right] + \tfrac{1}{16}$

15. $y(t) = \tfrac{1}{4}[2t \sin(2t + 1) + \cos(2t + 1)] + \tfrac{3}{4}$

17. a) $s(2) = 164$ ft   **b)** $v = -64$ ft/sec   **c)** $v = -102.45$ ft/sec

19. $M = \frac{4}{3}\pi r^3 \rho$, $S = 4\pi r^2$, $\dot{M} = -kS$ implies that $\dot{r} = -(k/\rho)$ and $r = r_0 - (k/\rho)t$. Hence $t_{\text{final}} = r_0\rho/k$. From $M(75) = M(0)/2$, $r(75) = (\frac{1}{2})^{1/3}r_0 = r_0 - (k/\rho)75$. Therefore $k/\rho = [1 - (\frac{1}{2})^{1/3}]r_0/75$ and $t_{\text{final}} = 75/[1 - (\frac{1}{2})^{1/3}] \simeq 365$.

20. b) $y(1) \simeq 1.463$

Section 1.4

1. $\dfrac{y^2}{2} = \dfrac{t^3}{3} + c$, $y(t) = \pm(\frac{2}{3}t^3 + 2c)^{1/2}$

3. $e^{-y} + e^t = c$, $y(t) = -\ln|c - e^{-t}|$

5. $\ln\left|\dfrac{y-2}{y-1}\right| = \ln|Kt|$, $y(t) = \dfrac{2 - Kt}{1 - Kt}$

7. $y^2 = (t - 1)e^t + c$, $y(t) = \pm[(t - 1)e^t + c]^{1/2}$

9. $\ln|y - 1| + y = -\dfrac{1}{t} + c$ or $y(t) \equiv 1$

11. $\dfrac{y^2}{2} = -\cos t + c$, $y(t) = -[-2\cos t + 3]^{1/2}$ all t **13.** $y(t) \equiv \frac{1}{2}$

15. $y^2 - 2y = t^3 + 2t + 3$ implies $y(0) = -1$ or $y(0) = 3$; $y(t) = 1 - (t^3 + 2t + 4)^{1/2}$

17. $\tan y = 2 - 1/t$, $y(t) = \arctan(2 - 1/t)$, $\lim\limits_{t\to\infty} y(t) = \arctan(2)$

19. $y^{-1/3} = -\cos t + 2$ if $y(\pi/2) = \frac{1}{8}$; $y(t) = (2 - \cos t)^{-3}$ is defined for all t; $y^{-1/3} = -\cos t + \frac{1}{2}$ if $y(\pi/2) = 8$; $y(t) = (\frac{1}{2} - \cos t)^{-3}$ if $\pi/3 < t < 5\pi/3$

21. c) $k \simeq \frac{1}{16}$, $s_{\max} \simeq 7900$ ft   **23.** 35%

24. a) $y' = -x/2y$ implies $x^2 + 2y^2 = K$ **c)** $y' = 4y/x$ implies $y = Kx^4$

25. a) $V = 16\pi h$, area $= \frac{1}{144}$, $2h^{1/2} = -kt + 2\sqrt{12}$, where $k = 4.8/(2304\pi)$; hence $T = 2\sqrt{12}/k$.

Section 1.5

1. $y(t) = ce^t$ **3.** $y(t) = ce^{-t^3/3}$ **5.** $y(t) = ce^{3t}$

7. $y(t) = ce^t - \frac{1}{5}(\sin 2t + 2\cos 2t)$ **9.** $y(t) = (c - \ln|t|)/t$

11. $y(t) = \sin^2 t + K \csc t$ **13.** $y(t) = 4e^{-2t}$

15. $y(t) = \dfrac{1}{\sqrt{2} + 2}(\tan t + \sec t)$ **17.** $y(t) = 2t + 3 + e^{-t}$

19. $y(t) = (2 - t)e^t$ **21.** $y(t) = \frac{3}{2}e^{t^2} - \frac{1}{2}$

23. $y(t) = (t^2 + 1)\left[\arctan(t) + \dfrac{\pi}{4}\right]$ **25.** $y(t) = (t + 1)e^{-t^2}$

27. $y(t) = \dfrac{\sin t}{t^2}$ **29.** $y(t) = 1/(ce^{-t} - t + 1)$ **31.** $y(t) = t^{-1}(c - t^2)^{-1/2}$

33. $t = 10\ln(0.05)/\ln(0.6)$ **35. b)** $t = 100\ln 5$

36. b) $y_p(t) = \frac{1}{4}(\sin 2t - \cos 2t) + \frac{1}{2}$ **d)** $T = \frac{1}{2}\ln(2 \cdot 10^5)$

Section 1.6

1. a) $h = \frac{1}{4}$

i	t	y	y'
0	0.	2.	0.
1	0.2500	2.	−1.
2	0.5000	1.750	−1.531
3	0.7500	1.367	−1.402
4	1.0000	1.017	

b)

i	t	y	s_1	s_2
0	0.	2.	0.	−1.
1	0.2500	1.875	−0.8789	−1.370
2	0.5000	1.594	−1.270	−1.221
3	0.7500	1.282	−1.233	−0.9482
4	1.0000	1.009		

c) $y(t) = \dfrac{2}{1 + t^2}$; $y(1) = 1$. The Euler difference is -0.017; the improved

Euler difference is -0.009.

2. a) $h = \frac{1}{4}$

i)

i	t	y
0	0.	1.
1	0.2500	0.5000
2	0.5000	0.3125
3	0.7500	0.2812
4	1.0000	0.3281

ii)

i	t	y
0	0.	1.
1	0.2500	0.6562
2	0.5000	0.4883
3	0.7500	0.4302
4	1.0000	0.4407

iii) $y(t) = \frac{5}{4}e^{-2t} + \frac{1}{2}t - \frac{1}{4}$; $y(1) - y_E(1) = 0.0911$; $y(1) - y_{IE}(1) = -0.0215$

c) $h = \frac{1}{4}$

i)

i	t	y
0	0.	0.
1	0.2500	0.
2	0.5000	0.0039
3	0.7500	0.0349
4	1.0000	0.1333

ii)

i	t	y
0	0.	0.
1	0.2500	0.0020
2	0.5000	0.0193
3	0.7500	0.0822
4	1.0000	0.2156

iii) $y(t) = \frac{1}{2}\ln(\frac{1}{2}t^4 + 1)$; $y(1) - y_E(1) = 0.0694$; $y(1) - y_{IE}(1) = -0.0129$

3. If $y' = g(t)$, $y(0) = y_0$, then $y_1 = y_0 + \dfrac{h}{2}[g(0) + g(h)]$. Here, $g(t) = t + 1$,

so $y_1 = y_0 + h + \dfrac{h^2}{2} = y(h)$, the exact solution.

5.

t	y (exact)	y ($h = 0.1$)	Error	y ($h = 0.2$)	Error	Δ
0.00	1.00000	1.00000	0.	1.00000	0.	0.
0.10	1.10551	1.10550	0.00001			
0.20	1.22421	1.22413	0.00008	1.22400	0.00021	0.00013
0.30	1.35958	1.35936	0.00022			
0.40	1.51547	1.51504	0.00043	1.51408	0.00140	0.00096
0.50	1.69616	1.69542	0.00074			
0.60	1.90636	1.90519	0.00117	1.90237	0.00398	0.00282
0.70	2.15126	2.14953	0.00173			
0.80	2.43662	2.43418	0.00244	2.42810	0.00853	0.00608
0.90	2.76881	2.76547	0.00334			
1.00	3.15484	3.15039	0.00445	3.13908	0.01577	0.01131

Here Δ is the difference of the two approximate solutions.

7. a) $y(t) = 1/(1 - t)$, $y(0.5) = 2$

b) $y_{(h=0.05)} = 1.884$ at $t = 0.5$; $y_{(h=0.025)} = 1.937$ at $t = 0.5$; $2.000 - 1.884 = 0.116$; $2.000 - 1.937 = 0.063$

c) $y_{(h=0.00315)} = 1.991$ at $t = 0.5$

9. With $h = \frac{1}{20}$, $y = 0.5053$ at $t = 1$; with $h = \frac{1}{40}$, $y = 0.5054$ at $t = 1$.

11. a) $y(t) = \arctan t$; $y(1) = \pi/4 \simeq 0.7854$; $y_{(h=1/10)} = 0.7850$

c) $y(t) = t^2$; $y(2) = 4$; $y_{(h=1/10)} = 3.987$

e) $y(t) = \ln\left[\dfrac{t^4}{4} + \dfrac{3}{4}\right]$; $y(2) = 1.5581$; $y_{(h=1/10)} = 1.5580$

g) $y(t) = -\ln(1 + e^{-1} - e^t)$; $y(0.25) = 2.4787$; $y_{(h=1/40)} = 2.4781$

i) $y(t) = \dfrac{19}{16}e^{4t} + \dfrac{4t - 3}{16}$; $y(0.5) = 8.712$; $y_{(h=1/80)} = 8.710$

13. For the Euler method: For the improved Euler method:

h	$[y(1) - y_E]/h$	h	$[y(1) - y_{IE}]/h^2$
$\frac{1}{2}$	0.236	$\frac{1}{2}$	-0.091
$\frac{1}{4}$	0.206	$\frac{1}{4}$	-0.074
$\frac{1}{8}$	0.194	$\frac{1}{8}$	-0.067
$\frac{1}{16}$	0.189	$\frac{1}{16}$	-0.064
$\frac{1}{32}$	0.186	$\frac{1}{32}$	-0.062
$\frac{1}{64}$	0.185	$\frac{1}{64}$	-0.058

Section 1.7

1. $ty^2 + t^2y = c$ **3.** $t^2y + y \cos t = c$ **5.** $\tan t \sin y + ty = c$

7. $y \ln|t| - t^2 \ln|y| = c$ **9.** $e^x \cos y = c$ **11.** $\dfrac{t^2}{y^2} + \dfrac{1}{t} = c$

13. $2 \ln|ty| + \dfrac{t}{y} = c$

15. a) yes   **c)** no   **e)** no

16. a) $y(t) = t \tan(\ln t)$

    c) $y(t) = t(\ln t^2 + 16)^{1/2}$ **e)** $\sin\left(\dfrac{y}{t}\right) = \dfrac{1}{4}t^{2/3}$

17. a) $y(t) = c(t + 1)^2 - \frac{1}{2}(3t - 1)$ **c)** $\dfrac{2y - t - 3}{(y - t - 1)^3} = K$

Section 1.8

1. $I(t) = 4(1 - e^{-3t})$, $|I(t) - 4| < 10^{-4}$ if $t > 3.1$

3. $I(t) = 4e^{-2t}$, $0 \le t < 1$, $I(t) = 1.5 + Ke^{-2t}$, $1 < t$, where $K \simeq -1.54$; $I'(t)$ is discontinuous at $t = 1$

Section 1.9

3. $P(t) = 1.535 + \dfrac{(6.336)(0.767)}{0.767 + (6.336 - 0.767)e^{-0.023t}}$

    $P(130) = 6.176$; $P(140) = 6.446$; $P(150) = 6.684$; $P(160) = 6.890$; $P(170) = 7.066$

5. a) $H_c = 4.80$ **b)** $K_2 = 157$ **c)** $20.0 < t < 20.5$

    d) $11.5 < t < 12.0$

Section 1.10

1. a) $500 \dfrac{dc}{dt} = -4c$ **b)** $t = 250 \ln 10$ **3.** $t = 200 \ln(99/80)$

5. $c(250) \simeq 0.320$ kg/l, $Q(250) \simeq 480$ kg; $c(500) \simeq 0.412$ kg/l, $Q(500) \simeq 824$ kg

7. $200 \dfrac{dc_1}{dt} = -2c_1 + 10$, $200 \dfrac{dc_2}{dt} = -2c_2 + 2c_1$; $c_1(t) = 5(1 - e^{-t/100})$, $c_2(t) =$

    $5(1 - e^{-t/100}) - \dfrac{t}{20} e^{-t/100}$

9. $c(t) = \frac{3}{100}e^{-t/125}$, $0 \le t \le t^*$, $0.026 = \frac{3}{100}e^{-t^*/125}$; $c(t) = 0.02 + 0.006e^{-(t - t^*)/125}$, $t^* < t$

11. $c(t) = \frac{3}{100}e^{-t/250}$, $0 \le t \le t^* + 2$, $0.026 = \frac{3}{100}e^{-t^*/250}$;

    $c(t) = 0.02 + (0.026e^{-1/125} - 0.02) \exp\left(-\dfrac{t - t^* - 2}{250}\right)$, $t^* + 2 < t$

12. Assume $\bar{s} > \bar{c}$: $c_0 > \bar{c}$ implies that $c(t) = c_0 e^{-t/125}$, $0 \le t \le t^*$, then it chatters, i.e., no solution; $c_0 = \bar{c}$ implies that $c(t) = \bar{s} + (c_0 - \bar{s})e^{-t/125}$, $0 \le t < t_*$, then it chatters, i.e., no solution; $c_0 = \bar{c}$ implies no solution since the differential equation has a discontinuity at $c = \bar{c}$. From $(\bar{s} - \bar{c})/(\bar{s} - c_0) = e^{-t_*/125}$, it follows that $dt_*/ds < 0$.

Chapter 2

Section 2.3

2. a) $y(t) = 4e^{-t} - 3e^{-2t}$ **c)** $y(t) = \dfrac{1}{3}\cos\left(3t - \dfrac{\pi}{2}\right)$

e) $y(t) = \frac{3}{8}e^{2t} + (\frac{3}{8}e^{\pi/2} - \frac{1}{2})\cos 2t - \frac{3}{8}e^{\pi/2}\sin 2t$

g) $y(t) = \sin t + \cos t - 4e^{(t-\pi/2)}\cos t$

Section 2.4

1. a) $y(t) = c_1 e^{5t} + c_2 e^{4t}$ **c)** $y(t) = c_1 e^{6t} + c_2 e^{-5t}$

e) $y(t) = \dfrac{1}{2\sqrt{5}}(e^{\sqrt{5}t} - e^{-\sqrt{5}t})$ **g)** $y(t) = \frac{1}{2}e^{6t} + \frac{1}{2}e^{-2t}$

i) $y(t) = 3e^{-t/4} - e^{t/4}$

2. a) $y(t) = c_1 e^{2t} + c_2 t e^{2t}$ **c)** $y(t) = c_1 + c_2 t$

e) $y(t) = e^{-t/2} + \frac{3}{2}te^{-t/2}$ **g)** $y(t) = 3te^{3t}$

i) $y(t) = \frac{3}{4}e^{-(t-1)/4} + te^{-(t-1)/4}$

4. b) $y(t) = \cosh\dfrac{t}{3} - 3\sinh\dfrac{t}{3}$

5. b) $y(t) = c_1 e^{5t} + c_2 t e^{5t}$

Section 2.5

1. a) $y(t) = e^{-3t/2}(k_1 \sin t + k_2 \cos t)$ **c)** $y(t) = e^{t}(k_1 \sin t + k_2 \cos t)$

e) $y(t) = k_1 \sin(\frac{1}{2}t) + k_2 \cos(\frac{1}{2}t)$ **g)** $y(t) = 2\cos t - 6\sin t$

i) $y(t) = \sqrt{2}\sin 2t$

2. a) $y(t) = \sqrt{2}\cos\left(t - \dfrac{\pi}{4}\right)$ **c)** $y(t) = \sqrt{2}\cos\left(2t - \dfrac{\pi}{4}\right)$

3. b) $y(t) = 5e^{-t}\cos(3t - \phi)$, where $\cos\phi = \frac{4}{5}$, $\sin\phi = \frac{3}{5}$

6. a) $y(t) = \cos t - \sin t$ **c)** $y(t) = 2e^{t}[1 - \frac{1}{8}(e^{-2} + 1)t]$

e) Only one solution when ω is not an integer; many solutions when ω is an integer; there is always a solution.

Section 2.6

1. a) $W = -4$; nonzero for all t **c)** $W = -\omega$; nonzero for all t

e) $W = t^3$; nonzero for $t \neq 0$ **g)** $W = -t^{-1}$; nonzero for $t > 0$

2. a) $\mathbf{Y}_1(t) = (e^{-3t})\mathbf{e}_1 + (-3e^{-3t})\mathbf{e}_2$, $\mathbf{Y}_2(t) = (e^{-2t})\mathbf{e}_1 + (-2e^{-2t})\mathbf{e}_2$

c) $\mathbf{Y}_1(t) = (e^{-t}\cos 2t)\mathbf{e}_1 + (-e^{-t}\cos 2t - 2e^{-t}\sin 2t)\mathbf{e}_2$; $\mathbf{Y}_2(t) = $
$(e^{-t}\sin 2t)\mathbf{e}_1 + (-e^{-t}\sin 2t + 2e^{-t}\cos 2t)\mathbf{e}_2$

e) $\mathbf{Y}_1(t) = (e^{2t})\mathbf{e}_1 + (2e^{2t})\mathbf{e}_2$; $\mathbf{Y}_2(t) = (e^{-5t})\mathbf{e}_1 + (-5e^{-5t})\mathbf{e}_2$

3. b) $x(t) = \dfrac{\sqrt{5}}{2}\cos(2t - \phi)$, where $\cos\phi = \dfrac{2}{\sqrt{5}}$, $\sin\phi = \dfrac{-1}{\sqrt{5}}$; $x^2 + \dfrac{y^2}{4} = \dfrac{5}{4}$

5. a) $y(t) = At^{-2} + Bt^{-3}$; yes   **c)** $y(t) = (A + B\ln t)t^{-3}$; yes

e) $y(t) = t^{-1}[A\cos(2\ln t) + B\sin(2\ln t)]$; yes

Section 2.7

1. a) $y(t) = \dfrac{e^{3t}}{4} + c_1 e^{t} + c_2 e^{-t}$ **c)** $y(t) = (5t - 5)e^{4t} + c_1 e^{4t} + c_2 e^{3t}$

e) $y(t) = -\frac{1}{32}(\cos 6t - 24t\sin 2t - 9\cos 2t) + c_1 \sin 2t + c_2 \cos 2t$

g) $y(t) = -e^{-3t}(\ln|t| + 1) + e^{-3t}(c_1 + tc_2)$

i) $y(t) = \dfrac{e^t}{2} \ln(1 + e^{-t}) - \dfrac{e^{3t}}{2}\left[\dfrac{e^{-2t}}{2} - e^{-t} - \ln(1 + e^{-t})\right] + c_1 e^t + c_2 e^{3t}$

2. a) $y(t) = t^2(c_1 + c_2 \ln|t|) + \dfrac{2t + 1}{2}$

c) $y(t) = -\dfrac{(4\ln|t| - 8t - 1)}{4t} + \dfrac{c_1}{t} + \dfrac{c_2}{t^5}$

3. a) $y(t) = \frac{1}{4}e^{3t} + e^t - \dfrac{e^{-t}}{4}$

c) $y(t) = \cos t \ln(\cos t) + t \sin t - \sin t + 2 \cos t$

e) $y(t) = -t \ln t + 2t^2 - 2t$

6. b) $y(t) = \dfrac{t}{2} e^t + c_1 t \sin t + c_2 t \cos t$

Section 2.8

1. a) $y(t) = c_1 e^{2t} + c_2 e^{-t} - \dfrac{(2t - 1)}{4}$ **c)** $y(t) = 2 \cos t + e^t(c_1 + c_2 t)$

e) $y(t) = \frac{1}{12}(t^4 + 2t^3 + 6t^2) + c_1 + c_2 t$

g) $y(t) = c_1 \sin 2t + c_2 \cos 2t + \frac{1}{4}(2t \cos 2t + t)$

i) $y(t) = e^{-t/2}\left[c_1 \sinh\left(\dfrac{\sqrt{17}}{2} t\right) + c_2 \cosh\left(\dfrac{\sqrt{17}}{2} t\right) + \dfrac{e^{-t}}{4} - \dfrac{e^t}{2}\right]$

2. a) $y(t) = \frac{5}{12}e^{2t} - \frac{2}{3}e^{-t} - \dfrac{(2t - 1)}{4}$ **c)** $y(t) = 2 \cos t + (t - 1)e^t$

e) $y(t) = \dfrac{\sin 2t}{8} + \dfrac{2t \cos 2t + t}{4} - \dfrac{3\pi \cos 2t}{4}$

4. a) $y(t) = t^3 + c_1 t^2 + c_2 t^{-1}$

c) $y(t) = c_1 t^{(\sqrt{3}-2)} + c_2 e^{-(\sqrt{3}+2)} + \dfrac{10t^3 - 39}{26t}$

e) $y(t) = \dfrac{At^k}{k^2 + 9} + c_1 \sin(\ln t^3) + c_2 \cos(\ln t^3), \ t > 0$

Section 2.9

1. $Q(t) = \cos 2t, \ I(t) = -2 \sin 2t$, not damped

3. $Q(t) = \cos t, \ I(t) = -\sin t$, not damped

5. $Q(t) = e^{-2t}(\frac{2}{3} \sin 3t + \cos 3t), \ I(t) = e^{-2t}(-\frac{13}{3} \sin 3t)$, underdamped

7. $Q(t) = (2t + 1)e^{-2t}, \ I(t) = -4te^{-2t}$, critically damped

9. $Q(t) = (2t + 1)e^{-t}, \ I(t) = -te^{-t}$, critically damped

11. $Q(t) = 2e^{-t} - e^{-2t}, \ I(t) = -2e^{-t} + 2e^{-2t}$, overdamped

13. a) Amplitude 2, period 2π, $Q^2 + I^2 = 4$

e) Amplitude 4, period π, $y^2 + \dfrac{y'^2}{4} = 16$

14. b) Damping factor $e^{-t/20}$; period $40\pi/\sqrt{799}$; $T = -20 \ln 0.00005$

15. a) $y(\tfrac{1}{2}) = 2e^{-1/2}$ **c)** $y(10) = 80e^{-1}$

16. b) ii) $T = -\tfrac{1}{2} \ln \dfrac{2 - \sqrt{4 - \epsilon}}{2}$

Section 2.10

1. Steady state solution $= \dfrac{28 \cos 2t - 3 \sin 2t}{39}$;

transient solution $= e^{-3t/2}\left(c_1 \sin \dfrac{\sqrt{43}}{2} t + c_2 \cos \dfrac{\sqrt{43}}{2} t \right)$;

$Q(2) = \arctan \tfrac{2}{3}; \; K(2) = 145^{-1/2}$

3. Steady state solution $= \dfrac{15 \sin t - \cos t}{8}$;

transient solution $= e^{-2t}(c_1 \sin t + c_2 \cos t)$;

$Q(1) = \dfrac{\pi}{4}; \; K(1) = 32^{-1/2}$

5. Steady state solution $= \dfrac{56 \sin 2t - 3 \cos 2t}{370}$; transient solution $= c_1 e^{-t/3} +$

$c_2 e^{-4t}; \; Q(2) = \arctan(-\tfrac{13}{2}); \; K(2) = \tfrac{1}{26}$

7. a) $Q(3) = 0; \; K(3) = \tfrac{1}{9}$ **c)** $Q(\tfrac{22}{7}) = 0; \; K(\tfrac{22}{7}) = [(\tfrac{22}{7})^2 - \pi^2]^{-1}$

Section 2.11

1. a)

t_i	y_i	$y(t_i)$	**c)**	t_i	y_i	$y(t_i)$
0.25	0.2500	0.1967		0.25	0.0000	0.0372
0.50	0.3750	0.3161		0.50	0.0625	0.1796
0.75	0.4375	0.3884		0.75	0.2188	0.4954
1.00	0.4688	0.4323		1.00	0.5156	1.0973

2. a)

t_i	y_i	$y(t_i)$	**c)**	t_i	y_i	$y(t_i)$
0.25	0.1875	0.1967		0.25	0.0313	0.0372
0.50	0.3047	0.3161		0.50	0.1602	0.1796
0.75	0.3779	0.3884		0.75	0.4476	0.4954
1.00	0.4237	0.4323		1.00	0.9936	1.0973

3. a)

t_i	y_i	y_i'
0.	1.00000	0.
0.10	1.00000	0.00499
0.20	1.00102	0.02094
0.30	1.00421	0.04884
0.40	1.01082	0.08969
0.50	1.02218	0.14450
0.60	1.03975	0.21427
0.70	1.06508	0.30004
0.80	1.09980	0.40284
0.90	1.14569	0.52376
1.00	1.20460	0.66391

c)

t_i	y_i	y_i'
0.	0.	0.
0.10	0.00500	0.07304
0.20	0.01511	0.11486
0.30	0.02824	0.14063
0.40	0.04338	0.15901
0.50	0.06011	0.17477
0.60	0.07836	0.19049
0.70	0.09822	0.20749
0.80	0.11987	0.22643
0.90	0.14351	0.24764
1.00	0.16940	0.27131

e)

t_i	y_i	y_i'
0.	0.	1.00000
0.10	0.10000	0.96980
0.20	0.19402	0.88251
0.30	0.27665	0.74674
0.40	0.34357	0.57557
0.50	0.39199	0.38473
0.60	0.42074	0.19059
0.70	0.43027	0.00826
0.80	0.42246	−0.15007
0.90	0.40023	−0.27609
1.00	0.36710	−0.36576

5.

t_i	y_i	y_i'	t_i	y_i	y_i'
0.	0.35503	1.00000	1.60	1.24900	−0.49276
0.10	0.45503	0.99772	1.70	1.18973	−0.69465
0.20	0.55457	0.98990	1.80	1.11015	−0.89660
0.30	0.65301	0.97455	1.90	1.01050	−1.09346
0.40	0.74949	0.94975	2.00	0.89155	−1.27957
0.50	0.84296	0.91365	2.10	0.75468	−1.44891
0.60	0.93222	0.86454	2.20	0.60187	−1.59522
0.70	1.01588	0.80092	2.30	0.43572	−1.71230
0.80	1.09241	0.72153	2.40	0.25948	−1.79414
0.90	1.16019	0.62543	2.50	0.07696	−1.83529
1.00	1.21752	0.51208	2.60	−0.10754	−1.83106
1.10	1.26264	0.38143	2.70	−0.28924	−1.77784
1.20	1.29383	0.23393	2.80	−0.46312	−1.67341
1.30	1.30946	0.07069	2.90	−0.62398	−1.51715
1.40	1.30802	−0.10659	3.00	−0.76665	−1.31032
1.50	1.28821	−0.29545			

a) The zero is between $t = 2.50$ and $t = 2.60$.

b) To improve accuracy, start at $t = 2.50$ with $h = 0.01$ and integrate to $t = 2.60$, then use linear interpolation.

Chapter 3

Section 3.3

1. a) $y(t) = \frac{1}{2}(e^{2t} + 1)$; $y(0.5) = 1.859$, $y_5 = 1.851$

d) $y(t) = e^{-t}$; $y(0.5) = 0.6065$, $y_5 = 0.6071$

e) $y(t) = \dfrac{2}{t^2 + 1}$; $y(0.5) = 1.600$, $y_5 = 1.595$

3. $y_{k+1} = y_k + \dfrac{h}{6}[6a_k + h(3b_k + 8c_k)]$, where $a_k = t_k\, y_k + 1$, $b_k = y_k + t_k a_k$, c_k

$= 2a_k + t_k b_k$; $y_4 = 1.407$

Section 3.4

2. a)

t	y
0.0	1.000
0.10	1.105
0.20	1.221
0.30	1.349
0.40	1.491
0.50	1.647
0.60	1.820
0.70	2.012
0.80	2.223
0.90	2.456
1.00	2.714

c)

t	y
0.0	0.0
0.10	0.1002
0.20	0.2025
0.30	0.3090
0.40	0.4222
0.50	0.5454
0.60	0.6826
0.70	0.8398
0.80	1.025
0.90	1.253
1.00	1.543

3. b)

t	y	k_1	k_2	k_3	k_4	Δy
0.00	1.0000	-1.0000	-0.9013	-0.9154	-0.8286	-0.0910
0.10	0.9090	-0.8300	-0.7584	-0.7678	-0.7039	-0.0764
0.20	0.8325	-0.7048	-0.6508	-0.6574	-0.6086	-0.0655
0.30	0.7670					

4. a)

t	y_{RK4}
0.5	0.4870
1.0	0.2821

c)

t	y_{RK4}	$y(t)$	y_{2c}
0.5	0.5463	0.5463	0.5454
1.0	1.557	1.557	1.543

e)

t	y_{RK4}
0.5	1.128
1.0	1.540

5. a)

t	y_{RK4}
1.5	0.06725
2.0	0.4286

c)

t	y_{RK4}
1.0	1.834
2.0	2.689
3.0	1.838

e)

t	y_{RK4}
0.5	1.095
1.0	1.040

6. a)

t	x	y
0.5	0.1115	−0.3347
1.0	0.04978	−0.09958

c)

t	x	y
0.5	−1.210	1.039
1.0	−0.6850	−0.01825

e)

t	x	y
0.5	17.54	−24.21
1.0	447.0	−864.5

9. a)

t	y_{RK4} ($h = 0.5$)
0.5	1.245
1.0	1.462
1.5	1.635
2.0	1.762

b) The results agree with the corresponding values of $y(t) = 2e^t/(1 + e^t)$.

11. a) $R(2.0) = 13.17$, $P(2.0) = 11.87$ **b)** $R(2.0) = 5.0$, $P(2.0) = 5.0$

c) $R(2.0) = 0.7268$, $P(2.0) = 3.197$

12.

t	y	y'
0.	1.000	1.000
0.20	1.263	1.637
0.40	1.658	2.323
0.60	2.191	2.992
0.80	2.848	3.551
1.00	3.598	3.913

14.

t	θ
0.90	0.0192
0.95	−0.0008

Chapter 4

Section 4.2

1. a) $\dfrac{1}{s-2}$ **c)** $\dfrac{1}{s-10}$ **e)** $\dfrac{4}{s-1}$

4. $L\{1\} = 1/s, \operatorname{Re}(s) > 0$

Section 4.3

1. a) $y(t) = -3e^{-3t} + 4e^{-2t}$ **c)** $y(t) = \frac{1}{3} + \frac{2}{3}e^{3t}$ **e)** $y(t) = e^{3t/2} - e^{t/2}$

3. a) $y(t) = -\frac{1}{2}e^{-2t} + 2e^t - \frac{1}{2}e^{4t}$ **c)** $y(t) = \frac{3}{4}e^{-t} + \frac{1}{4}e^t + \frac{1}{2}te^{-t}$

Section 4.4

1. a) $\dfrac{6}{s^4}, \operatorname{Re}(s) > 0$ **c)** $\dfrac{3}{s^2+9}, \operatorname{Re}(s) > 0$ **e)** $\dfrac{4}{(s-2)^2+1}, \operatorname{Re}(s) > 2$

g) $\dfrac{-4s}{(s^2+1)^2}, \operatorname{Re}(s) > 0$ **i)** $\dfrac{s+3}{(s+3)^2-1}, \operatorname{Re}(s) > -3$

k) $\dfrac{(s-2)^2-1}{[(s-2)^2+1]^2}, \operatorname{Re}(s) > 2$ **m)** $\dfrac{72(s^2-3)}{(s^2+9)^3}, \operatorname{Re}(s) > 0$

o) $\dfrac{3}{s+1} + \dfrac{8}{s^2+4}, \operatorname{Re}(s) > 0$

q) $\dfrac{27}{2}\left[\dfrac{1}{(s-1)^2+9} - \dfrac{1}{(s+1)^2+9}\right], \operatorname{Re}(s) > 1$

2. a) $2t^2$ **c)** $5e^{2t}$ **e)** $4\cos\sqrt{3}t$

g) $\frac{1}{2}e^t \sin 2t$ **i)** $2\sqrt{3}e^{-t} \sin\sqrt{3}t$ **k)** $3e^{3t}\cos 3t + \frac{10}{3}e^{3t}\sin 3t$

m) $-9te^{2t}$ **o)** $6e^{3t} + \cos 5t + \frac{4}{5}\sin 5t$

q) $\int_0^t r \sin r\, dr = \sin t - t\cos t$

7. a) $L\{y'(t)\} = sY(s), L\{y''(t)\} = s^2Y(s) - 1$

c) $L\{y'(t)\} = sY(s) - 1, L\{y''(t)\} = s^2Y(s) - s$

e) $L\{y'(t)\} = sY(s), L\{y''(t)\} = s^2Y(s)$

8. b) $L\{y'(t)\} = sY(s) - 1, L\{y''(t)\} = s^2Y(s) - s - 1,$

$L\{y^{(3)}(t)\} = s^3Y(s) - s^2 - s + 1,$

$L\{y^{(4)}(t)\} = s^4Y(s) - s^3 - s^2 + s + 1$

Section 4.5

1. a) $y(t) = \frac{2}{3}\sin 3t + \cos 3t$ **c)** $y(t) = -\frac{1}{2}\sin t + 2\cos t - \frac{1}{2}t\cos t$

e) $y(t) = 2e^{2t} + 2e^{-2t} - 2$

2. a) $x(t) = 3e^t\cosh 3t - e^t\sinh 3t, y(t) = -3e^t\cosh 3t + 9e^t\sinh 3t$

c) $x(t) = e^t\cosh 3t, y(t) = e^t\sinh 3t$

e) $x(t) = 4\cos 3t + \sin 3t, Y(t) = 3\cos 3t + 5\sin 3t$

4. a) $x(t) = \frac{1}{3}e^t\cosh 3t - \frac{1}{3}e^t, y(t) = \frac{1}{3}e^t\sinh 3t$

5. a) $x(t) = 10e^{-2t/25} \cosh \dfrac{t}{25}$ lb, $y(t) = 20e^{-2t/25} \sinh \dfrac{t}{25}$ lb (the change of

variables $\tau = t/50$ makes the algebra a little easier).

$x(t) = (10e^{-t/25} + 10e^{-2t/25})/2 \to 0$ as $t \to \infty$; same argument for $y(t)$.

c) i) $y(t) = \dfrac{8}{10\sqrt{17}} e^{-9t/100} \sinh \dfrac{\sqrt{17}t}{100}$ lb/gal

Section 4.6

1. a) $A_1 = \%$, $A_2 = \%$ **c)** $A_1 = 2$, $A_2 = 1$

e) $A_1 = 1$, $A_2 = 3$, $A_3 = -\%$ **g)** $A_1 = 4$, $A_2 = -4$, $A_3 = \sqrt{2}$

3. a) $5e^{-3t} - 3e^{-2t}$ **c)** $e^{-t} \cos 5t - \%e^{-t} \sin 5t$ **e)** $-\%t + \%e^{2t} - \%e^{-2t}$

g) $\%te^t \sin 2t + \%e^t \sin 2t - \%te^t \cos 2t$ **i)** $^{13}\!\%_6 \sin 2t - \%t \cos 2t$

5. a) $y(t) = \%e^{-t} - e^{-2t} + \%e^{2t}$

c) $y(t) = \% \sin 3t + \cos 3t + \%t \sin 3t$

e) $y(t) = \%_5 e^{-t} \cos 2t - \%_5 e^{-t} \sin 2t + \%t - \%_5$

g) $y(t) = 2e^{-2t} + 4te^{-2t} - \%t^2 e^{-2t}$

6. a) $x(t) = ^{23}\!\%_{224} e^{4t} - \%e^{-2t} + \%e^{-3t} - \%_2 + \%t$

$y(t) = ^{23}\!\%_{224} e^{4t} + \%e^{-2t} - \%e^{-3t} + \%_2 - \%t$

c) $x(t) = -\%e^t \cos 2t - \%e^t \sin 2t + \% - \% \cos t - \% \sin t$

$y(t) = \%e^t \cos 2t - \%e^t \sin 2t - \% + \%_0 \cos t - \%_0 \sin t$

Section 4.7

1. a) $\%t^3$ **c)** $\%(e^t - e^{-t}) = \sinh t$

2. a) $\%t - \%_6 + \%_6 e^{-4t}$

c) $\%_4 \sin t - \%_4 \sin t \cos 4t - \%_2 \sin t \cos 2t - \%_2 \cos t \sin 2t +$

$\%_4 \cos t \sin 4t = \% \sin t - \%_4 \sin 3t$

3. a) $P(s) = \dfrac{1}{s^2 - 3s + 10}$, $p(t) = \dfrac{2}{\sqrt{31}} e^{3t/2} \sin \dfrac{\sqrt{31}}{2} t$

c) $P(s) = \dfrac{1}{s^2 - 9}$, $p(t) = \dfrac{1}{3} \sinh 3t$

7. a) $y(t) = 4t + \%t^3$ **c)** Case $c < 1$: $y(t) = \dfrac{a}{\sqrt{1-c}} \sin(t\sqrt{1-c})$

Section 4.8

1. a) $u(t) - 2u(t-2) + u(t-4)$, $\dfrac{1}{s} - 2\dfrac{e^{-2s}}{s} + \dfrac{e^{-4s}}{s}$

c) $tu(t) - 2(t-1)u(t-1) + 2(t-3)u(t-3) - (t-4)u(t-4)$,

$\dfrac{1}{s^2} - 2\dfrac{e^{-s}}{s^2} + 2\dfrac{e^{-3s}}{s^2} - \dfrac{e^{-4s}}{s^2}$

e) $\sin(t-2\pi)u(t-2\pi)$, $e^{-2\pi s}/(s^2+1)$

2. a) $\dfrac{e^{-s}}{s} + 4\dfrac{e^{-2s}}{s}$ **c)** $e\dfrac{e^{-3s}}{s-1}$ **e)** $\dfrac{1}{s^2} + \dfrac{e^{-s}}{s^2}$

3. a) $y(t) = f(t) - 2f(t-2)u(t-2) + f(t-4)u(t-4)$, where $f(t) =$

$\% - e^{-t} + \%e^{-2t}$

c) $y(t) = \frac{1}{2}[\sin t - (t - 2\pi) \cos t] \, u(t - 2\pi)$; yes

e) $y(t) = n(t - \sin t) - n\left[\left(t - \dfrac{1}{n}\right) - \sin\left(t - \dfrac{1}{n}\right)\right] u\left(t - \dfrac{1}{n}\right)$; yes

4. a) $\dfrac{2(1 - e^{-s})}{s(1 - e^{-2s})} = \dfrac{2}{s(1 + e^{-s})}$ **b)** $\dfrac{e^{-s/2} - e^{-3s/2}}{s(1 - e^{-2s})} = \dfrac{1}{2s \cosh \dfrac{s}{2}}$

5. b) $y(t) = \frac{1}{2}\Sigma(-1)^n[1 - \cos 2(t - n)] \, u(t - n)$

Section 4.9

1. a) $y(t) = \frac{1}{2}(\sin 2t) \, u(t - \pi) + \cos 2t$

c) $y(t) = e^{-t} \sin 2t + 2e^{-t} + \frac{1}{2}[e^{-(t-2)} \sin 2(t - 2)] \, u(t - 2)$

e) $y(t) = \cos 3t + \dfrac{1}{3}u\left(t - \dfrac{\pi}{3}\right) \sin 3\left(t - \dfrac{\pi}{3}\right)$

2. b) $y(t) = 2 \cos 4t + \dfrac{1}{4}u\left(t - \dfrac{\pi}{8}\right) \sin 4\left(t - \dfrac{\pi}{8}\right)$

Chapter 5

Section 5.1

1. $y_1(t) = ae^{-k_1 t}$; $y_2(t) = ak_1 te^{-k_1 t} + be^{-k_1 t}$; $y_3(t) = c + a(1 - e^{-k_1 t} - k_1 te^{-k_1 t}) + b(1 - e^{-k_1 t})$

3. i) $y_1(0) = y_2(0) = a$, $Y_1(s) = Y_2(s) = sa(s^2 + \omega^2)^{-1}$, $y_1(t) = y_2(t) = a \cos \omega t$

ii) $y_1(0) = -y_2(0) = b$, $Y_1(s) = -Y_2(s) = sb(s^2 + 3\omega^2)^{-1}$, $y_1(t) = -y_2(t) = b \cos(\sqrt{3}\omega t)$

Section 5.2

1. a) $\mathbf{x} + \mathbf{y} = \begin{bmatrix} 3 \\ 0 \end{bmatrix}$, $\mathbf{x} - \mathbf{y} = \begin{bmatrix} -1 \\ 0 \end{bmatrix}$, $\mathbf{x} \cdot \mathbf{y} = 2$, dependent; $c_1 = 2$, $c_2 = -1$

c) $\mathbf{x} + \mathbf{y} = \begin{bmatrix} 3 \\ 6 \end{bmatrix}$, $\mathbf{x} - \mathbf{y} = \begin{bmatrix} -1 \\ -2 \end{bmatrix}$, $\mathbf{x} \cdot \mathbf{y} = 10$, dependent; $c_1 = 2$, $c_2 = -1$

e) $\mathbf{x} + \mathbf{y} = \begin{bmatrix} 3 \\ 3 \\ 0 \end{bmatrix}$, $\mathbf{x} - \mathbf{y} = \begin{bmatrix} -1 \\ -3 \\ 2 \end{bmatrix}$, $\mathbf{x} \cdot \mathbf{y} = 1$, independent

2. a) Independent   **c)** Independent

e) Dependent; $c_1 = -2$, $c_2 = 1$, $c_3 = -1$, $c_4 = 1$

3. a) $\mathbf{x}'(t) = \begin{bmatrix} 1 \\ 2t \end{bmatrix}$, $\displaystyle\int_0^t \mathbf{x}(s) \, ds = \begin{bmatrix} t^2/2 \\ t^3/3 \end{bmatrix}$

c) $\mathbf{x}'(t) = \begin{bmatrix} te^t + e^t \\ 2 \cos 2t \\ 2 \end{bmatrix}$, $\displaystyle\int_0^t \mathbf{x}(s) \, ds = \begin{bmatrix} te^t - e^t + 1 \\ -\frac{1}{2} \cos 2t + \frac{1}{2} \\ t^2 + t \end{bmatrix}$

e) $\mathbf{x}'(t) = \begin{bmatrix} \cos t - \sin t \\ \cos t + \sin t \end{bmatrix}$, $\displaystyle\int_0^t \mathbf{x}(s) \, ds = \begin{bmatrix} 1 - \cos t + \sin t \\ 1 - \cos t - \sin t \end{bmatrix}$

4. a) Independent   **c)** Independent

5. a) $\text{col}(s^{-2}, 2s^{-3})$ **c)** $\text{col}[(s - 1)^{-2}, 2(s^2 + 4)^{-1}, 2s^{-2} + s^{-1}]$

e) $\text{col}[(s + 1)(s^2 + 1)^{-1}, (-s + 1)(s^2 + 1)^{-1}]$

6. a) $A + B = \begin{bmatrix} 3 & 3 & 3 \\ 3 & 1 & 1 \\ 0 & 2 & 1 \end{bmatrix}$,   **c)** $A - 2B = \begin{bmatrix} -3 & 0 & -3 \\ 0 & 1 & -2 \\ 0 & -1 & 1 \end{bmatrix}$,

e) $AB = \begin{bmatrix} 4 & 2 & 4 \\ 5 & 2 & 5 \\ 1 & 1 & 1 \end{bmatrix}$

7. a) $3A + B = \begin{bmatrix} 7 & 4 & 4 \\ 11 & 6 & 1 \\ 1 & 6 & 4 \end{bmatrix}$,   **c)** $AB = \begin{bmatrix} 5 & 5 & 4 \\ 5 & 6 & 4 \\ 5 & 6 & 3 \end{bmatrix}$

9. a) $A^T = \begin{bmatrix} 1 & 3 \\ 2 & 4 \end{bmatrix}$,   **c)** $(A + B)^T = \begin{bmatrix} 6 & 10 \\ 8 & 12 \end{bmatrix}$,

e) $(AB)^T = \begin{bmatrix} 19 & 43 \\ 22 & 50 \end{bmatrix}$

11. a) -2, independent   **c)** -1, independent   **e)** 1, independent

13. a) $\mathbf{x} = \begin{bmatrix} -1 \\ 2 \end{bmatrix}$, $C = \begin{bmatrix} -1 & 1 \\ 2 & -1 \end{bmatrix}$

c) $\mathbf{x} = \begin{bmatrix} 2 \\ 3 \end{bmatrix}$, $C = \dfrac{1}{41} \begin{bmatrix} 3 & 4 \\ 5 & -7 \end{bmatrix}$

e) $\mathbf{x} = \begin{bmatrix} 1 \\ 0 \\ 1 \end{bmatrix}$, $C = \dfrac{1}{16} \begin{bmatrix} 1 & 2 & 1 \\ 3 & -10 & 3 \\ -12 & -8 & 4 \end{bmatrix}$

15. a) $A^{-1} = \begin{bmatrix} 2 & -5 \\ -1 & 3 \end{bmatrix}$ **c)** $A^{-1} = \begin{bmatrix} 3 & -2 \\ -1 & 1 \end{bmatrix}$

e) $A^{-1} = -\dfrac{1}{2} \begin{bmatrix} 4 & -2 \\ -3 & 1 \end{bmatrix}$

17. a) $\lambda^2 - 4\lambda + 3 = 0$; $\lambda = 1$, col$(1, 1)$; $\lambda = 3$, col$(1, -1)$

c) $\lambda^2 - 3\lambda = 0$; $\lambda = 0$, col$(1, -1)$; $\lambda = 3$, col$(1, 2)$

e) $-\lambda^3 + 7\lambda^2 - 11\lambda + 5 = 0$; $\lambda = 1$, col$(1, 0, -1)$, col$(0, 1, -2)$; $\lambda = 5$, col$(1, 1, 1)$

18. a) $A'(t) = \begin{bmatrix} e^t & -e^{-t} \\ 2e^t & -e^t \end{bmatrix}$

c) $A'(t) = e^{2t} \begin{bmatrix} 2\cos t - \sin t & 4\sin t + 2\cos t \\ 6\cos t - 3\sin t & 2\sin t + \cos t \end{bmatrix}$

e) $A'(t) = \begin{bmatrix} 0 & -2\sin t & 5\cos t \\ 2e^t & -2\sin t & 5\cos t \\ e^t & -3\sin t & -5\cos t \end{bmatrix}$

21. a) $\begin{bmatrix} \dfrac{1}{s-1} & \dfrac{1}{s+1} \\ \dfrac{2}{s-1} & \dfrac{-1}{s-1} \end{bmatrix}$

c) $\dfrac{1}{(s-2)^2+1}\begin{bmatrix} s-2 & 2 \\ 3(s-2) & 1 \end{bmatrix}$

e) $\begin{bmatrix} 0 & \dfrac{2s}{s^2+1} & \dfrac{5}{s^2+1} \\[2mm] \dfrac{2}{s-1} & \dfrac{2s}{s^2+1} & \dfrac{5}{s^2+1} \\[2mm] \dfrac{1}{s-1} & \dfrac{3s}{s^2+1} & \dfrac{-5}{s^2+1} \end{bmatrix}$

Section 5.3

1. $\mathbf{y}(t) = c_1 \begin{bmatrix} 1 \\ 2 \end{bmatrix} e^{-t} + c_2 \begin{bmatrix} 2 \\ 1 \end{bmatrix} e^{2t}$

3. $\mathbf{y}(t) = c_1 \begin{bmatrix} 1 \\ \dfrac{1+\sqrt{5}}{2} \end{bmatrix} e^{(-5+\sqrt{5})t/2} + c_2 \begin{bmatrix} 1 \\ \dfrac{1-\sqrt{5}}{2} \end{bmatrix} e^{(-5-\sqrt{5})t/2}$

5. $\mathbf{y}(t) = c_1 \begin{bmatrix} 1 \\ 1 \end{bmatrix} e^{-t} + c_2 \begin{bmatrix} 1 \\ -1 \end{bmatrix} e^{-5t}$

7. $\mathbf{y}(t) = \begin{bmatrix} 2e^{-t} - e^t \\ e^{-t} - e^t \end{bmatrix}$

9. $\mathbf{y}(t) = -\dfrac{2}{5}\begin{bmatrix} 2 \\ -1 \end{bmatrix} e^{-4t} + \dfrac{3}{5}\begin{bmatrix} 3 \\ 1 \end{bmatrix} e^t = \dfrac{1}{5}\begin{bmatrix} 9e^t - 4e^{-4t} \\ 3e^t + 2e^{-4t} \end{bmatrix}$

11. $\mathbf{y}(t) = \dfrac{1}{2}\begin{bmatrix} 3e^{-t} - e^t \\ 9e^{-t} - e^t \end{bmatrix}$

12. $\mathbf{y}(t) = e^{-t}\left\{ c_1 \begin{bmatrix} -2\sin 2t \\ \cos 2t \end{bmatrix} + c_2 \begin{bmatrix} 2\cos 2t \\ \sin 2t \end{bmatrix} \right\}$

14. $\mathbf{y}(t) = c_1 e^{3t}\begin{bmatrix} 2\cos 3t \\ \cos 3t + 3\sin 3t \end{bmatrix} + c_2 e^{3t}\begin{bmatrix} 2\sin 3t \\ \sin 3t - 3\cos 3t \end{bmatrix}$

16. $\mathbf{y}(t) = c_1 e^{t}\begin{bmatrix} 5\cos t \\ -2\cos t - \sin t \end{bmatrix} + c_2 e^{t}\begin{bmatrix} 5\sin t \\ \cos t - 2\sin t \end{bmatrix}$

18. $\mathbf{y}(t) = e^{-t}\left\{ \dfrac{1}{2}\begin{bmatrix} 2\cos t \\ -5\cos t - \sin t \end{bmatrix} + \dfrac{5}{2}\begin{bmatrix} 2\sin t \\ -5\sin t + \cos t \end{bmatrix} \right\}$

20. $\mathbf{y}(t) = e^{2(t-\pi)}\left\{ (-2)\begin{bmatrix} \sin 2t \\ \dfrac{-\cos 2t}{2} \end{bmatrix} + (1)\begin{bmatrix} \cos 2t \\ \dfrac{\sin 2t}{2} \end{bmatrix} \right\}$

22. $\mathbf{y}(t) = (-1)\begin{bmatrix} \cos t \\ -\sin t \end{bmatrix} + (-1)\begin{bmatrix} \sin t \\ \cos t \end{bmatrix}$

24. If $\mathbf{b} = c\mathbf{b}^*$, then $A\mathbf{b} = \lambda^*\mathbf{b}$ and $A\mathbf{b} = \lambda\mathbf{b}$. Hence $\lambda = \lambda^*$, a contradiction.

26. $\mathbf{y}(t) = c_1 \begin{bmatrix} 1 \\ 1 \end{bmatrix} e^{3t} + c_2 \left\{ \begin{bmatrix} 0 \\ 1 \end{bmatrix} + t\begin{bmatrix} 1 \\ 1 \end{bmatrix} \right\} e^{3t}$

28. $\mathbf{y}(t) = c_1 \begin{bmatrix} 2 \\ 1 \end{bmatrix} e^{t} + c_2 \left\{ \begin{bmatrix} 1 \\ 0 \end{bmatrix} + t\begin{bmatrix} 2 \\ 1 \end{bmatrix} \right\} e^{t}$

30. $\mathbf{y}(t) = c_1 \begin{bmatrix} 1 \\ -2 \end{bmatrix} e^{4t} + c_2 \left\{ \begin{bmatrix} 0 \\ -1 \end{bmatrix} + t\begin{bmatrix} 1 \\ -2 \end{bmatrix} \right\} e^{4t}$

32. $\mathbf{y}(t) = e^{-3t} \begin{bmatrix} 8t + 3 \\ 8t + 1 \end{bmatrix}$ **34.** $\mathbf{y}(t) = e^{-3t} \begin{bmatrix} 1 - t \\ t \end{bmatrix}$

36. $\mathbf{y}(t) = e^{-2(t-1)} \begin{bmatrix} 7 - 7t \\ 8 - 7t \end{bmatrix}$

38. $(A - \lambda I)^2 = \begin{bmatrix} p & a_{12} \\ a_{21} & -p \end{bmatrix}^2 = (p^2 + a_{12}a_{21})I = 0$, since $p =$

$(a_{11} - a_{22})/2$ and $p^2 + a_{12}a_{21} = 0$.

39. a) $\mathbf{y}(t) = \text{col}(2e^{4t}, -3(e^{4t} - e^{-2t}), -3(e^{4t} + e^{-2t}))$

b) $\mathbf{y}(t) = e^t \text{col}(2 \cos 2t, \sin 2t, 3 \sin 2t)$

41. $\mathbf{y}(t) = e^{3t} \text{col}(0, 1, 0)$ **43.** $\mathbf{y}(t) = \text{col}(5 \sin t, 5 \cos t, -2 \cos t + \sin t)$

Section 5.4

3. $\Phi(t) = \dfrac{e^{-t}}{6} \begin{bmatrix} 4e^{3t} + 2e^{-3t} & e^{3t} - e^{-3t} \\ 8e^{3t} - 8e^{-3t} & 2e^{3t} + 4e^{-3t} \end{bmatrix}$

5. $\Phi(t) = e^{2t} \begin{bmatrix} 1 + 3t & 3t \\ -3t & 1 - 3t \end{bmatrix}$

7. $\Phi(t) = \begin{bmatrix} \cos 2t + \frac{1}{2} \sin 2t & -\frac{1}{2} \sin 2t \\ \frac{5}{2} \sin 2t & \cos 2t - \frac{1}{2} \sin 2t \end{bmatrix}$

9. $\Phi(t) = e^{2t} \begin{bmatrix} \cos t + \sin t & \sin t \\ -2 \sin t & \cos t - \sin t \end{bmatrix}$

11. $\Phi(t) = e^{4t} \begin{bmatrix} 1 + t & -t \\ t & 1 - t \end{bmatrix}$

13. $\Phi(t) = e^t \begin{bmatrix} 1 - t & -t \\ t & 1 + t \end{bmatrix}$, $\mathbf{y}(t) = e^{-3t} \begin{bmatrix} 2 - 3t \\ 1 + 3t \end{bmatrix}$, $\mathbf{y}(t) = e^{-3(t-1)} \begin{bmatrix} 4 - 3t \\ 3t - 1 \end{bmatrix}$

15. $\Phi(t) = e^t \begin{bmatrix} \cos \sqrt{6}t + \dfrac{2}{\sqrt{6}} \sin \sqrt{6}t & \dfrac{-2}{\sqrt{6}} \sin \sqrt{6}t \\ \dfrac{5}{\sqrt{6}} \sin \sqrt{6}t & \cos \sqrt{6}t - \dfrac{2}{\sqrt{6}} \sin \sqrt{6}t \end{bmatrix}$,

$\mathbf{y}(t) = e^t \begin{bmatrix} \cos \sqrt{6}t + \dfrac{4}{\sqrt{6}} \sin \sqrt{6}t \\ \dfrac{7}{\sqrt{6}} \sin \sqrt{6}t - \cos \sqrt{6}t \end{bmatrix}$,

$\mathbf{y}(t) = e^{t-1} \begin{bmatrix} \cos \sqrt{6}(t - 1) \\ \cos \sqrt{6}(t - 1) + \dfrac{3}{\sqrt{6}} \sin \sqrt{6}(t - 1) \end{bmatrix}$

Section 5.5

1. $\mathbf{y}(t) = \frac{1}{4} \text{col}(-e^{-5t} + 8e^{2t} + 5e^t, -3e^{5t} - 8e^{2t} - 5e^t)$

3. $y_1(t) = -\frac{9}{2} - 5 \cos t + 10 \sin t + \frac{1}{2}e^{t-\pi} \cos t - \frac{27}{2}e^{t-\pi} \sin t$

$y_2(t) = \frac{3}{2} - \cos t - 3 \sin t + \frac{5}{2}e^{t-\pi} \cos t + \frac{11}{2}e^{t-\pi} \sin t$

5. $y(t) = e^{-t} \begin{bmatrix} \dfrac{-8}{25} \\ \dfrac{2}{25} \end{bmatrix} + e^{t} \begin{bmatrix} \dfrac{-1}{9} \\ \dfrac{-4}{9} \end{bmatrix}$

$+ e^{4(t-2)} \begin{bmatrix} t-1 & 2-t \\ t-2 & 3-t \end{bmatrix} \cdot \left\{ \begin{bmatrix} -3 \\ 2 \end{bmatrix} + e^{-2} \begin{bmatrix} \dfrac{8}{25} \\ \dfrac{-2}{25} \end{bmatrix} + e^{2} \begin{bmatrix} \dfrac{1}{9} \\ \dfrac{4}{9} \end{bmatrix} \right\}$

7. $\hat{\Phi}(s) = \begin{bmatrix} \dfrac{1}{s-3} & 0 \\ 0 & \dfrac{1}{s-1} \end{bmatrix}$, $\Phi(t) = \begin{bmatrix} e^{3t} & 0 \\ 0 & e^{t} \end{bmatrix}$

9. $\hat{\Phi}(s) = \dfrac{1}{s^2+9} \begin{bmatrix} s+1 & 5 \\ -2 & s-1 \end{bmatrix}$,

$\Phi(t) = \begin{bmatrix} \cos 3t + \frac{1}{3}\sin 3t & \frac{5}{3}\sin 3t \\ -\frac{2}{3}\sin 3t & \cos 3t - \frac{1}{3}\sin 3t \end{bmatrix}$

11. $\hat{\Phi}(s) = \begin{bmatrix} \dfrac{1}{s-1} & 0 & 0 \\ 0 & \dfrac{1}{s-2} & \dfrac{1}{(s-2)^2} \\ 0 & 0 & \dfrac{1}{s-2} \end{bmatrix}$, $\Phi(t) = \begin{bmatrix} e^{t} & 0 & 0 \\ 0 & e^{2t} & te^{2t} \\ 0 & 0 & e^{2t} \end{bmatrix}$

Chapter ⑥
Section 6.2

2. a) $a_2 = a_3 = 0$, $a_n = \dfrac{a_{n-4}}{n(n-1)}$, $n = 4, 5, \ldots$,

$y(x) = a_0 \left(1 - \dfrac{1}{4 \cdot 3} x^4 + \dfrac{1}{8 \cdot 7 \cdot 4 \cdot 3} x^8 + \cdots \right)$

$+ a_1 \left(x - \dfrac{1}{5 \cdot 4} x^5 + \dfrac{1}{9 \cdot 8 \cdot 5 \cdot 4} x^9 + \cdots \right)$

c) $a_0 = -2$, $a_1 = 2$, $a_2 = \dfrac{a_0}{2 \cdot 1}$, $a_3 = \dfrac{a_1}{3 \cdot 2}$, $a_n = \dfrac{a_{n-4} + a_{n-2}}{n(n-1)}$, $n = 4, 5, \ldots$,

$y(x) = -2 + 2x - x^2 + \dfrac{1}{3} x^3 - \dfrac{1}{4} x^4 + \dfrac{7}{5 \cdot 4 \cdot 3} x^5 + \cdots$

3. a) $y(x) = 1 - x - \frac{1}{2}x^2 + \frac{1}{12}x^4 + \frac{1}{24}x^5 + \cdots$

c) $y(x) = a_0 \left[1 - \dfrac{(\alpha + \beta)}{2} x^2 + \dfrac{(\alpha + \beta)^2 + 4\beta}{24} x^4 + \cdots \right]$

$+ a_1 \left[x - \dfrac{(\alpha + \beta)}{6} x^3 + \cdots \right]$

4. a) $y(x) = x + \frac{2}{3}x^3 + \frac{4}{15}x^4 + \cdots$

5. b) $y(x) = x - \frac{2}{3!}x^3 + \frac{9}{5!}x^5 - \frac{75}{7!}x^7 + \cdots$

Section 6.3

1. a) $r^2 + \frac{1}{2}r - \frac{1}{2} = 0, r_1 = \frac{1}{2}, r_2 = -1, y_1(x) = x^{1/2} \sum_0^\infty a_n x^n, y_2(x) = x^{-1} \sum_0^\infty b_n x^n$

c) $r^2 - \frac{1}{2}r = 0, r_1 = \frac{1}{2}, r_2 = 0, y_1(x) = x^{1/2} \sum_0^\infty a_n x^n, y_2(x) = \sum_0^\infty b_n x^n$

e) $r^2 + 4r + 4 = 0, r_1 = r_2 = 2, y_1(x) = x^2 \sum_0^\infty a_n x^n,$

$y_2(x) = y_1(x) \beta \ln x + x^2 \sum_0^\infty b_n x^n$

g) $r^2 + \frac{1}{2}r - \frac{3}{16} = 0, r_1 = \frac{1}{4}, r_2 = -\frac{3}{4},$

$y_1(x) = x^{1/4} \sum_0^\infty a_n x^n, y_2(x) = x^{-3/4} \sum_0^\infty b_n x^n$ or

$y_2(x) = y_1(x) \beta \ln x + x^{-3/4} \sum_0^\infty b_n x^n$

3. a) $a = -1, r^2 + \frac{1}{2}r = 0, r_1 = 0, r_2 = -\frac{1}{2}, y_1(x) = \sum_0^\infty a_n(x + 1)^n, y_2(x) =$

$(x + 1)^{-1/2} \sum_0^\infty b_n(x + 1)^n$

e) $a = 2, r^2 - \frac{5}{2}r + \frac{3}{2} = 0, r_1 = \frac{3}{2}, r_2 = 1, y_1(x) = (x - 2)^{3/2} \sum_0^\infty a_n(x - 2)^n,$

$y_2(x) = (x - 2) \sum_0^\infty b_n(x - 2)^n.$

A second singular point is $a = 0, r^2 + \frac{1}{2}r = 0, r_1 = 0, r_2 = -\frac{1}{2}, y_1(x) =$

$\sum_0^\infty a_n x^n, y_2(x) = x^{-1/2} \sum_0^\infty b_n x^n$

4. a) $r_1 = \frac{1}{2}; a_{n+1} = \frac{a_n}{(n + 1)(2n + 3)},$

$y_1(x) = a_0 x^{1/2}\left(1 + \frac{1}{3}x + \frac{1}{30}x^2 + \frac{1}{160}x^3 + \cdots\right)$

$= a_0 x^{1/2}\left(1 + \sum_{n=1}^\infty \frac{1}{n!(2n + 1)(2n - 1) \cdots 5 \cdot 3}x^n\right)$

$r_2 = 0: b_{n+1} = \frac{b_n}{(n + 1)(2n + 1)},$

$y_2(x) = b_0\left(1 + x + \frac{1}{6}x^2 + \frac{1}{90}x^3 + \cdots\right)$

$= b_0\left(1 + \sum_{n=1}^\infty \frac{1}{n!(2n - 1)(2n - 3) \cdots 3 \cdot 1}x^n\right)$

c) $r_1 = \dfrac{1}{2}$; $a_{n+1} = \dfrac{-a_n}{2(n+1)(2n+3)}$,

$$y_1(x) = a_0 x^{1/2}\left(1 - \frac{1}{6}x + \frac{1}{120}x^2 - \frac{1}{5040}x^3 + \cdots\right)$$

$$= a_0 x^{1/2}\left(1 + \sum_{n=1}^{\infty} \frac{(-1)^n}{2^n n!(2n+1)(2n-1)\cdots 5\cdot 3}x^n\right)$$

$r_2 = 0$: $b_{n+1} = \dfrac{-b_n}{2(n+1)(2n+1)}$,

$$y_2(x) = b_0\left(1 - \frac{1}{2}x + \frac{1}{24}x^2 - \frac{1}{720}x^3 + \cdots\right)$$

$$= b_0\left(1 + \sum_{n=1}^{\infty} \frac{(-1)^n}{2^n n!(2n-1)(2n-3)\cdots 3\cdot 1}x^n\right)$$

e) $r_1 = 3$: $y_1(x) = a_0 x^3(1 - \tfrac{5}{36}x^2 + \tfrac{19}{1404}x^4 + \cdots)$,

$r_2 = \tfrac{1}{2}$: $y_2(x) = b_0 x^{1/2}(1 + \tfrac{5}{24}x^2 - \tfrac{119}{3456}x^4 + \cdots)$

5. a) $r_1 = 0$, $r_2 = -1$: When r_2 is used, a_0 and a_1 are arbitrary, $a_2 = 0$, and

$$a_{n+3} = \frac{-a_n}{(n+3)(n+2)}, \quad n \geq 0;$$

$$y(x) = a_0 x^{-1}(1 - \tfrac{1}{6}x^3 + \tfrac{1}{180}x^6 - \tfrac{1}{12960}x^9 + \cdots)$$
$$+ a_1 x^{-1}(x - \tfrac{1}{12}x^4 + \tfrac{1}{504}x^7 - \tfrac{1}{45360}x^{10} + \cdots)$$
$$= a_0 y_1(x) + a_1 y_2(x)$$

c) $r_1 = 0$, $r_2 = -3$: When r_2 is used, a_0 and a_3 are arbitrary, $a_2 = 0$, and

$$a_{n+4} = \frac{-(n+2)}{(n+4)(n+1)}a_{n+3}, \quad n = -3, \text{ and } n \geq 0,$$

$$y(x) = a_0 x^{-3}\left(1 - \frac{1}{2}x\right) + a_3 x^{-3}\left(x^3 - \frac{1}{2}x^4 + \frac{3}{20}x^5 - \frac{1}{30}x^6 + \cdots\right)$$

$$= a_0 x^{-3}\left(1 - \frac{1}{2}x\right) + a_3\left[1 + \sum_{n=1}^{\infty} \frac{(-1)^n 6(n+1)!}{(n+3)!n!}x^n\right]$$

$$= a_0 y_1(x) + a_3 y_2(x)$$

e) $r_1 = 0$, $r_2 = -4$: a_0 and a_4 arbitrary, $a_n = 0$ for $n \geq 7$,

$$y(x) = a_0(x-1)^4[1 + 4(x-1) + 5(x-1)^2]$$
$$+ a_4[1 + \tfrac{4}{5}(x-1) + \tfrac{1}{5}(x-1)^2].$$

(It is useful to let $z = x - 1$, express x and $4 + x$ in terms of z, and let $y(z) = z^{-4}\sum_0^{\infty} a_n z^n$.)

Section 6.4

1. c) $y_2(x) = x\displaystyle\int^x \frac{1}{r^2}e^{r^2/2}\,dr$ **e)** $y_2(x) = x^3$

g) $y_2(x) = -(1 + x\tan x)$

2. a) $P(x)Q(x) = 1 - \tfrac{1}{4}x^2 + \tfrac{5}{36}x^3 - \tfrac{7}{144}x^4 + \cdots$

c) $\dfrac{1}{P(x)^2} = \dfrac{1}{1 + 2x + \tfrac{3}{2}x^2 + \tfrac{13}{18}x^3 + \cdots} = 1 - 2x + \dfrac{5}{2}x^2 - \dfrac{49}{18}x^3 + \cdots$

e) $P(x)Q^2(x) = 1 - x + \tfrac{1}{4}x^2 + \tfrac{5}{18}x^3 + \cdots$

3. a) $\int x^2(1 - x + \tfrac{1}{4}x^2 + \cdots)\,dx = \tfrac{1}{3}x^3 - \tfrac{1}{4}x^4 + \tfrac{1}{20}x^5 + \cdots$

c) $\displaystyle\int \dfrac{1}{x^3}\left(1 + \dfrac{2}{3}x^2 + \dfrac{2}{9}x^4 + \cdots\right)dx = -\dfrac{1}{2x^2} + \dfrac{2}{3}\ln x + \dfrac{1}{9}x^2 + \cdots$

4. a) $r_1 = r_2 = 0$: a_0 is arbitrary, $a_1 = 0$, $a_{n+2} = a_n/(n + 2)^2$,

$$y_1(x) = a_0\left(1 + \dfrac{1}{2^2}x^2 + \dfrac{1}{4^2 \cdot 2^2}x^4 + \dfrac{1}{6^2 \cdot 4^2 \cdot 2^2}x^6 + \cdots\right)$$

$$= a_0 \sum_{n=1}^{\infty} \dfrac{1}{2^{2n}(n!)^2}x^{2n},$$

$$y_2(x) = y_1(x)\int \dfrac{1}{x}\left(1 - \dfrac{1}{2}x^2 + \dfrac{5}{32}x^4 + \cdots\right)dx$$

$$= y_1(x)\left[\ln x - \dfrac{1}{4}x^2 + \dfrac{5}{128}x^4 + \cdots\right]$$

c) $r_1 = 1$, $r_2 = -1$: a_0 is arbitrary, $a_{n+1} = -a_n/[(n + 1)(n + 3)]$,

$$y_1(x) = a_0x\left(1 - \dfrac{1}{3}x + \dfrac{1}{24}x^2 - \dfrac{1}{360}x^3 + \cdots\right) = 2a_0x\sum_{n=0}^{\infty}\dfrac{(-1)^n}{(n + 2)!n!}x^n,$$

$$y_2(x) = y_1(x)\int \dfrac{1}{x^3}\left(1 + \dfrac{2}{3}x + \dfrac{1}{4}x^2 + \dfrac{19}{270}x^3 + \cdots\right)dx$$

$$= y_1(x)\left(-\dfrac{1}{2x^2} - \dfrac{2}{3x} + \dfrac{1}{4}\ln x + \dfrac{19}{270}x + \cdots\right)$$

e) $r_1 = r_2 = 2$: a_0 is arbitrary, $a_{n+1} = -4a_n/(n + 1)^2$,

$$y_1(x) = a_0x^2\left(1 - 4x + 4x^2 - \dfrac{16}{9}x^3 + \dfrac{4}{9}x^4 + \cdots\right)$$

$$= a_0x^2\sum_{n=0}^{\infty}\dfrac{(-1)^n 4^n}{(n!)^2}x^n,$$

$$y_2(x) = y_1(x)\int \dfrac{1}{x}\left(1 + 8x + 40x^2 + \dfrac{1472}{9}x^3 + \dfrac{5416}{9}x^4 + \cdots\right)dx$$

$$= y_1(x)\left(\ln x + 8x + 20x^2 + \dfrac{1472}{27}x^3 + \dfrac{1354}{9}x^4 + \cdots\right).$$

Section 6.5

1. a) $\Gamma(5.185) = 31.8304$, $\Gamma(-3.185) = -0.60453$

3. a) With 20 terms, the error is 41.275; with 30 terms, the error is $3.7805 \cdot 10^{-3}$ and since $e^{-10} \simeq 4.5399 \cdot 10^{-5}$, this is still unacceptable.

5. b) $J_{-3/2}(0.587) = -\left[\dfrac{1}{0.587}J_{-1/2}(0.587) + J_{1/2}(0.587)\right] = -2.0539$

9. a) Write as $x^2y'' + 4x^3y = 0$ with solutions $y(x) = c_1x^{1/2}J_{1/3}(\tfrac{2}{3}x^{3/2}) + c_2x^{1/2}J_{-1/3}(\tfrac{2}{3}x^{3/2})$

c) $y(x) = c_1 x^{3/4} J_{1/4}\left(\dfrac{\sqrt{2}}{2} x\right) + c_2 x^{3/4} J_{-1/4}\left(\dfrac{\sqrt{2}}{2} x\right)$

e) $y(x) = c_1 x^{1/2} J_1(2x^{1/2}) + c_2 x^{1/2} Y_1(2x^{1/2})$

Chapter 7

Section 7.1

1. Integrate the expression for $\overline{KE}$ by parts and use the fact that $x(0) = x(T)$ for any periodic motion.

3. Observe that for $0 < \alpha < \pi$ and $0 \leq x \leq \alpha$,

$$\int_0^\alpha \left(\sin^2 \frac{\alpha}{2} - \sin^2 \frac{x}{2}\right)^{-1/2} dx > \int_0^\alpha \left(1 - \sin^2 \frac{x}{2}\right)^{-1/2} dx$$

7. Since $\dot{\eta}^2 + \sin^2 \eta = 2E$, $E = \frac{1}{2}$ will do it. The equation $\dot{\eta} = \pm \cos \eta$ is separable and solving gives

$$\eta(t) = 2 \arctan(Ke^{\pm t}) - \frac{\pi}{2}, K = \text{const}; \dot{\eta}(t) = \pm 2Ke^{\pm t}/(1 + K^2 e^{\pm 2t})$$

Section 7.2

1. $(1, -1)$ 3. $(2, 4), (-1, 1)$ 5. $(2, 1), (2, -1), (-2, 1),(-2, -1)$

7. $(2, 3); x(t) = 2 - \frac{1}{2}e^{5t}(u_0 + v_0) + \frac{1}{2}e^{3t}(3u_0 + v_0),$

$\qquad y(t) = 3 + \frac{3}{2}e^{5t}(u_0 + v_0) - \frac{1}{2}e^{3t}(3u_0 + v_0), (u_0, v_0)$ near $(0, 0)$

9. $(0, 0); x(t) = u_0 \cos t + v_0 \sin t, y(t) = -u_0 \sin t + v_0 \cos t, (u_0, v_0)$ near $(0, 0)$

11. $(2, -4); x(t) = e^{2t}\left[u_0 \cosh \sqrt{7}t + \dfrac{1}{\sqrt{7}}(2u_0 + v_0) \sinh \sqrt{7}t\right],$

$\qquad y(t) = e^{2t}\left[v_0 \cosh \sqrt{7}t + \dfrac{1}{\sqrt{7}}(3u_0 - 2v_0) \sinh \sqrt{7}t\right], (u_0 v_0)$ near

$\qquad (0, 0)$

$\qquad (-1, -1); x(t) = -1 + u_0 e^{-t} \cos \sqrt{2}t - \dfrac{(u_0 - v_0)}{\sqrt{2}} e^{-t} \sin \sqrt{2}t,$

$\qquad y(t) = -1 + \dfrac{v_0}{\sqrt{2}} e^{-t} \cos \sqrt{2}t - \dfrac{(3u_0 - v_0)}{\sqrt{2}} e^{-t} \sin \sqrt{2}t, (u_0, v_0)$

$\qquad$ near $(0, 0)$

13. Note that $x \sin y = (\rho \cos \phi) \sin(\rho \sin \phi)$

$$= \rho \cos \phi \left[\rho \sin \phi - \frac{(\rho \sin \phi)^3}{3!} + \cdots\right]$$

$$= \rho^2 \left[\cos \phi \sin \phi - \frac{(\rho^2 \cos (\sin \phi)^3}{3!} + \cdots\right]$$

and similarly for the term $y \sin x$. Also, $\det \begin{bmatrix} -4 & 5 \\ 1 & 4 \end{bmatrix} = -21 \neq 0$.

15. $u(t) = 10^{-4}e^{-t} (\cos 3t + \sin 3t); v(t) = 10^{-4}e^{-t} (\sin 3t - \cos 3t)$

Section 7.3

1. Saddle point, $\lambda = 7, -5$; portrait similar to Fig. 7.8

3. Stable node, $\lambda = -1, -2$; portrait similar to Fig. 7.6

5. Unstable spiral, $\lambda = 1 \pm 2i$; portrait similar to Fig. 7.12(b)

7. Saddle point, $\lambda = 2, -1$; portrait similar to Fig. 7.8

9. Stable node, $\lambda = -1$, double root; portrait similar to Fig. 7.10

11. Unstable node, $\lambda = 2$, double root; portrait similar to Fig. 7.11

13. Center, $\lambda = \pm 4i$; portrait similar to Fig. 7.12(c)

15. Saddle point, $\lambda = (1 \pm \sqrt{17})/2$; portrait similar to Fig. 7.12(c)

17. Unstable node, $\lambda = 3, 4$; portrait similar to Fig. 7.7

19. Stable spiral, $\lambda = -2 \pm i$; portrait similar to Fig. 7.12(a)

21. $(1, 2)$, stable node, $\lambda = -3$, double root, $x(t) = 1 + u_0 e^{-3t} + 4te^{-3t}(u_0 - v_0)$,
    $y(t) = 2 + v_0 e^{-3t} + 4te^{-3t}(u_0 - v_0)$, (u_0, v_0) near $(0, 0)$

23. $(0, 0)$, center, $\lambda = \pm i$, $x(t) = u_0 \cos t + v_0 \sin t$, $y(t) = -u_0 \sin t + v_0 \cos t$,
    (u_0, v_0) near $(0, 0)$
    $(1, -1)$, saddle point, $\lambda = \pm\sqrt{3}$, $x(t) = 1 + u_0 \cosh \sqrt{3}t$
    $+ \dfrac{(2u_0 + v_0)}{\sqrt{3}} \sinh \sqrt{3}t, \; y(t) = -1 + v_0 \cosh \sqrt{3}t - \dfrac{(u_0 + 2v_0)}{\sqrt{3}} \sinh \sqrt{3}t,$
    (u_0, v_0) near $(0, 0)$

25. $(1, 1)$, unstable spiral, $\lambda = 1 \pm i$, $x(t) = 1 + e^t(u_0 \cos t - v_0 \sin t)$, $y(t) = 1 + e^t(u_0 \sin t + v_0 \cos t)$, (u_0, v_0) near $(0, 0)$
    $(-1, -1)$, saddle point, $\lambda = \pm\sqrt{2}$, $x(t) = -1 + u_0 \cosh \sqrt{2}t + \dfrac{(u_0 - v_0)}{\sqrt{2}} \sinh \sqrt{2}t, \; y(t) = -1 + v_0 \cosh \sqrt{2}t - \dfrac{(u_0 + v_0)}{\sqrt{2}} \sinh \sqrt{2}t, \; (u_0, v_0)$
    near $(0, 0)$

27. $(-1, -1)$, unstable node, $\lambda = 2 \pm \sqrt{2}$,
    $$x(t) = -1 + \frac{e^{2t}}{\sqrt{2}} [u_0 \sqrt{2} \cosh \sqrt{2}t - (u_0 + v_0) \sinh \sqrt{2}t],$$
    $$y(t) = -1 + \frac{e^{2t}}{\sqrt{2}} [v_0 \sqrt{2} \cosh \sqrt{2}t - (u_0 - v_0) \sinh \sqrt{2}t], \; (u_0, v_0) \text{ near } (0, 0)$$
    $(-3, -3)$, saddle point, $\lambda = 1 \pm \sqrt{3}$,
    $$x(t) = -3 + e^t \left[u_0 \cosh \sqrt{3}t - \frac{v_0}{\sqrt{3}} \sinh \sqrt{3}t \right],$$
    $$y(t) = -3 + e^t [v_0 \cosh \sqrt{3}t - u_0 \sqrt{3} \sinh \sqrt{3}t], \; (u_0, v_0) \text{ near } (0, 0)$$

29. See Fig. A.1.

31. $x(t) = Ae^{2t} - B, \; y(t) = Ae^{2t} + B, \; A, B$ arbitrary constants
    (see Fig. A.2)

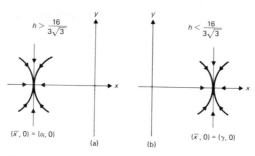

Figure A.1

Figure A.2

33. $dz/dt = 2z - z^2 = 2z(1 - z/2)$ (compare with logistic equation of Chapter 1), $dw/dt = wz$,

$$z(t) = \frac{2z_0}{z_0 + (2 - z_0)e^{-2t}} = \frac{2z_0e^{2t}}{(2 - z_0) + z_0e^{2t}} \text{ if } z(0) = z_0,$$

$$w(t) = w_0 \exp\left[\int_0^t z(s) \, ds\right] = \frac{w_0}{2}[(2 - z_0) + z_0e^{2t}] \text{ if } w(0) = w_0.$$

There is a line of equilibrium points ($z = 0$) and an isolated equilibrium point $(2, 0)$ which is a saddle point.

Section 7.4

1. $(0, 0)$, a saddle point, $x(t) = u_0\cosh \sqrt{2}t + \frac{v_0}{\sqrt{2}} \sinh \sqrt{2}t,$

$y(t) = v_0 \cosh \sqrt{2}t + \sqrt{2} u_0 \sinh \sqrt{2}t, (u_0, v_0)$ near $(0, 0)$;

$(1, 0)$, a center, $x(t) = 1 + u_0 \cos \sqrt{3}t + \frac{v_0}{\sqrt{3}} \sin \sqrt{3}t, y(t) = v_0 \cos \sqrt{3}t -$

$\sqrt{3}u_0 \sin \sqrt{3}t, (u_0, v_0)$ near $(0, 0)$;

$(-2, 0)$, a center, $x(t) = -2 + u_0 \cos \sqrt{2}t + \frac{v_0}{\sqrt{2}} \sin \sqrt{2}t, y(t) =$

$v_0 \cos \sqrt{2}t - \sqrt{2}u_0 \sin \sqrt{2}t, (u_0, v_0)$ near $(0, 0)$

3. $(0, 0)$, a center, $x(t) = u_0 \cos t + v_0 \sin t, y(t) = v_0 \cos t - u_0 \sin t, (u_0, v_0)$ near $(0, 0)$;

$(1, 0)$, a saddle point, $x(t) = 1 + u_0 \cosh \sqrt{2}t + \frac{v_0}{\sqrt{2}} \sinh \sqrt{2}t, y(t) =$

$v_0 \cosh \sqrt{2}t + \sqrt{2}u_0 \sinh \sqrt{2}t, (u_0, v_0)$ near $(0, 0)$;

$(-1, 0)$, a saddle point; same as $(1,0)$ with 1 replaced by -1.

5. See Fig. A.3.　　　**7.** See Fig. A.4.　　　**9.** See Fig. A.5.

11. The potential function $F(u) = \frac{u^2}{2} - u - \lambda\frac{u^3}{3}$ has a local minimum at $u =$

$\frac{1}{2\lambda}[1 - \sqrt{1 - 4\lambda}]$ and a local maximum at $u = \frac{1}{2\lambda}[1 + \sqrt{1 - 4\lambda}]$ for $0 <$

$\lambda < \frac{1}{4}$. For $\lambda > \frac{1}{4}$ it is a strictly decreasing function. (See Fig. A.6.)

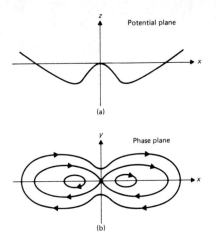

Figure A.3

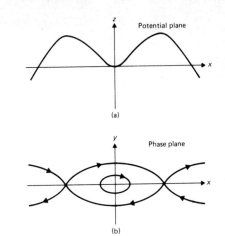

Figure A.4

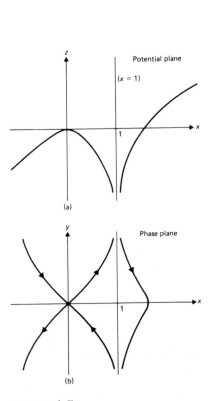

Figure A.5

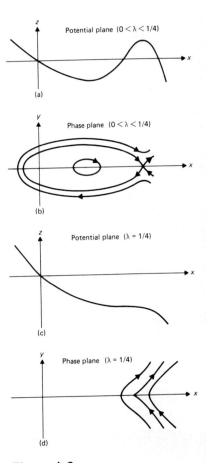

Figure A.6

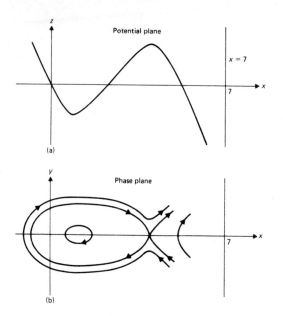

Figure A.7

13. See Fig. A.7.   **15.** See Fig. A.8.   **17.** See Fig. A.9.

19. See Fig. A.10.

21. If $E(x, \dot{x}) = e^{2x}\dot{x}^2/2 + e^{2x}F(x)$, where $F'(x) + 2F(x) = g(x)$, then

$$\frac{dE(x, \dot{x})}{dt} = e^{2x}\dot{x}[\ddot{x} + \dot{x}^2 + 2F(x) + F'(x)] = 0.$$

a) If $g(x) = x$, then $F(x) = Ae^{-2x} + x/2 - \frac{1}{4}$. Set $A = \frac{1}{4}$, then $F(0) = 0$ and

$$E(x, y) = \frac{e^{2x}y^2}{2} + h(x) = k, \quad \text{where} \quad h(x) = \frac{e^{2x}(2x - 1)}{4} + \frac{1}{4}.$$

The integral curve through $(b, 0)$ is given by $y^2 = 2e^{-2x}[h(b) - h(x)]$. If $0 < b \ll 1$, then $h(x) - h(b) = 0$ has a negative root $(x = a)$ near $x = -b$ and y^2 is positive on $a < x < b$ and vanishes at the endpoints of the interval. It follows that the integral curve through $(b, 0)$ is closed.

b) $E(x, y) = \dfrac{e^{2x}y^2}{2} + \dfrac{e^{2x}(2 \sin x - \cos x)}{5} = k.$

Since the points $(\pi, 0)$ and $(-\pi, 0)$ determine different values of k, they lie on different integral curves.

c) Let $G(x)$ be an antiderivative of $e^{2x}g(x)$; then $G(x) = e^{2x}F(x)$. Then, if $y = dx/dt$, $z = dx/ds$, $E(x, y) = e^{2x}y^2/2 + G(x) = k$ and $E^*(x, z) = z^2/2 + G(x) = k^*$ are the corresponding integrals. Therefore, the integral curves are different (see Fig. A.11).

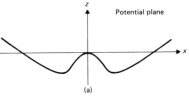

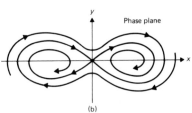

Figure A.8

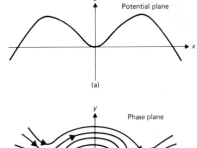

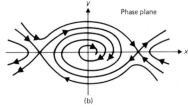

Figure A.9

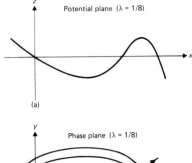

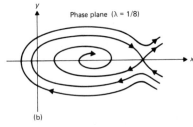

Figure A.10

Figure A.11

Section 7.5

1. **a)** $(0, 0)$, unstable node   **b)** $(s/\beta, r/\alpha)$, saddle point

4. **a)** $r = 2, K = 10, \alpha = \tfrac{1}{5}, s = 1, J = 10, \beta = \tfrac{1}{5}$; $K > s/\beta$ implies that $(K, 0)$ is a stable node; $J < r/\alpha$ implies that $(0, J)$ is a saddle point. Hence there is competitive exclusion with Y becoming extinct (see Fig. 7.33).

   c) $r = 1, K = 20, \alpha = \tfrac{1}{5}, s = 1, J = 12, \beta = \tfrac{1}{16}$; $K > s/\beta$ implies that $(K, 0)$ is a stable node; $J > r/\alpha$ implies that $(0, J)$ is a stable node. Hence (x_∞, y_∞) is a saddle point and initial conditions will determine which population wins.

5. An equilibrium point off the coordinate axes can occur only if $K > s/\beta$, in which case it is the point $\left(\dfrac{s}{\beta}, \dfrac{r}{\alpha}\left(1 - \dfrac{s}{K\beta}\right)\right)$. The matrix of the linearized system at this point is

$$\begin{bmatrix} -\dfrac{rs}{K\beta} & -\dfrac{\alpha}{\beta}s \\[2ex] -\dfrac{\beta}{\alpha}r\left(1 - \dfrac{s}{K\beta}\right) & 0 \end{bmatrix}$$

with eigenvalues

$$\lambda = \frac{1}{2}\left[\frac{-rs}{K\beta} \pm \sqrt{\left(\frac{rs}{K\beta}\right)^2 + 4rs\left(1 - \frac{s}{K\beta}\right)}\,\right].$$ Since $1 - s/K\beta > 0$, these

eigenvalues are real and of opposite sign, so that the equilibrium point is a saddle point and there can be no stable coexistence.

6. a) $H = 0$: competitive exclusion; $H = \frac{5}{4}$: ($\frac{10}{3}$, $\frac{35}{12}$) is a saddle point and unstable coexistence; $H = \frac{25}{12}$: extinction.

7. a) i)

t	x	y
0.	1.750	4.000
0.50	1.851	3.591
1.00	2.085	3.525
1.50	2.264	3.876
2.00	2.186	4.391
2.50	1.926	4.522
3.00	1.759	4.147
3.50	1.802	3.684

ii)

t	x	y
0.	2.000	2.000
0.50	3.137	2.646
1.00	3.291	5.334
1.50	1.748	6.944
2.00	1.048	4.918
2.50	1.079	3.016
3.00	1.575	2.112
3.50	2.569	2.147

10. a) $\beta = \frac{1}{5}$: (5, 6) is a stable spiral point; $\beta = \frac{1}{40}$: no equilibrium points in the interior of the first quadrant.

11. a) $K = 20$: equilibrium points are (0, 0), a saddle point; (20, 0), a saddle point; (10, 10), a stable spiral point.

c) $K = 25$: equilibrium points are (0, 0), a saddle point; (25, 0), a saddle point; $(20, 8(1 - e^{-2})^{-1}) = (20, 9.252)$, a stable node.

12. b)

$H = 0$			$H = 1.1$		
t	x	y	t	x	y
30.00	19.91	23.08	30.00	38.73	2.082
31.00	19.98	23.06	31.00	39.17	1.170
32.00	20.04	23.07	32.00	39.67	0.1471
33.00	20.07	23.08	33.00	40.21	−1.002
34.00	20.07	23.11	34.00	40.80	−2.294
35.00	20.06	23.13	35.00	41.45	−3.753

Chapter 8

Section 8.2

1.

t_i	y_i	Local error
0.25	0.0	0.01111
0.50	0.0156	0.1088
0.75	0.1172	0.4507
1.00	0.5508	1.6472

The actual solution is $y(t) = \dfrac{e^{10t}}{500} - \dfrac{50t^2 + 10t + 1}{500}$, so $y(1) = 43.9309$. The global error at $t = 1$ is 43.3801; the sum of local errors is 2.2178. Notice that even though the sum of local errors is small, the global error is not.

2.

h	Error	h	Error
$\frac{1}{2}$	0.027198	$\frac{1}{64}$	0.000016
$\frac{1}{4}$	0.005096	$\frac{1}{128}$	0.000006
$\frac{1}{8}$	0.001104	$\frac{1}{256}$	0.000005
$\frac{1}{16}$	0.000257	$\frac{1}{512}$	0.000009
$\frac{1}{32}$	0.000063	$\frac{1}{1024}$	0.000019

Index